Instructor's Guide
and
Solutions Manual

Elementary Statistics

Sixth Edition

Instructor's Guide
and
Solutions Manual

Elementary Statistics
Sixth Edition

Mario F. Triola
Dutchess Community College

Addison-Wesley Publishing Company

Reading, Massachusetts • Menlo Park, California • New York
Don Mills, Ontario • Wokingham, England • Amsterdam • Bonn
Sydney • Singapore • Tokyo • Madrid • San Juan • Milan • Paris

Reproduced by Addison-Wesley from camera-ready copy supplied by the author.

Copyright © 1995 by Addison-Wesley Publishing Company, Inc.

ISBN 0-201-57685-6
1 2 3 4 5 6 7 8 9 10-VG-97969594

CONTENTS

CONTENTS

INTRODUCTION

This guide and manual contains a variety of different materials designed to supplement the sixth edition of *Elementary Statistics* by Triola. Included are the following.

* *Three different sample **syllabi** to provide suggestions and help in planning a course outline

* *A comparison of the **STATDISK** and **Minitab** software packages to help in choosing a computer supplement, if one is desired

* *A listing (by text chapter) of the recommended programs from the **videotape series** *Against All Odds: Inside Statistics* and a separate listing of the recommended programs from the videotape series *Statistics: Decisions Through Data*.

* *Four different **quizzes** for each chapter. These quizzes, which may be photocopied for student use, are particularly helpful when you want to provide make-up quizzes for students who miss a regularly scheduled test. Four comprehensive **final examinations** are also included. **Answers** to the chapter quizzes and final examinations are also provided. (In addition to this printed test bank, a separate computerized test generator is also available.)

* *Several **transparency masters** for convenient reference

* ***Solutions** to exercises that show details not included with the answers in the textbook. These solutions were developed by Dr. Milton Loyer.

DESIGNING A COURSE SYLLABUS

Almost all instructors will find that the entire textbook cannot be covered in one semester. Since some topics will normally be omitted, we have included three samples of course syllabi. The "class hour" referred to in these outlines consists of approximately one hour. In addition to the three syllabi that follow, there are many other configurations that will work very successfully. The particular outline chosen will depend on individual needs and preferences and the following three outlines are provided only as a source of some suggestions. Specific comments about these three syllabi are given on the following page.

In planning a new course syllabus for the sixth edition of *Elementary Statistics*, users of the fifth edition should consider these major changes:

*New section in Chapter 1: **Statistics and Computers**

*New section in Chapter 3: **Probabilities through Simulations**

*The former Section 3-5 on Complements and Odds is deleted; complements are discussed earlier in the chapter and odds are included as B exercises.

*The former Sections 4-2 and 4-3 are now combined in one Section 4-2 (Random Variables) that includes calculations of the mean, variance, and standard deviation of a probability distribution.

*The former Sections 5-2, 5-3, and 5-4 are now reorganized in two sections: Section 5-2 (the Standard Normal Distribution) and Section 5-3 (Nonstandard Normal Distributions).

*Chapter 8 on Inferences from Two Samples has undergone major reorganization. The former section 8-3 on Inferences from Two Means has been divided into these new sections:
    8-2 Inferences about Two Means: Dependent Samples
    8-3 Inferences about Two Means: Independent and Large Samples
    8-5 Inferences about Two Means: Independent and Small Samples

*New Chapter 12: **Statistical Process Control**
    that includes control charts for monitoring variation and mean in a process, as well as control charts for monitoring process attributes.

SYLLABUS (Version A)

Version A does not include the simulation approach to probability (Section 3-5) or the permutations and combinations found in Section 3-6 (Counting). It also excludes Section 6-4 on confidence intervals and sample size determinations that relate to standard deviations or variances. All of Chapter 8 (hypothesis testing and confidence intervals with two populations) is omitted along with Sections 9-4 (Variation) and 9-5 (Multiple Regression), Chapter 11 (Analysis of Variance), and Chapter 13 (Nonparametric Statistics). This version does include the essential concepts of descriptive statistics, probability, estimating parameters, and hypothesis testing. The amount of material covered and the pace of this version make it suitable for many different colleges.

SYLLABUS (Version B)

Version B excludes the simulation approach to probability (Section 3-5) and the permutations and combinations of Section 3-6 (Counting). Also excluded is Section 6-4 on confidence intervals and sample size determinations that relate to standard deviations or variances. This version does not include Section 9-4 (Variation), Chapter 11 (Analysis of Variance), nor does it include anything from Chapter 13 on nonparametric methods. The content and pace of this version make it reasonable for many different colleges. While Version A excludes Chapter 8 (Inferences From Two Samples), that chapter is included in Version B with the result that this version covers more material and has a faster pace.

SYLLABUS (Version C)

Version C has a fast pace and it covers more material than the preceding two versions. This version would be suitable only for classes with well-prepared and highly-motivated students. The simulation approach to probability (Section 3-5) is excluded along with Section 3-6 on permutations and combinations. Also excluded is Section 6-4 on confidence intervals and sample size determinations that relate to standard deviations or variances. Several other sections from Chapters 8, 9, 11, and 13 are also excluded.

SYLLABUS (Version A)

Class Hour	Text Section	Topic
1	Chapter 1	Introduction to Statistics
2	2-1, 2-2, 2-3	Summarizing Data and Pictures of Data
3	2-4	Measures of Central Tendency
4	2-5	Measures of Dispersion
5	2-6	Measures of Position
6	2-7	Exploratory Data Analysis
7	Ch. 1, 2	Review
8	Ch. 1, 2	Test 1
9	3-1, 3-2, 3-3	Fundamentals of Probability and the Addition Rule
10	3-4	Multiplication Rule
11	4-1, 4-2	Random Variables
12	4-3	Binomial Experiments
13	4-4	Mean and Standard Deviation for the Binomial Distribution
14	Ch. 3, 4	Review
15	Ch. 3, 4	Test 2
16	5-1, 5-2	The Standard Normal Distribution
17	5-3	Nonstandard Normal Distributions
18	5-4	The Central Limit Theorem
19	5-5	Normal as Approximation to Binomial
20	6-1, 6-2	Estimates and Sample Sizes of Means
21	6-1, 6-2	Estimates and Sample Sizes of Means
22	6-3	Estimates and Sample Sizes of Proportions
23	Ch. 5, 6	Review
24	Ch. 5, 6	Test 3
25	7-1, 7-2	Fundamentals of Hypothesis Testing
26	7-3	Claims about a Mean: Large Samples
27	7-3	Claims about a Mean: Large Samples
28	7-4	Claims about a Mean: Small Samples
29	7-5	Claims about a Proportion
30	7-6	Claims about a Variance
31	Ch. 7	Review
32	Ch. 7	Test 4
33	9-1, 9-2	Correlation
34	9-2, 9-3	Correlation and Regression
35	9-3	Regression
36	10-1, 10-2	Multinomial Experiments
37	10-3	Contingency Tables
38	12-1, 12-2	Control Charts for Variation and Mean
39	12-3	Control Charts for Attributes
40	Ch. 9, 10, 12	Review
41	Ch. 9, 10, 12	Test 5
42		Review for Comprehensive Final Exam
43		Review for Comprehensive Final Exam

SYLLABUS (Version B)

Class Hour	Text Section	Topic
1	Chapter 1	Introduction to Statistics
2	2-1, 2-2, 2-3	Summarizing Data and Pictures of Data
3	2-4	Measures of Central Tendency
4	2-5	Measures of Dispersion
5	2-6, 2-7	Measures of Position; Exploratory Data Analysis
6	Ch. 1, 2	Review
7	Ch. 1, 2	Test 1
8	3-1, 3-2, 3-3	Fundamentals of Probability and the Addition Rule
9	3-4	Multiplication Rule
10	4-1, 4-2	Random Variables
11	4-3	Binomial Experiments
12	4-4	Mean and Standard Deviation for the Binomial Distribution
13	Ch. 3, 4	Review
14	Ch. 3, 4	Test 2
15	5-1, 5-2	The Standard Normal Distribution
16	5-3	Nonstandard Normal Distributions
17	5-4	Central Limit Theorem
18	5-5	Normal as Approximation to Binomial
19	6-1, 6-2	Estimates and Sample Sizes of Means
20	6-3	Estimates and Sample Sizes of Proportions
21	Ch. 5, 6	Review
22	Ch. 5, 6	Test 3
23	7-1, 7-2	Fundamentals of Hypothesis Testing
24	7-3	Claims about a Mean: Large Samples
25	7-4	Claims about a Mean: Small Samples
26	7-5	Claims about a Proportion
27	7-6	Claims about a Variance
28	12-1, 12-2	Control Charts for Variation and Mean
29	12-3	Control Charts for Attributes
30	Ch. 7, 12	Review
31	Ch. 7, 12	Test 4
32	8-1, 8-2	Two Means: Dependent Samples
33	8-3	Two Means: Independent/Large Samples
34	8-5	Inferences about Two Proportions
35	9-1, 9-2	Correlation
36	9-2, 9-3	Correlation and Regression
37	9-3	Regression
38	9-5	Multiple Regression
39	10-1, 10-2	Multinomial Experiments
40	10-3	Contingency Tables
41	Ch. 8, 9, 10	Review
42	Ch. 8, 9, 10	Test 5
43		Review for Comprehensive Final Exam
44		Review for Comprehensive Final Exam

SYLLABUS (Version C)

Class Hour	Text Section	Topic
1	Chapter 1	Introduction to Statistics
2	2-1, 2-2, 2-3	Summarizing Data and Pictures of Data
3	2-4	Measures of Central Tendency
4	2-5	Measures of Dispersion
5	2-6, 2-7	Measures of Position; Exploratory Data Analysis
6	3-1, 3-2, 3-3	Fundamentals of Probability and the Addition Rule
7	3-4	Multiplication Rule
8	4-1, 4-2	Random Variables
9	4-3	Binomial Experiments
10	4-4	Mean and Standard Deviation for the Binomial Distribution
11	Ch. 1-4	Review
12	Ch. 1-4	Test 1
13	5-1, 5-2	The Standard Normal Distribution
14	5-3	Nonstandard Normal Distributions
15	5-4	Central Limit Theorem
16	5-5	Normal as Approximation to Binomial
17	6-1, 6-2	Estimates and Sample Sizes of Means
18	6-3	Estimates and Sample Sizes of Prop's.
19	7-1, 7-2	Fundamentals of Hypothesis Testing
20	7-3	Claims about a Mean: Large Samples
21	7-4	Claims about a Mean: Small Samples
22	7-5	Claims about a Proportion
23	7-6	Claims about a Variance
24	Ch. 5, 6, 7	Review
25	Ch. 5, 6, 7	Test 2
26	8-1, 8-2	Two Means: Dependent Samples
27	8-3	Two Means: Independent/Large Samples
28	8-5	Inferences about Two Proportions
29	9-1, 9-2	Correlation
30	9-2, 9-3	Correlation and Regression
31	9-3	Regression
32	9-5	Multiple Regression
33	10-1, 10-2	Multinomial Experiments
34	10-3	Contingency Tables
35	Ch. 8, 9, 10	Review
36	Ch. 8, 9, 10	Test 3
37	11-1, 11-2	One-Way ANOVA with Equal Sample Sizes
38	11-3	One-Way ANOVA with Unequal Sample Sizes
39	12-1, 12-2	Control Charts for Variation and Mean
40	12-3	Control Charts for Attributes
41	13-1, 13-2	Nonparametric Methods; Sign Test
42	13-7	Runs Test for Randomness
43	Ch. 11, 12, 13	Review
44	Ch. 11, 12, 13	Test 4
45		Review for Comprehensive Final Exam

COMPUTER USAGE: STATDISK and Minitab

With the availability of microcomputers and statistics software packages, we have an excellent opportunity to use computers as a supplement to the introductory statistics course. We recommend two different levels of software for use with this text. **STATDISK**, developed specifically for this textbook, is provided free to colleges that adopt the book. It is very easy to use and is ideal for those students who have little or no prior computer experience. STATDISK is available for the IBM PC and compatible computers, as well as the Macintosh. A separate *STATDISK STUDENT LABORATORY MANUAL AND WORKBOOK* (4th edition) is available. This package is highly recommended for those who can spare little or no class time for the discussion of computer use.

For those who wish to incorporate computer usage as a major component of the course, we recommend **Minitab**, although other popular packages (such as BMDP, SAS, or SPSS) can also be used with the textbook. There is a *MINITAB STUDENT LABORATORY MANUAL AND WORKBOOK* (3rd edition) that is designed specifically for this textbook. An inexpensive *Student Edition of Minitab* is available from Addison-Wesley Publishing Company, 1 Jacob Way, Reading, MA 01867 (or call 800-447-2226). Instructors concerned about the cost of this package might consider buying a few copies that could be made available in a computer laboratory. This package consists of software and a user's manual.

The use of Minitab would normally require some classroom instruction in the formats required and the commands that are relevant to the different topics in statistics. However, the textbook includes Section 1-5, Statistics and Computers, which presents the fundamental commands used by Minitab. Also, many individual sections include the relevant Minitab commands along with sample displays.

The following page summarizes some of the basic differences between STATDISK and Minitab.

	STATDISK	**Minitab**
Cost of software:	Free to colleges that adopt the textbook.	As of this printing, less than $50 for the student version.
Difficulty:	Extremely easy to use. Students can be given assignments with little or no class time required for computer instruction.	Very easy to use when compared to other packages like SPSS or SAS. Does require some classroom time for computer instruction.
Data required:	Works with original data or with summary statistics, such as the sample mean and standard deviation.	Works only with original data. Does not work with summary statistics.
Number of scores that can be used:	Usually a maximum of 240 scores.	The student version allows up to 2000 scores, and the standard version allows much larger data sets.
Supplements:	*STATDISK Student Laboratory Manual and Workbook*, 4th edition; designed specifically as a supplement for this textbook.	*Minitab Student Laboratory Manual and Workbook*, 3rd edition; designed specifically as a supplement for this textbook. Also, several other reference books are available.
Scope:	The **STATDISK** program can be used for almost every major procedure in this textbook.	The Minitab program covers most of the major procedures in this textbook, as well as many other procedures not included in the textbook.

VIDEOTAPE SUPPLEMENTS

For videotape supplements we recommend either the videotape series *Against All Odds: Inside Statistics* or the videotape series *Statistics: Decisions Through Data*.

The series *Against All Odds: Inside Statistics* consists of 26 half-hour programs on 13 video-cassettes; see below for the listing of programs arranged according to the chapter from *Elementary Statistics*, 6th edition. For information about obtaining this series, call 1-800-LEARNER.

Text Chapter	Recommended program from *Against All Odds*
1	1: What is Statistics?
2	2: Picturing Distributions
	3: Describing Distributions
3	15: What is Probability?
4	16: Random Variables
	17: Binomial Distributions
5	4: Normal Distributions
	5: Normal Calculations
6	19: Confidence Intervals
7	20: Significance Tests
	21: Inference for One Mean
8	22: Comparing Two Means
	23: Inference for Proportions
9	8: Describing Relationships
	9: Correlation
	11: The Question of Causation
	25: Inference for Relationships
10	24: Inference for Two-Way Tables
	10: Multidimensional Data Analysis
11	13: Blocking and Sampling
12	18: The Sample Mean and Control Charts
	6: Time Series
13	7: Models for Growth
	12: Experimental Design
	14: Samples and Surveys
	26: Case Study

The series *Statistics: Decisions Through Data* consists of 21 video modules ranging between 12 min and 15 min in length. See below for the listing of programs arranged according to the chapter from *Elementary Statistics*, 6th edition. The scope of these video modules is much less than that of the modules from *Against All Odds: Inside Statistics*, so that fewer text chapters apply. For information about obtaining this series, call 1-800-LEARNER.

Text Chapter	Recommended program from *Decisions Through Data*
1	1: What is Statistics?
	17: Census and Sampling
	18: Sample Surveys
2	2: Stemplots
	3: Histograms and Distributions
	4: Measures of Center: Mean and Median
	5: Five-Number Summary and Boxplots
	6: The Standard Deviation
5	7: Normal Curves
	8: Normal Calculations
6	20: Confidence Intervals
7	21: Tests of Significance
8	15: Designing Experiments
9	9: Straight-Line Growth
	10: Exponential Growth
	11: Scatterplots
	12: Fitting Lines to Data
	13: Correlation
	14: Case Study: Save the Bay
	16: The Question of Causation
12	19: Sampling Distributions

Chapter 1   Quiz A   Name_____

1. In each of the following, determine which of the four
   levels of measurement (nominal, ordinal, interval,
   ratio) is most appropriate.
   (a) Eye colors of fashion models. _____
   (b) A paint manufacturer's ratings of "good, better, and
       best." _____
   (c) Times spent in prison by convicted embezzlers._____
   (d) Temperatures of newborn baby girls._____
   (e) Social security numbers. _____

2. Identify the type of sampling used in each case.
   (a) A quality control analyst inspects every 50th compact
       disk from the assembly line._____
   (b) A sample of 12 jurors is selected by first placing all
       names on cards that are mixed in a drum._____
   (c) A television news team samples reactions to an elec-
       tion by selecting adults who pass by the news studio
       entrance. _____
   (d) A political analyst first identifies three different
       economic groups, then selects 200 subjects from each of
       them. _____
   (e) News coverage includes exit polls of everyone from each
       of 80 randomly selected election precincts._____

3. Criticize this report: "An early morning raid netted
   police $475,325 worth of counterfeit designer pocket-
   books."

4. You plan to test a sample of M&M candies manufactured at a
   Mars, Inc. manufacturing plant. What is wrong with selecting the last 100 candies made in one particular day?

5. Define the terms *population, sample, parameter,* and
   statistic.

Chapter 1   Quiz B   **Name**_____

1. Identify the type of sampling used in each case.
   (a) A tax auditor selects every 500th income tax return
       that is received. _____
   (b) A market researcher selects 500 drivers under 30
       years of age and 500 drivers over 30 years of age.
       _____
   (c) A market researcher selects all adults in each of 10
       randomly selected villages. _____
   (d) A pollster uses a computer to generate 500 random num-
       bers, then interviews the voters corresponding to
       those numbers. _____
   (e) To avoid working late, a quality control analyst
       simply inspects the first 100 items produced in a day.

       _____

2. In each of the following, determine which of the four
   levels of measurement (nominal, ordinal, interval,
   ratio) is most appropriate.
   (a) Temperatures of the ocean at various depths._____
   (b) Salaries of statistics professors._____
   (c) Survey responses of "good, better, best."_____
   (d) Month of birth of survey respondents._____
   (e) Ages of survey respondents.   _____

3. Distinguish between a statistic and a parameter.

4. You plan to poll 16-year-old women about their attitudes
   concerning education. What is wrong with selecting a
   sample of women from different high schools?

5. A college conducts a survey of its alumni in an attempt
   to determine their median annual salary. Identify two
   reasons why the results may be misleading.

Chapter 1   Quiz C   **Name**_____

1. In each of the following, determine which of the four
   levels of measurement (nominal, ordinal, interval,
   ratio) is most appropriate.
   (a) Rankings of the top ten professional tennis players.

   (b) Weights of tennis players.   _____
   (c) Body temperatures of tennis players._____
   (d) Political party affiliations of tennis players._____
   (e) Telephone area codes.   _____

2. Identify the type of sampling used in each case.
   (a) A pollster selects drivers who are waiting to have
       their cars repaired at a local Sears Auto store.

   (b) A pollster selects every 500th name in a telephone
       book.
   (c) A pollster selects 100 men and 100 women._____
   (d) A pollster selects everyone in each of 12 randomly
       selected election precincts.   _____
   (e) A pollster writes the names of each voter on a card,
       shuffles the cards, then draws 25 names._____

3. Define the terms *population, sample, parameter,* and
   statistic.

4. You plan to conduct a poll of drivers in your region.
    What is wrong with using a sample that consists of
    drivers who are waiting to have their cars repaired
    at a local Sears Auto store?

5. Criticize this statement: "Last year, Americans consumed 121,358,399,451 gallons of gasoline."

Chapter 1   Quiz D   **Name**_____

1. You plan to conduct a poll by sampling fans from the population of all fans of the New York Giants. What is wrong
   with using a sample of fans selected as they leave the
   stadium upon completion of a game?

2. Identify the type of sampling used in each case.
   (a) The type of sampling used in problem 1._____
   (b) A quality control manager selects every 500th news-
       paper from the printing press.   _____
   (c) A researcher identifies a numbered list of 500 sub-
       jects, then selects 50 of them by using a computer to
       randomly generate 50 numbers.   _____
   (d) A market researcher selects all customers in each of 15
       randomly selected stores.   _____
   (e) A political analyst selects 1000 Democrats and 1000
       Republicans.   _____

3. In each of the following, determine which of the four
   levels of measurement (nominal, ordinal, interval,
   ratio) is most appropriate.
   (a) Consumer ratings of poor, fair, good, and excellent.
                                             _____
   (b) Times required by different students to learn how to
       drive a car.   _____
   (c) Heights of fashion models.   _____
   (d) Temperatures of the 50 state capitols._____
   (e) Zip codes   _____

4. Distinguish between a statistic and a parameter.

5. (a) Describe one way in which a graph may be constructed so
       as to be misleading.

   (b) A researcher randomly selects 7000 owners of American
       cars, then mails a survey to each of them. Five hun-
       dred responses are received. What's wrong with this
       survey?

Chapter 2   Quiz A   **Name**_____

1. Use the given sample data to find each of the listed
   values.   49   52   52   52   74   67   55   55

   (a) mean   _____   (b) median   _____

   (c) mode   _____   (d) midrange_____

   (e) range   _____   (f) variance_____

   (g) st. dev._____   (h) Q_3 _____

   (i) Σx^2 _____   (j) $(\Sigma x)^2$ _____

2. Construct the stem-and-leaf plot and a boxplot for the
   data given in problem 1.

3. Which is better: A score of 580 on a test with a mean of
   430 and a standard deviation of 130, or a score of 93
   on a test with a mean of 75 and a standard deviation of
   15? Explain your choice.

4. Use the frequency table below
   to find the values listed at
   the right.   (a) mean   _____
                                       (b) standard
   | x | f | deviation_____
   |-------|----|
   | 0- 8 | 2 |
   | 9-17 | 8 |
   | 18-26 | 5 |
   | 27-35 | 10 |
   | 36-44 | 4 |
   | 45-53 | 1 |

5. (a) What is the class width for the data summarized in
       the frequency table of problem 4?   _____

   (b) If a set of data has a mean of 9.7, what is the mean
       after 20 has been added to each score?   _____

   (c) If a set of data has a standard deviation of 5.8,
       what is the standard deviation after 20 has been
       added to each score?   _____

Chapter 2   Quiz B   **Name**_____

1. Use the given sample data to find each of the listed
   values.   63   76   88   75   82   82   94   81   61

   (a) mean   _____   (b) median   _____

   (c) mode   _____   (d) midrange_____

   (e) range   _____   (f) variance_____

   (g) st. dev._____   (h) Q_3 _____

   (i) Σx^2 _____   (j) $(\Sigma x)^2$ _____

2. Use the frequency table below
   to find the values listed at
   the right.   (a) mean   _____

x	f
1-15	2
16-30	0
31-45	5
46-60	8
61-75	12
76-90	20

   (b) standard
       deviation_____

3. Referring to the frequency table given in problem 2,
   answer the following questions.

   (a) What is the lower class limit for the first class?
                                       (a)_____
   (b) What is the class mark of the first class?
                                       (b)_____
   (c) What is the lower class boundary of the first class?
                                       (c)_____
   (d) What is the sample size n?   (d)_____
   (e) What is the class width?   (e)_____

4. Construct the histogram for the data summarized in the
   frequency table for problem 2.

5. Construct the stem-and-leaf plot and a boxplot for the
   data given in problem 1.

6. Which is better: A score of 66 on a test with a mean of
   80 and a standard deviation of 6, or a score of 70 on a
   test with a mean of 80 and a standard deviation of 3?
   Explain your choice.

Chapter 2   Quiz C   **Name**_____

1. Use the given sample data to find each of the listed
   values.   68   59   59   59   57   56   56   47   58   80   50

   (a) mean   _____   (b) median   _____

   (c) mode   _____   (d) midrange_____

   (e) range   _____   (f) variance_____

   (g) st. dev._____   (h) Q_3 _____

   (i) Σx^2 _____   (j) $(\Sigma x)^2$ _____

2. Construct the stem-and-leaf plot and a boxplot for the
   data given in problem 1.

3. Which is better: A score of 82 on a test with a mean of
   70 and a standard deviation of 8, or a score of 82 on a
   test with a mean of 75 and a standard deviation of 4?
   Explain your choice.

4. Use the frequency table below
   to find the values listed at
   the right.

x	f
21– 35	2
36– 50	5
51– 65	8
66– 80	12
81– 95	11
96–110	6

   (a) mean   _____
   (b) standard
       deviation_____
   (c) class width_____
   (d) lowery boundary of
       the first class___
   (e) class mark of the
       first class_____

Chapter 2   Quiz D   **Name**_____

1. Use the given sample data to find each of the listed
   values.   62   52   52   52   64   69   69   76

   (a) mean   _____   (b) median   _____

   (c) mode   _____   (d) midrange_____

   (e) range   _____   (f) variance_____

   (g) st. dev._____   (h) Q_1 _____

   (i) Σx^2 _____   (j) $(\Sigma x)^2$ _____

2. Use the frequency table below
   to find the values listed at
   the right.   (a) mean   _____

x	f
41- 50	2
51- 60	1
61- 70	5
71- 80	12
81- 90	8
91-100	4

   (b) standard
       deviation_____

3. Referring to the frequency table given in problem 2,
   answer the following questions.

   (a) What is the lower class limit for the first class?
                                            (a)_____
   (b) What is the class mark of the first class?
                                            (b)_____
   (c) What is the lower class boundary of the first class?
                                            (c)_____
   (d) What is the sample size n?   (d)_____
   (e) What is the class width?   (e)_____

4. Construct the histogram for the data summarized in the
   frequency table for problem 2.

5. Construct the stem-and-leaf plot and a boxplot for the
   data given in problem 1.

6. Which is better: A score of 80 on a test with a mean of
   70 and a standard deviation of 5, or a score of 50 on a
   test with a mean of 40 and a standard deviation of 8?
   Explain.

Chapter 3   Quiz A   Name_____

1. A process of manufacturing clocks has a 90% yield, meaning
   that 90% of the products are acceptable and 10% are defective. If three of the clocks are randomly selected, find
   the probability that all of them are acceptable.

                                                    _____

2. A doctor knows that one particular treatment has a 30%
   cure rate. If that treatment is used on six different
   patients, what is the probability that at least one
   patient is cured?
                                                    _____

3. (a) Evaluate $_{12}C_9$.   a._____
   (b) For a doubles tennis match, one man and one woman
       must be chosen from a pool of five men and seven
       women. If both selections are random, how many
       different pairings are possible?   b._____

   (c) A computer is used to randomly generate five differ-
       ent numbers. What is the probability that those
       numbers are arranged in order from lowest to high-
       est?   c._____

4. A batch consists of 12 defective clocks and 88 good ones.
   Find the probability of getting two good clocks when two
   coils are randomly selected if ...
   (a) the first selection is replaced before the second
       selection is made.   a._____
   (b) the first selection is NOT replaced.   b._____

5. A fund raising committee is to be selected from a group
   of 25 members, including nine college graduates (6 of
   whom are women) and 16 people who did not graduate from
   college (12 of whom are women).
   (a) If the chairperson is randomly selected, find the
       probability of getting a woman.   a._____
   (b) If the chairperson is randomly selected, find the
       probability of getting a man or a college graduate.
                                                  b._____
   (c) If two members are randomly selected for a special
       project, find the probability that they are both
       women.   c._____
   (d) At each meeting, one of the 25 members is randomly
       chosen to be secretary. Find the probability that
       the first two secretaries are both men.   d._____
   (e) If 1 of the 25 members is randomly selected, find the
       probability of getting a college graduate, given that
       the person selected is a woman.   e._____

Chapter 3   Quiz B   Name_____

1. A batch consists of 15 defective capacitors and 85 good
   capacitors. Find the probability of getting two good
   capacitors when two are selected if ...
   (a) the first selection is replaced before the second
       selection is made.   a._____
   (b) the first selection is NOT replaced.   b._____

2. A multiple choice test allows answers of a, b, c, d, e
   for each question. A student answers ten questions by
   making random guesses.
   (a) What is the probability that the first answer is
       correct?   a._____
   (b) Find P(all ten answers are wrong).   b._____
   (c) Find P(at least one correct answer).   c._____

3. (a) Evaluate $_{12}C_8$.   a._____
   (b) If five women are selected for a lineup, how many
       different ways can they be arranged?   b._____
   (c) One doctor and one nurse must be chosen from a pool
       of twelve doctors and 8 nurses. If both selections
       are random, how many different doctor-nurse pairings
       are possible?   c._____

4. A citizen's action committee against nuclear power is
   comprised of 6 Democrats (4 of whom are men), 8 Republicans (5 of whom are men), and 3 Conservatives (1 of whom
   is a man).
   (a) If 1 member is randomly selected, find the probability
       of getting a Conservative.   a._____
   (b) If 1 member is randomly selected, find the probability
       of getting a Republican or a Conservative.
                                                       b._____
   (c) If 2 different people are randomly selected from
       this committee, find the probability that they are
       both Democrats.   c._____
   (d) If 1 member is randomly selected, find the probability
       of getting a Democrat or a man.   d._____
   (e) If 1 member is randomly selected, find the probability
       of getting a Democrat, given that the selected person
       is a man.   e._____

Chapter 3   Quiz C   **Name**_____

1. The governors of three states appoint a crime commission
   that includes 12 Floridians (8 of whom are women), 9 resi-
   dents of Alabama (6 of whom are women), and 10 Georgians
   (7 of whom are women).

   (a) If the chairperson is randomly selected, find the
       probability of getting a woman.   a._____

   (b) If the chairperson is randomly selected, find the
       probability of getting a man or a Floridian.
                                                 b._____

   (c) If two different members are randomly selected for a
       special project, find the probability that they are
       both women.   c._____

   (d) At each meeting, one of the members is randomly
       chosen to be secretary. Find the probability that
       the first two secretaries are both men.   d._____

   (e) If one of the commission members is randomly selected,
       find the probability of getting a Floridian, given that
       the selected person is a woman.   e._____

2. Based on previous studies, one particular treatment has
   a 70% cure rate. If 6 patients are randomly selected and
   given this treatment, find the probability that they are
   all cured.
                                                 _____

3. Assume that boys and girls are equally likely and a
   couple will have four children.

   (a) Find P(at least one girl).   a._____

   (b) Find P(all girls).   b._____

4. (a) Five different people apply for a job and the appli-
       cations arrive on different days. What is the prob-
       ability that they arrive in alphabetical order?____

   (b) Evaluate $_{10}C_7$.   b._____

   (c) If $P(A) = 0.174$, find $P(\overline{A})$. c._____

Chapter 3   Quiz D   **Name**_____

1. In one region, 30% of all residential telephone numbers
   are unlisted. If 4 residential housing units are randomly selected, find the probability that all of them have
   unlisted numbers.   1._____

2. (a) Evaluate $_{16}P_3$.   _____

   (b) A supplier must fill four large orders. How
       many different ways can those orders be
       arranged in a queue?   _____

   (c) A state lottery involves the random selection
       of six different numbers between 1 and 44. If
       you select one six-number combination, what is
       the probability it will be the winning combination?

                                                      _____

3. A legislative advisory committee consists of 20 Democrats
   (8 of whom are college graduates) and 10 Republicans (3 of
   whom are college graduates).

   (a) Two of the committee members are randomly selected
       for a special research project. What is the proba-
       bility that they are both Democrats?   _____
   (b) If the chairperson is randomly selected, find the
       probability of getting a Democrat or a college grad-
       uate.   _____
   (c) At each meeting of this committee, one person is
       randomly chosen from the 30 members and that person
       must act as a secretary for the meeting. Find the
       probability that the first two meetings have Democrat
       secretaries.   _____
   (d) If one of the committee members is randomly selected
       as treasurer, find the probability that a college grad-
       uate is chosen.   _____
   (e) If one of the committee members is randomly selected,
       find the probability of getting a college graduate,
       given that the selected person is a Democrat._____

4. An unprepared student makes random guesses for the ten
   true/false questions on a quiz.

   (a) Find the probability that there is at least one
       correct answer.   a._____

   (b) Find the probability that the answers are either all
       correct or all wrong.   b._____

Chapter 4   Quiz A   **Name**_____

1. A company manufactures calculators in batches of 64 and there is a 5% rate of defects.

   (a) Find the mean number of defects per batch._____

   (b) Find the standard deviation for the number of defects per batch.   b._____

   (c) Find the probability of getting exactly three defects in a batch.   c._____

2.

x	P(x)
0	0.23
1	0.38
2	0.31
3	0.08

   Find the mean, standard deviation, and variance for the accompanying probability distribution. The variable x represents the number of students (among 3) who were able to preregister in less than 1 hour.

   mean:_____

   standard deviation:_____

   variance:_____

3. (a) Find the probability of getting three defective items in a batch of ten items if the overall percentage of defects is 40%.   _____

   (b) Find the probability of getting at least three defective items in a batch of ten items if the overall percentage of defects is 40%.   _____

4. Does $P(x) = (x+2)/6$ for $x = 0$, 1, 2 determine a probability distribution? Explain.   5._____

5. For one particular policy, an insurance company has a 0.998 probability of gaining $160 and a 0.002 probability of losing $9840. What is the company's expected value?

_____

Chapter 4   Quiz B   **Name**_____

1. An IRS agent randomly selects 15 income tax returns for audit. (She audits 15 returns each day.) In the past, the agent has found that 20% of her audited returns contain errors.

   (a) Find the mean number of returns with errors per day.

   a._____

   (b) Find the standard deviation for the number of returns with errors (per day).   b._____

   (c) Find the probability that for one particular day, there are exactly two returns with errors.   c._____

   (d) Find the probability that for one particular day, there are at least two returns with errors.   d._____

   (e) If the agent works on a weekend and audits 30 returns, find the probability that there will be exactly 4 returns with errors.   e._____

2.

x	P(x)
0	0.05
2	0.17
4	0.43
6	0.35

Find the mean, standard deviation, and variance for the accompanying probability distribution.

mean:_____
standard deviation:_____
variance:_____

3. A contractor has a 0.80 probability of making $70,000, a 0.15 probability of losing $20,000, and a 0.05 probability of breaking even. What is the contractor's expected value?   _____

4. Does $P(x) = x/2$ for $x = 0, 1, 2$ determine a probability distribution? Explain.

Chapter 4   Quiz C   **Name**_____

1. A question on a proficiency test is multiple choice with four possible answers, one of which is correct. Assuming that all responses are random guesses, find the probability that among 12 test subjects, exactly five answer the question correctly.   _____

2.
x	P(x)
0	0.18
1	0.21
2	0.23
3	0.17
4	0.21

Find the mean, standard deviation, and variance for the accompanying probability distribution.

mean:_____
standard deviation:_____
variance:_____

3. A small company manufactures a new product and it is estimated that there is a 0.65 probability of making $45,000, a 0.20 probability of losing $30,000, and a 0.15 probability of breaking even. What is the expected value for this company?   _____

4. In conducting certain stress analysis tests, forces of 5000 pounds are applied to cables. The cables are tested in batches of 12 and an average of 5% of the cables fail. Find ...

   (a) the mean number of failures per batch.   a._____

   (b) the standard deviation for the numbers of failures per batch.   b._____

   (c) the probability that in a batch of 12 cables, exactly one will fail.   c._____

   (d) the probability that in a batch of 12 cables, at least two will fail.   d._____

Chapter 4   Quiz D   **Name**_____

1. The Electronic Gallery is a retail store specializing in video games. Prospective employees are separated into groups of 12 and are given a test of reasoning ability. Based on past results, 70% of the prospective employees pass the test. Find ...

   (a) the mean number of prospective employees per group who pass the test.   a._____

   (b) the standard deviation for the number of prospective employees per group who pass the test.   b._____

   (c) the probability of having exactly 8 prospective employees pass in a given group.   c._____

   (d) the probability of having at least 8 prospective employees pass in a given group.   d._____

2. The Winston Public Relations Company finds that an average of 20% of all new employees resign during the first year. Find the probability that among the next 30 employees hired, exactly 5 resign during the first year. _____

3.

x	P(x)
0	0.04
1	0.16
2	0.45
3	0.20
4	0.15

Find the mean and standard deviation for the accompanying probability distribution for the number of aircraft operations per minute at an airport.

mean:_____

standard deviation:_____

4. A contractor has a 0.60 probability of making $80,000, a 0.30 probability of losing $10,000, and a 0.10 probability of breaking even. What is the expected value?

4._____

Chapter 5   Quiz A   Name_____

1. The Worthington Pottery Company manufactures beer mugs in batches of 120 and the overall rate of defects is 5%. Find the probability of having more than 6 defects in a batch.

_____

2. A bank's loan officer rates applicants for credit. The ratings are normally distributed with a mean of 200 and a standard deviation of 50.

   (a) If an applicant is randomly selected, find the probability of a rating that is between 200 and 275.

   a._____

   (b) If an applicant is randomly selected, find the probability of a rating that is below 250.   b._____

   (c) If an applicant is randomly selected, find the probability of a rating above 300.   c._____

   (d) If an applicant is randomly selected, find the probability of a rating between 170 and 220. d._____

   (e) If an applicant is randomly selected, find the probability of a rating above 178.   e._____

   (f) Find D_6, the score which separates the lower 60% from the top 40%.   f._____

   (g) If 40 different applicants are randomly selected, find the probability that their mean is above 215.

   g._____

Chapter 5   Quiz B   **Name**_____

1. An educational testing corporation has designed a standard test of mechanical aptitude. Scores on this test
   are normally distributed with a mean of 75 and a standard deviation of 15.

   (a) If a subject is randomly selected and tested, find
       the probability that his score will be between 75
       and 100.   a._____

   (b) If a subject is randomly selected and tested, find
       the probability that his score will be between 70
       and 90.   b._____

   (c) If a subject is randomly selected and tested, find
       the probability that his score will be below 85.
                                                    c._____

   (d) If a subject is randomly selected and tested, find
       the probability that his score will be greater than
       80.   d._____

   (e) If a subject is randomly selected and tested, find
       the probability that his score will be between 60
       and 70.   e._____

   (f) Find the 33rd percentile.   f._____

   (g) Eighty different subjects are randomly selected and
       tested. Find the probability that their mean is be-
       low 71.   g._____

2. In one county, the conviction rate for speeding is 85%.
   Find the probability that of the next 100 speeding sum-
   monses issued, there will be at least 90 convictions.

                                                    _____

Chapter 5   Quiz C   Name_____

1. In one region, the September energy consumption levels for single-family homes are found to be normally distributed with a mean of 1050 kwh and a standard deviation of 218 kwh.

   (a) For a randomly selected home, find the probability that the September energy consumption level is between 1050 kwh and 1250 kwh.   a._____

   (b) For a randomly selected home, find the probability that the September energy consumption level is below 1200 kwh.   b._____

   (c) For a randomly selected home, find the probability that the September energy consumption level is between 1100 kwh and 1225 kwh.   c._____

   (d) For a randomly selected home, find the probability that the September energy consumption level is above 1175 kwh.   d._____

   (e) For a randomly selected home, find the probability that the September energy consumption level is between 1000 kwh and 1200 kwh.   e._____

   (f) Find the 45th percentile.   f._____

   (g) If 50 different homes are randomly selected, find the probability that their mean energy consumption level for September is greater than 1075 kwh.
   g._____

2. In a study of the causes of death, it was found that 52% of all Americans die from heart disease. Find the probability that in a group of 500 randomly selected Americans, at least 275 die from heart disease.   _____

Chapter 5   Quiz D   Name_____

1. A physical fitness researcher devises a test of strength and finds that the results are normally distributed with a mean of 110.0 pounds and a standard deviation of 10.4 pounds.

   (a) If a subject is randomly selected and measured, find the probability of a score between 110.0 pounds and 125.0 pounds.   a._____

   (b) If a subject is randomly selected and measured, find the probability of a score between 90.0 pounds and 115.0 pounds.   b._____

   (c) If a subject is randomly selected and measured, find the probability of a score below 95.0 pounds.
                                              c._____

   (d) If a subject is randomly selected and measured, find the probability of a score greater than 85.0 pounds.
                                              d._____

   (e) If a subject is randomly selected and measured, find the probability of a score between 80.0 pounds and 100.0 pounds.   e._____

   (f) Find the 20th percentile.   f._____

   (g) Sixty different subjects are randomly selected and tested. Find the probability that their mean is below 108.0 pounds.   g._____

2. A certain question on a test is answered correctly by 22% of the respondents. Find the probability that among the next 150 respondents, there will be more than 40 correct answers.   _____

Chapter 6   Quiz A   **Name**_____

1. Of 346 vending machines tested, 12 are found to be defective. Construct the 98% confidence interval for the proportion of defective vending machines.

                                        _____

2. The Consumer Protection Corporation wants to estimate the proportion of Chevrolet Corvettes that have crashed. In order to obtain a sample proportion that is in error by no more than 0.04 (with 98% confidence), how many cars must be sampled?   _____

3. A sociologist develops a test to measure attitudes about public transportation, and 27 randomly selected subjects are given the test. Their mean is 76.2 and their standard deviation is 21.4. Construct the 95% confidence interval for the mean score of all such subjects.

                                        _____

4. Using the same sample data in problem 3, construct the 95% confidence interval for the standard deviation of the scores of all subjects. _____

5. (a) Evaluate $z_{\alpha/2}$ for a confidence level of 92%.

                                        a._____

    (b) Evaluate $t_{\alpha/2}$ for a confidence level of 99% and a

         sample size of 17.   b._____

    (c) Evaluate χ_L^2 for a confidence level of 99% and a

         sample size of 20.   c._____

    (d) Evaluate χ_R^2 for a confidence level of 99% and a

         sample size of 20.   d._____

    (e) For the sample data described in problem 3, what is the best point estimate of the population mean?_____

Chapter 6   Quiz B   **Name**_____

1. A survey of 865 voters in California reveals that 408 favor approval of an issue before the legislature. Construct the 95% confidence interval for the true proportion of all voters in California who favor approval.

   _____

2. We want to estimate the proportion of voters who would
   again vote for the current president. We want a confidence level of 97% that our sample proportion is in
   error by no more than 0.03. How large should the sample
   be?   _____

3. Twenty cars were randomly selected and weighed. This
   sample has a mean of 3650 pounds and a standard deviation of 420 pounds. Construct the 99% confidence interval for the population mean.   _____

4. Use the same sample data given in problem 3 to construct
   the 99% confidence interval for the population standard
   deviation.   _____

5. (a) Evaluate $z_{\alpha/2}$ for $\alpha = 0.08$.   a._____

   (b) Evaluate $t_{\alpha/2}$ for $\alpha = 0.02$ and $n = 8$.   b._____

   (c) Given the sample data in problem 3, what is the best
       point estimate of the population mean?   c._____

   (d) Given the sample data in problem 1, what is the best
       point estimate of the population proportion of vot-
       ers who favor approval?   d._____

   (e) Evaluate χ_L^2 for $\alpha = 0.05$ and $n = 10$.   e._____

Chapter 6   Quiz C   **Name**_____

1. We want to estimate the mean energy consumption level
   for a home in one region. We want to be 90% confident
   that our sample mean is within 25 kwh of the true population mean, and past data strongly suggest that the
   population standard deviation is 137 kwh. How large
   must our sample be?   _____

2. A study of shark attacks on humans showed that 15 of
   200 attacks occurred in deep water. Construct the 99%
   confidence interval for the true proportion of shark
   attacks that occur in deep water.   _____

3. A botanist measures the heights of 24 seedlings and ob-
   tains a mean and standard deviation of 41.6 cm and 4.8
   cm, respectively. Construct the 98% confidence interval
   for the population mean.   _____

4. Use the same sample data given in problem 3 to construct
   the 95% confidence interval for the population standard
   deviation σ.   _____

5. (a) Evaluate $t_{\alpha/2}$ for $\alpha = 0.05$ and $n = 18$.   a._____

   (b) Evaluate $z_{\alpha/2}$ for $\alpha = 0.06$.   b._____

   (c) Evaluate χ_L^2 for $\alpha = 0.05$ and $n = 14$.   c._____

   (d) Given the sample data in problem 3, what is the best
       point estimate of the population mean?   d._____

   (e) Given the sample data in problem 2, what is the best
       point estimate of the population proportion of shark
       attacks that occur in deep water?   e._____

Chapter 6   Quiz D   **Name**_____

1. Eighteen subjects are randomly selected and given proficiency tests. The mean for this group is 492.3 and the standard deviation is 37.6. Construct the 95% confidence interval for the population mean.

_____

2. Use the same sample data given in problem 1 to construct the 98% confidence interval for the population standard deviation.   _____

3. We want to estimate, with a maximum error of 0.025, the true proportion of adult Americans who are opposed to nuclear power, and we want 95% confidence in our results. How many people should we survey?

_____

4. A survey of 300 union members in New York State reveals that 37.3% favor the Republican candidate for governor. Construct the 98% confidence interval for the true population proportion of all New York State union members who favor the Republican candidate.

_____

5. (a) Given the sample data in problem 1, what is the best point estimate of the population mean?   a._____

   (b) Given the sample data in problem 4, what is the best point estimate of the population proportion p of voters favoring the Republican?   b._____

   (c) Evaluate $t_{\alpha/2}$ for $\alpha = 0.01$ and $n = 12$.   c._____

   (d) Evaluate $z_{\alpha/2}$ for $\alpha = 0.005$.   d._____

   (e) Evaluate χ_L^2 for $\alpha = 0.05$ and $n = 10$.   e._____

Chapter 7   Quiz A   **Name**_____

1. The Chandler Department Store accepts only its own credit card. Among 36 randomly selected card holders, it was found that the mean amount owed was $175.37, while the standard deviation was $84.77. Use a 0.05 level of significance to test the claim that the mean amount charged by all customers is greater than $150.00.

2. A machine dispenses a liquid drug into bottles in such a way that the standard deviation of the contents is 81 milliliters. A new machine is tested on a sample of 24 containers and the standard deviation for this sample group is found to be 26 milliliters. At the 0.05 level of significance, test the claim that the new machine produces a lower standard deviation.

3. A supplier of 3.5" disks claims that no more than 1% of the disks are defective. In a random sample of 600 disks, it is found that 3% are defective, but the supplier claims that this is only a sample fluctuation. At the 0.01 level of significance, test the supplier's claim that no more than 1% are defective.

4. For the hypothesis test of problem 3, identify the ...

   (a) *P*-value

   (b) type I error

   (c) type II error

Chapter 7   Quiz B   **Name**_____

1. A poll of 1068 adult Americans reveals that 48% of the voters surveyed prefer the Democratic candidate for the presidency. At the 0.05 level of significance, test the claim that at least half of all voters prefer the Democrat.

2. For the hypothesis test of problem 1, identify the ...

   (a) *P*-value

   (b) type I error

   (c) type II error

3. A test of sobriety involves measuring of the subject's motor skills. Twenty randomly selected sober subjects take the test and produce a mean of 41.0 with a standard deviation of 3.7. At the 0.01 level of significance, test the claim that the mean for this test is equal to 35.0.

4. When 12 bolts are tested for hardness, their indexes have a standard deviation of 41.7. Test the claim that the standard deviation of the hardness indexes for all such bolts is greater than 30.0. Use a 0.025 level of significance.

Chapter 7   Quiz C   **Name**_____

1 A large software company gives job applicants a test of
   programming ability and the mean for that test has been
   160.   Twenty-five job applicants are randomly selected
   from one large university and they produce a mean and
   standard deviation of 183 and 12, respectively. Use a
   0.05 level of significance to test the claim that this
   sample comes from a population with an aptitude level
   greater than 160.

2. A manufacturer considers her production process to be
    out of control when defects exceed 3%. In a random sam-
    ple of 170 items, the defect rate is 5.9% but the manager
    claims that this is only a sample fluctuation and pro-
    duction is not really out of control. At the 0.01 level
    of significance, test the manager's claim.

3. For the hypothesis test of problem 2, identify the ...

    (a) *P*-value

    (b) type I error

    (c) type II error

4. With several different waiting lines, a bank finds that
    the standard deviation of waiting times on Monday morn-
    ings is 3.2 minutes. The bank experiments with a single
    main waiting line and finds that for a random sample of
    25 customers, the waiting times have a standard devia-
    tion of 0.3 minutes. At the 0.05 level of significance,
    test the bank manager's claim that the single line pro-
    duces a lower standard deviation.

Chapter 7   Quiz D   Name_____

1. An educational testing company has been using a standard test of verbal ability and the mean and standard deviation have been 430 and 130, respectively. In analyzing a new version of that test, it is found that a sample of 100 randomly selected subjects produces a mean and a standard deviation of 424 and 155, respectively. At the 0.05 level of significance, test the claim that the new version has a standard deviation equal to that of the past version.

2. Before endorsing a candidate for political office, a newspaper editor surveys 200 randomly selected readers and finds that 120 favor the candidate in question. At the 0.05 level of significance, test the editor's claim that the candidate is favored by 2/3 of the readers.

3. For the hypothesis test of problem 2, identify the ...

   (a) P-value

   (b) type I error

   (c) type II error

4. In tests of a computer component, it is found that the mean time between failures is 937 hours. A modification is made which is supposed to increase reliability by increasing the time between failures. Tests on a sample of 36 modified components produce a mean time between failures of 983 hours, with a standard deviation of 52 hours. At the 0.01 level of significance, test the claim that the modified components have a longer mean time between failures.

Chapter 8   Quiz A   **Name**_____

1. A marketing survey involves product recognition in New
   York and California. Of 558 New Yorkers surveyed, 193
   knew the product while 196 out of 614 Californians knew
   the product. At the 0.05 significance level, test the
   claim that the recognition rates are the same in both
   states.

2. Construct a 99% confidence interval for the difference
   between the two population proportions referred to in
   problem 1.

3. A test of abstract reasoning is given to a random sample
   of students before and after they completed a formal
   logic course. The results are given below. At the 0.05
   significance level, test the claim that the mean score
   is not affected by the course.

Before	74 83 75 88 84 63 93 84 91 77
After	73 77 70 77 74 67 95 83 84 75

4. Sample data are collected from   Brand X   Brand Y
   two brands of a product and the
   results are given at the right. $n = 25$ $n = 16$
   At the 0.05 significance level, $\bar{x} = 48.8$ $\bar{x} = 39.5$
   test the claim that these samples $s = 1.5$ $s = 3.7$
   come from populations with equal
   standard deviations.

5. Refer to the same sample data given in problem 4 and
   test the claim that the two populations have equal
   means. Use a 0.05 significance level.

6. Construct a 99% confidence interval for the difference
   $\mu_x - \mu_y$ based on the sample data from problem 4.

Chapter 8   Quiz B   **Name**_____

1. Two types of flares are tested   Brand X   Brand Y
   for their burning times (in min-
   utes) and sample results are given $n = 35$ $n = 40$
   at the right. Use a 0.05 sig- $\bar{x} = 19.4$ $\bar{x} = 15.1$
   nificance level to test the claim $s = 1.4$ $s = 0.8$
   that the two brands have equal
   variances.

2. Refer to the same sample data given in problem 1 and
   test the claim that the two populations have equal
   means. Use a 0.05 significance level.

3. Construct a 95% confidence interval for the difference
   $\mu_x - \mu_y$ based on the sample data from problem 1.

4. A radio station manager conducts an audience survey and
   finds that among 275 men surveyed, 18 were listeners
   while there were 31 listeners among the 325 women surveyed. At the 0.05 level of significance, test the
   claim that the proportions of male and female listeners
   are the same.

5. Construct a 99% confidence interval for the difference
   between the two population proportions referred to in
   problem 4.

6. A course is designed to improve scores on a college entrance examination. A randomly selected experimental
   group is used to test the effectiveness of the course
   and the results are given below. At the 0.05 level of
   significance, test the claim that the course has no effect on the grades.

Subject	A	B	C	D	E	F	G	H	I	J
Before	469	496	529	527	484	446	534	531	539	565
After	480	509	529	538	487	464	545	542	551	572

Chapter 8   Quiz C   **Name**_____

1. A survey of consumer preferences involves two different
   market areas. Of 150 consumers surveyed in area A, 20
   preferred a certain product. There were 350 consumers
   surveyed in market area B, and 52 of them preferred the
   same product. At the 0.01 level of significance, test
   the claim that the product is preferred by the same market shares.

2. Construct a 95% confidence interval for the difference
   between the two population proportions described in
   problem 1.

3. A test of writing ability is given to a random sample
   of students before and after they completed a formal
   writing course. The results are given below. At the
   0.05 significance level, test the claim that the mean
   score is not affected by the course.

Before	70 80 92 99 93 97 76 63 68 71 74
After	69 79 90 96 91 95 75 64 62 64 76

4. Sample data are collected from   Brand X   Brand Y
   two brands of a product and the
   results are given at the right. $n = 10$ $n = 20$
   At the 0.05 significance level, $\bar{x} = 77.3$ $\bar{x} = 71.8$
   test the claim that these samples $s = 5.4$ $s = 12.9$
   come from populations with equal
   standard deviations.

5. Refer to the same sample data given in problem 4 and
   test the claim that the two populations have equal
   means. Use a 0.05 significance level.

6. Construct a 95% confidence interval for the difference
   $\mu_x - \mu_y$ based on the sample data from problem 4.

Chapter 8   Quiz D   **Name**_____

1. A study is made of the defect rates of two machines used
   in manufacturing. Of 200 randomly selected items produced by the first machine, 5 are defective. Of 250
   randomly selected items produced by the second machine,
   15 are defective. At the 0.05 level of significance,
   test the claim that the two machines have the same rate
   of defects.

2. Construct a 95% confidence interval for the difference
   between the two population proportions referred to in
   problem 1.

3. Two different firms design their own versions of the
   same aptitude test, and a psychologist administers both
   versions to randomly selected subjects with the results
   given below. At the 0.05 level of significance, test
   the claim that both versions produce the same mean.

   | Test A | 109 118 104 127 126 99 104 108 113 |
   | Test B | 102 115 107 116 104 91 113 112 112 |

4. Sample data are collected from   Brand X   Brand Y
   two brands of a product and the
   results are given at the right. $n = 20$ $n = 27$
   At the 0.05 significance level, $x = 125.4$ $x = 119.3$
   test the claim that these samples $s = 8.2$ $s = 2.6$
   come from populations with equal
   standard deviations.

5. Refer to the same sample data given in problem 4 and
   test the claim that the two populations have equal
   means. Use a 0.05 significance level.

6. Construct a 99% confidence interval for the difference
   $\mu_x - \mu_y$ based on the sample data from problem 4.

Chapter 9   Quiz A   Name_____

1. <u>Productivity</u>| 23 25 28 21 21 25 26 30 34 36
   <u>Dexterity | 49 53 59 42 47 53 55 63 67 75</u>

   Listed above are results from two different tests designed
   to measure productivity and dexterity for randomly selected employees.
   (a) Plot the scatter diagram on the back of this page.
   (b) Find the value of the linear correlation
       coefficient r.   b._____
   (c) Assuming a 0.05 significance level, find the crit-
       ical value.   c._____
   (d) What do you conclude about the correlation between
       the two variables?_____
   (e) Use the given sample data to find the estimated
       equation of the regression line.   e._____
   (f) Plot the regression line with the equation given in
       part (e). Plot that line on the same scatter dia-
       gram given in part (a).
   (g) What percentage of the total variation can be ex-
       plained by the regression line?   g._____

2. (a) What do you conclude at the 0.05 significance level
       if 18 pairs of data yield $r = -0.456$?_____

   _____
   (b) What do you conclude at the 0.01 significance level
       if 73 pairs of data yield $r = 0.308$?_____

   _____
   (c) Find the best predicted value of y for $x = 5$ given 6
       pairs of data that yield $r = 0.789$, $\bar{y} = 19.0$, and the
       regression equation:
       $$\hat{y} = -2 + 4x$$ c._____
   (d) (True or false) "The centroid $(\bar{x}, \bar{y})$ must lie on the
       regression line, even if there is no significant
       linear correlation."   d._____
   (e) The data set given below has $\hat{y} = 3 + 2x$ for its re-
       gression equation. Find the standard error of est-
       imate s_e.   <u>x | 8 2 3 5</u>   e._____
                           y |19 7 9 13

3. A measure of job satisfaction was developed and results
   are obtained for the same eight employees sampled in
   problem 1. Here are the corresponding results:

   Job satisfaction: 56 58 60 50 54 61 59 63 67 69

   Find the multiple regression equation that expresses the
   job satisfaction scores in terms of the productivity and
   dexterity scores. Also find the value of the multiple
   coefficient of determination.

Chapter 9   Quiz B   Name_____

1. Managers rate employees according to job performance and attitude. The results for several randomly selected employees are given below.

Performance	59	63	65	69	58	77	76	69	70	64
Attitude	72	67	78	82	75	87	92	83	87	78

(a) Plot the scatter diagram on the back of this page.
(b) Find the value of the linear correlation coefficient r.   b._____
(c) Assuming a 0.05 significance level, find the critical value.   c._____
(d) What do you conclude about the correlation between the two variables?

_____

(e) Use the given sample data to find the estimated equation of the regression line.   e._____
(f) Plot the regression line with the equation given in part (e). Plot that line on the same scatter diagram given in part (a).
(g) What percentage of the total variation can be explained by the regression line?   g._____

2. (a) If the linear correlation coefficient r is negative, must the slope of regression line be negative?_____
(b) What do you conclude at the 0.01 significance level if 42 pairs of data yield $r = 0.362$?_____

_____

(c) Find the best predicted value of y given that $x = 2$ for 10 pairs of data that yield $r = 0.003$, $\bar{y} = 5.0$, and the following regression equation:
$$\hat{y} = 2 + 3x$$ c._____
(d) What is known about the standard error of estimate s_e for a set of paired data with a linear correlation coefficient of $r = 1$?_____
(e) What do you conclude at the 0.05 significance level if 20 pairs of data yield $r = -0.456$?_____

_____

3. In addition to the performance and attitude sample data in problem 1, managers also rate the same employees according to adaptability and these are the results that correspond to those given in problem 1:

   Adaptability: 50 52 54 60 46 67 66 59 62 55

Find the multiple regression equation that expresses performance in terms of attitude and adaptability.

_____

Chapter 9   Quiz C   **Name**_____

1. Two separate tests are designed to measure student's ability to solve problems. Several students are randomly selected to take both tests and the results are given below.

Test A	48 52 58 44 43 43 40 51 59
Test B	73 67 73 59 58 56 58 64 74

   (a) Plot the scatter diagram on the back of this page.
   (b) Find the value of the linear correlation
       coefficient r.   b._____
   (c) Use a 0.05 significance level and find the critical
       value.   c._____
   (d) What do you conclude about the correlation between
       the two variables?_____
   (e) Use the given sample data to find the estimated
       equation of the regression line.   e._____
   (f) Plot the regression line with the equation given in
       part (e). Plot that line on the same scatter dia-
       gram given in part (a).
   (g) What percentage of the total variation can be ex-
       plained by the regression line?   g._____

2. (a) What do you conclude at the 0.05 significance level
       if 90 pairs of data yield $r = -0.209$?_____

   _____
   (b) What do you conclude at the 0.01 significance level
       if 42 pairs of data yield $r = 0.375$?_____

   _____
   (c) Find the best predicted value of y given that $x = 5$
       for 6 pairs of data that yield $r = 0.444$, $\bar{y} = 18.3$, and
       the regression equation
           $\hat{y} = 2 + 5x$ c._____
   (d) (True or false) "The centroid $(\bar{x}, \bar{y})$ must lie on the
       regression line, even if there is no significant
       linear correlation."   d._____
   (e) The data set given below has $\hat{y} = -5 + 3x$ for its re-
       gression equation. Find the standard error of est-
       imate s_e. | x | 3 2 5 8 | e._____
                            | y | 4 1 10 19 |

3. In addition to the test results given in problem 1, a
   third test was designed to measure the same problem-
   solving ability and the following results correspond
   to the same students.

Test C:	48 41 59 45 42 44 40 58 60

   (a) Find the multiple regression equation that expresses
       the test C results in terms of those for tests A and
       B._____
   (b) Find the multiple coefficient of determination._____

Chapter 9   Quiz D   **Name**_____

1. Two separate tests are designed to measure a student's ability to solve problems. Several students are randomly selected to take both tests and the results are given below.

Test A	64 48 51 59 60 43 41 42 35 50 45
Test B	91 68 80 92 91 67 65 67 56 78 71

   (a) Plot the scatter diagram on the back of this page.
   (b) Find the value of the linear correlation
       coefficient r.   b._____
   (c) Assuming a 0.05 significance level, find the crit-
       ical value.   c._____
   (d) What do you conclude about the correlation between
       the two variables?_____
   (e) Use the given sample data to find the estimated
       equation of the regression line.   e._____
   (f) Plot the regression line with the equation given in
       part (e). Plot that line on the same scatter dia-
       gram given in part (a).
   (g) What percentage of the total variation can be ex-
       plained by the regression line?   g._____

2. (a) Find the best predicted value of y given that $x = 3$
       for 10 pairs of data that yield $r = 0.555$, $\bar{y} = 23.0$,
       and the regression equation
       $$\hat{y} = 8 + 5x$$ a._____
   (b) What do you conclude at the 0.05 significance level
       if 11 pairs of data yield $r = 0.707$?_____
       _____
   (c) What do you conclude at the 0.01 significance level
       if 27 pairs of data yield $r = -0.413$?_____
       _____
   (d) What is known about the standard error of estimate
       s_e for a set of paired data with a linear correla-
       tion coefficient of $r = -1$?_____
   (e) (True or false) "Even if there is no correlation,
       the centroid $(\bar{x}, \bar{y})$ must lie on the regression line."
                                                           e._____

3. A third test is designed to measure a student's problem-
   solving ability and it is administered to the same students referred to in problem 1. These results correspond to those given in problem 1:

   Test C: 88   64   76   90   88   65   63   62   64   75   70

   Find the multiple regression equation that expresses the
   test C results in terms of those for tests A and B.

Chapter 10   Quiz A   **Name**_____

1. In studying the occurrence of genetic characteristics, the following sample data were obtained. At the 0.05 significance level, test the claim that the characteristics occur with the same frequency.

Characteristic	A	B	C	D	E	F
Frequency	28	30	45	48	38	39

2. Among the four northwestern states, Washington has 51% of the total population, Oregon has 30%, Idaho has 11%, and Montana has 8%. A market researcher selects a sample of 1000 subjects, with 450 in Washington, 340 in Oregon, 150 in Idaho, and 60 in Montana. At the 0.05 significance level, test the claim that the sample of 1000 subjects has a distribution that agrees with the distribution of state populations.

3. Responses to a survey question are broken down according to employment and the sample results are given below. At the 0.10 significance level, test the claim that the response and employment status are independent.

	Yes	No	Undecided
Employed	30	15	5
Unemployed	20	25	10

Chapter 10   Quiz B   **Name**_____

1. The correct and incorrect responses to a test question
   are listed according to major with the sample results
   given below. At the 0.10 significance level, test the
   claim that response and major are independent.

	Correct	Incorrect
Math	27	53
English	43	37

2. A company manager wishes to test a union leader's claim
   that absences occur on the different week days with the
   same frequencies. Test this claim at the 0.05 level of
   significance if the following sample data have been compiled.

Day	Mon	Tue	Wed	Thur	Fri
Absences	37	15	12	23	43

3. Using the same sample data and significance level from
   problem 2, test the claim that absences on the five days
   occur with percentages of 30%, 10%, 10%, 15%, and 35%
   respectively.

Chapter 10   Quiz C   **Name**_____

1. Responses to a survey question are broken down according to sex and the sample results are given below. At the 0.05 significance level, test the claim that the response and sex are independent.

	Yes	No	Undecided
Male	25	50	15
Female	20	30	10

2. In studying the responses to a multiple-choice test question, the following sample data were obtained. At the 0.05 significance level, test the claim that the responses occur with the same frequency.

Response	A	B	C	D	E
Frequency	12	15	16	18	19

3. Using the same sample data and significance level from problem 2, test the claim that the responses occur with percentages of 15%, 20%, 25%, 25%, and 15% respectively.

Chapter 10   Quiz D   **Name**_____

1. Responses to a survey question are broken down according
   to employment and the sample results are given below.
   At the 0.10 significance level, test the claim that the
   response and employment status are independent.

	Yes	No	Undecided
Employed	15	35	20
Unemployed	25	25	10

2. In analyzing the random number generator of a certain
   computer, the following sample results were obtained.
   At the 0.05 level of significance, test the claim that
   the outcomes occur with equal frequency.

Outcome	1	2	3	4	5	6
Frequency	18	12	14	16	21	15

3. Using the same sample data and significance level from
   problem 2, test the claim that the outcomes occur with
   percentages of 20%, 10%, 15%, 15%, 20%, and 20% respectively.

Chapter 11   Quiz A   Name_____

1. At the 0.01 significance level, test the claim that the
   three brands have the same mean if the following sample
   results have been obtained.

Brand A	Brand B	Brand C
44	30	28
47	32	27
44	34	31
40	36	32
39	38	36
	40	
	42	

2. A manager records the production output of three employees
   who each work on three different machines for three different days. The sample results are given below and the Minitab results follow.

		Employee		
		A	B	C
Machine	I	23, 27, 29	30, 27, 25	18, 20, 22
	II	25, 26, 24	24, 29, 26	19, 16, 14
	III	28, 25, 26	25, 27, 23	15, 11, 17

   ANALYSIS OF VARIANCE ITEMS

SOURCE	DF	SS	MS
MACHINE	2	34.67	17.33
EMPLOYEE	2	504.67	252.33
INTERACTION	4	26.67	6.67
ERROR	18	98.00	5.44
TOTAL	26	664.00	

   (a) Using a 0.05 significance level, test the claim that
       the interaction between employee and machine has no
       effect on the number of items produced.
   (b) Assume that the results of part (a) showed that there
       is no interaction effect. Using a 0.05 significance
       level, test the claim that the machine has no effect
       on the number of items produced.
   (c) Assume that the results of part (a) showed that there
       is no interaction effect. Using a 0.05 significance
       level, test the claim that the choice of employee has
       no effect on the number of items produced.

Chapter 11   Quiz B   **Name**_____

1. At the 0.025 significance level, test the claim that the four brands have the same mean if the following sample results have been obtained.

Brand A	Brand B	Brand C	Brand D
17	18	21	22
20	18	24	25
21	23	25	27
22	25	26	29
21	26	29	35
		29	36
			37

2. A manager records the production output of three employees who each work on three different machines for three different days. The sample results are given below and the Minitab results follow.

		Employee		
		A	B	C
	I	16, 18, 19	15, 17, 20	14, 18, 16
Machine	II	20, 27, 29	25, 28, 27	29, 28, 26
	III	15, 18, 17	16, 16, 19	13, 17, 16

ANALYSIS OF VARIANCE ITEMS

SOURCE	DF	SS	MS
MACHINE	2	588.74	294.37
EMPLOYEE	2	2.07	1.04
INTERACTION	4	15.48	3.87
ERROR	18	98.67	5.48
TOTAL	26	704.96	

(a) Using a 0.05 significance level, test the claim that the interaction between employee and machine has no effect on the number of items produced.
(b) Assume that the results of part (a) showed that there is no interaction effect. Using a 0.05 significance level, test the claim that the machine has no effect on the number of items produced.
(c) Assume that the results of part (a) showed that there is no interaction effect. Using a 0.05 significance level, test the claim that the choice of employee has no effect on the number of items produced.

Chapter 11   Quiz C   Name_____

1. At the 0.025 significance level, test the claim that the
   three brands have the same mean if the following sample
   results have been obtained.

Brand A	Brand B	Brand C
32	27	22
34	24	25
37	33	32
33	30	22
36		21
39		

2. A manager records the production output of three employees
   who each work on three different machines for three different days. The sample results are given below and the Minitab results follow.

		Employee		
		A	B	C
Machine	I	31, 34, 32	29, 23, 22	21, 20, 24
	II	19, 26, 22	35, 33, 30	25, 19, 23
	III	21, 18, 26	20, 23, 24	36, 37, 31

ANALYSIS OF VARIANCE ITEMS

SOURCE	DF	SS	MS
MACHINE	2	1.19	0.59
EMPLOYEE	2	5.85	2.93
INTERACTION	4	710.81	177.70
ERROR	18	160.00	8.89
TOTAL	26	877.85	

(a) Using a 0.05 significance level, test the claim that
    the interaction between employee and machine has no
    effect on the number of items produced.
(b) Assume that the results of part (a) showed that there
    is no interaction effect. Using a 0.05 significance
    level, test the claim that the machine has no effect on
    the number of items produced.
(c) Assume that the results of part (a) showed that there
    is no interaction effect. Using a 0.05 significance
    level, test the claim that the choice of employee has
    no effect on the number of items produced.

Chapter 11 Quiz D Name______________________

1. At the 0.025 significance level, test the claim that the four brands have the same mean if the following sample results have been obtained.

Brand A	Brand B	Brand C	Brand D
15	20	21	15
25	17	22	15
21	22	20	14
23	23	19	23
22		18	22
20			28
			28

2. A manager records the production output of three employees who each work on two different machines for four different days. The sample results are given below and the Minitab results follow.

		Employee A	Employee B	Employee C
Machine	I	21 22 19 24	15 12 11 10	19 23 25 22
	II	9 11 13 10	19 23 22 22	25 18 21 23

ANALYSIS OF VARIANCE ITEMS

SOURCE	DF	SS	MS
MACHINE	1	2.04	2.04
EMPLOYEE	2	166.58	83.29
INTERACTION	2	410.08	205.04
ERROR	18	90.25	5.01
TOTAL	23	668.96	

(a) Using a 0.05 significance level, test the claim that the interaction between employee and machine has no effect on the number of items produced.

(b) Assume that the results of part (a) showed that there is no interaction effect. Using a 0.05 significance level, test the claim that the machine has no effect on the number of items produced.

(c) Assume that the results of part (a) showed that there is no interaction effect. Using a 0.05 significance level, test the claim that the choice of employee has no effect on the number of items produced.

Chapter 12   Quiz A   Name_____

1.   The Collins Computer Manufacturing Company makes laser
      printer cartridges that are supposed to contain 350 g
      of toner. During production, a sample of 5 cartridges
      is randomly selected each hour and the results are list-
      ed below for 20 hours. Construct a control chart for R,
      a control chart for $\bar{x}$, then write a conclusion about the
      statistical stability of this manufacturing process.

Sample			Weight of Toner (mg)			Mean	Range
1	350.1	343.3	335.9	343.7	342.8	343.16	14.2
2	343.0	344.8	345.0	337.5	344.6	342.98	7.5
3	347.9	347.5	347.0	343.8	349.5	347.14	5.7
4	343.3	353.1	348.8	345.4	345.8	347.28	9.8
5	351.5	343.5	340.8	342.5	341.9	344.04	10.7
6	345.7	350.0	352.6	341.0	342.8	346.42	11.6
7	352.4	348.3	342.8	348.3	342.0	346.76	10.4
8	344.2	342.5	340.0	347.1	340.1	342.78	7.1
9	335.7	341.7	345.4	343.6	355.5	344.38	19.8
10	335.2	346.8	349.9	341.2	349.6	344.54	14.7
11	355.9	334.7	348.2	352.1	353.3	348.84	21.2
12	341.8	353.5	350.5	359.2	348.2	350.64	17.4
13	341.0	349.0	347.9	341.9	342.3	344.42	8.0
14	348.0	345.2	348.2	343.5	343.6	345.70	4.7
15	341.9	345.1	340.1	340.1	339.2	341.28	5.9
16	346.8	351.2	344.3	351.6	339.9	346.76	11.7
17	334.7	336.2	345.5	346.8	344.3	341.50	12.1
18	345.4	356.8	346.1	343.3	347.6	347.84	13.5
19	340.9	346.7	341.2	352.7	353.9	347.08	13.0
20	342.6	344.1	344.0	348.8	337.3	343.36	11.5

2.   The Collins Computer Manufacturing Company makes printer
      cables that are either acceptable or defective. During
      each day, 200 cables are randomly selected and tested.
      Listed below are the numbers of defects for each of 15
      consecutive work days. Construct a control chart for p,
      and then determine whether the process is within statis-
      tical control.

          6   7   7   6   9   7   8   10   11   9   12   13   10   16   17

Chapter 12   Quiz B   Name_____

1. The Telektronic Company manufactures outdoor lamps that
    automatically turn on when motion is detected. The man-
    ufacturing process is monitored by recording times re-
    quired to assemble the lamps, and results are listed be-
    low for 5 lamps randomly selected during each of 20 con-
    secutive work days. Construct a control chart for R,
    a control chart for $\bar{x}$, then write a conclusion about the
    statistical stability of this manufacturing process.

Sample	Time (min)					Mean	Range
1	32	27	26	32	21	27.6	11
2	31	27	23	26	24	26.2	8
3	21	32	34	28	35	30.0	14
4	23	20	33	28	30	26.8	13
5	33	28	28	27	26	28.4	7
6	26	31	30	27	30	28.8	5
7	31	27	29	20	32	27.8	12
8	24	26	17	27	26	24.0	10
9	26	23	25	26	24	24.8	3
10	27	28	26	31	31	28.6	5
11	17	26	22	25	17	21.4	9
12	24	25	24	33	26	26.4	9
13	21	27	23	26	13	22.0	14
14	22	21	28	19	30	24.0	11
15	31	23	23	28	27	26.4	8
16	25	35	25	29	28	28.4	10
17	32	33	35	34	30	32.8	5
18	28	26	30	21	25	26.0	9
19	24	31	24	31	21	26.2	10
20	23	29	25	24	29	26.0	6

2. The same lamps described in problem 1 are monitored for
    the proportion of defects. During each of 15 work days,
    200 lamps are randomly selected and tested. The numbers
    of defects are listed below. Construct a control chart
    for p, then determine whether the process is within sta-
    tistical control.

         9 9 8 10 11 11 8 7 12 12 13 6 13 5 14

Chapter 12   Quiz C   Name_____

1.   The Thomas Tire Company monitors a process of changing
     and mounting sets of tires on different cars. For each
     of 24 consecutive work days, the times are recorded for
     4 randomly selected cars and the results are listed be-
     low. Construct a control chart for R, a control chart for
     $\bar{x}$, then write a conclusion about the statistical stabil-
     ity of this process.

Sample	Time	(min)			Mean	Range
1	38	37	38	41	38.50	4
2	38	37	38	35	37.00	3
3	37	40	43	36	39.00	7
4	37	36	38	39	37.50	3
5	38	33	37	37	36.25	5
6	35	33	39	38	36.25	6
7	33	37	37	41	37.00	8
8	39	40	37	38	38.50	3
9	37	36	37	35	36.25	2
10	38	35	38	37	37.00	3
11	37	37	39	34	36.75	5
12	36	36	33	36	35.25	3
13	33	34	38	36	35.25	5
14	36	36	37	37	36.50	1
15	40	36	37	37	37.50	4
16	41	37	36	38	38.00	5
17	30	38	29	40	34.25	11
18	39	39	31	32	35.25	8
19	30	35	42	43	37.50	13
20	32	37	40	45	38.50	13
21	29	27	46	44	36.50	19
22	28	47	30	46	37.75	19
23	38	35	37	33	35.75	5
24	37	34	39	41	37.75	7

2.   The Collins Computer Company manufactures switch boxes
     that allow computers to share printers. The manufactur-
     ing process is monitored by recording the numbers of de-
     fects among 200 switch boxes produced on each of 15 con-
     secutive days and the results are listed below. Construct
     a control chart for p, then determine whether the process
     is within statistical control.

          8 8 7 7 9 8 7 6 10 8 8 9 7 9 8

Chapter 12   Quiz D   Name_____

1. The Swiss Mountain Candy Company makes chocolate bars
    that are supposed to weigh 6 oz. The production pro-
    cess is monitored by recording the weights of the 4 bars
    randomly selected each hour. Results are listed below
    for 24 consecutive working hours. Construct a control
    chart for R, a control chart for $\bar{x}$, then write a conclu-
    sion about the statistical stability of this process.

Sample	Weight (oz)				Mean	Range
1	6.1	6.0	6.0	5.8	5.975	0.3
2	5.9	6.3	6.1	6.1	6.100	0.4
3	5.8	6.3	6.6	6.1	6.200	0.8
4	6.4	5.9	6.1	5.8	6.050	0.6
5	6.2	5.7	6.1	5.9	5.975	0.5
6	6.1	6.1	6.4	5.6	6.050	0.8
7	5.9	6.1	5.8	6.0	5.950	0.3
8	6.1	5.9	5.7	5.6	5.825	0.5
9	5.5	5.6	5.4	5.5	5.500	0.2
10	6.2	6.1	6.1	6.0	6.100	0.2
11	5.8	5.8	6.2	5.9	5.925	0.4
12	5.8	6.0	6.4	5.9	6.025	0.6
13	6.0	6.2	6.1	6.0	6.075	0.2
14	6.0	6.0	5.9	6.1	6.000	0.2
15	6.0	6.0	5.8	6.0	5.950	0.2
16	5.6	6.0	5.9	5.8	5.825	0.4
17	6.0	6.1	6.1	5.8	6.000	0.3
18	5.8	6.0	5.7	5.9	5.850	0.3
19	5.5	6.1	6.1	5.9	5.900	0.6
20	5.8	5.9	5.9	5.6	5.800	0.3
21	6.2	6.0	5.8	6.1	6.025	0.4
22	6.0	5.6	6.2	6.0	5.950	0.6
23	6.0	6.3	5.8	6.1	6.050	0.5
24	5.9	5.7	6.1	6.2	5.975	0.5

2. The Newport Fan Company makes air circulating fans and
    monitors the manufacturing process by testing 400 fans
    made each week. Listed below are the numbers of defects
    for 15 consecutive weeks. Construct a control chart for
    p, then determine whether the process is within statis-
    tical control.

         6 7 8 7 8 6 8 18 6 7 7 6 9 8 8

Chapter 13   Quiz A   **Name**_____

1. A fire science specialist tests three different brands
   of flares for their burning times (in minutes) and the
   results are given below for the sample data. At the
   0.05 significance level, test the claim that the three
   brands have the same mean burn time. Use the Kruskal-
   Wallis test.

Brand X	16.4	17.6	18.3	17.0	17.1	17.3	
Brand Y	17.9	18.0	17.8	18.4	17.6	19.0	19.1
Brand Z	17.3	16.4	16.5	16.0	15.8	16.3	17.1

2. Use the Wicoxon rank-sum approach to test the claim that
   students at two colleges achieve the same distribution
   of grade averages. The sample data are listed below.
   Use a 0.05 level of significance.

   College A 3.2 4.0 2.4 2.6 2.0 1.8 1.3 0.0 0.5 1.4 2.9
   College B 2.4 1.9 0.3 0.8 2.8 3.0 3.1 3.1 3.1 3.5 3.5

3. A pollster interviews voters and claims that her selection process is random. Listed below is the sequence of
   voters identified according to sex. At the 0.05 level
   of significance, test her claim that the sequence is random according to the criterion of sex.

            M, M, M, M, M, M, M, M, M, M, M, M, F, F, F, F,
            M, M, M, M, M, M, M, M, M, M, F, F, F, F, F, F

Chapter 13   Quiz B   **Name**_____

1 Use a 0.05 level of significance to test the claim that
  the sequence of computer-generated numbers is random.
  Test for randomness above and below the mean.

  8 7 5 7 3 9 1 8 0 4 3 8 4 6 2 3 9 7 5

2. A standard aptitude test is given to several randomly
   selected programmers, and the scores are given below for
   the mathematics and verbal portions of the test. Use
   the sign test to test the claim that programmers do better on the mathematics portion of the test. Use a 0.05
   level of significance.

Mathematics	347	440	327	456	427	349	377	398	425
Verbal	285	378	243	371	340	271	294	322	385

3. Use the Wilcoxon signed-ranks test and the sample data
   from problem 2. At the 0.05 significance level, test
   the claim that math and verbal scores are the same.

4. Use the sample data given in problem 2 to find the rank
   correlation coefficient, the critical value, and determine whether there is a significant correlation. Use a
   0.05 significance level.

Chapter 13   Quiz C   **Name**_____

1. The sequence of state lottery numbers given below was recorded for the last 20 consecutive weeks. Use a 0.05 level of significance to test for randomness above and below the mean. (The data are in order by row.)

        674 321 452 550 978 473 993 564 772 284
        531 585 952 450 184 305 385 389 221 002

2. An instructor gives a test before and after a lesson and results from randomly selected students are given below. At the 0.05 level of significance, test the claim that the lesson has no effect on the grade. Use the sign test.

Before	54	61	56	41	38	57	42	71	88	42	36	23	22	46	51
After	82	87	84	76	79	87	42	97	99	74	85	96	67	84	79

3. Do problem 2 using Wilcoxon's signed-ranks test.

4. Use the sample data given in problem 2 to find the rank correlation coefficient, the critical value, and determine whether there is a significant correlation. Use a 0.05 significance level.

Chapter 13   Quiz D   **Name**_____

1. Listed below are grade averages for randomly selected
   students with three different categories of high school
   background. At the 0.05 level of significance, test the
   claim that the three groups come from identical populations. (Use the appropriate *nonparametric* test.)

                   HIGH SCHOOL RECORD
        Good   Fair   Poor
        3.21   2.87   2.01
        3.65   3.05   2.31
        1.00   2.00   2.98
        3.12   0.00   0.50
        2.75   1.98   2.36

2. A highway traffic commission studies speeds on inter-
   state highways and records the following sample data
   for randomly selected cars in two different states.
   At the 0.05 level of significance, use the Wilcoxon
   rank-sum approach to test the claim that there is no
   difference between the speeds in the two states.

Florida	70	62	65	65	67	75	58	58	63	76	71
Montana	73	78	67	60	60	76	69	68	68	68	82

3. Two different tests are designed to measure verbal aptitude and the sample results are listed below for randomly selected subjects. Use the rank correlation coefficient to test for a correlation between the scores
   on the two tests. Use a 0.05 level of significance.

Test A	164	179	151	217	227	184	162	194	265
Test B	277	301	234	384	387	314	271	336	498

Final Examination A   Name _____

Show all work. In tests of hypotheses, be sure to include the null and alternative hypotheses, the test statistics, critical values, graphs, and conclusions.

1. Scores on a visual perception test are normally distributed with a mean of 2020 and a standard deviation of 250.
   (a) If one subject is randomly selected and tested, find the probability of a score greater than 1800.

      _____

   (b) If 50 subjects are randomly selected and tested, find the probability that their mean is between 2000 and 2500.

      _____

   (c) A job requirement stipulates that all applicants for employment must score in the top 80% of this test. What test score separates the top 80% of all scores from the bottom 20%?

      _____

2. A pollster wants to determine whether or not there are regional differences in the way people feel about nuclear power and sample survey results follow. At the 0.05 significance level, test the claim that the geographic region is independent of the opinion expressed.

	West	Central	East
Opposed	80	20	50
In favor	60	80	10

3. An educator wants to determine the proportion of college graduates who have had the good sense to take a statistics course. She wants to be 95% confident that her sample survey produces a proportion with an error no larger than 0.04. How large must the sample be?

      _____

Page 2 of Final Examination A

4. At the 0.05 level of significance, test the following
sequence of state lottery numbers for randomness above
and below the mean.
365 669 263 173 390 055 782 217
547 521 349 573 296 803 218

5. (a) A two-tailed hypothesis test produces a test statistic of $z = 2.05$. What is the P-value?_____
(b) What do you conclude about the null hypothesis if
the P-value is 0.001 in a test conducted with a
0.05 significance level?_____
(c) Evaluate $_{10}C_3$._____
(d) Find $P(4)$ in a binomial experiment for which $n = 5$
and $p = 3/4$._____
(e) If $r = 0.900$ for a set of paired data, what percentage of the total variation can be explained by
the regression line?_____
(f) A multiple choice test question has five possible
answers of which one is correct and four students
make random guesses. What is the probability that
all four guesses are correct?_____
(g) What do you know about a collection of scores having a standard deviation of zero?_____
(h) A set of scores has a mean of 34.5. What is the
mean after 10 has been added to each score?_____
(i) A set of scores has a standard deviation of 15.6.
What is the standard deviation after 10 has been
added to each score?_____
(j) If a <u>statistic</u> describes a sample characteristic,
what describes a population characteristic?_____

6. As part of an experiment in parapsychology, the subject
must guess a number from 1 through 5. If this test is
repeated on 36 different subjects, find the probability
of getting more than 4 correct answers. _____

Page 3 of Final Examination A

7. Samples of similar car models from three different manu-
facturers are tested for fuel consumption and the results (in mph) follow. At the 0.05 significance level,
test the claim that the means are equal. (Use a *parametric* test.)

Model A	Model B	Model C
21	25	25
19	24	28
19	27	23
23	28	27
24	23	29

8. Two teaching methods are used on two separate but similar groups of students and the test results follow. At
the 0.05 significance level, test the claim that both
methods will produce the same mean score for both populations.

Method A	Method B
$n = 10$	$n = 25$
$\bar{x} = 79.6$	$\bar{x} = 73.3$
$s = 5.6$	$s = 11.3$

9. Use the sample data given for Method A in problem 8 and
test the claim that Method A produces a variance equal
to 50. Use a 0.05 level of significance.

Page 4 of Final Examination A

10. Use the sample data given for Method A in problem 8 to construct the 95% confidence interval for the mean.

_____

11. Randomly selected subjects are given an I.Q. test and a college entrance examination and the results follow.

I.Q.	112	114	123	104	96	119	138	105
Entrance Exam	52	57	60	53	44	58	68	54

(a) Find the value of the linear correlation coefficient r.

_____

(b) Find the equation of the regression line. _____

(c) Is there a significant linear correlation between the I.Q. score and the exam score? Explain.

_____

(d) Find the value of the rank correlation coefficient.

_____

Final Examination B   **Name**_____

*Show all work. In tests of hypotheses, be sure to include
the null and alternative hypotheses, the test statistics,
critical values, graphs, and conclusions.*

1. Nine randomly selected volunteers agreed to test the
   effectiveness of a new diet and their results follow.
   At the 0.05 significance level, test the claim that the
   diet has no effect on the subject's weight. (Use a
   parametric test.)

Weight before	128 192 169 146 176 168 195 152 138 146
Weight after	118 170 158 133 168 163 195 138 118 134

2. Do problem 1 using the sign test.

3. Find the 95% confidence interval for the mean weight of
    people *before* they undergo the diet referred to in
    problem 1.   _____

4. Use the sample data summarized in the frequency table
    to find the indicated statistics.

x	f
0- 4	2
5- 9	4
10-14	8
15-19	16
20-24	12

    (a) $\bar{x}$ = _____

    (b) s = _____

Page 2 of Final Examination B

5. At the 0.05 level of significance, test the claim of a college dean who states that the mean grade point average of students is greater than 2.50. A random sample of 75 students yields a sample with mean $\bar{x} = 2.66$ and with a sample standard deviation of $s = 0.42$.

6. The table below summarizes the time spent taking a quiz and the corresponding scores for a randomly selected group of students.

Time (min)	34	41	41	36	29	25
Score	81	75	70	82	92	84

(a) Find the value of the linear correlation coefficient r. _____

(b) Find the equation of the regression line. _____

(c) Is there a significant linear correlation? Explain. _____

(d) Find the value of the rank correlation coefficient. _____

7. An appliance manufacturer plans to conduct a study to determine the proportion of their appliances that require no repairs during the first year. How large of a sample is needed if we want to be 95% confident that our sample proportion differs from the true population proportion by no more than two percentage points? _____

Page 3 of Final Examination B

8. A company president wanted to determine whether the rate of defective items was independent of the factory in which the items are made. At the 0.05 significance level, test the claim that the rate of defects is independent of the factory in which the products are made.

	Defective	Acceptable
Factory A	15	45
Factory B	25	35
Factory C	20	30

9. Find the probability that of 100 test questions, there are more than 25 correct answers. Assume that each question is multiple choice with five possible answers of which only one is correct. Also assume that all answers are random guesses. _____

10. A manufacturer finds that the daily numbers of items produced by a machine are normally distributed with a mean of 560 and a standard deviation of 12.

   (a) For a randomly selected day, find the probability that more than 575 items are produced. _____

   (b) For a randomly selected day, find the probability that the number of items produced is between 565 and 580. _____

   (c) For 36 different randomly selected days, find the probability that the mean number of items produced is less than 565. _____

   (d) Find the value of P_{20}, the twentieth percentile. _____

Page 4 of Final Examination B

11. (a) If six students line up in a random order, what is the probability that they will be in alphabetical order? _____

(b) A left-tailed hypothesis test produces a test statistic of $z = -1.23$. What is the P-value? _____

(c) True or false: "The distribution of sample variances is a normal distribution." _____

(d) Each time a particular hypothesis test is conducted there is a 0.025 probability of making a type I error. What is the probability of making three consecutive type I errors in each of three separate and independent hypothesis tests? _____

(e) What is a type I error? _____

_____

(f) A set of data has a mean of 75.0 and a standard deviation of 10.4. What is the standard deviation after 15 has been added to each score? _____

(g) A data collection consists of the party affiliation of registered voters. This data collection is at the (nominal, ordinal, interval, ratio) level of measurement. _____

(h) A committee consists of 30 Democrats (10 of whom are men) and 20 Republicans (5 of whom are men). If one person is randomly selected from this committee, find the probability of getting a man. _____

(i) If two different people are randomly selected from the same committee in part (h), find the probability that they are both men. _____

(j) For a binomial experiment in which $n = 8$ and $p = 1/4$, find $P(3)$. _____

Final Examination C   Name_____

Show all work. In tests of hypotheses, be sure to include the null and alternative hypotheses, the test statistics, critical values, graphs, and conclusions.

1. Batteries of a certain type have voltages that are normally distributed with a mean of 13.9 volts and a standard deviation of 1.2 volts.
   (a) If one battery is randomly selected, find the probability that it yields more than 13.0 volts._____

   (b) If 36 batteries are randomly selected, find the probability that the mean voltage of this sample group is less than 13.5 volts.   _____

   (c) Batteries that rank in the bottom 5% according to voltage are saved for further testing. What is the level separating these batteries from the others?

                                                  _____

2. Two separate tests are designed to measure the same personality characteristic and sample results for randomly selected subjects follow.

Test X	57	47	47	77	87	67	79	37
Test Y	89	57	47	92	99	82	98	29

   (a) Find the value of the linear correlation coefficient r.   _____

   (b) Find the equation of the regression line. _____

   (c) Is there a significant linear correlation at the 0.05 significance level? Explain.   _____

   (d) Find the value of the rank correlation coefficient

                                                  _____

Page 2 of Final Examination C

3. Students rate the quality of a book, and the results are broken down according to sex. Sample results are listed below. At the 0.05 significance level, test the claim that rating is independent of the sex of the student.

	Superior	Good	Average	Fair	Poor
Male	15	25	30	20	10
Female	15	30	20	10	5

4. (a) Find the median of the N.Y. times in problem 9._____

(b) A right-tailed hypothesis test produces a test statistic of $z = 1.46$. What is the P-value?_____

(c) If a committee of five people must be selected from nine available candidates, how many different combinations are possible? _____

(d) What is a type II error?_____

(e) If a couple plans to have three children, what is the probability that they will all be girls?_____

(f) If $P(A) = p$, what is the maximum value that the product $p(1 - p)$ can assume? _____

(g) In testing a claim that two populations have equal variances, what distribution applies? _____

(h) Find $P(2)$ in a binomial experiment for which $n = 5$ and $p = 2/3$. _____

(i) If $P(A) = 0.35$ and event B is the complement of A, what do you know about $P(B)$? _____

(j) In probability, what term is used to describe two events that cannot occur simultaneously? _____

(k) If a data set has a mean of 125.7, what is the mean after 50 has been subtracted from each score?_____

(l) If a data set has a standard deviation of 8.9, what is the standard deviation after 50 has been added to each score? _____

(m) If data set A has a range smaller than the range of data set B, must the standard deviation for data set A also be smaller than that of data set B?_____

(n) True or false: The median equals $(Q_1 + Q_3)/2$. _____

(o) Identify a point that must always lie on the regression line. _____

Page 3 of Final Examination C

5. A telephone solicitor finds that when calling someone, the probability of a sale is 1/25.

    (a) Find the probability of getting exactly three sales among 50 randomly selected calls. _____

    (b) Find the probability of at least 15 sales among 400 randomly selected calls. _____

6. Randomly selected men and women are given tests of memory and the results are given below. At the 0.05 significance level, test the claim that the populations of men and women have the same mean.

Men	Women
$n = 40$	$n = 50$
$\bar{x} = 79.8$	$\bar{x} = 81.2$
$s = 6.8$	$s = 8.4$

7. Use the sample data given for men in problem 6 to construct the 90% confidence interval for the mean memory score of all men. _____

8. Use the sample data given in problem 6 to test the claim that the variance of the scores of all men is equal to the variance of the scores for all women. Use a 0.05 significance level.

Page 4 of Final Examination C

9. Use the Wilcoxon rank-sum test at the 0.05 significance level to test the claim that the times of runners from New York and Colorado are equal. High school athletes from the two states are randomly selected for a cross-country race and their times (in minutes) are given below.

New York	Colorado
41.8	44.7
45.6	45.6
35.2	37.4
35.1	34.4
50.3	46.7
35.2	37.2
35.4	43.6
39.6	44.5
42.7	44.9
43.4	45.2
48.2	46.2
49.4	46.8
	48.3

10. Samples are randomly selected from three normally distributed populations and the results are given below. At the 0.05 significance level, test the claim that the three populations have the same mean. (Use a *parametric* test.)

Sample A	Sample B	Sample C
32.3	37.8	38.1
32.6	42.4	40.7
37.5	35.6	42.1
30.2	36.3	43.2
31.9	37.3	44.6

Final Examination D   **Name**_____

*Show all work. In tests of hypotheses, be sure to include
the null and alternative hypotheses, the test statistics,
critical values, graphs, and conclusions.*

1. To test the effectiveness of a physical training program, subjects are given a test of fitness before and
   after the program. Sample results are listed below.
   At the 0.01 significance level, test the claim that the
   program is effective in raising scores. Use a parametric
   test.

Before program	50 44 62 68 49 43 55 76 66 54 58
After program	69 66 75 94 64 59 78 91 79 77 86

2. Use the Wilcoxon signed-ranks test and the sample data
   from problem 1. At the 0.05 significance level, test
   the claim that the before and after scores are the same.

3. A measurement of depth perception among adult males produces normally distributed results with a mean of 120
   and a standard deviation of 8.
   (a) For a randomly selected adult male, find the prob-
       ability that his score is between 125 and 130._____

   (b) For a randomly selected adult male, find the prob-
       ability that his score is below 135.   _____

   (c) For 60 randomly selected adult males, find the prob-
       ability that their mean score is greater than 122.

       _____

   (d) Find the score which is P_{35}, the 35th percentile.

       _____

Page 2 of Final Examination D

4. Three different diets are tested by using randomly selected subjects with the same characteristic. The resulting weight losses (in pounds) are shown below. At
the 0.05 level of significance, test the claim that the
diets produce the same mean weight loss. (Use a *parametric* test.)

Diet 1	Diet 2	Diet 3
12	7	8
14	6	16
11	10	19
9	12	9
14	5	21

5. An algebra test is given to students at the beginning of
a calculus course. The algebra test scores and the final calculus grades are collected for several randomly
selected students and the results are given below.

Algebra	36 48 45 43 35 30 20 47
Calculus	77 92 87 81 74 63 52 88

(a) Find the value of the linear correlation coefficient
r. _____

(b) Find the equation of the regression line. _____

(c) Is there a significant linear correlation at the
0.05 significance level? Explain. _____

(d) Find the value of the rank correlation coefficient.
_____

6. Use the sample algebra scores from problem 5 to find the
95% confidence interval for the mean algebra scores of
all students. _____

Page 3 of Final Examination D

7. Use the sample algebra scores in problem 5 to test the claim that the variance of all such scores is equal to 15. Use a 0.05 significance level.

8. Use the sample algebra scores in problem 5 to test the claim that the mean of all such scores is equal to 18.0. Use a 0.05 significance level.

9. A company produces a certain component on Monday, Tuesday, and Wednesday of each week. Sample results are tested and the results are listed below. At the 0.05 level of significance, test the claim that the day of the week makes no difference in the rate of defects.

	Mon	Tues	Wed
Acceptable components	80	105	130
Defective components	20	15	10

Page 4 of Final Examination D

10. (a) Find the mean of the data summarized in the
accompanying stem-and-leaf plot. _____
30 | 004
31 | 123569
32 | 447

(b) A two-tailed hypothesis test produces a test statistic of $z = -1.80$. What is the P-value?_____

(c) The transcripts of six students are randomly selected and mixed in a pile. What is the probability that they are in alphabetical order from top to
bottom? _____

(d) Two delegates must be selected from ten available
people. How many different pairs of delegates are
possible? _____

(e) Find the median of the data given in the stem-and-
leaf plot from part (a). _____

(f) If a committee of four people must be selected from
nine available candidates, how many different combinations are possible? _____

(g) What is a type I error?_____

(h) Find the probability that of the next six children
born, there are exactly four girls. _____

(i) A newspaper editor routinely conducts polls by
publishing a question along with a telephone number
that respondents can call. What's wrong with this?
_____

(j) In testing a claim that two populations have equal
variances, what distribution applies? _____

(k) True or false: The median equals $(Q_1 + Q_3)/2$.
_____

(l) If a data set has a mean of 430.7, what is the mean
after 30 has been subtracted from each score?_____

(m) If a data set has a standard deviation of 5.0, what
is the standard deviation after 5 has been added to
each score? _____

(n) If data set A has a range smaller than the range of
data set B, must the standard deviation for data
set A also be smaller than that of data set B?_____

(o) Find $P(3)$ in a binomial experiment for which $n = 5$
and $p = 1/3$. _____

ANSWERS TO CHAPTER QUIZZES

ANSWERS TO CHAPTER 1 (QUIZ A)

1. (a) nominal   (b) ordinal   (c) ratio
   (d) interval   (e) nominal
2. (a) systematic   (b) random   (c) convenience
   (d) stratified   (e) cluster
3. The number seems to suggest that it is very accurate and precise, when it is actually an estimate and should be reported as being "about $475,000." Also, the value should be identified as the retail value, wholesale value, cost, or whatever.
4. The sample is not necessarily representative because of tired employees and machinery that has been running for hours.
5. A population is the complete collection of elements to be studied. A sample is a subcollection of elements drawn from a population. A parameter is a numerical measurement describing some characteristic of a population. A statistic is a numerical measurement describing some characteristic of a sample.

ANSWERS TO CHAPTER 1 (QUIZ B)

1. (a) Systematic   (b) Stratified   (c) Cluster
   (d) Random   (e) Convenience
2. (a) Interval   (b) Ratio   (c) Ordinal
   (d) Nominal   (e) Ratio
3. A parameter is a numerical measurement describing some characteristic of a population, whereas a statistic is a numerical measurement describing some characteristic of a sample.
4. The sample will not be representative of 16-year-old women because it excludes dropouts.
5. Those with embarrassingly low incomes are less likely to respond; there may be a tendency to exaggerate incomes.

ANSWERS TO CHAPTER 1 (QUIZ C)

1. (a) ordinal   (b) ratio   (c) interval
   (d) nominal   (e) nominal
2. (a) convenience   (b) systematic   (c) stratified
   (d) cluster   (e) random
3. A population is the complete collection of elements to be studied. A sample is a subcollection of elements drawn from a population. A parameter is a numerical measurement describing some characteristic of a population. A statistic is a numerical measurement describing some characteristic of a sample.
4. The sample may be biased in that it tends to exclude drivers living in the more distant rural areas, drivers who do their own repairs, drivers who own new cars that are under dealer warranties, and so on.

5. The number suggests that a very accurate and precise count was taken, but it is actually an estimate. The precision of the number is unwarranted and deceptive. It would be better to say that "121 billion gallons of gasoline were consumed."

ANSWERS TO CHAPTER 1 (QUIZ D)

1. The sample may be biased since it includes only those who could afford tickets. Also, their attitudes and responses may be affected by the outcome of the game they saw. In general, the fans who actually saw the game tend to be more fanatical than those who were not present.
2. (a) convenience   (b) systematic   (c) random
   (d) cluster   (e) stratified
3. (a) ordinal   (b) ratio   (c) ratio   (d) interval   (e) nominal
4. A parameter is a numerical measurement describing some characteristic of a population, whereas a statistic is a numerical measurement describing some characteristic of a sample.
5. (a) One answer: Elongate the vertical scale and cut off the bottom part of the graph so that differences are exaggerated.
   (b) The responses are self-selected. There is such a serious potential for bias that such sampling methods are not considered valid for statistical purposes.

ANSWERS TO CHAPTER 2 (QUIZ A)

1. (a) 57.0   (b) 53.5   (c) 52   (d) 61.5   (e) 25.0
   (f) 76.6   (g) 8.8   (h) 61.0   (i) 26,528   (j) 207,936
2. 4|9
   5|22255
   6|7
   7|4

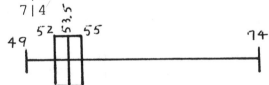

3. The score of 93 since its z score of 1.20 is greater than the z score of 1.15.
4. (a) 24.7   (b) 11.6
5. (a) 9   (b) 29.7   (c) 5.8

ANSWERS TO CHAPTER 2 (QUIZ B)

1. (a) 78.0   (b) 81.0   (c) 82   (d) 77.5   (e) 33.0
   (f) 115.5   (g) 10.7   (h) 82.0   (i) 55,680   (j) 492,804
2. (a) 66.1   (b) 19.7
3. (a) 1   (b) 8   (c) 0.5   (d) 47   (e) 15

4.

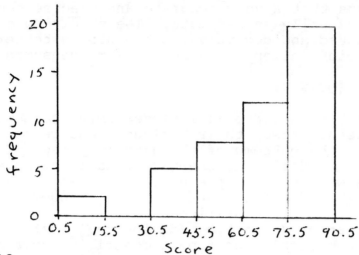

5. 6|13
    7|56
    8|1228
    9|4

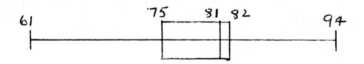

6. The score of 66 since its *z* score of −2.33 is greater than the
    z score of −3.33.

ANSWERS TO CHAPTER 2 (QUIZ C)

1. (a) 59.0   (b) 58.0   (c) 59   (d) 63.5   (e) 33.0
    (f) 77.0   (g) 8.8   (h) 59.0   (i) 39,061   (j) 421,201
2. 4|7
    5|06678999
    6|8
    7|
    8|0

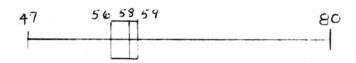

3. The second 82 since its *z*
    score of 1.75 is greater
    than the *z* score of 1.5.

4. (a) 72.7   (b) 20.6   (c) 15
   (d) 20.5   (e) 28

ANSWERS TO CHAPTER 2 (QUIZ D)

1. (a) 62.0   (b) 63.0   (c) 52   (d) 64.0   (e) 24.0
   (f) 85.4   (g) 9.2   (h) 52.0   (i) 31,350   (j) 246,016
2. (a) 76.4   (b) 12.8
3. (a) 41   (b) 45.5   (c) 40.5   (d) 32   (e) 10
4.

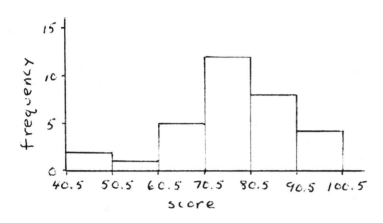

5. 5|222
   6|2499
   7|6

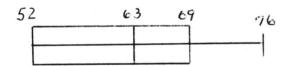

6. The score of 80 since its
   z score of 2 is greater
   than the z score of 1.25.

ANSWERS TO CHAPTER 3 (QUIZ A)

1. 0.729   2. 0.882
3. (a) 220   (b) 35   (c) 1/120
4. (a) 0.774   (b) 0.773
5. (a) 18/25   (b) 13/25   (c) 51/100   (d) 49/625   (e) 1/3

ANSWERS TO CHAPTER 3 (QUIZ B)

1. (a) 0.723   (b) 0.721
2. (a) 1/5   (b) 0.107   (c) 0.893
3. (a) 495   (b) 120   (c) 96
4. (a) 3/17   (b) 11/17   (c) 0.110   (d) 12/17   (e) 0.4

ANSWERS TO CHAPTER 3 (QUIZ C)

1. (a) 21/31   (b) 18/31   (c) 14/31   (d) 0.104   (e) 8/21
2. 0.118
3. (a) 15/16   (b) 1/16
4. (a) 1/120   (b) 120   (c) 0.826

ANSWERS TO CHAPTER 3 (QUIZ D)

1. 0.0081
2. (a) 3360   (b) 24   (c) 1/7,059,052
3. (a) 0.437   (b) 23/30   (c) 4/9   (d) 11/30   (e) 0.4
4. (a) 1023/1024   (b) 1/512

ANSWERS TO CHAPTER 4 (QUIZ A)

1. (a) 3.2   (b) 1.7   (c) 0.228
2. (a) 1.2   (b) 0.9   (c) 0.8
3. (a) 0.215   (b) 0.833
4. No, since $P(0) + P(1) + P(2) = 1.5$ and does not equal 1.
5. $140

ANSWERS TO CHAPTER 4 (QUIZ B)

1. (a) 3.0   (b) 1.5   (c) 0.231   (d) 0.833   (e) 0.133
2. 4.2; 1.7; 2.9   3. $53,000
4. No, because $P(0) + P(1) + P(2) = 1.5$, which exceeds 1.

ANSWERS TO CHAPTER 4 (QUIZ C)

1. 0.103   2. 2.0; 1.4; 2.0   3. $23,250
4. (a) 0.6   (b) 0.8   (c) 0.341   (d) 0.118

ANSWERS TO CHAPTER 4 (QUIZ D)

1. (a) 8.4   (b) 1.6   (c) 0.231   (d) 0.724
2. 0.172
3. 2.3; 1.0
4. $45,000

ANSWERS TO CHAPTER 5 (QUIZ A)

1. 0.4168
2. (a) 0.4332   (b) 0.8413   (c) 0.0228   (d) 0.3811
      (e) 0.6700   (f) 212.5   (g) 0.0287

ANSWERS TO CHAPTER 5 (QUIZ B)

1. (a) 0.4525   (b) 0.4706   (c) 0.7486   (d) 0.3707
      (e) 0.2120   (f) 68.4   (g) 0.0084
2. 0.1038

ANSWERS TO CHAPTER 5 (QUIZ C)

1. (a) 0.3212   (b) 0.7549   (c) 0.1971   (d) 0.2843
      (e) 0.3459   (f) 1021.7   (g) 0.2090
2. 0.0968

ANSWERS TO CHAPTER 5 (QUIZ D)

1. (a) 0.4251   (b) 0.6570   (c) 0.0749   (d) 0.9918
      (e) 0.1665   (f) 101.3   (g) 0.0681
2. 0.0694

ANSWERS TO CHAPTER 6 (QUIZ A)

1. $0.0118 < p < 0.0576$
2. 849
3. $67.7 < \mu < 84.7$
4. $16.9 < \sigma < 29.3$
5. (a) 1.75   (b) 2.921   (c) 6.844   (d) 38.582   (e) 76.2

ANSWERS TO CHAPTER 6 (QUIZ B)

1. $0.438 < p < 0.505$ 2. 1309
3. $3381 < \mu < 3919$ 4. $294.7 < \sigma < 699.8$
5. (a) 1.75   (b) 2.998   (c) 3650 lb   (d) 0.472   (e) 2.700

ANSWERS TO CHAPTER 6 (QUIZ C)

1. 82   2. $0.027 < p < 0.123$
3. $39.2 < \mu < 44.0$ 4. $3.7 < \sigma < 6.7$
5. (a) 2.110   (b) 1.88   (c) 5.009   (d) 41.6 cm   (e) 0.075

ANSWERS TO CHAPTER 6 (QUIZ D)

1. $473.6 < \mu < 511.0$ 2. $26.8 < \sigma < 61.2$
3. 1537   4. $0.308 < p < 0.438$
5. (a) 492.3   (b) 0.373   (c) 3.106   (d) 2.81   (e) 2.700

ANSWERS TO CHAPTER 7 (QUIZ A)

1. $H_0: \mu \le 150.00$. $H_1: \mu > 150.00$.
   Test statistic: $z = 1.796$. Critical value: $z = 1.645$.
   Reject the null hypothesis. There is sufficient evidence to
   support the claim that the mean is greater than $150.00.

2. $H_0: \sigma \ge 81$. $H_1: \sigma < 81$.
   Test statistic: $\chi^2 = 2.370$. Critical value: $\chi^2 = 13.091$.
   Reject the null hypothesis. There is sufficient evidence to
   support the claim that the new machine produces a lower standard deviation.

3. $H_0: p \le 0.01$. $H_1: p > 0.01$.   Test statistic: $z = 4.92$.
   Critical value: $z = 2.33$. Reject the null hypothesis. There
   is sufficient evidence to warrant rejection of the claim that
   no more than 1% are defective.

4. (a) Less than 0.0001
   (b) Rejecting the claim that no more than 1% are defective
       when, in reality, the rate of defects is no more than 1%.
   (c) Fail to reject the claim that no more than 1% are defec-
       tive when the rate of defects is actually greater than 1%.

ANSWERS TO CHAPTER 7 (QUIZ B)

1. $H_0: p \ge 0.5$. $H_1: p < 0.5$.
   Test statistic: $z = -1.29$. Critical value: $z = -1.645$.
   Fail to reject the null hypothesis. There is not sufficient
   evidence to warrant rejection of the claim that at least half
   prefer the Democrat.

2. (a) 0.0985
   (b) Reject the claim that at least half favor the Democrat
       when the Democrat is really favored by at least half of
       all voters.
   (c) Fail to reject the claim that at least half favor the
       Democrat when fewer than half of all voters actually
       favor the Democrat.

3. $H_0: \mu = 35.0$. $H_1: \mu \ne 35.0$.
   Test statistic: $t = 7.252$. Critical values: $t = -2.861, 2.861$.
   Reject the null hypothesis. There is sufficient evidence to
   warrant rejection of the claim that the mean is equal to 35.0.

4. H_0: $\sigma \leq 30.0$. H_1: $\sigma > 30.0$.
   Test statistic: $\chi^2 = 21.253$.   Critical value: $\chi^2 = 21.920$.
   Fail to reject the null hypothesis.   There is not sufficient
   evidence to support the claim that the standard deviation is
   greater than 30.0.

ANSWERS TO CHAPTER 7 (QUIZ C)

1. H_0: $\mu \leq 160$. H_1: $\mu > 160$.
   Test statistic: $t = 9.583$.   Critical value: $t = 1.711$.
   Reject the null hypothesis.   There is sufficient evidence
   to support the claim that the mean is greater than 160.

2. H_0: $p \leq 0.03$. H_1: $p > 0.03$.
   Test statistic: $z = 2.22$.   Critical value: $z = 2.33$.
   Fail to reject the null hypothesis.   There is not sufficient
   evidence to warrant rejection of the manager's claim that
   production is not really out of control.

3. (a) 0.0594
   (b) Concluding that the manager's claim is incorrect when it
       is actually correct.
   (c) Failure to reject the manager's claim when it is actually
       wrong.

4. H_0: $\sigma \geq 3.2$. H_1: $\sigma < 3.2$.
   Test statistic: $\chi^2 = 0.211$. Critical value: $\chi^2 = 13.848$.
   Reject the null hypothesis.   There is sufficient evidence to
   support the claim of a lower standard deviation.

ANSWERS TO CHAPTER 7 (QUIZ D)

1. H_0: $\sigma = 130$. H_1: $\sigma \neq 130$.
   Test statistic: $\chi^2 = 140.738$.   Critical values: $\chi^2 =$
   74.222, 129.561.   Reject the null hypothesis.   There is sufficient evidence to warrant rejection of the claim that the
   new version has a standard deviation equal to that of the old
   version.

2. H_0: $p = 2/3$. H_1: $p \neq 2/3$.
   Test statistic: $z = -2.00$.   Critical values: $z = -1.96$, 1.96.
   Reject the null hypothesis.   There is sufficient evidence to
   warrant rejection of the claim that the candidate is favored
   by two-thirds of the readers.

3. (a) 0.0456
   (b) Rejecting the editor's claim when the candidate really is
       favored by two-thirds of the readers.
   (c) Fail to reject the editor's claim when the candidate is
       NOT favored by two-thirds of the readers.

4. H_0: $\mu \leq 937$. H_1: $\mu > 937$.
   Test statistic: $z = 5.31$. Critical value: $z = 2.33$.
   Reject the null hypothesis. There is sufficient evidence to
   support the claim that the modified components have a longer
   mean time between failures.

ANSWERS TO CHAPTER 8 (QUIZ A)

1. H_0: $p_1 = p_2$. H_1: $p_1 \neq p_2$.
   Test statistic: $z = 0.97$. Critical values: $z = -1.96, 1.96$.
   Fail to reject the null hypothesis. There is not sufficient
   evidence to warrant rejection of the claim that the recognition rates are the same in both states.

2. $-0.0443 < p_1 - p_2 < 0.0976$

3. H_0: $\mu_1 = \mu_2$. H_1: $\mu_1 \neq \mu_2$.
   Test statistic: $t = 2.366$. Critical values: $t = -2.262, 2.262$.
   Reject the null hypothesis. There is sufficient evidence to
   warrant rejection of the claim that the mean is not affected
   by the course.

4. H_0: $\sigma_1 = \sigma_2$. H_1: $\sigma_1 \neq \sigma_2$.
   Test statistic: $F = 6.0844$. Critical value: $F = 2.4374$.
   Reject the null hypothesis. There is sufficient evidence to
   warrant rejection of the claim that the samples come from
   populations with equal standard deviations.

5. Preliminary F results: See the solution to problem 4 above.
   H_0: $\mu_1 = \mu_2$. H_1: $\mu_1 \neq \mu_2$.
   Test statistic: $t = 9.564$. Critical values: $t = -2.132, 2.132$.
   Reject the null hypothesis. There is sufficient evidence to
   warrant rejection of the claim that the two populations have
   equal means.

6. $6.4 < \mu_1 - \mu_2 < 12.2$

ANSWERS TO CHAPTER 8 (QUIZ B)

1. H_0: $\sigma_1^2 = \sigma_2^2$. H_1: $\sigma_1^2 \neq \sigma_2^2$.
   Test statistic: $F = 3.0625$. Critical value: $F = 1.9429$.
   Reject the null hypothesis. There is sufficient evidence to
   warrant rejection of the claim that the two brands have equal
   variances.

2. Preliminary F test results: See solution to problem 1 above.
   H_0: $\mu_1 = \mu_2$. H_1: $\mu_1 \neq \mu_2$.
   Test statistic: $z = 16.025$. Critical values: $z = -1.96, 1.96$.
   Reject the null hypothesis. There is sufficient evidence to
   warrant rejection of the claim that the two populations have

equal means.

3. $3.8 < \mu_1 - \mu_2 < 4.8$

4. $H_0: p_1 = p_2$. $H_1: p_1 \neq p_2$.
   Test statistic: $z = -1.33$. Critical values: $z = -1.96, 1.96$.
   Fail to reject the null hypothesis. There is not sufficient
   evidence to warrant rejection of the claim that both proportions are the same.

5. $-0.0868 < p_1 - p_2 < 0.0269$

6. $H_0: \mu_1 = \mu_2$. $H_1: \mu_1 \neq \mu_2$.
   Test statistic: $t = -5.964$. Crit. values: $t = -2.262, 2.262$.
   Reject the null hypothesis. There is sufficient evidence to
   warrant rejection of the claim that the course has no effect
   on grades.

ANSWERS TO CHAPTER 8 (QUIZ C)

1. $H_0: p_1 = p_2$. $H_1: p_1 \neq p_2$.
   Test statistic: $z = -0.44$. Critical values: $z = -2.575, 2.575$.
   Fail to reject the null hypothesis. There is not sufficient
   evidence to warrant rejection of the claim that the product is
   preferred by the same market shares.

2. $-0.0812 < p_1 - p_2 < 0.0507$

3. $H_0: \mu_1 = \mu_2$. $H_1: \mu_1 \neq \mu_2$.
   Test statistic: $t = 2.507$. Critical values: $t = -2.228, 2.228$.
   Reject the null hypothesis. There is sufficient evidence to
   warrant rejection of the claim that the mean is not affected
   by the course.

4. $H_0: \sigma_1 = \sigma_2$. $H_1: \sigma_1 \neq \sigma_2$.
   Test statistic: $F = 5.7068$. Critical value: $F = 3.6669$.
   Reject the null hypothesis. There is sufficient evidence
   to warrant rejection of the claim that the populations have
   equal standard deviations.

5. Preliminary F test: See the solution to problem 4 above.
   $H_0: \mu_1 = \mu_2$. $H_1: \mu_1 \neq \mu_2$.
   Test statistic: $t = 1.641$. Critical values: $t = -2.262, 2.262$.
   Fail to reject the null hypothesis. There is not sufficient
   evidence to warrant rejection of the claim that the two populations have equal means.

6. $-2.1 < \mu_1 - \mu_2 < 13.1$

ANSWERS TO CHAPTER 8 (QUIZ D)

1. H_0: $p_1 = p_2$. H_1: $p_1 \neq p_2$.
   Test statistic: $z = -1.79$. Critical values: $z = -1.96$, 1.96.
   Fail to reject the null hypothesis. There is not sufficient evidence to warrant rejection of the claim that the two machines have the same rate of defects.

2. $-0.0733 < p_1 - p_2 < 0.0033$

3. H_0: $\mu_1 = \mu_2$. H_1: $\mu_1 \neq \mu_2$.
   Test statistic: $t = 1.292$. Critical values: $t = -2.306$, 2.306.
   Fail to reject the null hypothesis. There is not sufficient evidence to warrant rejection of the claim that both versions produce the same mean.

4. H_0: $\sigma_1 = \sigma_2$. H_1: $\sigma_1 \neq \sigma_2$.
   Test statistic: $F = 9.9467$. Critical value: $F = 2.2759$.
   Reject the null hypothesis. There is sufficient evidence to warrant rejection of the claim that the samples come from populations with equal standard deviations.

5. Preliminary F test results: See solution to problem 4 above.
   H_0: $\mu_1 = \mu_2$. H_1: $\mu_1 \neq \mu_2$.
   Test statistic: $t = 3.209$. Critical values: $t = -2.093$, 2.093.
   Reject the null hypothesis. There is sufficient evidence to warrant rejection of the claim of equal means.

6. $0.7 < \mu_1 - \mu_2 < 11.5$

ANSWERS TO CHAPTER 9 (QUIZ A)

1. (b) 0.986   (c) 0.632
   (d) Significant linear correlation   (e) $\hat{y} = 5.05 + 1.91x$
   (g) 97.2%
2. (a) No significant linear correlation
   (b) Significant linear correlation
   (c) 19.0   (d) True   (e) 0
3. $S = 28.279 + 0.086P + 0.516D$; 0.940

ANSWERS TO CHAPTER 9 (QUIZ B)

1. (b) 0.863   (c) 0.632
   (d) Significant linear correlation   (e) $\hat{y} = 11.66 + 1.02x$
   (g) 74.5%
2. (a) Yes   (b) No significant linear correlation
   (c) 5.0   (d) 0
   (e) Significant linear correlation
3. $P = 14.087 + 0.013(\text{Att.}) + 0.907(\text{Adapt.})$

ANSWERS TO CHAPTER 9 (QUIZ C)

1. (b) 0.867   (c) 0.666
   (d) Significant linear correlation   (e) $\hat{y} = 19.40 + 0.93x$
   (g) 75.2%
2. (a) Significant linear correlation
   (b) No significant linear correlation
   (c) 18.3   (d) True   (e) 0
3. (a) $C = 3.40 + 1.12A - 0.144B$ (b) 0.678

ANSWERS TO CHAPTER 9 (QUIZ D)

1. (b) 0.974   (c) 0.602
   (d) Significant linear correlation   (e) $\hat{y} = 10.59 + 1.32x$
   (g) 95.0%
2. (a) 23.0   (b) Significant linear correlation
   (c) No significant linear correlation
   (d) 0   (e) True
3. $C = 7.336 - 0.173A + 0.989B$

ANSWERS TO CHAPTER 10 (QUIZ A)

1. H_0: The proportions of occurrences are all equal.
   H_1: Those proportions are not all equal.
   Test statistic: $\chi^2 = 8.263$. Critical value: $\chi^2 = 11.071$.
   Fail to reject the null hypothesis. There is not sufficient
   evidence to warrant rejection of the claim that the character-
   istics occur with the same frequency.

2. H_0: The sampling distribution agrees with the population
   distribution. H_1: It doesn't agree.
   Test statistic: $\chi^2 = 31.938$. Critical value: $\chi^2 = 7.815$.
   Reject the null hypothesis. There is sufficient evidence to
   warrant rejection of the claim that the sampling distribution
   agrees with the distribution of the state populations.

3. H_0: Employment and response are independent.
   H_1: Employment and response are dependent.
   Test statistic: $\chi^2 = 5.942$. Critical value: $\chi^2 = 4.605$.
   Reject the null hypothesis. There is sufficient evidence to
   warrant rejection of the claim that the response and employ-
   ment status are independent.

ANSWERS TO CHAPTER 10 (QUIZ B)

1. H_0: Major and response are independent.
   H_1: Major and response are dependent.
   Test statistic: $\chi^2 = 6.502$. Critical value: $\chi^2 = 2.706$.
   Reject the null hypothesis. There is sufficient evidence to

warrant rejection of the claim that response and major are independent.

2. H_0: The proportions of absences are all the same.
   H_1: The proportions of absences are not all the same.
   Test statistic: $\chi^2 = 28.308$. Critical value: $\chi^2 = 9.488$.
   Reject the null hypothesis. There is sufficient evidence to warrant rejection of the claim that absences occur on the different week days with the same frequency.

3. H_0: The absences occur according to the stated percentages.
   H_1: The absences do not occur according to the stated percentages. (That is, at least one of the days has an occurrence rate different than the stated percentage.)
   Test statistic: $\chi^2 = 1.253$. Critical value: $\chi^2 = 9.488$.
   Fail to reject the null hypothesis. There is not sufficient evidence to warrant rejection of the claim that the absences occur with the stated percentages.

ANSWERS TO CHAPTER 10 (QUIZ C)

1. H_0: Sex and response are independent.
   H_1: Sex and response are dependent.
   Test statistic: $\chi^2 = 0.579$. Critical value: $\chi^2 = 5.991$.
   Fail to reject the null hypothesis. There is not sufficient evidence to warrant rejection of the claim that the response and sex are independent.

2. H_0: The proportions of responses are all equal.
   H_1: The proportions of responses are not all equal.
   Test statistic: $\chi^2 = 1.875$. Critical value: $\chi^2 = 9.488$.
   Fail to reject the null hypothesis. There is not sufficient evidence to warrant rejection of the claim that the responses occur with the same frequency.

3. H_0: The responses occur according to the stated percentages.
   H_1: The responses do not occur according to the stated percentages.
   Test statistic: $\chi^2 = 5.146$. Critical value: $\chi^2 = 9.488$.
   Fail to reject the null hypothesis. There is not sufficient evidence to warrant rejection of the claim that the responses occur according to the stated percentages.

ANSWERS TO CHAPTER 10 (QUIZ D)

1. H_0: Employment and response are independent.
   H_1: Employment and response are dependent.
   Test statistic: $\chi^2 = 6.771$. Critical value: $\chi^2 = 4.605$.
   Reject the null hypothesis. There is sufficient evidence to warrant rejection of the claim that response and employment

status are independent.

2. H_0: The proportions of outcomes are all equal.
   H_1: The proportions of outcomes are not all equal.
   Test statistic: $\chi^2 = 3.125$. Critical value: $\chi^2 = 11.071$.
   Fail to reject the null hypothesis. There is not sufficient
   evidence to warrant rejection of the claim that the outcomes
   occur with the same frequency.

3. H_0: The outcomes occur according to the stated percentages.
   H_1: The outcomes do not occur according to the stated percent-
       ages.
   Test statistic: $\chi^2 = 1.951$. Critical value: $\chi^2 = 11.071$.
   Fail to reject the null hypothesis. There is not sufficient
   evidence to warrant rejection of the claim that the outcomes
   occur according to the given percentages.

ANSWERS TO CHAPTER 11 (QUIZ A)

1. H_0: $\mu_1 = \mu_2 = \mu_3$. H_1: The means are not all equal.
   Test statistic: $F = 12.3465$. Critical value: $F = 6.5149$.
   Reject the null hypothesis. There is sufficient evidence
   to warrant rejection of the claim that the three brands have
   the same mean.

2. (a) H_0: There is no interaction effect.
       H_1: There is an interaction effect.
       Test statistic: $F = 1.2261$. Critical value: $F = 2.9277$.
       Fail to reject the null hypothesis. There does not appear
       to be an interaction effect.
   (b) H_0: There is no machine effect.
       H_1: There is a machine effect.
       Test statistic: $F = 3.1857$. Critical value: $F = 3.5546$.
       Fail to reject the null hypothesis. The type of machine
       does not appear to have an effect on the items produced.
   (c) H_0: There is no employee effect.
       H_1: There is an employee effect.
       Test statistic: $F = 46.3842$. Critical value: $F = 3.5546$.
       Reject the null hypothesis. The employee does appear to
       have an effect on the items produced.

ANSWERS TO CHAPTER 11 (QUIZ B)

1. H_0: $\mu_1 = \mu_2 = \mu_3$. H_1: The means are not all equal.
   Test statistic: $F = 6.6983$. Critical value: $F = 3.9034$.
   Reject the null hypothesis. There is sufficient evidence to
   warrant rejection of the claim that the four brands have the
   same mean.

2. (a) H_0: There is no interaction effect.
       H_1: There is an interaction effect.
       Test statistic: $F = 0.7062$. Critical value: $F = 2.9277$.
       Fail to reject the null hypothesis. There does not appear
       to be an interaction effect.
   (b) H_0: There is no machine effect.
       H_1: There is a machine effect.
       Test statistic: $F = 53.7172$. Critical value: $F = 3.5546$.
       Reject the null hypothesis. There does appear to be a
       machine effect.
   (c) H_0: There is no employee effect.
       H_1: There is an employee effect.
       Test statistic: $F = 0.1898$. Critical value: $F = 3.5546$.
       Fail to reject the null hypothesis. There does not appear
       be an employee effect.

ANSWERS TO CHAPTER 11 (QUIZ C)

1. H_0: $\mu_1 = \mu_2 = \mu_3$. H_1: The means are not all equal.
   Test statistic: $F = 12.1267$. Critical value: $F = 5.0959$.
   Reject the null hypothesis. There is sufficient evidence to
   warrant rejection of the claim that the three brands have the
   same mean.

2. (a) H_0: There is no interaction effect.
       H_1: There is an interaction effect.
       Test statistic: $F = 19.9888$. Critical value: $F = 2.9277$.
       Reject the null hypothesis. There does appear to be an
       interaction effect.
       Note: Because there does appear to be an interaction ef-
       fect, we would normally not proceed with parts (b) and
       (c), but we are directed to assume that there is no inter-
       action effect. With that assumption, we get the following.
   (b) H_0: There is no machine effect.
       H_1: There is a machine effect.
       Test statistic: $F = 0.0664$. Critical value: $F = 3.5546$.
       Fail to reject the null hypothesis. There does not appear
       to be a machine effect.
   (c) H_0: There is no employee effect.
       H_1: There is an employee effect.
       Test statistic: $F = 0.3296$. Critical value: $F = 3.5546$.
       Fail to reject the null hypothesis. There does not appear
       be an employee effect.

ANSWERS TO CHAPTER 11 (QUIZ D)

1. H_0: $\mu_1 = \mu_2 = \mu_3 = \mu_4$. H_1: The means are not all equal.
   Test statistic: $F = 0.0555$. Critical value: $F = 3.9539$.
   Fail to reject the null hypothesis. There is not sufficient
   evidence to warrant rejection of the claim that the four

brands have the same mean.
2. (a) H_0: There is no interaction effect.
   H_1: There is an interaction effect.
   Test statistic: $F = 40.9261$. Critical value: $F = 3.5546$.
   Reject the null hypothesis. There does appear to be an
   interaction effect.
   Note: Because there does appear to be an interaction ef-
   fect, we would normally not proceed with parts (b) and
   (c), but we are directed to assume that there is no inter-
   action effect. With that assumption, we get the following.
   (b) H_0: There is no machine effect.
   H_1: There is a machine effect.
   Test statistic: $F = 0.4072$. Critical value: $F = 4.4139$.
   Fail to reject the null hypothesis. There does not appear
   to be a machine effect.
   (c) H_0: There is no employee effect.
   H_1: There is an employee effect.
   Test statistic: $F = 16.6248$. Critical value: $F = 3.5546$.
   Reject the null hypothesis. There does appear be an
   employee effect.

ANSWERS TO CHAPTER 12 (QUIZ A)

1. The control chart for R indicates statistical stability of the
   process variation. The control chart for $\bar{x}$ indicates
   statistical stability of the process mean. Control charts for
   R and $\bar{x}$ are as follows.

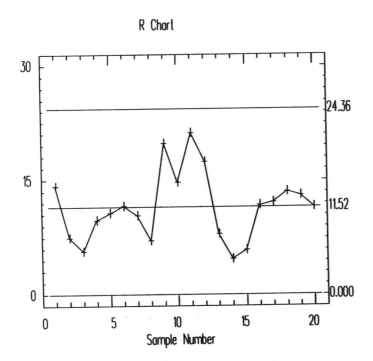

R Chart

X-bar Chart

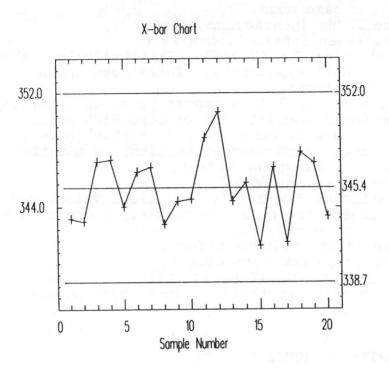

2. The control chart for *p* shows a pattern of an upward trend, so the process does not appear to be within statistical control.

P Chart

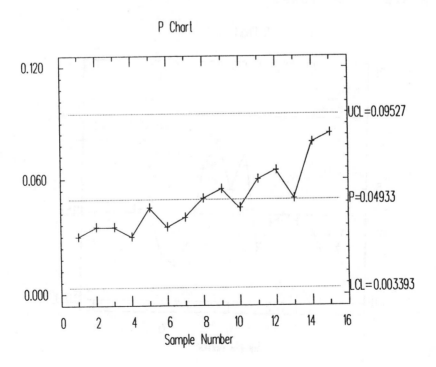

ANSWERS TO CHAPTER 12 (QUIZ B)

1. The control chart for R indicates statistical stability of the
   process variation, but the control chart for $\bar{x}$ has a point
   beyond the upper control limit, so the process mean is not
   within statistical control.

R Chart

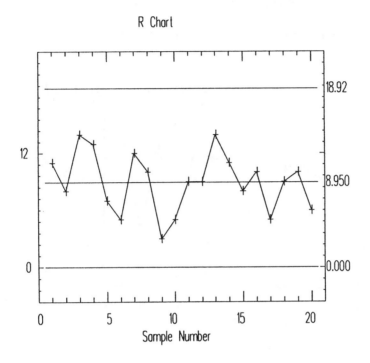

X-bar Chart

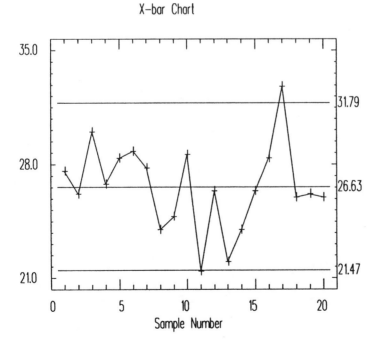

2. The *p* chart shows that the process is within statistical control.

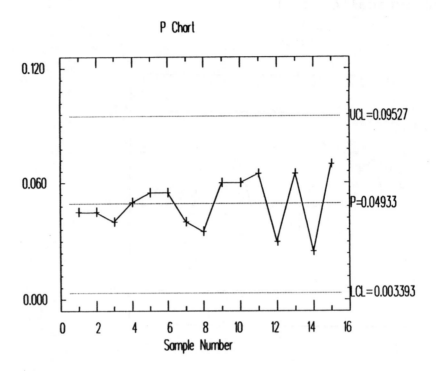

ANSWERS TO CHAPTER 12 (QUIZ C)

1. The control chart for R includes points beyond the upper
   control limit and there are 8 consecutive points below the
   center line, so the process variation is not within
   statistical control. The control chart for $\bar{x}$ indicates that
   the process mean is within statistical control.

R Chart

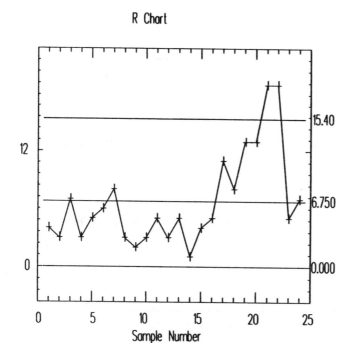

X-bar Chart

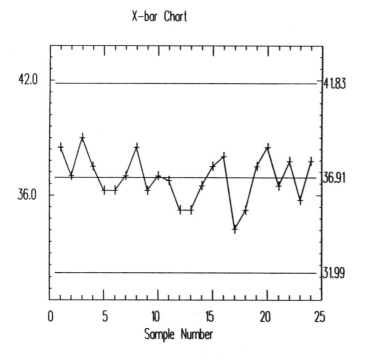

2. The control chart for *p* indicates that the process is within
   statistical control.

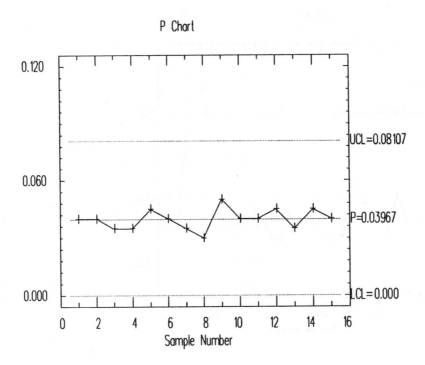

ANSWERS TO CHAPTER 12 (QUIZ D)

1. The control chart for R indicates that the process variation
   is within statistical control. The control chart for $\bar{x}$
   includes a point below the lower control limit, so the process
   mean is not within statistical control.

R Chart

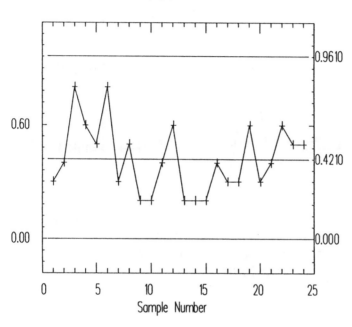

X-bar Chart for C1

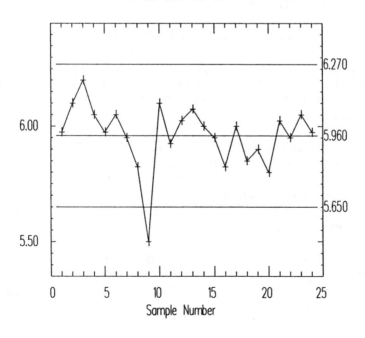

2. The control chart for *p* includes a point beyond the upper control limit, so the process does not appear to be within statistical control.

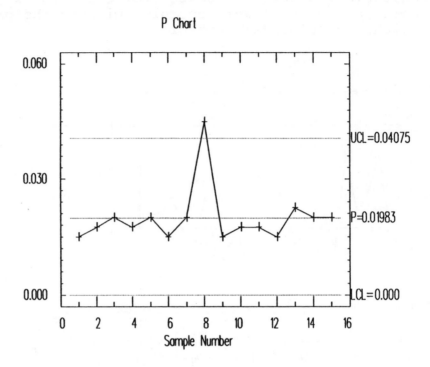

P Chart

ANSWERS TO CHAPTER 13 (QUIZ A)

1. H_0: The three brands have identical populations.
   H_1: The three populations are not identical.
   Test statistic: $H = 12.7979$. The critical value is 5.991. Reject the null hypothesis. There is sufficient evidence to warrant rejection of the claim that the three brands have the same mean burn time.

2. H_0: The two populations are identical.
   H_1: The two populations are not identical.
   Test statistic: $z = -1.18$. Critical values: $z = -1.96, 1.96$. Fail to reject the null hypothesis. There is not sufficient evidence to warrant rejection of the claim that the two populations are identical.

3. $n_1 = 22$, $n_2 = 10$, $G = 4$, $\mu_G = 14.75$, $\sigma_G = 2.38$.
   Test statistic: $z = -4.52$. Critical values: $z = -1.96, 1.96$. Reject the null hypothesis of randomness. The sequence does not appear to be random.

ANSWERS TO CHAPTER 13 (QUIZ B)

1. $n_1 = 9$, $n_2 = 10$, $G = 14$.
   Test statistic: $G = 14$. Critical values: 5, 16.
   Fail to reject the null hypothesis of randomness.

2. H_0: The math scores are equal to or less than the verbal
   scores. H_1: The math scores are greater than the verbal
   scores. Test statistic: $x = 0$. Critical value: $x = 1$.
   Reject the null hypothesis. There is sufficient evidence
   to support the claim that the math scores are greater than
   the verbal scores.

3. H_0: There is no difference between math and verbal scores.
   H_1: There is a difference between math and verbal scores.
   Test statistic: $T = 0$. Critical value: $T = 6$.
   Reject the null hypothesis of no difference.

4. $r_s = 0.867$. Critical value: 0.683. There is a significant
   correlation.

ANSWERS TO CHAPTER 13 (QUIZ C)

1. $n_1 = 9$, $n_2 = 11$, $G = 8$.
   Test statistic: $G = 8$. Critical values: 6, 16.
   Fail to reject the null hypothesis of randomness above
   and below the mean.

2. H_0: There is no difference between before and after grades.
   H_1: There is a difference between before and after grades.
   Test statistic: $x = 0$. Critical value: $x = 2$.
   Reject the null hypothesis of no difference.

3. H_0: There is no difference between before and after grades.
   H_1: There is a difference between before and after grades.
   Test statistic: $T = 0$. Critical value: $T = 21$.
   Reject the null hypothesis of no difference.

4. $r_s = 0.559$. Critical value: 0.525. There is a significant
   correlation.

ANSWERS TO CHAPTER 13 (QUIZ D)

1. H_0: The three populations are identical.
   H_1: The three populations are not identical.
   Test statistic: $H = 2.9600$. Critical value is 5.991.
   Fail to reject the null hypothesis. There is not sufficient
   evidence to warrant rejection of the claim that the three
   groups come from identical populations.

2. H_0: The two populations of speeds are identical.
   H_1: The two populations are not identical.
   Test statistic: $z = -1.28$. Critical values: $z = -1.96$, 1.96.
   Fail to reject the null hypothesis. There is not sufficient
   evidence to warrant rejection of the claim that there is no
   difference between the speeds in the two states.

3. $r_s = 1.000$. Critical value: 0.683. There is a significant
   correlation.

ANSWERS TO FINAL EXAMINATION A

1. (a) 0.8106   (b) 0.7156   (c) 1810

2. H_0: Opinion and geographic region are independent.
   H_1: Opinion and geographic region are dependent.
   Test statistic: $\chi^2 = 65.524$. Crit. value: $\chi^2 = 5.991$.
   Reject the null hypothesis. There is sufficient evidence to
   warrant rejection of the claim that opinion and geographic
   region are independent.

3. 601

4. $n_1 = 9$, $n_2 = 6$, $G = 11$.
   Test statistic: $G = 11$. Critical values: 4, 13.
   Fail to reject the null hypothesis of randomness.

5. (a) 0.0404   (b) Reject the null hypothesis.
   (c) 120   (d) 0.396   (e) 81%   (f) 1/625
   (g) All scores are the same.
   (h) 44.5   (i) 15.6   (j) parameter

6. 0.8708

7. H_0: $\mu_1 = \mu_2 = \mu_3$. H_1: The means are not all equal.
   Test statistic: $F = 7.4641$. Critical value: $F = 3.8853$.
   Reject the null hypothesis. There is sufficient evidence to
   warrant rejection of the claim that the means are equal.

8. F test results: Test statistic is $F = 4.0717$ and the critical
   value is $F = 3.6142$. Reject equality of the standard dev's.
   T-test results: Test statistic is $t = 2.194$ and the critical
   values are $t = -2.262$, 2.262. Fail to reject the null hypothesis. There is not sufficient evidence to warrant rejection of the claim that both methods produce the same mean.

9. H_0: $\sigma^2 = 50$. H_1: $\sigma^2 \neq 50$.
   Test statistic: $\chi^2 = 5.6448$. Critical values: $\chi^2 = $
   2.700, 19.023. Fail to reject the null hypothesis. There
   is not sufficient evidence to warrant rejection of the claim
   that the variance equals 50.

10. $75.6 < \mu < 83.6$

11. (a) 0.955   (b) $\hat{y} = -2.714 + 0.509x$
    (c) Yes. At a 0.05 or 0.01 significance level, the linear correlation coefficient of 0.955 exceeds the critical value.
    (d) 0.929

ANSWERS TO FINAL EXAMINATION B

1. H_0: The before and after weights are the same.
   H_1: The before and after weights are not the same.
   Test statistic: $t = 5.593$. Crit. values: $t = -2.262, 2.262$.
   Reject the null hypothesis. There is sufficient evidence to warrant rejection of the claim that the diet has no effect on weight.

2. H_0: The before and after weights are the same.
   H_1: The before and after weights are not the same.
   Test statistic: $x = 0$. Critical value: $x = 1$.
   Reject the null hypothesis. There is sufficient evidence to warrant rejection of the claim that the diet has no effect on weight.

3. $144.8 < \mu < 177.2$

4. (a) 15.8   (b) 5.6

5. H_0: $\mu \leq 2.50$. H_1: $\mu > 2.50$.
   Test statistic: $z = 3.30$. Critical value: $z = 1.645$.
   Reject the null hypothesis. There is sufficient evidence to support the claim that the mean is greater than 2.50.

6. (a) −0.813   (b) $\hat{y} = 113.510 - 0.957x$
   (c) If the significance level is 0.05, there is a significant linear correlation because the critical value is 0.811. If the significance level is 0.01, there is no significant linear correlation because the critical value is 0.917.
   (d) −0.843

7. 2401

8. H_0: The rate of defects and the factory are independent.
   H_1: The rate of defects and the factory are dependent.
   Test statistic: $\chi^2 = 4.336$. Critical value: $\chi^2 = 5.991$.
   Fail to reject the null hypothesis. There is not sufficient evidence to warrant rejection of the claim that the factory and the rate of defects are independent.

9. 0.0838

10. (a) 0.1056   (b) 0.2897   (c) 0.9938   (d) 549.9

11. (a) 1/720   (b) 0.1093   (c) False   (d) 0.0000156
    (e) It is the mistake of rejecting a true null hypothesis.
    (f) 10.4   (g) Nominal   (h) 3/10
    (i) 3/35   (j) 0.208

ANSWERS TO FINAL EXAMINATION C

1. (a) 0.7734   (b) Q.0228   (c) 11.9 volts
2. (a) 0.921   (b) $y = -9.83 + 1.35x$
    (c) Yes. The value of $r = 0.921$ exceeds the critical value
        at the 0.05 or 0.01 significance level.
    (d) 0.970
3. H_0: Sex and rating are independent.
    H_1: Sex and rating are dependent.
    Test statistic: $\chi^2 = 5.298$. Critical value: $\chi^2 = 9.488$.
    Fail to reject the null hypothesis. There is not sufficient
    evidence to warrant rejection of the claim that rating is
    independent of the sex of the student.
4. (a) 42.25   (b) 0.0721   (c) 126
    (d) The mistake of failing to reject a false null hypothesis.
    (e) 1/8   (f) 0.25   (g) F distribution
    (h) 0.165   (i) 0.65   (j) mutually exclusive
    (k) 75.7   (l) 8.9   (m) No
    (n) False   (o) Centroid $(\bar{x}, \bar{y})$
5. (a) 0.184   (b) 0.6480
6. H_0: $\mu_1 = \mu_2$. H_1: $\mu_1 \neq \mu_2$.
    Test statistic: $z = -0.87$. Critical values: $z = -1.96, 1.96$.
    Fail to reject the null hypothesis. There is not sufficient
    evidence to warrant rejection of the claim that the two pop-
    ulations have the same mean.
7. $78.0 < \mu < 81.6$
8. H_0: $\sigma_1^2 = \sigma_2^2$. H_1: $\sigma_1^2 \neq \sigma_2^2$.
    Test statistic: $F = 1.5260$. Critical value: $F = 1.8752$.
    Fail to reject the null hypothesis. There is not sufficient
    evidence to warrant rejection of the claim that the two pop-
    ulation variances are the same.
9. H_0: The times are the same. H_1: The times are not the same.
    Test statistic: $z = -0.84$. Critical values: $z = -1.96, 1.96$.
    Fail to reject the null hypothesis. There is not sufficient
    evidence to warrant rejection of the claim that the times of
    runners from New York and Colorado are the same.
10. H_0: $\mu_1 = \mu_2 = \mu_3$. H_1: The means are not all equal.
    Test statistic: $F = 14.1598$. Critical value: $F = 3.8853$.
    Reject the null hypothesis. There is sufficient evidence
    to warrant rejection of the claim that the three populations
    have the same mean.

ANSWERS TO FINAL EXAMINATION D

1. $H_0: \mu_1 \geq \mu_2$. $H_1: \mu_1 < \mu_2$.
   Test statistic: $t = -12.082$. Critical value: $t = -2.764$.
   Reject the null hypothesis. There is sufficient evidence to support the claim that the program is effective.

2. H_0: Before and after scores are the same.
   H_1: Before and after scores are not the same.
   Test statistic: $T = 0$. Critical value: $T = 11$.
   Reject the null hypothesis. There is sufficient evidence to warrant rejection of the claim that before and after scores are the same.

3. (a) 0.1587   (b) 0.9699   (c) 0.0262   (d) 116.9

4. $H_0: \mu_1 = \mu_2 = \mu_3$. H_1: The means are not all equal.
   Test statistic: $F = 3.5053$. Critical value: $F = 3.8853$.
   Fail to reject the null hypothesis. There is not sufficient evidence to warrant rejection of the claim that the diets produce the same mean weight loss.

5. (a) 0.989   (b) $\hat{y} = 24.04 + 1.39x$
   (c) Yes, $r = 0.989$ exceeds the critical value of 0.707.
   (d) 1.000

6. $29.9 < \mu < 46.1$

7. $H_0: \sigma^2 = 15$. $H_1: \sigma^2 \neq 15$.
   Test statistic: $\chi^2 = 43.733$. Critical values: $\chi^2 = 1.690, 16.013$. Reject the null hypothesis. There is sufficient evidence to warrant rejection of the claim that the variance is equal to 15.

8. $H_0: \mu = 18.0$. $H_1: \mu \neq 18.0$.
   Test statistic: $t = 5.843$. Crit. values: $t = -2.365, 2.365$.
   Reject the null hypothesis. There is sufficient evidence to warrant rejection of the claim that the mean is 18.0.

9. H_0: The rate of defects and the day of the week are independent. H_1: They are dependent.
   Test statistic: $\chi^2 = 8.816$. Critical value: $\chi^2 = 5.991$.
   Reject the null hypothesis. There is sufficient evidence to warrant rejection of the claim that the rate of defects and the day of the week are independent.

10. (a) 313.8 (b) 0.0718   (c) 1/720   (d) 45 (e) 314   (f) 126
    (g) The mistake of rejecting a true null hypoth.   (h) 0.234
    (i) The respondents are self-selected and the results have an extremely high chance of being biased.   (j) F dist.
    (k) False   (l) 400.7   (m) 5.0   (n) No   (o) 0.165

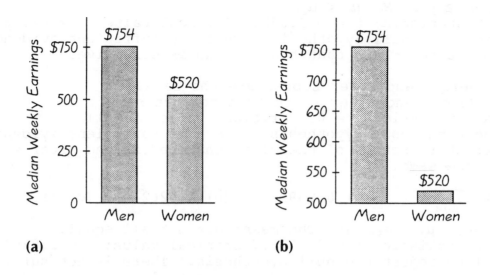

(a) **(b)**

Figure 1–1

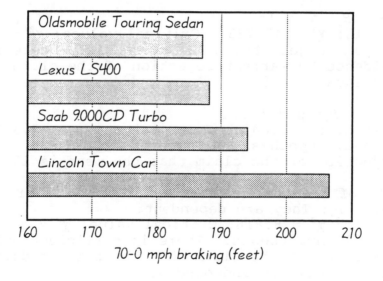

Figure 1–2

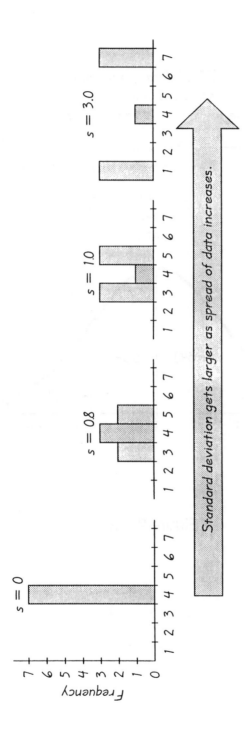

Frequency

s = 0

s = 0.8

s = 1.0

s = 3.0

Standard deviation gets larger as spread of data increases.

Figure 2–10

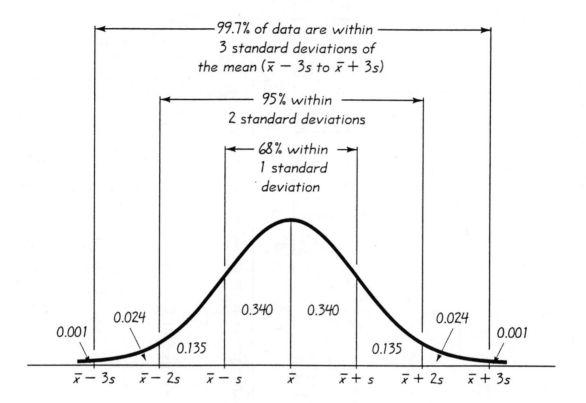

Figure 2–12

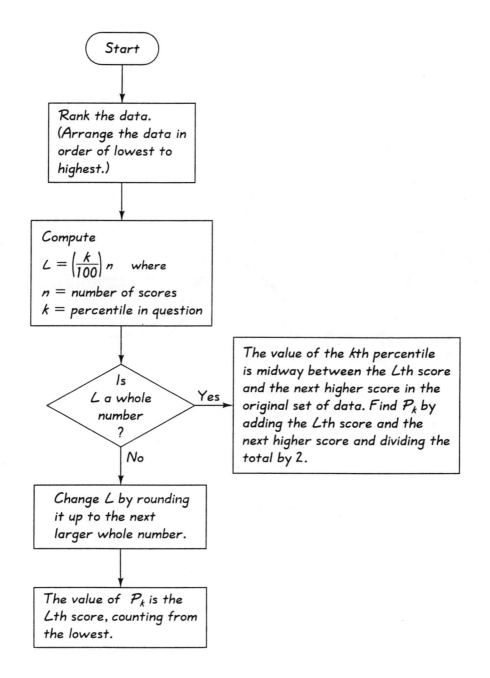

Figure 2–14

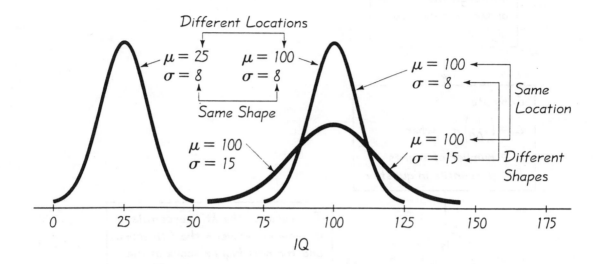

Figure 5-4

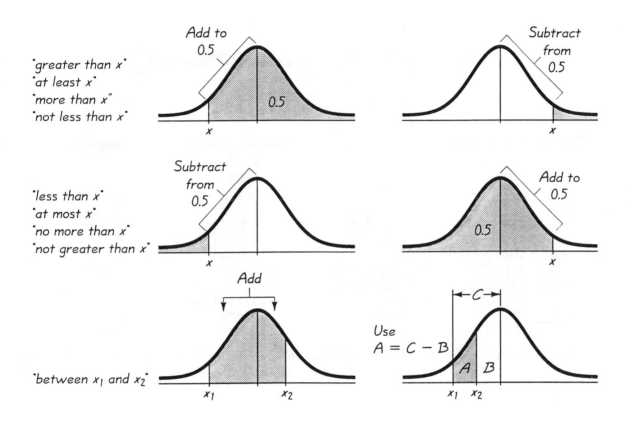

Figure 5-10

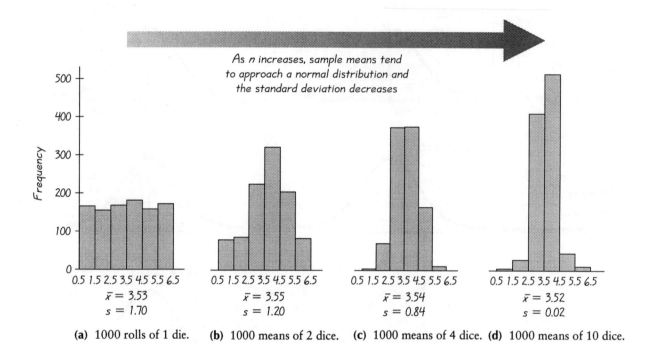

(a) 1000 rolls of 1 die.   (b) 1000 means of 2 dice.   (c) 1000 means of 4 dice.   (d) 1000 means of 10 dice.

Figure 5-23

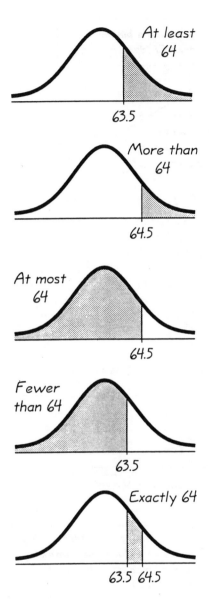

Figure 5-27

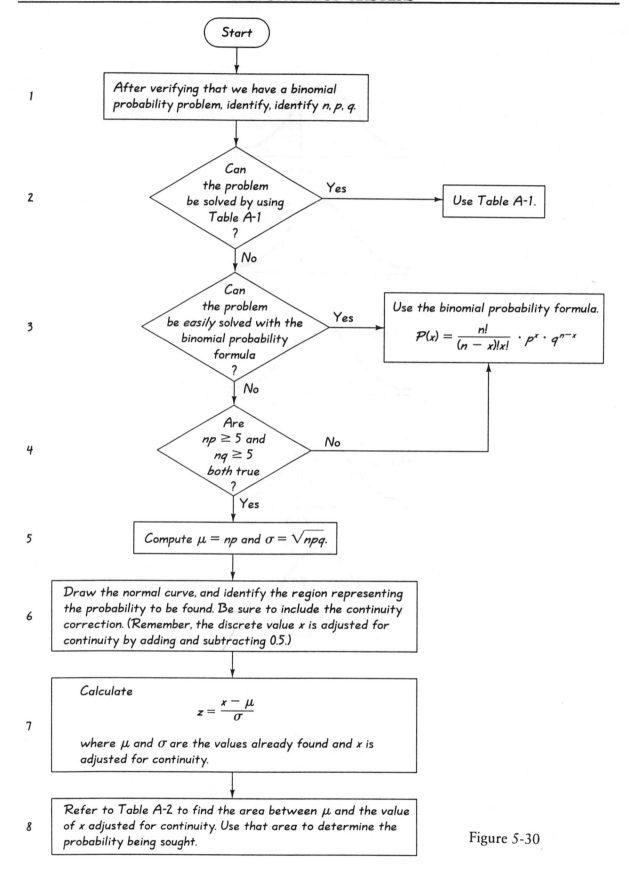

Figure 5-30

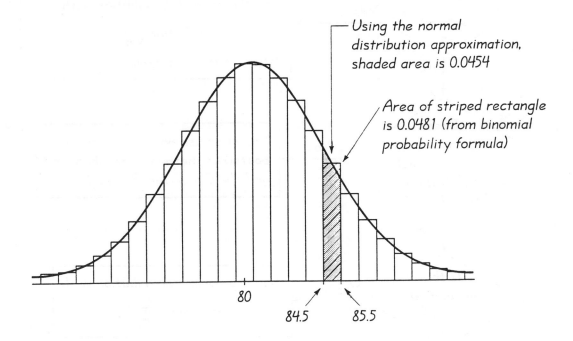

Figure 5-31

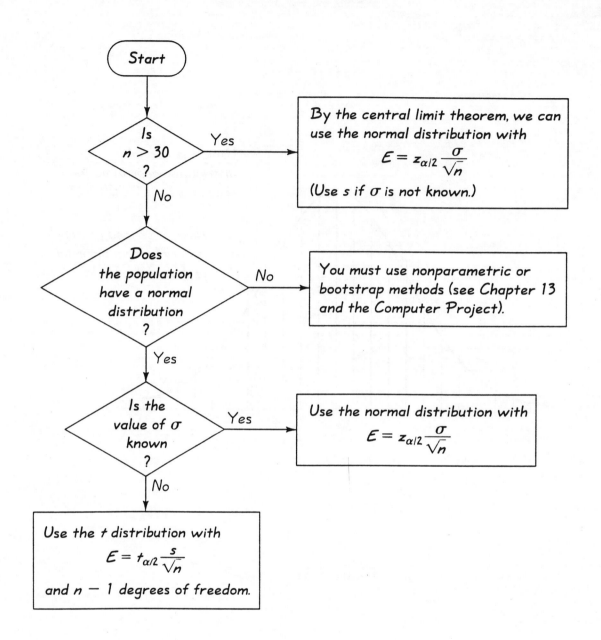

Figure 6-5

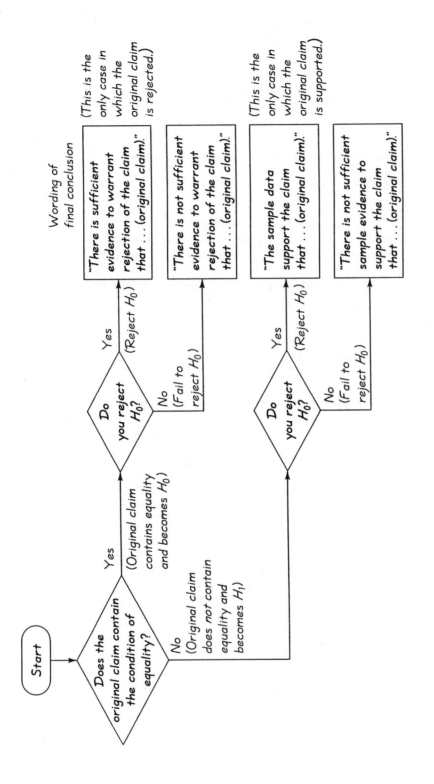

Figure 7-2

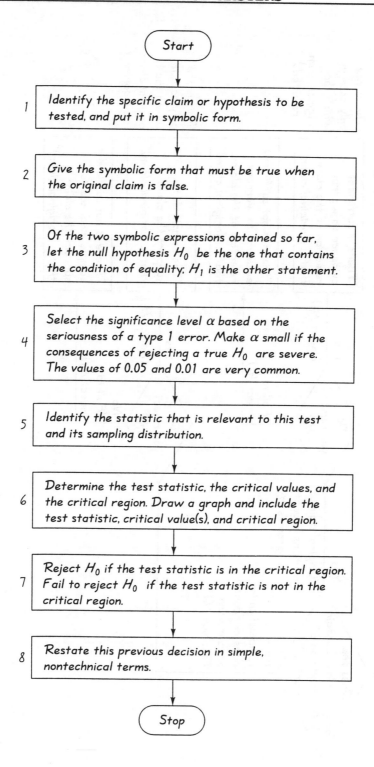

Figure 7-4

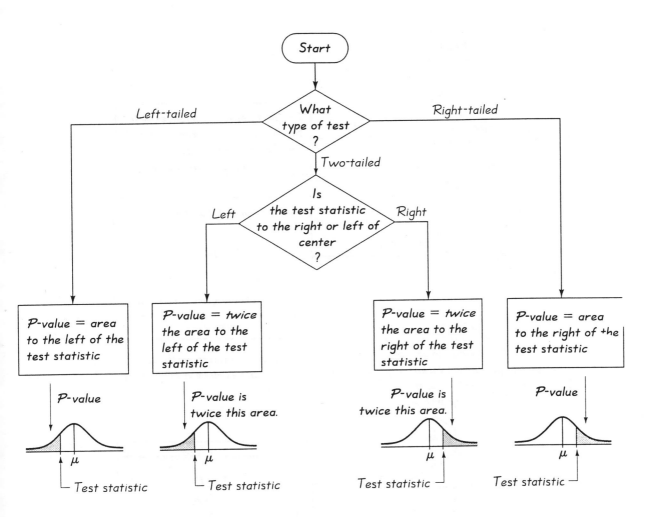

Figure 7-7

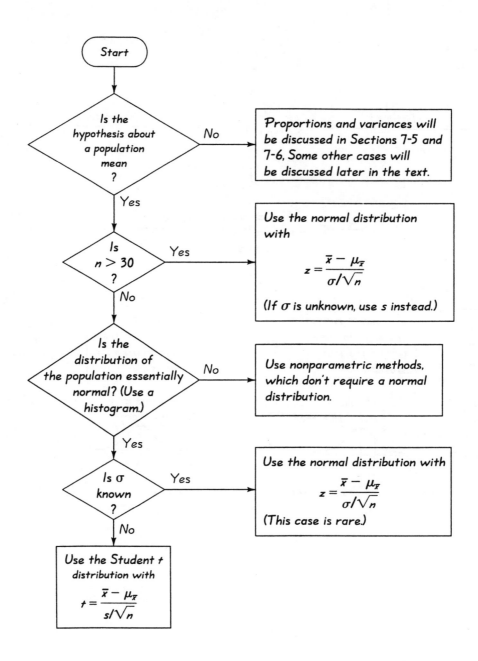

Figure 7-10

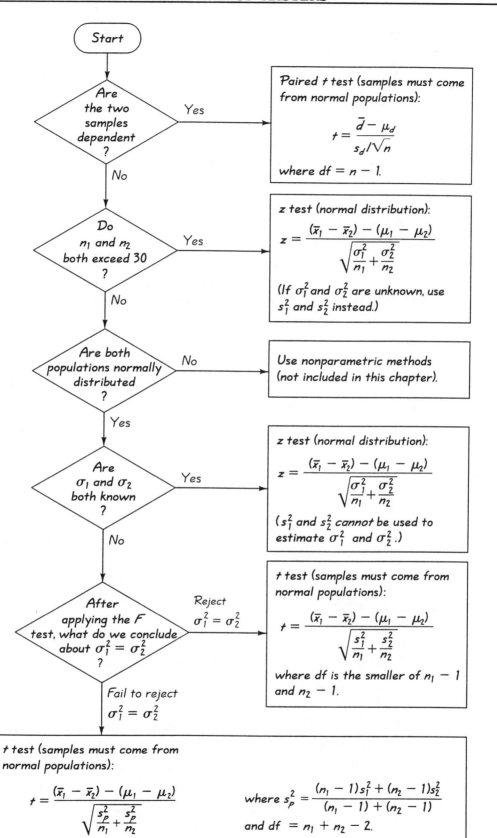

Figure 8-6

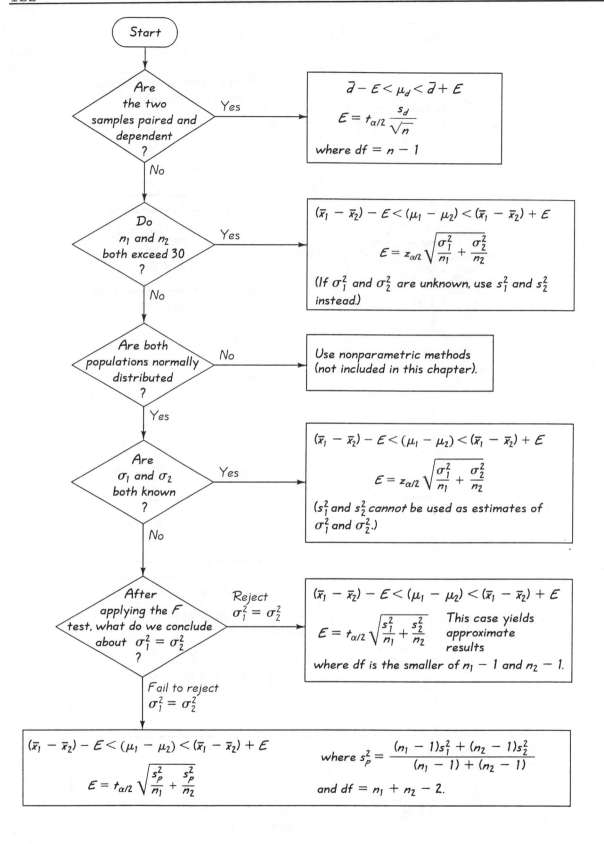

Figure 8-7

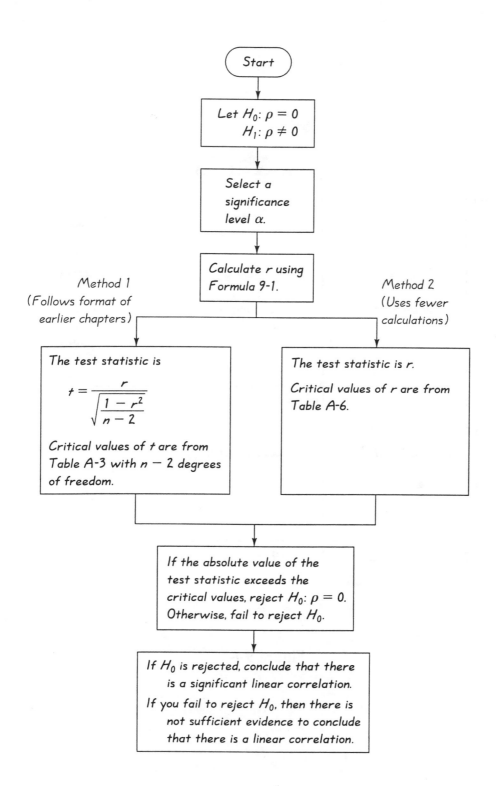

Figure 9-2

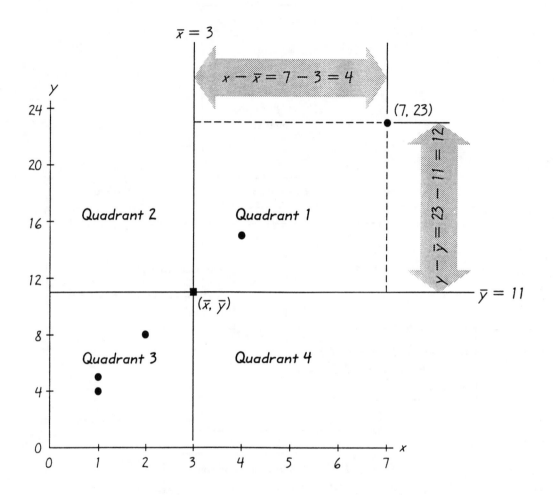

Figure 9-6

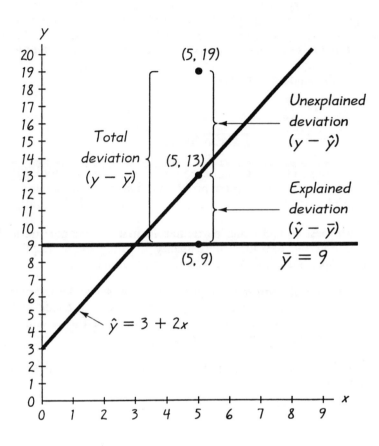

Figure 9-9

INSTRUCTOR'S SOLUTIONS

by Milton Loyer

The following pages contain solutions to all the exercises. For convenience, we include all information (i.e., PREFACE, solutions for the odd-numbered exercises, NOTES, illustrations, etc.) given in the **Student's Solutions Manual** as it appears in that volume -- except that in some cases the information has been arranged with less "white space" into a somewhat less user-friendly but much more efficient space-saving format. In general, the even-numbered solutions added to form the **Instructor's Solutions** do not include the intermediate steps and non-required illustrations provided for the student's manual. Bold-face type identifies the exercise numbers and NOTE designations for material appearing only in the instructor's manual. Accordingly, even numbers not in bold-face type identify solutions appearing for one reason or another in the student's manual.

PREFACE

This manual contains the solutions to the odd-numbered problems of the textbook Elementary Statistics, Sixth Edition, by Mario Triola. In the worked problems, intermediate steps are provided for the calculations. When appropriate, additional hints and comments are included and prefaced by NOTE.

Many statistical problems are best solved using particular formats. Recognizing and following these patterns promotes understanding and develops the capacity to apply the concepts to other problems. This manual identifies and employs such formats whenever practicable.

For best results, read the text carefully before attempting the exercises and attempt the exercises before consulting the solutions. This manual has been prepared to provide a check and extra insights for exercises that have already been completed and to provide guidance for solving exercises that have already been attempted but have not been successfully completed.

I would like to thank Mario Triola for writing an excellent elementary statistics book and for inviting me to prepare this solutions manual.

Chapter 1

Introduction to Statistics

1-2 The Nature of Data

1. discrete, since the number of defectives must be an integer

2. discrete, since the number of employees absent must be an integer
   NOTE: In situations involving identifying/counting characteristics, we follow the convention that the designation (e.g., defective, absent) either applies or it doesn't -- i.e., there is no such thing as being partly defective or partly absent.

3. continuous, since weight can be any value on a continuum

4. discrete, since the number having perfect scores must be an integer

5. continuous, since speed can be any value on a continuum

6. continuous, since time can be any value on a continuum

7. discrete, since the number recognizing the name must be an integer

8. continuous, since weight gain/loss can be any value on a continuum

9. ordinal, since the labels give relative position in a hierarchy

10. ratio, since differences are meaningful and there is an inherent zero

11. nominal, since the color is used only to distinguish the pieces

12. interval, since differences are meaningful but ratios are not

13. nominal, since the numbers are used for identification only. Even though they are assigned alphabetically within regions, zip codes are merely numerical names for post offices.
    NOTE: Suppose the numbers had resulted from placing all the post offices in one large list alphabetically (or by city size, or by mail volume, etc.), however, so that 17356 [Red Lion, PA] was the $17,356^{th}$ post office in the list; then zip codes, like the order of finishers in a race, would illustrate the ordinal level of measurement.

14. nominal, since the numbers are used for identification only

15. ratio, since differences are meaningful, and there is an inherent zero

16. ordinal, since the labels give relative position in a hierarchy but do not necessarily guarantee uniform differences between grades -- e.g., grades might be assigned from percents so that $95 \leq A \leq 100, 85 \leq B < 95, 70 \leq C < 85, 50 \leq D < 70, 0 \leq F < 50$

17. interval, since differences are meaningful but ratios are not

18. ordinal, since the labels give relative position in a hierarchy but do not necessarily guarantee uniform differences between levels

19. a. If a person with an IQ of 150 is twice as intelligent as a person with an IQ of 75, then IQ is being measured on a ratio scale.
    b. If a person with an IQ of 150 is not twice as intelligent as a person with an IQ of 75, then IQ is not being measured on a ratio scale. If the difference in intelligence between persons with IQ's of 75 and 100 is the same as the difference between persons with IQ's of 150 and 175, then IQ is being measured on an interval scale.

20. While 1 hour is half of 2 hours (since time is measured on a ratio scale), 600° is not twice 300° (since temperature is measured on an interval scale and not a ratio scale). Even if both quantities were measured on a ratio scale, the result may be different -- e.g., going 30 mph for 2 hours or 60 mph for 1 hour may get you to the same point, but they will not necessarily use the same amount of gasoline or produce the same wear and tear on the car (or the passengers).

21. Years are not data at the ratio level of measurement because the year zero has been arbitrarily assigned so that 1900 does not represent twice as much time as 950 -- consider, for example, the Chinese numerical representations for the given years. Since the time difference between 1900 and 1920 is the same as the time difference between 1920 and 1940, however, years are data at the interval level of measurement.

1-3 Uses and Abuses of Statistics

1. Studies sponsored by groups with a desire to show a particular result are sometimes biased toward that result. Such bias is not necessarily intentional and may owe to well-intentioned efforts to control the experiment by eliminating subjects and/or situations not conducive to the desired result, or to a subconscious tendency to interpret unclear data in the hoped-for direction. In addition, studies that do not support the desired conclusion will probably be re-done and/or not be reported.

2. The population from which the sample is drawn is not the same as (or even representative of) the population of consumers. People with unlisted numbers and people without phones would not be included in the survey. Homes with multiple listings would be more likely to be sampled than homes listed only once.

3. Because the graph does not start at zero, the given sizes are not in the same ratio as the numbers they represent and give an exaggerated impression of the differences.

4. The respondents were not selected randomly, but they self-selected themselves. Those who choose to respond are not necessarily representative of the entire population -- e.g., they may be more conscientious, less busy, etc., etc.

5. When there appears to be a cause-and-effect relationship between two factors, that relationship may be created by an unobserved third factor; or the cause-and-effect may even work in the reverse direction. There may, for example, be a genetic factor that tends to produce individuals that are both lower-scoring and inclined to physically addictive behaviors; or it may be that getting lower grades causes students to become more nervous and more inclined toward physically addictive behaviors.

6. The number that say they've committed a crime is $(.08)(1875) = 150$; the number that say they did so under the influence of alcohol or drugs is $(.62)(150) = 93$.

7. The study showed that Corvette drivers have a higher death rate than Volvo drivers, not necessarily that Corvettes are inherently more deadly and unsafe than Volvos. It could be, for example, that people with reckless personalities tend to purchase Corvettes and that they would have just as many accidents no matter what type of car they drove.

8. Students leaving the cafeteria at mealtime are not necessarily representative of the student body. Those who skip meals would be under-represented. Commuters and others who do not eat at the cafeteria would not be represented at all. Workers in the kitchen/dishroom might be missed.

9. Alumni with low (or no) annual salaries would probably feel discouraged and be less inclined to respond than would those with high annual salaries. This means that the responses would not represent all alumni and would tend to overstate the true typical annual salary. Other factors that might affect the results include (1) alumni wanting to make themselves look good might overstate annual salary, (2) alumni with income from unreputable sources and/or who understated income on tax forms might understate annual salary, (3) alumni wishing to avoid pleas for donations might understate annual incomes.

10. Since the poll was sponsored by a shoe polish company, the questions (especially if they were multiple choice and not open ended) may have been designed to encourage and/or suggest responses relating specifically to the appearance of shoes. Similarly, a necktie company might make certain there were questions and/or suggested responses about neck wear but fail to include items about foot wear.

11. a. The pay cut was 20% of $400, or $80, leaving a weekly salary of $320.
    b. The pay increase was 20% of $320, or $64, producing a weekly salary of $384.
    c. The percentages were calculated on the various current salaries and did not produce cuts and raises to return the employee to her original salary.

12. a. correct, since $1000 - 100 + 300 = 1200$
    b. not correct, since $(1200-1000)/1000 = .20 \neq .33$
    c. correct, since $(1200-1000)/1000 = .20$
    d. correct, since $300/1200 = .25$

13. Statement (a) claims a specific cause-and-effect relationship, while the more general statement (b) merely states that the factors are related. Because statement (a) goes beyond the facts to reach a conclusion, it is open to challenge. It could be argued, for example, that people under the influence of alcohol don't actually cause crashes but might be too impaired to avoid involvement in accidents caused by others.

14. A person who eats three meals a day without prolonged mealtime conversation might spend $20 + 30 + 40 = 90$ minutes a day eating meals. Adding 20 minutes a day for snacks/coffee/ soda produces 110 minutes a day. Ignoring life cycle adjustments (infancy, etc.), a life of 75 years translates into $110 \times 365.25 \times 75 = 3013312.5$ minutes $= 50221.875$ hours $= 2092.578$ days $= 5.729$ years. The estimate is not unreasonable.

15. The purpose of a graph is to convey information visually. Since there is no correspondence between the lengths of the bars and the amounts they represent (the 983.5 bar, for example, is over twice the size of the 643.3 bar), the bars are misleading and actually convey misinformation that hinders rather than helps the reader to receive the proper values.

16. Assuming a male can't become a symphony conductor until about age 30, one should eliminate from consideration all males who die before 30. The average age at death for male symphony conductors should then be compared to the average age at death of those males who survived past 30 -- an age that would certainly be higher than 69.5 and possibly very close to 73.4.

17. Assuming that each of the 20 individual subjects is ultimately counted as a success or not (i.e., that there are no "dropouts" or "partial successes"), the success rates in fraction form must be one of 0/20, 1/20, 2/20,..., 19/20, 20/20. In percentages, these rates are multiples of 5 (0%, 5%, 10%,..., 95%, 100%), and values such as 53% and 58% are not mathematical possibilities.

18. let S = salary at the start
    $1 - x/100 = (100-x)/100$ = proportion of salary left after a cut in pay of x percent
    $S \cdot (100-x)/100$ = salary left after a cut in pay of x percent
    to restore the salary, find the multiplier y such that $y \cdot S \cdot (100-x)/100 = S$
    and so, $y = 100/(100-x)$ = the necessary multiplier
    $y - 1 = 100/(100-x) - 1 = x/(100-x)$ = the proportion increase
    $100x/(100-x)$ = the percent increase

19. a. Since 100% is the totality of whatever is being measured, removing 100% of some quantity means that none of it is left.

    b. Reducing plaque by over 300% would mean removing three times as much plaque as is there, and then removing even more!

20. One possible listing of four major flaws is as follows.

    (1) The question is not objective, but worded to encourage negative answers. Consider the question, "Do you support the development of atomic weapons whose presence could discourage an enemy attack?"

    (2) The sample size of 20 is probably too small.

    (3) The respondents are not necessarily a random sample, but merely those who chose to reply.

    (4) In a sample of size 20, the percentages of "yes" and "no" responses must be multiples of 5 -- 87% and 13% are not mathematically possible.

1-4 Methods of Sampling

1. systematic, since pagers were selected at regular intervals

2. random, since each U.S. senator has an equal chance of being selected

3. cluster, since the entire set of interest (assumed to be all Ohio State University students) was divided into sections, and all the members in the selected sections were surveyed. NOTE: An added complication is the fact that the clusters overlap; any student in more than one class could conceivably be in two different selected clusters and be counted twice -- whether or not that creates a problem depends on the nature of the survey and may require advanced techniques. Notice also the stated assumption; if the set of interest is all U.S. college students, for example, the dean is using convenience sampling (because he is using a readily available group -- his own university), with cluster sub-sampling to reduce the sample to a manageable size.

4. stratified, since the entire set of interest was divided into 8 subpopulations (fr males, fr females, so males, so females, jr males, jr females, sr males, sr females) from which the actual sampling was done

5. convenience, since the sample was simply those choosing to respond

6. stratified, since the set of interest was divided into subpopulations from which the actual sampling was done

7. stratified, since the set of interest was divided into subpopulations from which the actual sampling was done

8. cluster, since the entire set of interest (assumed to be all the drivers in a particular city) was divided into sections (i.e., city blocks), and all the drivers in the randomly selected sections were surveyed. **NOTE:** The preceding assumption and interpretation of the problem are not the only ones possible. If the set of interest is all city drivers in the U.S., for example, there are at least three other possibilities: (1) If the city blocks were selected at random from the all the city blocks of all the cities, the answer is cluster sampling for the reasons previously stated. (2) If the city blocks were selected from the city blocks of the city in which the researcher is located, the answer is convenience sampling with cluster sub-sampling to reduce the sample to a manageable size. (3) If the city blocks were selected from the city blocks of one city selected at random, the answer is cluster sampling with additional cluster sub-sampling to reduce the sample to a manageable size.

9. cluster, since the entire set of interest (all leukemia patients) was divided into counties, and all the appropriate persons in each selected county were interviewed

10. systematic, since CEO's were selected from the list at regular intervals. **NOTE:** This answer the one following assume that the population of interest is the 1000 companies (and/or their CEO's) with the highest stock market values.

11. random, since each company has an equal chance of being selected

12. convenience, since the sample was those who happened to be leaving the cafeteria at the time. **NOTE:** If the population of interest is all students leaving the cafeteria, and the 40 students are from one randomly chosen time slot into which the psychology student has divided the exit times, then the answer is cluster sampling.

13. People probably don't know their waist size precisely and would respond with estimates and/or rounded values. In addition, people would probably tend to give themselves the benefit of the doubt and respond with values smaller than the true measurement.

14. a. No; the following types of vehicles are counted as passenger cars in some contexts but not others -- station wagons, mini-vans, full-sized vans, jeeps. Even pick-up trucks are classified as passenger cars for toll purposes on some bridges and highways.
   b. Yes; there are probably more jeeps and four-wheel drive vehicles in the Rocky Mountain states, and more convertibles in the southern states.
   c. If the states would cooperate, a sample could be obtained from title or registration records. Since the exercise suggests cars "in use," using sales information from car dealerships would not take into consideration the fact the some cars last longer than others.

15. a. Open questions elicit the respondent's true feelings without putting words or ideas into his mind. In addition, open questions might produce responses the pollster failed to consider. Unfortunately open questions sometimes produce responses that are rambling, unintelligible or not relevant.
   b. Closed questions help to focus the respondent and prevent misinterpretation of the question. Sometimes, however, closed questions reflect only the wording and opinions of the pollster and do not allow respondents to express legitimate alternatives.
   c. Closed questions are easier to analyze because the pollster can control the number of possible responses to each question and word the responses to establish relationships between questions.

1.5 Statistics and Computers

1. *MTB > SET C1*
   DATA> 2 4 1 2 3 2 3 1
   DATA> END
   MTB > PRINT C1

   C1
     2   4   1   2   3   2   3   1

2. [Even though it bears an even number, this problem is included because it is used in subsequent odd problems. In general, such will be the procedure throughout this manual.]
   MTB > SET C2
   DATA> 3.2 22.6 23.1 16.9 0.4 6.6 12.5 22.8
   DATA> END
   MTB > PRINT C2

   C2
     3.2   22.6   23.1   16.9   0.4   6.6   12.5   22.8

3. a. *MTB > LET C3 = C1 + C2*
   MTB > PRINT C3

   C3
     5.2 26.6 24.1 18.9 3.4 8.6 15.5 23.8

   Each value in C3 (column 3) is the sum of the corresponding values in C1 and C2. NOTE: Even though each set of values is designated as a column, Minitab prints the values of a single column in row format in order to save space. As the next exercises indicate, Minitab uses the column format when printing more than one column. Since C1 and C2 each contain 8 values, C3 contains 8 sums; when C1 and C2 do not contain the same number of values, Minitab finds as many sums as possible and provides an appropriate message.

   b. No, the values in C3 are of no practical significance. Calculators and computers give numerical values without considering whether they make any sense; it is the always the user's responsibility to consider the operations being performed and the units involved (e.g., feet, pounds, mpg, etc.).

4. *READ> C4 C5 C6*
   DATA> 623 509 2.6
   DATA> 454 471 2.3
   DATA> 643 700 2.4
   DATA> 585 719 3.0
   DATA> 719 710 3.1
   DATA> END
     5 ROWS READ
   MTB > PRINT C4 C5 C6

ROW	C4	C5	C6
1	623	509	2.6
2	454	471	2.3
3	643	700	2.4
4	585	719	3.0
5	719	710	3.1

5. *MTB > SAVE 'EXER'*
   MTB > STOP
   [The computer will print a statement about ending Minitab.]

   MTB > RETRIEVE 'EXER'
   [The computer will print details about retrieving the file.]
   MTB > PRINT C1-C6

ROW	C1	C2	C3	C4	C5	C6
1	2	3.2	5.2	623	509	2.6
2	4	22.6	26.6	454	471	2.3
3	1	23.1	24.1	643	700	2.4
4	2	16.9	18.9	585	719	3.0
5	3	0.4	3.4	719	710	3.1
6	2	6.6	8.6			
7	3	12.5	15.5			
8	1	22.8	23.8			

   NOTE: In general, Minitab is very user-friendly and follows normal human conventions. Within a PRINT command, Minitab assumes the hyphen means "through"; within a LET command, Minitab assumes the hyphen means "minus."

6. **NOTE:** The method for gaining access to data sets already installed in Minitab will vary. At the facility used for writing this manual, the sequence
>   *MTB > RETRIEVE 'TREES'*
>   ** ERROR * REQUESTED FILE DOES NOT EXIST*

resulted because the data sets were located in a DATA directory within a MINITAB directory within the Minitab application. The following sequence [CD = "change directory"] indicates how access was gained to the data sets. If students are to use the data sets already installed in Minitab, either the instructor or a computer technician should move the appropriate sets to a location from which they can be easily accessed but not student-edited and re-saved in a different form.

MTB > CD MINITAB
MTB > CD DATA
MTB > RETRIEVE 'TREES'
   [The computer will print details about retrieving the file.]
MTB > PRINT C1-C3

ROW	DIAMETER	HEIGHT	VOLUME
1	8.3	70	10.3
2	8.6	65	10.3
3	8.8	63	10.2
:	:	:	:

[There were 31 trees in this data file.]

7. *MTB > SET C7*
DATA > 25(12.345)
DATA > END
MTB > PRINT C7

C7
12.345	12.345	12.345	12.345	12.345	12.345	12.345
12.345	12.345	12.345	12.345	12.345	12.345	12.345
12.345	12.345	12.345	12.345	12.345	12.345	12.345
12.345	12.345	12.345	12.345			

Minitab prints the 25 12.345 values located in column 7.

8. *MTB > SET C8*
DATA > 2 4 7 85 90 102.4
DATA > END
MTB > PRINT C8

C8
2.0   4.0   7.0   85.0   90.0   102.4

NOTE: Since one of the numbers placed in column 8 was accurate to tenths (i.e., had one decimal place), all of C8 is stored and printed with that accuracy.

a. *MTB > LET C9 = C8/10*
   PRINT C9

C9
0.20   0.40   0.70   8.50   9.00   10.24

The results are the original numbers divided by 10. Each number is stored and printed with two decimal place accuracy.

b. *LET C10 = C8**2*
   PRINT C10

   C10
   4.0   16.0   49.0   7225.0   8100.0   10485.8

The results are the squares (i.e., raised to the power 2) of the original numbers. Since most versions of Minitab store and print only six significant digits, the final value (which should be 10485.76) is reported as 10485.8.

c. LET C11 = 50*C8

9. *MTB > SET C11*
   DATA> 2 4 1 2 3 2 3 1
   DATA> END
   a. *MTB > MEAN C11*
      MEAN   =   2.2500
   b. *MTB > MAXIMUM C11*
      MAXIMUM =   4.0000
   c. *MTB > MINIMUM C11*
      MINIMUM =   1.0000
   d. *MTB > SUM C11*
      SUM   =   18.000
   e. *MTB > SSQ C11*
      SSQ   =   48.000

The commands in parts (a)-(e) above produce the arithmetic average of the values, the largest value, the smallest value, the sum of the values, and the sum of the squares of the values. NOTE: Statisticians use the word "mean" for "arithmetic average." This is actually a mathematical version of an historic use of the word, for in very early English "mean" was defined as "average" in the sense of "commonplace" or "not superior." The Christmas song *What Child Is This?* asks, for example, "Why lies he in such mean estate, where ox and ass are feeding?" The SSQ stands for Sum of SQuares and is $2^2 + 4^2 + \ldots 1^2 = 48$.

10. *MTB > SET C11*
    DATA> 2 4 1 2 3 2 3 1
    DATA> END
    a. *MTB > MEAN C11*
       MEAN   =   225.00
    b. *MTB > MAXIMUM C11*
       MAXIMUM =   400.00
    c. *MTB > MINIMUM C11*
       MINIMUM =   100.00
    d. *MTB > SUM C11*
       SUM   =   1800.0
    e. *MTB > SSQ C11*
       SSQ   =   480000

Each value in column 11 has been multiplied by 100. The mean, maximum, minimum and sum in parts (a)-(d) are each multiplied by 100. The sum of the squares in part (e) is multiplied by $100^2 = 10,000$.

Review Exercises

1. a. ordinal, the categories give relative rankings but differences between categories are not necessarily consistent
    b. ratio, differences between values are consistent and there is a meaningful zero
    c. ordinal, the categories give relative rankings but differences between categories are not necessarily consistent
    d. nominal, the categories give names only and not any ordering
    e. interval, differences between values are consistent but there is no meaningful zero

2. a. continuous, since any h.p. rating on a continuum is possible
    b. ratio, since differences between values are consistent and there is a meaningful zero
    c. stratified, since the mowers were organized by manufacturer in order to make selections

3. The police had to estimate the quantity of albums seized; assuming they did not count each album, this could have been done by volume, weight, number of boxes, etc. They also had to estimate the value per album; this could have been done using retail price, "street" price, wholesale price, actual cost to produce, etc. They might tend to exaggerate the value of the albums and/or use the method that produces the largest value in order to appear more effective or in hopes that the magnitude of the criminal activity might justify an increase in manpower, salaries, etc.

4. a. discrete, since the number of small business owners surveyed must be an integer
    b. continuous, since times can be any value on a continuum
    c. continuous, since volume can be any value on a continuum

5. a. systematic, since products are selected at regular intervals
    b. random, since each car has the same chance of being selected
    c. cluster, since the stocked items were organized in stores and all the items in randomly selected stores were chosen
    d. stratified, since drivers were classified by sex and age in order to make the selections
    e. convenience, since the sample was composed of those who chose to take a test drive.
       NOTE: If the car maker first selected the particular dealership at random from among all its dealerships, then cluster sampling was used. Since the apparent set of interest is all potential customers and not just all who take a test drive, the selections at the local dealership were made by convenience sampling.

6. Even though the word "about" is not used, 1250 is probably an estimate. Because 1250 sounds more precise than 1200 or 1300, the article suggests an accuracy that is probably unwarranted. In addition, the subjective adjective "angry" with the figure raises questions -- e.g. how was it determined who was angry and who was not? how many non-angry protesters attended?

7. When giving their own personal data (height, weight, income, age, etc.) people sometimes give round numbers as a subconscious way to avoid invasions of privacy (and they often round in the most favorable direction). When answering for someone else, people give round numbers because they may not know the exact value.

8. a. The average number of deaths per day is $46,000/365 = 126$.
    b. Evenly distributed deaths would suggest about $126 \times 4 = 504$ during each 4-day period.
    c. As 630 is a 25% increase over 504, it is the expected number of deaths if travel increases 25%. No; it does not appear that driving is proportionately more dangerous over the holiday weekend. **NOTE:** Even if the proportions are the same, increased counts still warrant avoiding peak travel times. If the probability a given oncoming car crashes into you is constant, facing more oncoming cars increases the probability of at least one crash. [Analogy: Tossing a die more times increases the probability of getting at least one 6.]

Chapter 2

Descriptive Statistics

2-2 Summarizing Data

1. Subtracting two consecutive lower class limits indicates that the class width is 88 - 80 = 8. Since there is a gap of 1.0 between the upper class limit of one class and the lower class limit of the next, class boundaries are determined by increasing or decreasing the appropriate class limits by (1.0)/2 = 0.5. The class boundaries and class marks are given in the following table.

IQ	class boundaries	class mark	frequency
80- 87	79.5 - 87.5	83.5	16
88- 95	87.5 - 95.5	91.5	37
96-103	95.5 - 103.5	99.5	50
104-111	103.5 - 111.5	107.5	29
112-119	111.5 - 119.5	115.5	14
			146

NOTE: Although they often contain extra decimal points and may involve consideration of how the data were obtained, class boundaries are the key to tabular and pictorial data summaries. Once the class boundaries are obtained, everything else falls into place. In this case, the first class width is readily seen to be 87.5 - 79.5 = 8.0 and the first class mark is (79.5 + 87.5)/2 = 83.5. In this manual, class boundaries will typically be calculated first and then be used to determine other values. In addition, since the sum of the frequencies (i.e., the total number of values) is an informative number and used in many subsequent calculations, it will typically be shown as an integral part of each table.

2. Subtracting two consecutive lower class limits indicates that the class width is 7.6 - 0.0 = 7.6. Since there is a gap of .1 between the upper class limit of one class and the lower class limit of the next, class boundaries are determined by increasing or decreasing the appropriate class limits by (.1)/2 = .05. The class boundaries and class marks are given in the following table.

time (hrs)	class boundaries	class mark	frequency
0.0 - 7.5	-0.05 - 7.55	3.75	16
7.6 - 15.1	7.55 - 15.15	11.35	18
15.2 - 22.7	15.15 - 22.75	18.95	17
22.8 - 30.3	22.75 - 30.35	26.55	15
30.4 - 37.9	30.35 - 37.95	34.15	19
			85

3. Since the gap between classes is 0.1, the appropriate class limits are increased or decreased by (0.1)/2 = .05 to obtain the class boundaries and the following table.

Weight (kg)	class boundaries	class mark	frequency
16.2-21.1	16.15 - 21.15	18.65	16
21.2-26.1	21.15 - 26.15	23.65	15
26.2-31.1	26.15 - 31.15	28.65	12
31.2-36.1	31.15 - 36.15	33.65	8
36.2-41.1	36.15 - 41.15	38.65	3
			54

The class width is 21.15 - 16.15 = 5.00 and the first class mark is (16.15 + 21.15)/2 = 18.65.

4. Since the gap between classes is 1.0, the appropriate class limits are increased or decreased by $(1.0)/2 = .5$ to obtain the class boundaries and the following table.

Sales ($'s)	class boundaries	class mark	frequency
0 - 21	-0.5 - 21.5	10.5	2
22 - 43	21.5 - 43.5	32.5	5
44 - 65	43.5 - 65.5	54.5	8
66 - 87	65.5 - 87.5	76.5	12
88 - 109	87.5 - 109.5	98.5	14
110 - 131	109.5 - 131.5	120.5	20
			61

The class width is $21.5 - (-0.5) = 22.0$ and the first class mark is $(21.5 + -0.5)/2 = 10.5$.

5. The relative frequency for each class is found by dividing its frequency by 146, the sum of the frequencies. **NOTE:** As before, the sum is included as an integral part of the table. For relative frequencies, this should always be 1.000 (i.e., 100%) and serves as a check for the calculations.

IQ	relative frequency
80- 87	.110
88- 95	.253
96-103	.342
104-111	.199
112-119	.096
	1.000

6. The relative frequency for each class is found by dividing its frequency by 85, the sum of the frequencies. **NOTE:** As before, the sum is included as an integral part of the table. For relative frequencies, this should always be 1.000 (i.e., 100%) and serves as a check for the calculations.

time (hrs)	relative frequency
0.0 - 7.5	.188
7.6 - 15.1	.212
15.2 - 22.7	.200
22.8 - 30.3	.176
30.4 - 37.9	.224
	1.000

7. The relative frequency for each class is found by dividing its frequency by 54, the sum of the frequencies. NOTE: In #5, the relative frequencies were expressed as decimals; here they are expressed as percents The choice is arbitrary.

Weights (kg)	relative frequency
16.2-21.1	29.6%
21.2-26.1	27.8%
26.2-31.1	22.2%
31.2-36.1	14.8%
36.2-41.1	5.6%
	100.0%

8. The relative frequency for each class is found by dividing its frequency by 61, the sum of the frequencies. **NOTE:** In #6, the relative frequencies were expressed as decimals; here they are expressed as percents The choice is arbitrary.

Sales ($'s)	relative frequency
0 - 21	3.28%
22 - 43	8.20%
44 - 65	13.11%
66 - 87	19.67%
88 - 109	22.95%
110 - 131	32.79%
	100.00%

9. The cumulative frequencies are determined by repeated addition of successive frequencies to obtain the combined number in each class and all previous classes. NOTE: Consistent with the emphasis that has been placed on class boundaries, we choose to use upper class boundaries in the "less than" column. It is assumed that intelligence occurs on a continuum and that IQ points are the nearest whole number representation of a person's true measure of intelligence. An IQ of 87.9, for example, would be reported as 88 and fall in the second class. The 16 IQ's in the first class, therefore, are better described as being "less than 87.5" (using the upper class boundary) than as being "less than 88." This distinction becomes crucial in the construction of pictorial representations in the next section. In addition, the fact that the final cumulative frequency must equal the total number (i.e, the sum of the frequency column) serves as a check for calculations. The sum of cumulative frequencies, however, has absolutely no meaning and is not included.

IQ	cumulative frequency
less than 87.5	16
less than 95.5	53
less than 103.5	103
less than 111.5	132
less than 119.5	146

10. The cumulative frequencies are determined by repeated addition of successive frequencies to obtain the combined number in each class and all previous classes.

Time (hours)	cumulative frequency
less than 7.55	16
less than 15.15	34
less than 22.75	51
less than 30.35	66
less than 37.95	85

11. The cumulative frequencies are determined by repeated addition of successive frequencies to obtain the combined number in each class and all previous classes.

Weight (kg)	cumulative frequency
less than 21.15	16
less than 26.15	31
less than 31.15	43
less than 36.15	51
less than 41.15	54

12. The cumulative frequencies are determined by repeated addition of successive frequencies to obtain the combined number in each class and all previous classes.

Sales ($'s)	cumulative frequency
less than 21.5	2
less than 43.5	7
less than 65.5	15
less than 87.5	27
less than 109.5	41
less than 131.5	61

13. There is more than one acceptable solution. One possibility is to note that for a range of .95 - 0.26 = 4.69 to be covered with 10 classes, there must be at least (4.69)/10 = .469 units per class. Rounding up to a class width of .50 and starting at .25, for example, produces a first class with lower and upper class limits of .25 and .74. NOTE: The second class would then have lower and upper class limits of .75 and 1.24.

14. There is more than one acceptable solution. One possibility is to note that for a range of 157,800 - 3,600 = 154,200 to be covered with 16 classes, there must be at least (154,200)/16 = 9,637.5 units per class. Rounding up to a class width of 10,000 and starting at 0, for example, produces a first class with lower and upper class limits of 0 and 9,999. NOTE: The second class would then have lower and upper class limits of 10,000 and 19,999.

15. There is more than one acceptable solution. One possibility is to note that for a range of 4367 - 1650 = 2717 to be covered with 7 classes, there must be at least (2717)/7 = 388 units per class. Rounding up to a class width of 400 and starting at 1600, for example, produces a first class with lower and upper limits of 1600 and 1999. NOTE: The second class would then have lower and upper class limits of 2000 and 2399.

16. There is more than one acceptable solution. One possibility is to note that for a range of 1.027 - .802 = .225 to be covered with 12 classes, there must be at least (.225)/12 = .01875 units per class. Rounding up to a class width of .020 and starting at .800, for example, produces a first class with lower and upper class limits of .800 and .819. NOTE: The second class would then have lower and upper class limits of .820 and .839.

17. The class limits determined in exercise 13 produce the table given at the right.

Weights (lbs)	frequency
0.25 - 0.74	2
0.75 - 1.24	8
1.25 - 1.74	17
1.75 - 2.24	8
2.25 - 2.74	8
2.75 - 3.24	7
3.25 - 3.74	7
3.75 - 4.24	1
4.25 - 4.74	3
4.75 - 5.24	7
	62

18. The class limits determined in exercise 14 produce the table given at the right.

Income ($,s)	frequency
0- 9,999	6
10,000- 19,999	27
20,000- 29,999	31
30,000- 39,999	25
40,000- 49,999	21
50,000- 59,999	4
60,000- 69,999	4
70,000- 79,999	2
80,000- 89,999	3
90,000- 99,999	0
100,000-109,999	1
110,000-119,999	0
120,000-129,999	0
130,000-139,999	0
140,000-149,999	0
150,000-159,999	1
	125

19. The class limits determined in exercise 15 produce the table given at the right.

Weight (lbs)	frequency
1600 - 1999	1
2000 - 2399	1
2400 - 2799	12
2800 - 3199	6
3200 - 3599	7
3600 - 3999	4
4000 - 4399	1
	32

NOTE: Starting with 1600 instead of the lowest score of 1650 makes the lower class limits in 100's instead of 50's and makes the data more "centered" --i.e., the lowest and highest values of 1650 and 4367 are 50 and 32 units in from the limits of the table.

20. The class limits determined in exercise 16 produce the table given at the right.

Weight (g)	frequency
.800 - .819	3
.820 - .839	3
.840 - .859	5
.860 - .879	7
.880 - .899	14
.900 - .919	12
.920 - .939	26
.940 - .959	19
.960 - .979	7
.980 - .999	2
1.000 - 1.019	1
1.020 - 1.039	1
	100

21. The total numbers (i.e., sums of the frequency columns) for the men and women were 13,055 and 721 respectively. The relative frequency tables are as follows.

ethanol consumed by men (oz.)	relative frequency	ethanol consumed by women (oz.)	relative frequency
0.0 - 0.9	.019	0.0 - 0.9	.010
1.0 - 1.9	.071	1.0 - 1.9	.072
2.0 - 2.9	.118	2.0 - 2.9	.173
3.0 - 3.9	.171	3.0 - 3.9	.265
4.0 - 4.9	.087	4.0 - 4.9	.042
5.0 - 9.9	.273	5.0 - 9.9	.279
10.0 - 14.9	.142	10.0 - 14.9	.060
15.0 or more	.118	15.9 or more	.100
	1.000		1.000

The distributions seem very similar except that there are proportionately more women in the 2.0-3.9 range and proportionately less women in the 10.0-14.9 range. NOTE: Due to rounding, the relative frequencies given actually sum to 0.999 for the men and to 1.001 for the women. Discrepancies of 1 or 2 at the last decimal place are probably due to rounding; larger discrepancies should not be so attributed, and such work should be carefully checked.

23. a. classes mutually exclusive? yes
    b. all classes included? yes
    c. same width for all classes? no
    d. convenient class limits? yes
    e. between 5 and 20 classes? no
    Two of the five guidelines were not followed. NOTE: This does not mean the table is in error.
    The guidelines are only suggestions that make most presentations more readable; depending on
    the context, the given table may be the best way to present the ages.

24. NOTE: The ratio (log n)/(log 2) is the same regardless of the base of the log.
    a. 1 +   (log 50)/(log 2) = 1 +   5.64 =   6.64 =   7
    b. 1 +   (log 100)/(log 2) = 1 +   6.64 =   7.64 =   8
    c. 1 +   (log 150)/(log 2) = 1 +   7.23 =   8.23 =   8
    d. 1 +   (log 500)/(log 2) = 1 +   8.97 =   9.97 = 10
    e. 1 +   (log 1000)/(log 2) = 1 +   9.97 = 10.97 = 11
    f. 1 + (log 50,000)/(log 2) = 1 + 15.61 = 16.61 = 17

2-3 Pictures of Data

1. See the figure below. The bars extend from class boundary to class boundary. Each axis is
   labeled numerically <u>and</u> with the name of the quantity represented.

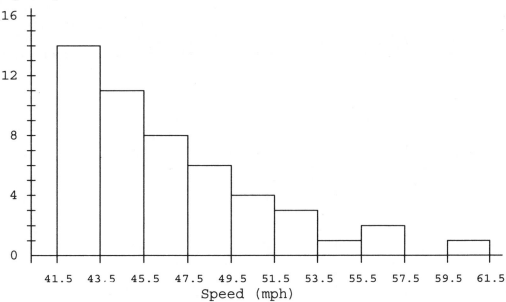

2. See the figure below. **NOTE:** Assuming the ages were reported in whole years (and not nearest years), the boundaries are 0, 10, 20,... (i.e., not -0.5, 9.5, 19.5,...).

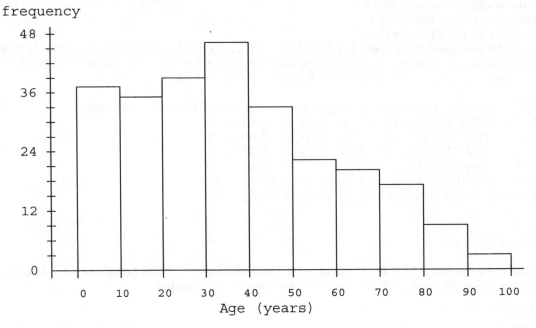

3. See the figure below. The bars extend from class boundary to class boundary. The relative frequencies are the original frequencies divided by 40, the sum of the frequencies. Boundaries were determined assuming the reported ages were rounded to the nearest tenth of a year.

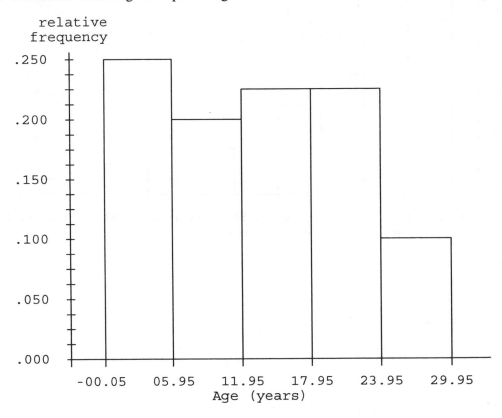

4. See the figure below. **NOTE:** Assuming the ages were reported in whole years (and not nearest years), the boundaries are 16, 26, 36,… (i.e., not 15.5, 25.5, 35.5,…).

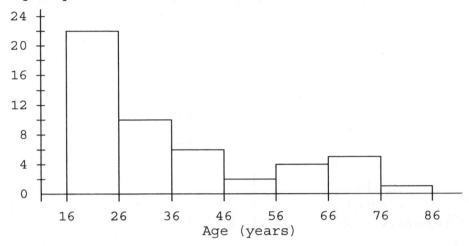

5. See the figure below, with bars arranged in order of magnitude.

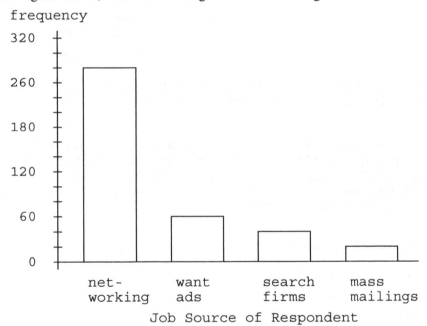

6. The pie chart is given at the right. The Pareto chart appears to be more effective in showing the relative importance of job sources.

Job Source of Respondent

7. See the figure below **at the left**. The sum of the frequencies is 50; the relative frequencies are 23/50 = 46%, 9/50 = 18%, 12/50 = 24%, and 6/50 = 12%. The corresponding central angles are (.46)360° = 165.6°, (.18)360° = 64.8°, (.24)360° = 86.4°, and (.12)360° = 43.2°. NOTE: To be complete, the figure needs to be titled with the name of the quantity being measured.

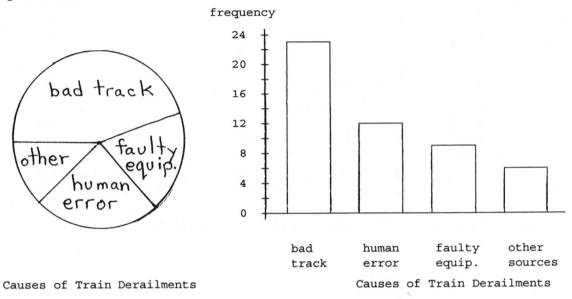

Causes of Train Derailments   Causes of Train Derailments

8. See the figure above at the right, with bars arranged in order of magnitude. The Pareto chart appears to be more effective in showing the relative importance of the causes of train derailments.

9. There is more than one acceptable solution. For a range of 193 - 72 = 121 to be covered with 12 classes, there must be at least (121)/12 = 10.08 units per class. One possibility is to round down to a class width of 10 and use 70 as the first lower class limit. While this produces 13 classes instead of 12, the added readability created by such round numbers justifies that choice and results in the following frequency distribution.

length (minutes)	frequency
70 - 79	1
80 - 89	2
90 - 99	18
100 - 109	17
110 - 119	7
120 - 129	6
130 - 139	3
140 - 149	1
150 - 159	2
160 - 169	2
170 - 179	0
180 - 189	0
190 - 199	1
	60

a. See the figure below. The bars extend from class boundary to class boundary.

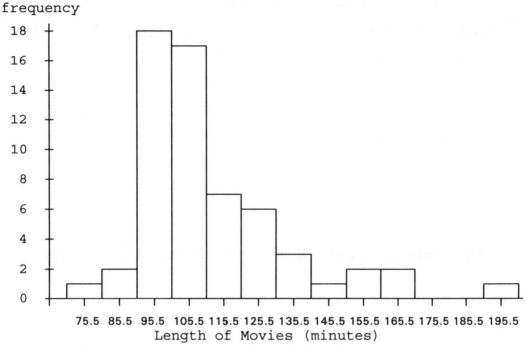

NOTE: The bars always extend from class boundary to class boundary, but the numerical labels along the horizontal axis may be placed at the class midpoints (as above) or at any other convenient points.

b. The figure shows that the data are positively skewed.

10. a. Refer to exercises 16 and 20 of section 2-2 to construct the frequency distribution below at the left. The histogram is given below at the right.

Weight (g)	frequency
.800 - .819	3
.820 - .839	3
.840 - .859	5
.860 - .879	7
.880 - .899	14
.900 - .919	12
.920 - .939	26
.940 - .959	19
.960 - .979	7
.980 - .999	2
1.000 - 1.019	1
1.020 - 1.039	1
	100

b. The figure is approximately bell-shaped.

11. There is more than one acceptable solution. One possibility is to note that for a range of 20.58 - 1.65 = 18.93 to be covered with 10 classes, there must be at least (18.93)/10 = 1.893 units per class. Rounding up to 2.00 and using 1.00 for the first lower class limit produces first class limits from 1.00 to 2.99. The resulting frequency distribution is given below.

weight (lbs)	frequency
1.00 - 2.99	4
3.00 - 4.99	4
5.00 - 6.99	12
7.00 - 8.99	9
9.00 - 10.99	12
11.00 - 12.99	10
13.00 - 14.99	5
15.00 - 16.99	3
17.00 - 18.99	1
19.00 - 20.99	2
	62

a. See the figure below. The bars extend from class boundary to class boundary.

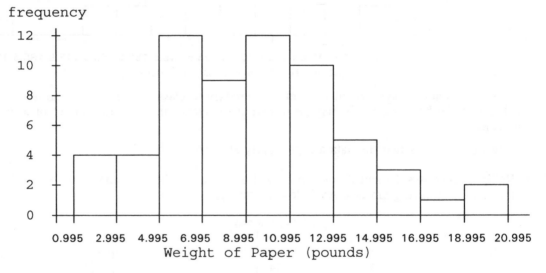

b. The figure shows that the data have a slight positive skew.

12. a. Since the variable assumes only integer values from 1 to 11, and since the directions require 11 classes, the figure at the right is the only solution.

   b. The figure shows that the data have a positive skew.

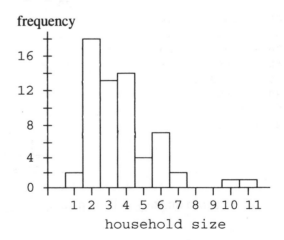

13. There is more than one acceptable solution. One possibility is to group the numbers naturally by fives to produce the frequency distribution given below. NOTE: Since zero was not one of the possible numbers, the first class contains only 4 achievable values; each of the other classes contains 5 achievable values.

number	frequency
0 - 4	23
5 - 9	34
10 - 14	26
15 - 19	30
20 - 24	33
25 - 29	31
30 - 34	30
35 - 39	34
40 - 44	31
45 - 49	28
	300

a. See the figure below. The bars extend from class boundary to class boundary.

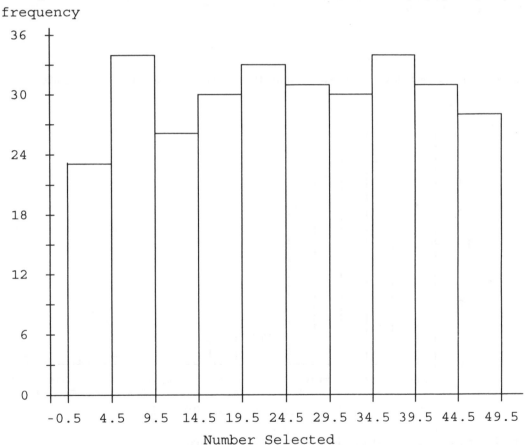

b. The figure shows that the data are approximately uniform.

15. There is more than one acceptable solution. One possibility is to note that for a range of 157.8 - 3.6 = 154.2 to be covered with 16 classes, there must be at least (154.2)/16 = 9.64 units per class. Rounding up to a class width of 10 and using 0 for the first lower class limit produces the frequency distribution given below.

income			frequency
0.0	-	9.9	6
10.0	-	19.9	27
20.0	-	29.9	31
30.0	-	39.9	25
40.0	-	49.9	21
50.0	-	59.9	4
60.0	-	69.0	4
70.0	-	79.9	2
80.0	-	89.9	3
90.0	-	99.9	0
100.0	-	109.9	1
110.0	-	119.9	0
120.0	-	129.9	0
130.9	-	139.9	0
140.0	-	149.9	0
150.0	-	159.9	1
			125

a. See the figure below. The bars extend from class boundary to class boundary.

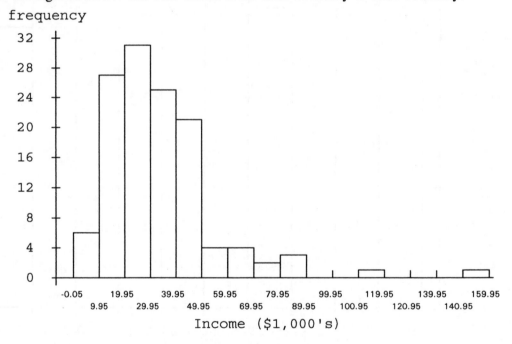

b. The figure shows that the data are positively skewed.

16. a. There is more than one acceptable solution. One possibility is to note that the ages range from 18 to 38. Beginning with 17 and using a class width of 2 years, produces the frequency distribution given below at the left. The histogram is given below at the right. Since age is reported in whole number of years achieved (and not rounded to the nearest year), the class boundaries are 17, 19,... (and not 16.5, 18.5,...).

age	frequency
17 - 18	6
19 - 20	13
21 - 22	16
23 - 24	11
25 - 26	14
27 - 28	15
29 - 30	11
31 - 32	12
33 - 34	5
35 - 36	3
37 - 38	1
	107

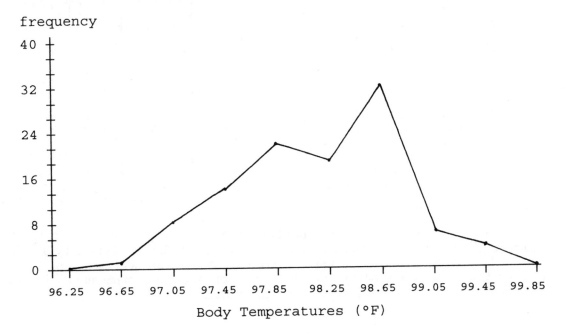

b. While there is an approximate uniform distribution (except at the ends -- and the small frequency at the lower end could be eliminated by starting at 18 instead of 17), the figure indicates that none of the given categorizations applies to these data.

17. See the figure below. NOTE: The graph returns to a frequency of zero at the midpoints of the classes before and after the given data.

18. The cumulative frequency distribution and ogive are given below.

temperature	cumulative frequency
less than 96.45	0
less than 96.85	1
less than 97.25	9
less than 97.65	23
less than 98.05	45
less than 98.45	64
less than 98.85	96
less than 99.25	102
less than 99.65	106

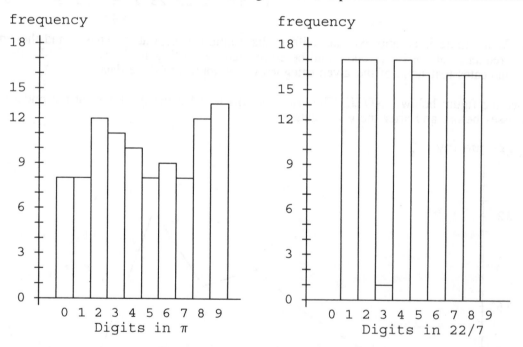

NOTE: The directions for exercise 18 specified that the cumulative frequency distribution include a "beginning" row with cumulative frequency 0, while the cumulative frequency distribution as discussed in the previous section did not include such a row. Most presentations of cumulative frequency distributions <u>do</u> include such a row as an integral part of the table.

19. a. The two figures are given below. The histogram for π represents a more even distribution.

b. The number π is irrational (i.e., it cannot be represented as the quotient of two whole numbers). The decimal representation of the rational number 22/7 is obtained by dividing 7 into 22; because there are a maximum of 6 possible remainder digits at each step (1,2,3,4,5,6) that keep generating digits in the quotient, there are a maximum of six possible quotient digits that will repeat for ever in a regular cycle.

20. Doubling the number of classes gives each class about half of its original frequency. If the original vertical scale is maintained, the figure will reach about half of its original height. If the original figure followed the guideline that the height be about three-fourths of the width, the new figure will be "short and fat" and may not give an accurate visual impression.

21. a. Here is the bar chart.

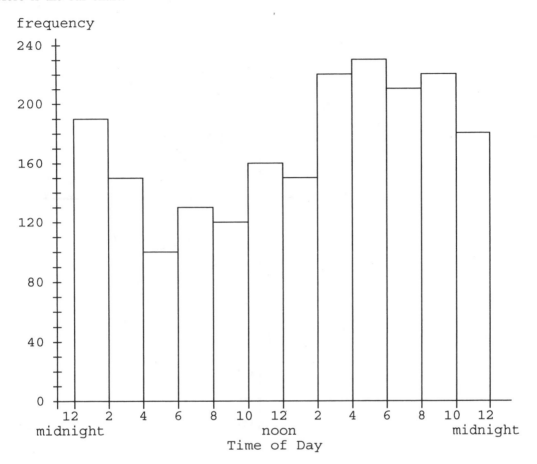

The circular bar chart is given on the following page.
NOTE: Such figures are difficult to construct and interpret.
When constructing the figure, make certain that...
    * there is a clear zero circle, so that one bar having twice the height of another
      corresponds to one category having twice the frequency of another.
    * the sides of the each bar are parallel to each other and perpendicular to the tangent line at
      the point where they intersect the zero circle.
    * the bars have the same width.
    * the tops of each bar are perpendicular to its sides.
When interpreting the figure, remember that ...
    * because the bars emanate from a zero circle and not a zero line, the area of each bar will
      not correspond exactly to the frequency it represents.
    * because the tops of the bars are straight and the height lines are concentric circles, reading
      the height with any precision is more difficult than for a normal bar chart.
If approximate information can be portrayed with added insight and interest, however, the use of circular bar charts and other "creative" figures is justified and appropriate -- especially in informal settings.

Number of Traffic Fatalities

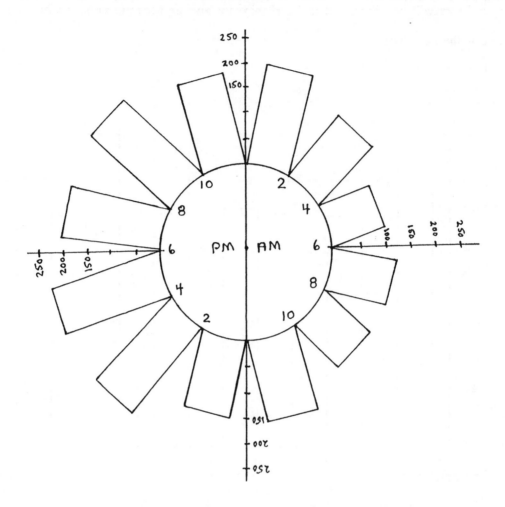

b. Because the circular bar chart has no beginning and no end, it does not artificially center on noon and better conveys the pattern in the numbers of death across the midnight hours. It might be argued, however, that the traditional bar chart is easier to read, more accurate, much less difficult to construct, and that adequate labeling allows the reader to appreciate the continuity in the time scale.

c. The fact that there are the least number of fatal crashes between 4 and 6 am does not necessarily make it the safest time to drive. That's like saying the best place to sit when you're sick is on the top of a piano because very, very few ever die while sitting on a piano. There are fewer cars on the road during those hours, and it that sense it may be considered a wise and safe time to travel. One must consider the <u>rates</u> of fatal accidents, however, as well as the <u>numbers</u> of such occurrences -- it could be, for example, that fatal crashes involve 1.0% of all the cars on the road during the 4-6 am slack period but only 0.1% of the many cars on the road during the 4-6 pm rush hour.

2-4 Measures of Central Tendency

NOTE: Of the four measures of central tendency discussed in the text, only the mean is given a particular symbol. As it is common in mathematics and statistics to use symbols instead of words to represent quantities that are used often and/or that may appear in equations, this manual also employs symbols for the other three measures of central tendency to produce the following notation:

mean = $\bar{x}$ mode = M
median = m   midrange = m.r.

1. Arranged in order, the 10 scores are: 65 66 67 68 71 73 74 77 77 77
   a. $\bar{x}$ = (Σx)/n = (715)/10 = 71.5   c. M = 77
   b. m = (71 + 73)/2 = 72.0   d. m.r. = (65 + 77)/2 = 71.0
   NOTE: Since the median is the middle score <u>when the scores are arranged in order</u> and the midrange is halfway between the first and last score <u>when the scores are arranged in order</u>, it is usually helpful to begin by placing the scores in order -- this will not affect the mean, and it may also aid in identifying the mode. In addition, no measure of central tendency can have a value lower than the smallest score or higher than the largest score -- remembering this helps to protect against gross errors, which most commonly occur when calculating the mean.

2. Arranged in order, the 10 scores are: 42 54 58 62 67 77 77 85 93 100
   a. $\bar{x}$ = (Σx)/n = (715)/10 = 71.5   c. M = 77
   b. m = (67 + 77)/2 = 72.0   d. m.r. = (42 + 100)/2 = 71.0

3. Arranged in order, the 15 scores are:
   .12 .13 .14 .16 .16 .16 .17 .17 .17 .18 .21 .24 .24 .27 .29
   a. $\bar{x}$ = (Σx)/n = (2.81)/15 = .187   c. M = .16 and .17 (bimodal)
   b. m = (71 + 73)/2 = .170   d. m.r. = (.12 + .29)/2 = .205

4. Arranged in order, the 20 scores are:
   0.0 0.0 1.0 1.5 1.7 1.8 2.0 2.0 2.1 2.1 2.3 2.4 2.4 2.9 3.3 3.4 3.7 4.4 4.4 4.5
   a. $\bar{x}$ = (Σx)/n = (47.9)/20 = 2.395   c. M = 0.0, 2.0, 2.1, 2.4, 4.4 (multimodal)
   b. m = (2.1 + 2.3)/2 = 2.20   d. m.r. = (0.0 + 4.5)/2 = 2.25
   NOTE: See the note given with exercise 7.

5. Arranged in order, the 27 scores are:
   14 16 17 17 18 18 19 20 21 23 23 24 25 25
   27 28 28 29 30 31 33 34 37 38 40 42 51
   a. $\bar{x}$ = (Σx)/n = (728)/27 = 27.0   c. M = 17,18,23,25,28 (multimodal)
   b. m = 25.0   d. m.r. = (14 + 51)/2 = 32.5
   NOTE: The purpose of descriptive statistics in general and measures of central tendency in particular is to provide a meaningful summary of the data, and the above determination of the mode followed the definition quite literally. From a practical point of view, most statisticians would probably say there was no meaningful mode. Grouping by ten's, one could say the mode was in the 20's -- i.e., more people died in their 20's than in any other 10-year group.

6. Arranged in order, the 42 scores are:
   42 43 46 46 47 48 49 49 50 51 51 51 51 51 52 52 54 54 54 54 55
   55 55 55 56 56 56 57 57 57 57 58 60 61 61 61 62 64 64 65 68 69
   a. $\bar{x}$ = (Σx)/n = (2304)/42 = 54.9   c. M = 51
   b. m = (55 + 55)/2 = 55.0   d. m.r. = (42 + 69)/2 = 55.5

7. Arranged in order, the 8 scores are: 0.63 0.92 1.40 1.41 1.74 2.10 2.19 2.87
a. $\bar{x} = (\Sigma x)/n = (13.26)/8 = 1.6575$ c. M = [none]
b. m = $(1.41 + 1.74)/2 = 1.575$ d. m.r. = $(0.63 + 2.87)/2 = 1.750$
NOTE: The mean was 1.6575; according to the rule given in the text, this value should be rounded to three decimal places. While the text describes how many decimal places to present in an answer, it does not describe the actual rounding process. When the figure to be rounded is <u>exactly</u> half-way between two values (i.e., the digit in the position to be discarded is a 5, and there are no further digits because the calculations have "come out even"), there is no universally accepted rule. Some authors say to always round up a 5; others correctly note that always rounding up a value exactly half-way between introduces a consistent bias -- the value should be rounded up half the time and rounded down half the time. In such cases, some authors suggest rounding toward the even value (e.g., .65 becomes .6 and .75 becomes .8) and other simply suggest flipping a coin. In this manual, the results of calculations producing values <u>exactly</u> half-way between will be reported without rounding (i.e., stated to one more decimal than usual).

8. Arranged in order, the 8 scores are: 8.08 9.46 10.58 11.03 11.42 13.61 15.09 16.39
a. $\bar{x} = (\Sigma x)/n = (95.66)/8 = 11.9575$ c. M = [none]
b. m = $(11.03 + 11.42)/2 = 11.225$ d. m.r. = $(8.08 + 16.39)/2 = 12.235$

9. Arranged in order, the 12 scores are:
654.2 661.3 662.2 662.7 667.0 667.4 669.8 670.7 672.2 672.2 672.6 679.2
a. $\bar{x} = (\Sigma x)/n = (8011.5)/12 = 667.625$ c. M = 672.2
b. m = $(667.4 + 669.8)/2 = 668.60$ d. m.r. = $(654.2 + 679.2)/2 = 666.70$

10. Arranged in order, the 11 scores are: .842 .886 .912 .913 .914 .920 .925 .939 .947 .957 .958
a. $\bar{x} = (\Sigma x)/n = (10.113)/11 = .9194$ c. M = [none]
b. m = .9200   d. m.r. = $(.842 + .958)/2 = .9000$

11. Arranged in order, the 10 scores are:
2360 2526 2784 2863 2992 3253 3315 3536 3766 3784
a. $\bar{x} = (\Sigma x)/n = (31179)/10 = 3117.9$ c. M = [none]
b. m = $(2992 + 3253)/2 = 3122.5$ d. m.r. = $(2360 + 3784)/2 = 3072.0$

12. Arranged in order, the 15 scores are: 6,400   7,400   7,800 10,400 15,000 17,200 22,000
23,600   25,400   26,400   26,800   34,800   40,000   47,400   67,000
a. $\bar{x} = (\Sigma x)/n = (377,600)/15 = 25,173$ c. M = [none]
b. m = 23,600   d. m.r. = $(6,400 + 67,000)/2 = 36,700$

13. Construct a stem-and-leaf plot [see section 2.7] to arrange the 38 scores in order.

96.	2 6
97.	0 2 3 4 4 4 4 5 6 7 8
98.	0 0 0 0 1 2 2 2 2 4 4 6 7 7 8 8 8 8 9 9
99.	0 0 0 2 4

a. $\bar{x} = (\Sigma x)/n = (3728.8)/38 = 98.13$
b. m = $(98.2 + 98.2)/2 = 98.20$
c. M = 97.4, 98.0, 98.2, 98.8 (multi-modal)
d. m.r. = $(96.2 + 99.4)/2 = 97.80$

14. Construct a stem-and-leaf plot [see section 2.7] to arrange the 62 scores in order.

0.	26 41
0.	85 94 99
1.	00 04 05 09 22 25 31 32 33 36 36 37 41 42 49
1.	50 50 52 55 67 70 73 76 86 89 93
2.	02 04 10 11 32 35 43
2.	50 57 58 62 74 83 96 97
3.	02 09 20 22 26 27 29
3.	57 60 61 69 99
4.	41 48
4.	63 95

a. $\bar{x} = (\Sigma x)/n = (137.50)/62 = 2.218$
b. m = $(1.93 + 2.02)/2 = 1.975$
c. M = 1.36, 1.50 (bi-modal)
d. m.r. = $(0.26 + 4.95)/2 = 2.605$

15. Construct a stem-and-leaf plot [see section 2.7] to arrange the 62 scores in order.

```
 1. | 65
 2. | 41 61 80
 3. | 26 27 45 69
 4. |
 5. | 86 87 88
 6. | 05 16 33 38 44 67 68 96 98
 7. | 57 72 98
 8. | 08 26 72 78 82 96
 9. | 09 19 41 45 46 55 64 83 92
10. | 00 58 99
11. | 03 08 36 42
12. | 29 32 43 45 56 73
13. | 05 11 31 61
14. | 33
15. | 09
16. | 08 39
17. | 65
18. |
19. |
20. | 12 58
```

a. $\bar{x} = (\Sigma x)/n = (584.54)/62 = 9.428$
b. $m = (9.19 + 9.41)/2 = 9.300$
c. $M = $ [none]
d. m.r. $= (1.65 + 20.58)/2 = 11.115$

16. Construct a stem-and-leaf plot [see section 2.7] to arrange the 35 scores in order.

```
 7 | 2
 8 | 2
 9 | 0 2 3 5 6 6 7 8
10 | 0 1 2 4 4 4 5 6 6 7
11 | 0 1 4 5 7 9
12 | 0 5 9
13 | 3 4 9
13 | 3 4 9
14 |
15 | 5 9
16 | 0
```

a. $\bar{x} = (\Sigma x)/n = (3890)/35 = 111.1$
b. $m = 106$
c. $M = 104$
d. m.r. $= (72 + 160)/2 = 116.0$

17.

x	f	x·f
42.5	14	595.0
44.5	11	489.5
46.5	8	372.0
48.5	6	291.0
50.5	4	202.0
52.5	3	157.5
54.5	1	54.5
56.5	2	113.0
58.5	0	0.0
60.5	1	60.5
	50	2335.0

$\bar{x} = (\Sigma x \cdot f)/n$
$= (2335.0)/50$
$= 46.7$

18.

x	f	x·f
4.5	37	166.5
14.5	35	507.5
24.5	39	955.5
34.5	46	1587.0
44.5	33	1468.5
54.5	22	1199.0
64.5	20	1290.0
74.5	17	1266.5
84.5	9	760.5
94.5	3	283.5
	261	9484.5

$$\bar{x} = (\Sigma x \cdot f)/n$$
$$= (9484.5)/261$$
$$= 36.3$$

NOTE: Since ages are reported as of one's last birthday, and not to the nearest year, care must be taken when working with ages. If n people each report an age of 14, for example, the mean reported age is clearly 14.0; since those n people were probably evenly spread out between 14.0 and 14.999 years old, however, their true mean age is about 14.5. The mean calculated above is the mean reported age and not necessarily the true mean time in years since birth. The latter (i.e., true mean time in years since birth) is found by adding .5 to the mean calculated above <u>OR</u> by recognizing that the true midpoint for the first interval is 5.0 (since it includes ages from 0.000 to 9.999), the true midpoint for the second interval is 15.0 (since it includes ages from 10.000 to 19.999), etc.

19.

x	f	x·f
2.95	10	29.50
8.95	8	71.60
14.95	9	134.55
20.95	9	188.55
26.95	4	107.80
	50	532.00

$$\bar{x} = (\Sigma x \cdot f)/n$$
$$= (532.00)/50$$
$$= 13.30$$

20.

x	f	x·f
20.5	22	451.0
30.5	10	305.0
40.5	6	243.0
50.5	2	101.0
60.5	4	242.0
70.5	5	352.5
80.5	1	80.5
	50	1775.0

$$\bar{x} = (\Sigma x \cdot f)/n$$
$$= (1775.0)/50$$
$$= 35.5$$

NOTE: Refer to the NOTE given in exercise #18.

21. a. Arranged in order, the 5 scores are: 108,000   179,000   206,000   236,000   236,000
   $\bar{x} = (\Sigma x)/n = (965,000)/5 = 193,000$
   m = 206,000
   M = 236,000
   m.r. = (108,000 + 236,000)/2 = 172,000

   b. Arranged in order, the 5 scores are: 128,000   199,000   226,000   256,000   256,000
   $\bar{x} = (\Sigma x)/n = (1065,000)/5 = 213,000$
   m = 226,000
   M = 256,000
   m.r. = (128,000 + 256,000)/2 = 192,000

   c. In general, adding (or subtracting) a constant k from each score will add (or subtract) k from each measure of central tendency.

d. Arranged in order, the 5 scores are: 108   179   206   236   236
   $\bar{x} = (\Sigma x)/n = (965)/5 = 193$
   $m = 206$
   $M = 236$
   m.r. $= (108 + 236)/2 = 172$

e. In general, dividing (or multiplying) each score by a constant k will divide (or multiply) each measure of central tendency by k.

22. a.

x	w	w·f
70	.20	14.0
65	.20	13.0
90	.20	18.0
85	.40	34.0
	1.00	79.0

$\bar{x} = (\Sigma w \cdot f)/\Sigma w$
$\phantom{\bar{x}} = (79.0)/1.00$
$\phantom{\bar{x}} = 79.0$

b.

x	w	w·f
4	4	16
2	3	6
3	1	3
4	2	8
2	4	8
	14	41

$\bar{x} = (\Sigma w \cdot f)/\Sigma w$
$\phantom{\bar{x}} = (41)/14$
$\phantom{\bar{x}} = 2.929$

23. a. The answers to exercises 1 and 2 are identical. Since the measures of central tendency attempt to summarize the data, the logical inference would be that the two data sets are identical -- or at least very similar.
   b. The measures of central tendency alone do not necessarily reflect differences in spread (the data in exercise 2 is much more spread out; all the scores in exercise 1 were in the 60's or 70's) or difference in skewness (the data in exercise 2 is more symmetric and has a mode near the center of the scores; the scores in exercise 2 "pile up" at the high end and have the highest value for a mode).

24. Arranged in order, the 10 scores are: 65 66 67 68 71 73 77 77 77 740
   a. $\bar{x} = (\Sigma x)/n = (1381)/10 = 138.1$
   b. $m = (71 + 73)/2 = 72.0$
   c. $M = 77$
   d. m.r. $= (65 + 740)/2 = 402.5$
   The extreme value had a dramatic effect on the mean and the midrange, but the median and mode were unchanged.

25. In general, $\log \bar{x}$ does not equal $(\Sigma \log x)/n$. In words, the log of the average is not the same as the average of the logs. This is analogous to the more straightforward fact that the square of the average is not the same as the average of the squares and illustrates the principle that the order in which operations are applied can make a difference. Consider the following example, in which the log base 10 is used.

x	log x	x^2
1	0	1
10	1	100
100	2	10000
111	3	10101

$\bar{x} = (\Sigma x)/n = 111/3 = 37$

$(\Sigma \log x)/n = 3/3 = 1$

$(\Sigma x^2)/n = 10101/3 = 3367$

$\log \bar{x} = \log 37 = 1.568 \qquad 1 = (\Sigma \log x)/n$
$\bar{x}^2 = 37^2 = 1369 \qquad 3367 = (\Sigma x^2)/n$

26. Let $\bar{x}_h$ stand for the harmonic mean.

  a. $\bar{x}_h = n/[\Sigma(1/x)]$

      $= 10/[1/2 + 1/3 + 1/6 + 1/7 + 1/7 + 1/8 + 1/9 + 1/9 + 1/9 + 1/10]$

      $= 10/[1.844]$

      $= 5.4$

  b. $\bar{x}_h = n/[\Sigma(1/x)] = 2/[1/40 + 1/60] = 2/[.0417] = 48$

  c. $\bar{x}_h = n/[\Sigma(1/x)]$

      $= 14/[1/42.6 + 1/41.3 + 1/38.2 + 1/42.9 + 1/43.4 + 1/43.7 + 1/40.8 +$

          $1/34.2 + 1/40.1 + 1/41.2 + 1/40.5 + 1/41.7 + 1/39.8 + 1/39.6]$

      $= 14/[.3451]$

      $= 40.57$

27. The geometric mean of the five values is the fifth root of their product.

$$\sqrt[5]{(1.10)(1.08)(1.09)(1.12)(1.07)} = 1.092$$

28. R.M.S. $= \sqrt{\Sigma x^2/n} = \sqrt{[(151)^2 + (162)^2 + (0)^2 + (81)^2 + (-68)^2]/5} = \sqrt{60230/5} = 109.8$

29. a. The class mark of the last class is $(20.5 + 25.5)/2 = 23$

x	f	x·f
0	5	0
3	96	288
8	57	456
13	25	325
18	11	198
23	6	138
	200	1405

$\bar{x} = (\Sigma x \cdot f)/n$
$= (1405)/200$
$= 7.0$

  b. The class mark of the last class is $(20.5 + 30.5)/2 = 25.5$

x	f	x·f
0.0	5	0.0
3.0	96	288.0
8.0	57	456.0
13.0	25	325.0
18.0	11	198.0
25.5	6	153.0
	200	1420.0

$\bar{x} = (\Sigma x \cdot f)/n$
$= (1420)/200$
$= 7.1$

  c. The class mark of the last class is $(20.5 + 40.5)/2 = 30.5$

x	f	x·f
0.0	5	0.0
3.0	96	288.0
8.0	57	456.0
13.0	25	325.0
18.0	11	198.0
30.5	6	183.0
	200	1450.0

$\bar{x} = (\Sigma x \cdot f)/n$
$= (1450)/200$
$= 7.25$

In this case, because it involves relatively few scores and there is a reasonable upper limit (e.g, we know no student spends 100 hours a week studying) the interpretation of the open-ended class seems to make little difference and justifies the use of such classes when they make for significantly simpler presentations.

30. The following chart applies.

temp.	f.	c.f.
96.5-96.8	1	1
96.9-97.2	8	9
97.3-97.6	14	23
97.7-98.0	22	45
98.1-98.4	19	64
98.5-98.8	32	96
98.9-99.2	6	102
99.3-99.6	4	106

Since there are 106 scores, the median of $(x_{53} + x_{54})/2$ is in the 98.1 to 98.4 class and the key values are:
lower limit of median class = 98.1
class width = .40
n = 106 total scores
m = 45 scores entering the median class
frequency of the median class = 19

median = lower boundary + (class width)$\cdot$[(n+1)/2 -(m+1)]/(median class frequency)
$= 98.1 + (.40)[53.5 - 46]/19$
$= 98.1 + (.40)[7.5/19]$
$= 98.26$

31. Arranged in order, the original 20 scores are:
0.0 0.0 1.0 1.5 1.7 1.8 2.0 2.0 2.1 2.1
2.3 2.4 2.4 2.9 3.3 3.4 3.7 4.4 4.4 4.5
$\bar{x} = (\Sigma x)/n = (47.9)/20 = 2.395$
a. After trimming the highest and lowest 10%, the remaining 16 scores are:
1.0 1.5 1.7 1.8 2.0 2.0 2.1 2.1 2.3 2.4 2.4 2.9 3.3 3.4 3.7 4.4
$\bar{x} = (\Sigma x)/n = (39.0)/16 = 2.44$
b. After trimming the highest and lowest 20%, the remaining 12 scores are:
1.7 1.8 2.0 2.0 2.1 2.1 2.3 2.4 2.4 2.9 3.3 3.4
$\bar{x} = (\Sigma x)/n = (28.4)/12 = 2.37$

32. No, the result is not the national average teacher's salary. Since there are many more teachers in California, for example, than Nevada, those state averages should not be counted equally -- a weighted average (where the weights are the numbers of teachers in each state) should be used.

2-5 Measures of Variation

1. NOTE: Although not given in the text, the symbol R will be used for the range throughout this manual. Remember that the range is the difference between the highest and the lowest scores, and not necessarily the difference between the last and the first values as they are listed.

x	$x - \bar{x}$	$(x-\bar{x})^2$	x^2
65	-6.5	42.25	4225
66	-5.5	30.25	4356
67	-4.5	20.25	4489
68	-3.5	12.25	4624
71	-0.5	0.25	5041
73	1.5	2.25	5329
74	2.5	6.25	5476
77	5.5	30.25	5929
77	5.5	30.25	5929
77	5.5	30.25	5929
715	0	204.50	51327

$\bar{x} = (\Sigma x)/n$
$= 715/10$
$= 71.5$

R = 77 - 65 = 12
by formula 2-4, $s^2 = \Sigma(x-\bar{x})^2/(n-1) = (204.50)/(9) = 22.7$
$s = 4.8$

by formula 2-6, $s^2 = [n(\Sigma x^2) - (\Sigma x)^2]/[n(n-1)]$
$$= [10(51327) - (715)^2]/[10(9)]$$
$$= 2045/90 = 22.7$$
$$s = 4.8$$

NOTE: When using formula 2-4, constructing a table having the first three columns shown above will help to organize the calculations and make errors less likely; in addition, verify that $\Sigma(x-\overline{x}) = 0$ before proceeding -- if such is not the case, there is an error and further calculation is fruitless. For completeness, and as a check, both formulas 2-4 and 2-6 were used above. In general, only formula 2-6 will be used throughout the remainder of this manual for the following reasons:
(1) When the mean does not "come out even," formula 2-4 involves round-off error and/or many messy decimal calculations.
(2) The quantities Σx and Σx^2 needed for formula 2-6 can be found directly and conveniently on the calculator from the original data without having to construct a table like the one above.

2. preliminary values: $n = 10$, $\Sigma x = 715$, $\Sigma x^2 = 54109$
$R = 100 - 42 = 58$
$s^2 = [n(\Sigma x^2) - (\Sigma x)^2]/[n(n-1)]$
$$= [10(54109) - (715)^2]/[10(9)]$$
$$= (29865)/90 = 331.8$$
$s = 18.2$

3. preliminary values: $n = 15$, $\Sigma x = 2.81$, $\Sigma x^2 = .5631$
$R = .29 - .12 = .17$
$s^2 = [n(\Sigma x^2) - (\Sigma x)^2]/[n(n-1)]$
$$= [15(.5631) - (2.81)^2]/[15(14)]$$
$$= (.5504)/210 = .003$$
$s = .051$
NOTE: The quantity $[n(\Sigma x^2) - (\Sigma x)^2]$ cannot be less than zero. A negative value indicates that there is an error and that further calculation is fruitless. In addition, find the value for s by taking the square root of the precise value of s^2 showing on the calculator display before it is rounded to one more decimal place than the original data.

4. preliminary values: $n = 20$, $\Sigma x = 47.9$, $\Sigma x^2 = 146.53$
$R = 4.5 - 0.0 = 4.5$
$s^2 = [n(\Sigma x^2) - (\Sigma x)^2]/[n(n-1)]$
$$= [20(146.53) - (47.9)^2]/[20(19)]$$
$$= (636.19)/380 = 1.67$$
$s = 1.29$

5. preliminary values: $n = 27$, $\Sigma x = 728$, $\Sigma x^2 = 21,786$
$R = 51 - 14 = 37$
$s^2 = [n(\Sigma x^2) - (\Sigma x)^2]/[n(n-1)]$
$$= [27(21,786) - (728)^2]/[27(26)]$$
$$= 58238/702 = 83.0$$
$s = 9.1$

6. preliminary values: $n = 42$, $\Sigma x = 2304$, $\Sigma x^2 = 128,014$
$R = 69 - 42 = 27$
$s^2 = [n(\Sigma x^2) - (\Sigma x)^2]/[n(n-1)]$
$$= [42(128,014) - (2304)^2]/[42(41)]$$
$$= (68172)/1722 = 39.6$$
$s = 6.3$

7. preliminary values: $n = 8$, $\Sigma x = 13.26$, $\Sigma x^2 = 25.6620$
   $R = 2.87 - 0.63 = 2.24$
   $s^2 = [n(\Sigma x^2) - (\Sigma x)^2]/[n(n-1)]$
   $\quad = [8(25.6620) - (13.26)^2]/[8(7)]$
   $\quad = (29.4684)/56 = .526$
   $s = .725$

8. preliminary values: $n = 8$, $\Sigma x = 95.66$, $\Sigma x^2 = 1200.3640$
   $R = 16.39 - 8.08 = 8.31$
   $s^2 = [n(\Sigma x^2) - (\Sigma x)^2]/[n(n-1)]$
   $\quad = [8(1200.3640) - (95.66)^2]/[8(7)]$
   $\quad = (452.0764)56 = 8.073$
   $s = 2.841$

9. preliminary values: $n = 12$, $\Sigma x = 8011.5$, $\Sigma x^2 = 5,349,166.83$
   $R = 679.2 - 654.2 = 25.0$
   $s^2 = [n(\Sigma x^2) - (\Sigma x)^2]/[n(n-1)]$
   $\quad = [12(5,349,166.83) - (8011.5)^2]/[12(11)]$
   $\quad = (5869.71)/132 = 44.47$
   $s = 6.67$

10. preliminary values: $n = 11$, $\Sigma x = 10.113$, $\Sigma x^2 = 9.308837$
   $R = .958 - .842 = .116$
   $s^2 = [n(\Sigma x^2) - (\Sigma x)^2]/[n(n-1)]$
   $\quad = [11(9.308837) - (10.113)^2]/[11(10)]$
   $\quad = (.124438)/110 = .0011$
   $s = .0336$

11. preliminary values: $n = 10$, $\Sigma x = 31179$, $\Sigma x^2 = 99,425,707$
   $R = 3784 - 2360 = 1424$
   $s^2 = [n(\Sigma x^2) - (\Sigma x)^2]/[n(n-1)]$
   $\quad = [10(99,425,707) - (31179)^2]/[10(9)]$
   $\quad = 22127029/90 = 245855.9$
   $s = 495.8$

12. preliminary values: $n = 15$, $\Sigma x = 377600$, $\Sigma x^2 = 13,433,680,000$
   $R = 67,000 - 6,400 = 60,600$
   $s^2 = [n(\Sigma x^2) - (\Sigma x)^2]/[n(n-1)]$
   $\quad = [15(13,433,680,000) - (377,600)^2]/[15(14)]$
   $\quad = (58,923,440,000)/210 = 280,587,810$
   $s = 16,751$

13. Refer also to exercise #13 of section 2-4.
   preliminary values: $n = 38$, $\Sigma x = 3728.8$, $\Sigma x^2 = 365914.52$
   a. $R/4 = (99.4 - 96.2)/4 = .80$
   b. $s^2 = [n(\Sigma x^2) - (\Sigma x)^2]/[n(n-1)]$
   $\qquad = [38(365914.52) - (3728.8)^2]/[38(37)]$
   $\qquad = (802.32)/1406 = .57$
   $\quad s = .76$

14. Refer also to exercise #14 of section 2-4.
   preliminary values: n = 62, Σx = 137.50, Σx^2 = 377.5532
   a. R/4 = (4.95 - .26)/4 = 1.1725
   b. $s^2 = [n(\Sigma x^2) - (\Sigma x)^2]/[n(n-1)]$
       $= [62(377.5532) - (137.50)^2]/[62(61)]$
       $= (4502.0484)/3782 = 1.190$
     s = 1.091

15. Refer also to exercise #15 of section 2-4.
   preliminary values: n = 62, Σx = 584.54, Σx^2 = 6570.8216
   a. R/4 = (20.58 - 1.65)/4 = 4.7325
   b. $s^2 = [n(\Sigma x^2) - (\Sigma x)^2]/[n(n-1)]$
       $= [62(6570.8216) - (584.54)^2]/[62(61)]$
       $= (65703.9276)/3782 = 17.373$
     s = 4.168

16. Refer also to exercise #16 of section 2-4.
   preliminary values: n = 35, Σx = 3890, Σx^2 = 446620
   a. R/4 = (160 - 72)/4 = 22.0
   b. $s^2 = [n(\Sigma x^2) - (\Sigma x)^2]/[n(n-1)]$
       $= [35(446620) - (3890)^2]/[35(34)]$
       $= (499600)/1190 = 419.8$
     s = 20.5

17.

x	f	f·x	f·x^2
42.5	14	595.0	25287.50
44.5	11	489.5	21782.75
46.5	8	372.0	17298.00
48.5	6	291.0	14113.50
50.5	4	202.0	10201.00
52.5	3	157.5	8268.75
54.5	1	54.5	2970.25
56.5	2	113.0	6384.50
58.5	0	0.0	0.00
60.5	1	60.5	3660.25
	50	2334.0	109966.50

$s^2 = [n(\Sigma f \cdot x^2) - (\Sigma f \cdot x)^2]/[n(n-1)]$
   $= [50(109966.50) - (2335.0)^2]/[50(49)]$
   $= (46100.00)/2450 = 18.8$
s = 4.3

18.

x	f	f·x	f·x^2
4.5	37	166.5	749.25
14.5	35	507.5	7358.75
24.5	39	955.5	23409.75
34.5	46	1587.0	54751.50
44.5	33	1468.5	65348.25
54.5	22	1199.0	65345.50
64.5	20	1290.0	83205.00
74.5	17	1266.5	94354.25
84.5	9	760.5	64262.25
94.5	3	283.5	26790.75
	261	9484.5	485475.25

$s^2 = [n(\Sigma f \cdot x^2) - (\Sigma f \cdot x)^2]/[n(n-1)]$
$= [261(485475.25) - (9484.5)^2]/[261(260)]$
$= (36753300)/67860 = 541.6$
$s = 23.3$

19.

x	f	f·x	f·x²
2.95	10	29.50	87.0250
8.95	8	71.60	640.8200
14.95	9	134.55	2011.5225
20.95	9	188.55	3950.1225
26.95	4	107.80	2905.2100
	40	532.00	9594.7000

$s^2 = [n(\Sigma f \cdot x^2) - (\Sigma f \cdot x)^2]/[n(n-1)]$
$= [40(9594.7000) - (532.00)^2]/[40(39)]$
$= (100764.0000)/1560 = 64.59$
$s = 8.04$

20.

x	f	f·x	f·x²
20.5	22	451.0	9245.50
30.5	10	305.0	9302.50
40.5	6	243.0	9841.50
50.5	2	101.0	5100.50
60.5	4	242.0	14641.00
70.5	5	352.5	24851.25
80.5	1	80.5	6480.25
	50	1775.0	79462.50

$s^2 = [n(\Sigma f \cdot x^2) - (\Sigma f \cdot x)^2]/[n(n-1)]$
$= [50(79462.50) - (1775.0)^2]/[50(49)]$
$= (822500)/2450 = 335.7$
$s = 18.3$

21. NOTE: Because it allows for better appreciation of the concepts involved, formula 2-4 (and the suggested table) is employed for the next two exercises.

a.

x	x - x̄	(x-x̄)²
80	-6.5	42.25
81	-5.5	30.25
82	-4.5	20.25
83	-3.5	12.25
86	-0.5	0.25
88	1.5	2.25
89	2.5	6.25
92	5.5	30.25
92	5.5	30.25
92	5.5	30.25
865	0	204.50

$\bar{x} = (\Sigma x)/n$
$= 865/10$
$= 86.5$

$R = 92 - 80 = 12$
By formula 2-4, $s^2 = \Sigma(x-\bar{x})^2/(n-1) = (204.50)/(9) = 22.7$ and $s = 4.8$.
Adding a constant to each score does not affect the spread of the scores or the measures of dispersion.

b.

x	x - $\bar{x}$	$(x-\bar{x})^2$
60	-6.5	42.25
61	-5.5	30.25
62	-4.5	20.25
63	-3.5	12.25
66	-0.5	0.25
68	1.5	2.25
69	2.5	6.25
72	5.5	30.25
72	5.5	30.25
72	5.5	30.25
665	0	204.50

$$\bar{x} = (\Sigma x)/n$$
$$= 665/10$$
$$= 66.5$$

$R = 72 - 60 = 12$

By formula 2-4, $s^2 = \Sigma(x-\bar{x})^2/(n-1) = (204.50)/(9) = 22.7$ and $s = 4.8$.

Subtracting a constant from each score does not affect the spread of the scores or the measures of dispersion.

c.

x	x - $\bar{x}$	$(x-\bar{x})^2$
650	-65	4225
660	-55	3025
670	-45	2025
680	-35	1225
710	-5	025
730	15	225
740	25	625
770	55	3025
770	55	3025
770	55	3025
7150	0	20450

$$\bar{x} = (\Sigma x)/n$$
$$= 7150/10$$
$$= 715$$

$R = 770 - 650 = 120$

By formula 2-4, $s^2 = \Sigma(x-\bar{x})^2/(n-1) = (20450)/(9) = 2272.2$ and $s = 47.7$.

Multiplying each score by a constant also multiplies the distances between scores by that constant, and hence the measures of dispersion in the original units are multiplied by that constant (and the variance is multiplied by the square of that constant).

d.

x	x - $\bar{x}$	$(x-\bar{x})^2$
32.5	-3.25	10.5625
33.0	-2.75	7.5625
33.5	-2.25	5.0625
34.8	-1.75	3.0625
35.0	-0.25	0.0625
36.5	0.75	0.5625
37.0	1.25	1.5625
38.5	2.75	7.5625
38.5	2.75	7.5625
38.5	2.75	7.5625
357.5	0	51.1250

$$\bar{x} = (\Sigma x)/n$$
$$= 357.5/10$$
$$= 35.75$$

$R = 38.5 - 32.5 = 6.0$

By formula 2-4, $s^2 = \Sigma(x-\bar{x})^2/(n-1) = (51.1250)/(9) = 5.68$ and $s = 2.38$.

Dividing each score by a constant also divides the distances between scores by that constant, and hence the measures of dispersion in the original units are divided by that constant (and the variance is divided by the square of that constant).

22. a. Yes; if the scores are identical (and, hence, equal to $\bar{x}$), $s^2 = \Sigma(x-\bar{x})^2/(n-1) = 0/(n-1) = 0$.
b. No; defined as the positive square root of s^2, s measures the "typical distance" from the mean.

23.

x	x $-\bar{x}$	$(x-\bar{x})^2$
65	-73.1	5343.61
66	-72.1	5198.41
67	-71.1	5055.21
68	-70.1	4914.01
71	-67.1	4502.41
73	-65.1	4238.01
77	-61.1	3733.21
77	-61.1	3733.21
77	-61.1	3733.21
740	601.9	362283.61
1381	0	402734.90

$\bar{x} = (\Sigma x)/n$
$= 1381/10$
$= 138.1$

R = 740 - 65 = 675
by formula 2-4, $s^2 = \Sigma(x-\bar{x})^2/(n-1)$
$= (402734.90)/(9)$
$= 44748.3$
$s = 211.5$

The inclusion of one outlier has changed the measures of dispersion considerably.

24. Within k standard deviations of the mean, there must be at least $1 - 1/k^2$ of the scores.
Since 55 to 145 is $100 \pm 3 \cdot 15$ (i.e., $\bar{x} \pm 3 \cdot s$), k = 3.
There must be at least $1 - 1/3^2 = 1 - 8/9 = 8/9$ of the scores within those limits.
Since 85 to 115 is $100 \pm 1 \cdot 15$ (i.e., $\bar{x} \pm 1 \cdot s$), k = 1.
As stated in the text, Chebyshev's theorem requires $k > 1$ and does not say anything about the number of scores within those limits.
NOTE: If applied, the theorem would say there must be at least $1 - 1/1^2 = 1 - 1/1 = 0$ of the scores within those limits. Mathematically, there do not have to be <u>any</u> scores within one standard deviation of the mean.

25. a. The limits 52.0 and 60.0 are 1 standard deviation from the mean. The empirical rule for bell-shaped data states that about 68% of the scores should fall within those limits.
b. A distance of 8.0 is 2 standard deviations from the mean. The empirical rule for bell-shaped data states that about 95% of the scores should fall within those limits.
c. The empirical rule for bell-shaped data states that about 99.7% of the scores should fall within 3 standard deviations of the mean. In this case, that would be within 3(4.0) = 12.0 of the mean of 56.0 -- i.e., from 44.0 to 68.0.

26. a. For n = 100, $\sigma = \sqrt{(100^2-1)/12} = 28.9$.
b. The derivation below uses the following two algebraic facts
** $1 + 2 + 3 + ... + n = n(n+1)/2$
** $1^2 + 2^2 + 3^2 + ... + n^2 = n(n+1)(2n+1)/6$
$s^2 = [n \cdot \Sigma x^2 - (\Sigma x)^2]/[n(n-1)]$
$= [n \cdot n(n+1)(2n+1)/6 - n^2(n+1)^2/4]/[n(n-1)]$
$= n^2(n+1) \cdot [(2n+1)/6 - (n+1)/4]/[n(n-1)]$
$= n^2(n+1) \cdot [(n-1)/12]/[n(n-1)]$
$= n(n+1)/12$

27. It is known from algebra that $1+2+3+...+N = (N/2)(N+1)$. The mean of a population of those N values would then be $\mu = (\Sigma x)/N = [(N/2)(N+1)]/N = (N+1)/2$.
Exercise 26 gives The variance of the population $1,2,3,...,N$ as
$$\sigma^2 = (N^2 - 1)/12.$$
For N=165, the population of $1,2,3,...,164$ has
$\mu = (N+1)/2 = (165+1)/2 = 83.0$
$\sigma^2 = (N^2 - 1)/12 = (165^2 - 1)/12 = 2268.7$
$\sigma = 47.6$
Exercise 21 of section 2-4 indicates that adding a constant to every score adds that constant to each measure of central tendency. Exercise 21 of section 2-5 indicates that adding a constant to every score does not affect the measures of dispersion. Since the population $18,19,20,...,182$ is the population analyzed above but with 17 added to each score,
$\mu = 83.0 + 17 = 100.0$
$\sigma = 47.6$

28. From exercise #26, for $1,2,3,...,n$
$\mu = [\Sigma x]/n = [n(n+1)/2]/n = (n+1)/2$
$\sigma^2 = (n^2-1)/12$
Subtracting 1 from each score reduces μ by 1 but does not affect σ^2. For $0,1,2,...,n-1$
$\mu = (n+1)/2 - 1 = (n-1)/2$
$\sigma^2 = (n^2-1)/12$
For $0,1,2,...,99999999$ [i.e., $n = 100,000,000$]
$\mu = (n-1)/2 = 99999999/2 = 49999999.5$
$\sigma^2 = (n^2-1)/12 = (100,000,000^2-1)/12 = 8.33 \cdot 10^{14}$
$\sigma = 28867513.5$
Dividing each score by 100,000,000 divides both μ and σ by 100,000,000.
For .00000000, .00000001, .00000002, ..., .99999999
$\mu = .499999995$ rounded to .500
$\sigma = .288675135$ rounded to .289

29. Fahrenheit temperatures are converted to Celsius by $C = (5/9) \cdot F - 160/9$.
Exercise 21 of section 2-4 indicates that multiplying each score by a constant multiplies the mean by that constant and that subtracting the same amount from each score subtracts that amount from the mean. The mean of the Celsius temperatures is therefore
$\bar{x}_C = (5/9) \cdot \bar{x}_F - 160/9$
$= (5/9)(98.20) - 160/9$
$= 36.78$
Exercise 21 of section 2-5 indicates that multiplying each score by a constant multiplies the standard deviation by that constant and that subtracting the same amount from each score does not change the standard deviation of the scores. The standard deviation of the Celsius temperatures is therefore $s_C = (5/9) \cdot s_F = (5/9)(0.62) = .34$

30. $n = 10$, $\Sigma x = 105$, $\Sigma x^2 = 1999$, $\bar{x} = 10.5$, $s = 9.98$
a. $CV = (s/\bar{x}) \cdot 100\% = (9.98/10.5) \cdot 100\% = 95.1\%$
b. $\bar{x}/s = 10.5/9.98 = 1.052$

31. $I = 3(\bar{x} - m)/s$
$= 3(98.20 - 98.4)/.62$
$= -.6/.62$
$= -.968$
The data are not *significantly* skewed.

2-6 Measures of Position

1. In general, $z = (x - \bar{x})/s$.
    a. $z_{98.6} = (98.6 - 98.20)/.62 = .65$
    b. $z_{98.2} = (98.2 - 98.20)/.62 = 0$
    c. $z_{100.0} = (100.0 - 98.20)/.62 = 2.90$

2. In general, $z = (x - \bar{x})/s$.
    a. $z_{115} = (115 - 100)/15 = 1.00$
    b. $z_{90} = (90 - 100)/15 = -.67$
    c. $z_{127} = (127 - 100)/15 = 1.80$

3. $z = (x - \bar{x})/s$
    $z_{10.00} = (10.00 - 7.06)/5.32 = .55$

4. $z = (x - \bar{x})/s$
    $z_{15.0} = (15.0 - 10.7)/11.2 = .38$

5. $z = (x - \mu)/\sigma$
    a. $z_{70} = (70 - 63.6)/2.5 = 2.56$
    b. Yes, the height is considered unusual since $2.56 > 2.00$.

6. $z = (x - \mu)/\sigma$
    a. $z_{106.8} = (106.8 - 117.8)/5.52 = -1.99$
    b. No, the height is not considered unusual since $-2.00 < -1.99 < 2.00$.

7. $z = (x - \mu)/\sigma$
    a. $z_{10.00} = (10.00 - 22.83)/8.55 = -1.50$
    b. No, the score is not considered unusual since $-2.00 < -1.50 < 2.00$.

8. $z = (x - \mu)/\sigma$
    a. $z_{275.2} = (275.2 - 178.1)/40.7 = 2.39$
    b. Yes, the level is considered unusually high since $2.39 > 2.00$.

9. In general $z = (x - \bar{x})/s$.
    a. $z_{53} = (53 - 50)/10 = .30$
    b. $z_{53} = (53 - 50)/5 = .60$
    The score in part b has the better relative position since $.60 > .30$.

10. In general $z = (x - \bar{x})/s$.
    a. $z_{60} = (60 - 70)/10 = -1.00$
    b. $z_{480} = (480 - 500)/50 = -.40$
    The score in part b has the higher relative level since $.60 > .30$.

11. In general, $z = (x - \bar{x})/s$.
    a. $z_{60} = (60 - 50)/5 = 2.00$
    b. $z_{230} = (230 - 200)/10 = 3.00$
    c. $z_{540} = (540 - 500)/15 = 2.67$
    The score in part b has the highest relative position.

12. In general, $z = (x - \bar{x})/s$.
    a. $z_{3.6} = (3.6 - 4.2)/1.2 = -.50$
    b. $z_{72} = (72 - 84)/10 = -1.20$
    c. $z_{255} = (255 - 300)/30 = -1.50$
    The score in part a is the highest relative score.

13. Let L = # of scores less than x; let n = total number of scores.
   In general, the percentile of score x is $(L/n) \cdot 100$.
   The percentile of score 97.2 is $(8/106) \cdot 100 = 8$.

14. Let L = # of scores less than x; let n = total number of scores.
   In general, the percentile of score x is $(L/n) \cdot 100$.
   The percentile of score 98.6 is $(68/106) \cdot 100 = 64$.

15. Let L = # of scores less than x; let n = total number of scores.
   In general, the percentile of score x is $(L/n) \cdot 100$.
   The percentile of score 99.0 is $(98/106) \cdot 100 = 92$.

16. Let L = # of scores less than x; let n = total number of scores.
   In general, the percentile of score x is $(L/n) \cdot 100$.
   The percentile of score 97.6 is $(17/106) \cdot 100 = 16$.

17. To find P_{80}, L = $(80/100) \cdot 106 = 84.8$ rounded up to 85.
   Since the 85th score is 98.7, $P_{80} = 98.7$.

18. To find P_{40}, L = $(40/100) \cdot 106 = 42.4$ rounded up to 43.
   Since the 43rd score is 98.0, $P_{40} = 98.0$.

19. To find $Q_3 = P_{75}$, L = $(75/100) \cdot 106 = 79.5$ rounded up to 80.
   Since the 80th score is 98.6, $Q_3 = 98.6$.

20. To find $Q_1 = P_{25}$, L = $(25/100) \cdot 106 = 26.5$ rounded up to 27.
   Since the 27th score is 97.8, $Q_1 = 97.8$.

21. To find $D_3 = P_{30}$, L = $(30/100) \cdot 106 = 31.8$ rounded up to 32.
   Since the 32nd score is 97.9, $D_3 = 97.9$.

22. To find $D_9 = P_{90}$, L = $(90/100) \cdot 106 = 95.4$ rounded up to 96.
   Since the 96th score is 98.8, $D_9 = 98.8$.

23. To find $D_7 = P_{70}$, L = $(70/100) \cdot 106 = 74.2$ rounded up to 75.
   Since the 75th score is 98.6, $D_7 = 98.6$.

24. To find P_{37}, L = $(37/100) \cdot 106 = 39.2$ rounded up to 40.
   Since the 40th score is 98.0, $P_{37} = 98.0$.

25. Let L = # of scores less than x
       n = total number of scores
   In general, the percentile of score x is $(L/n) \cdot 100$.
   The percentile of score 120 is $(45/60) \cdot 100 = 75$.

26. Let L = # of scores less than x
       n = total number of scores
   In general, the percentile of score x is $(L/n) \cdot 100$.
   The percentile of score 90 is $(3/60) \cdot 100 = 5$.

27. Let L = # of scores less than x
       n = total number of scores
   In general, the percentile of score x is $(L/n) \cdot 100$.
   The percentile of score 100 is $(21/60) \cdot 100 = 35$.

NOTE: For exercises 25-36, refer to the following ordered list of the 60 film lengths.

			length (min.)		
1	72	94	99	104	111	129
2	82	94	100	105	114	133
3	88	94	100	105	115	134
4	90	95	100	106	117	139
5	90	96	101	106	119	144
6	91	96	102	107	120	155
7	92	96	103	108	121	159
8	92	97	104	108	123	160
9	93	98	104	110	123	168
0	93	99	104	111	125	193

28. Let L = # of scores less than x
  n = total number of scores
  In general, the percentile of score x is $(L/n) \cdot 100$.
  The percentile of score 123 is $(47/60) \cdot 100 = 78$.

29. To find P_{15}, $L = (15/100) \cdot 60 = 9$ -- a whole number.
  The mean of the 9th and 10th scores, $P_{15} = (93 + 93)/2 = 93$

30. To find P_{30}, $L = (30/100) \cdot 60 = 18$ -- a whole number.
  The mean of the 18th and 19th scores, $P_{30} = (97 + 98)2 = 97.5$.

31. To find P_{80}, $L = (80/100) \cdot 60 = 48$ -- a whole number.
  The mean of the 48th and 49th scores, $P_{80} = (123 + 123)2 = 123$.

32. To find P_{10}, $L = (10/100) \cdot 60 = 6$ -- a whole number.
  The mean of the 6th and 7th scores, $P_{10} = (91 + 92)2 = 91.5$.

33. To find $Q_1 = P_{25}$, $L = (25/100) \cdot 60 = 15$ -- a whole number.
  The mean of the 15th and 16th scores, $Q_1 = (96 + 96)/2 = 96$.

34. To find $Q_3 = P_{75}$, $L = (75/100) \cdot 60 = 45$ -- a whole number.
  The mean of the 45th and 46th scores, $Q_3 = (119 + 120)/2 = 119.5$.

35. To find $D_9 = P_{90}$, $L = (90/100) \cdot 60 = 54$ -- a whole number.
  The mean of the 54th and 55th scores, $D_9 = (139 + 144)/2 = 141.5$.

36. To find $D_3 = P_{30}$, $L = (30/100) \cdot 60 = 18$ -- a whole number.
  The mean of the 18th and 19th scores, $D_3 = (97 + 98)/2 = 97.5$

37. a. The interquartile range is $Q_3 - Q_1$.
      For $Q_3 = P_{75}$, $L = (75/100) \cdot 106 = 79.5$ rounded up to 80.
        Since the 80th score is 98.6, $Q_3 = 98.6$.
      For $Q_1 = P_{25}$, $L = (25/100) \cdot 106 = 26.5$ rounded up to 27.
        Since the 27th score is 97.8, $Q_1 = 97.8$.
      The interquartile range is $98.6 - 97.8 = 0.8$
   b. The midquartile is $(Q_1 + Q_3)/2 = (98.6 + 97.8)/2 = 98.2$
   c. The 10-90 percentile range is $P_{90} - P_{10}$.
      For P_{90}, $L = (90/100) \cdot 106 = 95.4$ rounded up to 96.
        Since the 96th score is 98.8, $P_{90} = 98.8$.
      For P_{10}, $L = (10/100) \cdot 106 = 10.6$ rounded up to 11.
        Since the 11th score is 97.3, $P_{10} = 97.3$.
      The 10-90 percentile range is $98.8 - 97.3 = 1.5$
   d. Yes, $Q_2 = P_{50}$ by definition. They are always equal.
   e. For $Q_2 = P_{50}$, $L = (50/100) \cdot 106 = 53$ exactly; we use the mean of the 53rd and 54th score.
      $Q_2 = (98.4 + 98.4)/2 = 98.4$.
      In this case $Q_2 \neq (Q_1 + Q_3)/2$, demonstrating that the median does not necessarily equal the midquartile.

38. To find $Q_1 = P_{25}$, $L = (25/100) \cdot 14 = 3.5$.
  The 3rd score is 53, and $53 + .5 \cdot (58-53) = 55.5$.
  To find $Q_3 = P_{75}$, $L = (75/100) \cdot 14 = 10.5$.
  The 10th score is 80, and $80 + .5 \cdot (84-80) = 82$.
  To find P_{33}, $L = (33/100) \cdot 14 = 4.62$
  The 4th score is 58, and $58 + .62 \cdot (60-58) = 59.24$.

39. Unusual values are those more than two standard deviations from the mean. In this case, that would be any values such that $x < \bar{x} - 2 \cdot s$ or $x > \bar{x} + 2 \cdot s$

$x < 98.20 - 2(.62)$ or $x > 98.20 + 2(.62)$

$x < 96.96$ or $x > 99.44$

NOTE: The practical interpretation of this is that a temperature below 96.96 or above 99.44 is unusual for a well person. The logical conclusion is that such a person either is not well (typically arrived at in the presence of other symptoms) or has an unusual "normal" temperature.

40. For the original x values: $n = 5$, $\Sigma x = 40$, $\Sigma x^2 = 430$, $\bar{x} = 8$, $s_x = 5.24$
    The z values are: -1.144155, -.572078, 0, .190693, 1.525540
    For the z values: $n = 5$, $\Sigma z = 0$, $\Sigma z^2 = 4$, $\bar{z} = 0$, $s_z = 1$
    As shown below, $\bar{z} = 0$ and $s_z = 1$ will be true for *any* set of z scores.
    $\Sigma z = \Sigma(x-\bar{x})s = (1/s) \cdot [\Sigma(x-\bar{x})] = (1/s) \cdot [\Sigma x - \Sigma\bar{x}] = (1/s) \cdot [n\bar{x} - n\bar{x}] = (1/s) \cdot 0 = 0$
    $z^2 = \Sigma[(x-\bar{x})/s]^2 = (1/s^2) \cdot [\Sigma(x-\bar{x})^2] = (1/s^2) \cdot [(n-1)s^2] = n-1$
    $\bar{z} = (\Sigma z)/n = 0/n = 0$
    $s_z^2 = [n \cdot \Sigma z^2 - (\Sigma z)^2]/[n(n-1)] = [n \cdot (n-1) - 0^2]/[n(n-1)] = 1;\ s_z = \sqrt{1} = 1$

2-7 Exploratory Data Analysis

1. 406 406 407 408 410 419 419 419 419
   421 423 424 426 426 430 438 438

2. 6845 6845 6847 6886 6933 6938 6989
   7052 7059 7093 7127

3. First go through the numbers in the order they appear and place the one's digit behind the appropriate ten's digit. Going through the numbers by rows produces the following.

```
Temperature
  6 84
  7 7659
  8 046519233441400
  9 3473024021
```

Now order the one's digits within each row to produce the following final form.

```
Temperature
  6 48
  7 5679
  8 000112334444569
  9 0012233447
```

4. The first pass through the numbers by rows produces the figure at the left. Sorting the digits within each row produces the final figure at the right.

```
Pulse Rate                    Pulse Rate
  5 884                         5 488
  6 2650                        6 0256
  7 478415143                   7 113444578
  8 2502                        8 0225
```

5. Make the stem accurate to tenths, and make two rows for each tenth -- one for low hundredth
   digits (0-4) and one for high hundredth digits (5-9). The final form, with the hundredth digits in
   order, is given below.

```
Weight (grams)
   5.4│9
   5.5│2333
   5.5│677777888889999
   5.6│00000001222333
   5.6│5666778
   5.7│01123334
   5.7│
   5.8│4
```

6. The first pass through the data produces the figure at the left. Sorting the digits within each row
   produces the final figure at the right. **NOTE:** Since the directions to "round to the nearest tenth"
   are ambiguous when the rounded hundredth place is a 5, and since always rounding such numbers
   up introduces a consistent bias, the figure was constructed rounding such numbers toward the even
   digit -- e.g., 1.75 rounds to 1.8 and 2.25 rounds to 2.2. Such numbers rounded down are
   identified in bold, and may be increased by one unit by those who prefer to always round up.

```
Weight (lbs)                    Weight (lbs)
   0.│324                          0.│234
   0.│89669987967                  0.│66677889999
   1.│44444444123                  1.│12344444444
   1.│855757566                    1.│555566778
   2.│221420013431                 2.│001112223344
   2.│8977989                      2.│7788999
   3.│04040                        3.│00044
   3.│5                            3.│5
   4.│4                            4.│4
   4.│7                            4.│7
   5.│3                            5.│3
```

7. The 10 scores are given in order in the text.
   The minimum score is .02 and the maximum score is .19.
   The median score is .095.
   The left hinge is .08, median of .02 .04 .08 .08 .09 .095.
   The right hinge is .11, median of .095 .10 .10 .12 .13 .19.

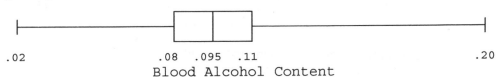

```
    .02                  .08  .095  .11                     .20
                    Blood Alcohol Content
```

8. The 15 scores are given in order in the text.
   The minimum score is 12.9 and the maximum score is 52.5.
   The median score is 32.1.
   The left hinge is 27.95, median of 12.9 13.4 18.3 24.7 31.2 31.3 32.0 32.1
   The right hinge is 35.15, median of 32.1 33.4 33.8 34.1 36.2 41.7 41.9 52.5

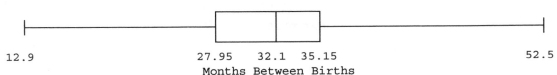

```
 12.9              27.95   32.1  35.15                   52.5
                  Months Between Births
```

9. Arranged in order, the 20 scores are
    70 130 138 142 157 157 159 162 164 170
    173 173 175 180 181 183 190 193 195 198
The minimum score is 70 and the maximum score is 198.
The median score is 171.5.
The left hinge is 157, median of 70 130 ... 164 170 171.5.
The right hinge is 181, median of 171.5 173 173 ... 195 198.

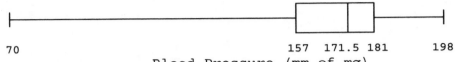

70   157 171.5 181   198
              Blood Pressure (mm of mg)

10. Arranged in order, the 25 scores are
    106 111 143 160 166 174 181 183 185 190 190 193 198
    207 213 213 219 220 221 229 243 257 281 308 350
The minimum score is 106 and the maximum score is 350.
The median score is 198.
The left hinge is 181, median of 106 111 ... 193 198.
The right hinge is 221, median of 198 207 ... 308 350.

106   181 198   221   350
              Words Per Minute

11. Arranged in order, the 30 scores are
     2   4   7   7   9   9 10 12 14 15 18 21 22 25 26
    27 29 30 31 31 32 34 35 36 36 38 40 41 43 45
The minimum score is 2 and the maximum score is 45.
The median score is 26.5.
The left hinge is 13, median of 2 4 ... 25 26 26.5.
The right hinge is 34.5, median of 26.5 27 29 ... 43 45.

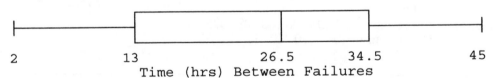

2   13   26.5   34.5   45
         Time (hrs) Between Failures

12. The 24 scores are given in order in the stem-and-leaf plot.
The minimum score is 50 and the maximum score is 95.
The median score is 78.5.
The left hinge is 72, median of 50 66 ... 74 78 78.5.
The right hinge is 86, median of 78.5 79 83 ... 92 95.

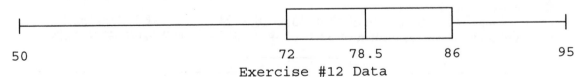

50   72   78.5   86   95
              Exercise #12 Data

13. Lottery numbers should be uniformly distributed, in which case the hinges and the center line in the boxplot would divide the figure into four equal lengths. The given boxplot indicates a concentration of numbers near the middle, with very low and very high numbers occurring less frequently.

14. IQ scores should be bell-shaped -- i.e., there should be a "bunching up" of the scores near the middle with fewer scores near the extremes. The scores given in the stem-and-leaf plot are even spread across the range from 90 to 128.

15. a. The final form of the back-to-back stem-and-leaf plot is given below. NOTE: This is another example of adapting a standard visual form in order to better communicate the data (see exercise 2.3 #21). While such decisions are arbitrary, we choose to display "outward" from the central stem but to keep the actor's ages in increasing order from left to right.

```
    actor's age     actress' age
                   2|146667
      122357899    3|00113344455778
   0012233456788   4|11129
           1566    5|
            012    6|011
              6    7|4
                   8|0
```

b. For the actors,
    the median age is 43.
    the left hinge is 39, median of 31 31 ... 42 43 43.
    the right hinge is 49.5, median of 43 43 44 ... 62 67.
For the actresses,
    the median age is 34.5.
    the left hinge is 30.5, median of 21 24 ... 34 34 34.5.
    the right hinge is 41, median of 34.5 35 35 ... 74 80.
The boxplots, using the same scale, are shown below.

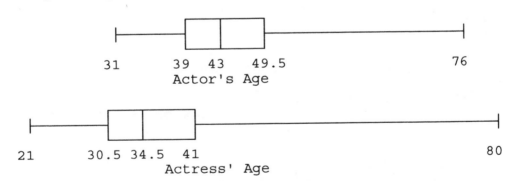

```
        31        39  43   49.5                76
                  Actor's Age
```

```
   21        30.5 34.5  41                      80
                  Actress' Age
```

c. Female Oscar winners tend to be younger than male Oscar winners. Assuming that acting ability doesn't peak differently for females and males, the data may reveal a difference in the standards by which way females and males are judged.

16. The 13 scores are given in order in the text.
The minimum score is 3 and the maximum score is 68.
The median score is 22.
The left hinge is 18, median of 3 15 17 18 21 21 22
The right hinge is 30, median of 22 25 27 30 38 49 68
D = 30 - 18 = 12
  1.5·D = (1.5)(12) = 18
  3·D = (3)(12) = 36
The lines extend to include points within 1.5·D = 18 of the hinges
  -- i.e., from 3 (15 below) to 38 (16 above).
Solid dots are used for values within 18 to 36 of the hinges
  -- i.e., for 49 (27 above).
Hollow dots are used for values more than 3·D = 36 beyond the hinges
  -- i.e., for 68 (46 above).

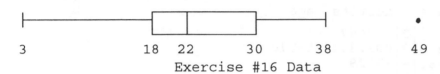

3   18   22   30   38   49   68

Exercise #16 Data

Review Exercises

1. Arranged in order, the scores are:
   4.10 4.42 4.95 5.01 7.24   7.51 7.98 8.60 11.53 12.30 14.59 14.67
   preliminary values: n = 12, Σx = 102.90, Σx^2 = 1044.7150
   a. $\bar{x}$ = $(\Sigma x)/n$ = (102.90)/12 = 8.575
   b. m = (7.51 + 7.98)/2 = 7.745
   c. M = [none]
   d. m.r. = (4.10 + 14.67)/2 = 9.385
   e. R = 14.67 - 4.10 = 10.57
   f. s^2 = $[n(\Sigma x^2) - (\Sigma x)^2]/[n(n-1)]$
      = $[12(1044.7150) - (102.90)^2]/[12(11)]$
      = (1948.1700)/132
      = 14.759
   g. s = 3.842

2. Use the class midpoint of 9.5 as the x value for the last class.

x	f	f·x	f·x^2
4	147	588	2352
5	81	405	2025
6	27	162	972
7	15	105	735
9.5	30	285	2707.5
	300	1545	8791.5

   NOTE: n = Σf = 300

   $\bar{x}$ = $(\Sigma f \cdot x)/n$ = 1545/300 = 5.15
   s^2 = $[n(\Sigma f \cdot x^2) - (\Sigma f \cdot x)^2]/[n(n-1)]$
      = $[300(8791.5) - (1545)^2]/[300(299)]$
      = (250425)/89700 = 2.79
   s = 1.67

3. The relative frequency for each class is obtained by dividing that class' frequency by
   $n = \Sigma f = 300$.

class boundaries	relative frequency
3.5 - 4.5	.49
4.5 - 5.5	.27
5.5 - 6.5	.09
6.5 - 7.5	.05
7.5 - 11.5	.10
	1.00

   The relative frequency histogram below is constructed from the above relative frequency distribution. NOTE: The purpose of a histogram is to show the relative amounts in each class using areas of bars -- i.e., the proportion of the total area in each bar should represent the proportion of the sample in that class. Since the last class is 4 times as wide as the other classes, its bar must be made 1/4 of its normal height in order to preserve proper proportions. The <u>area</u> of the last bar should be about equal to the <u>area</u> of the 5.5-6.5 bar and twice the <u>area</u> of the 6.5-7.5 bar; this conveys the information that 10% of the data were spread over the years 8,9,10,11 -- assumed to be about 2.5% of the data for each of those years.

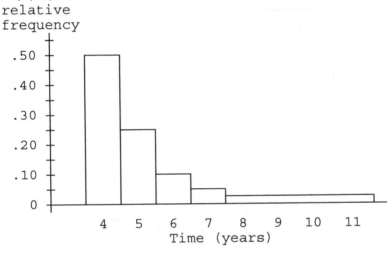

4. In general, $z = (x-\mu)/\sigma$.
   For the first test, $z_{66} = (66-75)/15 = -.60$.
   For the second test, $z_{223} = (223-250)/25 = -1.08$
   Assuming higher scores are better, the subject did better relative to others taking the test on the first test since $-.60 > -1.08$.

5. Subtracting the smallest value from the largest value and dividing by the desired number of classes determines that the class width must be at least $(1128 - 235)/9 = 99.22$. Rounding up to a convenient number, we choose 100. The exercise specifies that the first lower class limit is to be 235, the smallest data value. NOTE: In practice, the first lower class limit and the class width many be any convenient values. Since there are 60 seconds in a minute, for example, it might be reasonable to start at 180 or 210 and use a class width of 90 or 120.

time (seconds)	frequency
235 - 334	4
335 - 434	9
435 - 534	11
535 - 634	9
635 - 734	9
735 - 834	6
835 - 934	8
935 - 1034	2
1035 - 1134	2
	60

6. Obtain the relative frequencies by dividing each frequency by 60, the total number of cases.

time (seconds)	relative frequency
235 - 334	.067
335 - 434	.150
435 - 534	.183
535 - 634	.150
635 - 734	.150
735 - 834	.100
835 - 934	.133
935 - 1034	.033
1035 - 1134	.033
	1.000

7. frequency

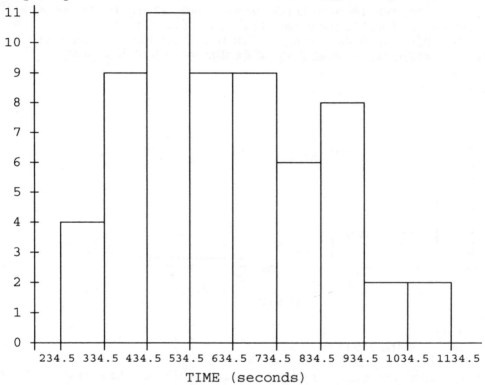

TIME (seconds)

8. The scores are given in order in the final stem-and-leaf plot of exercise #11.
Let L = # of scores less than x
    n = total number of scores
a. To find $Q_1 = P_{25}$, L = (25/100)·60 = 15 -- a whole number.
   The mean of the 15th and 16th scores, P_{25} = (447 + 448)/2 = 447.5.
b. To find P_{45}, L = (45/100)·60 = 27 -- a whole number.
   The mean of the 27th and 28th scores, P_{45} = (564 + 587)/2 = 575.5.
c. In general, the percentile of score x is (L/n)·100.
   The percentile of score 375 is (4/60)·100 = 7.

9. According to the range rule of thumb, the standard deviation is usually about 1/4 of the range -- i.e., we estimate s to be about (1/4)(893) = 223.25.

10.

x	f	f·x	f·x^2
284.5	4	1138.0	323761.00
384.5	9	3460.5	1330562.25
484.5	11	5329.5	2582142.75
584.5	9	5260.5	3074762.25
684.5	9	6160.5	4216862.25
784.5	6	4707.0	3692641.50
884.5	8	7076.0	6258722.00
984.5	2	1969.0	1938480.50
1084.5	2	2169.0	2352280.50
	60	37270.0	25770215.00

$\bar{x} = (\Sigma f\cdot x)/n = 37270.0/60 = 621.2$
$s^2 = [n(\Sigma f\cdot x^2) - (\Sigma f\cdot x)^2]/[n(n-1)]$
$= [60(25770215.00) - (37270.0)^2]/[60(59)]$
$= (157160000)/3540$
$= 44395.5$
$s = 210.7$

11. Use the hundred's digits for the stem and the last two digits -- with spaces between values -- for the leaves.

```
first pass, by columns               final form
 2|92 40 35                           2|35 40 92
 3|25 37 35 78 63 96 96 45            3|25 35 37 45 63 78 96 96
 4|48 43 04 57 47 95 94 20 74 83      4|04 20 43 47 48 57 74 83 94 95
 5|87 14 06 40 03 64 52               5|03 06 14 40 52 64 87
 6|26 70 15 88 09 25 76 70 93 66 27   6|09 15 25 26 27 66 70 70 76 88 93
 7|56 23 00 94 93 04 78 48            7|00 04 23 48 56 78 93 94
 8|61 71 20 53 52 60 62               8|20 52 53 60 61 62 71
 9|91 29 15                           9|15 29 91
10|23 70                             10|23 70
11|28                               11|28
```

12. The scores are given in order in the final stem-and-leaf plot of exercise #11.
The minimum score is 235 and the maximum score is 1128.
The median is $(615 + 625)/2 = 620$
The left hinge is 448, the median of 235,240,...,615,620.
The right hinge is 778, the median of 620,625,...,1070,1128.

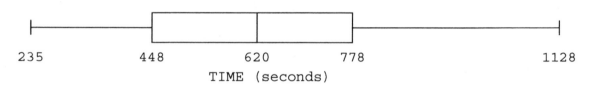

235   448   620   778   1128

TIME (seconds)

13. Arranged in order, the scores are: 8 8 12 16 18 19 22 25
preliminary values: $n = 8$, $\Sigma x = 128$, $\Sigma x^2 = 2322$
a. $\bar{x} = (\Sigma x)/n = (128)/8 = 16.0$
b. $m = (16 + 18)/2 = 17.0$
c. $M = 8$
d. m.r. $= (8 + 25)/2 = 16.5$
e. $R = 25 - 8 = 17$
f. $s^2 = [n(\Sigma x^2) - (\Sigma x)^2]/[n(n-1)]$
$= [8(2322) - (128)^2]/[8(7)]$
$= (2192)/56$
$= 39.1$
g. $s = 6.3$

14. a. 77.3, since adding 10 points to each score adds 10 points to the mean
   b. 14.2, since adding 10 points to each score does not affect the spread of the scores
   c. an error has been made, since $s^2 = \Sigma(x-\bar{x})^2/(n-1)$ cannot be less than zero
   d. false, since the standard deviation measures the "typical spread" and not just the difference between the extreme scores. **NOTE:** Consider the following, with $R_A = 50$ and $R_B = 100$.
   set A: n/2 0's and n/2 50's
   set B: one 0 and n-2 50's and one 100
   $s_A^2 > s_B^2$ whenever $n \cdot (25)^2 > 2 \cdot (50)^2$
   $$625n > 5000$$
   $$n > 8$$
   --e.g., for n=10, A is 0,0,0,0,0,50,50,50,50,50 with $s_A^2 = 10(25)^2/9 = 6250/9$
   and B is 0,50,50,50,50,50,50,50,50,100 with $s_B^2 = 2(50)^2/9 = 5000/9$
   e. the scores are equal, so that each score is $\bar{x}$ and $\Sigma(x-\bar{x})^2 = \Sigma 0^2 = 0$

15. Arranged in order, the scores are: 2 4 4 10 12 15 17 17 22 25 27 28 29 29 29 30 31 32 32
   preliminary values: $n = 19$, $\Sigma x = 395$, $\Sigma x^2 = 10137$
   a. $\bar{x} = (\Sigma x)/n = (395)/19 = 20.8$
   b. m = 25.0
   c. M = 29
   d. m.r. = (2 + 32)/2 = 17.0
   e. R = 32 - 2 = 30
   f. $s^2 = [n(\Sigma x^2) - (\Sigma x)^2]/[n(n-1)] = [19(10137) - (395)^2]/[19(18)] = (36578)/342 = 107.0$
   g. s = 10.3

16. Arranging the categories in decreasing order by frequency produces the following figure.

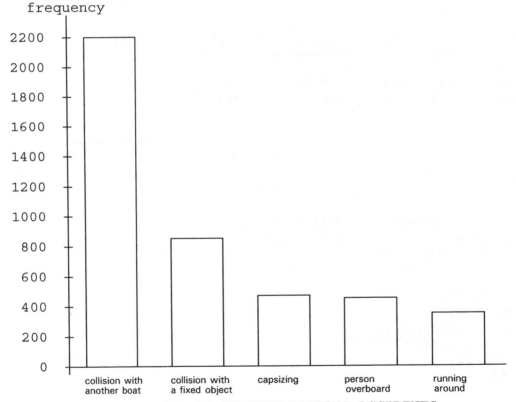

CAUSES OF SERIOUS BOATING ACCIDENTS

Chapter 3

Probability

3-2 Fundamentals

1. Since $0 \leq P(A) \leq 1$ is always true, the following values less than 0 or greater than 1 cannot be probabilities.
   values less than 0: -0.2
   values greater than 1: 4/3   1.001   2 $\sqrt{2}$

2. a. The probability of a certain event is 1.00.
   b. The probability of an impossible event is 0.
   c. For each equally likely event, let $P(A_i) = a$. Since $\Sigma P(A_i) = 200a = 1$, $a = 1/200$.
   d. Since only 1 of the 5 equally likely choices is correct, P(correct response) = 1/5.

3. let C = the selected student cheated
   P(C) = 1162/3630 = .320

4. let A = such a taxpayer is audited
   P(A) = 20/750 = .0267

5. let O = a selected person has group O blood
   P(O) = 36/80 = .45

6. let B = a person's birthday is October 18
   P(B) = 1/365 = .00274

7. let V = a selected 12-24 year old voted
   P(V) = (9,230,000)/(25,569,000) = .361

8. There were 831 + 18 = 849 consumers surveyed.
   let C = a consumer recognizes Campbell's Soup
   P(C) = 831/849 = .979

9. There were 288 + 962 = 1250 households surveyed.
   let C = a selected household has a computer
   P(C) = 288/1250 = .230

10. There were 36 + 114 = 150 derailments studied.
    let H = the derailments was caused by human error
    P(H) = 36/150 = .240

11. let S = a selected 18-25 year old smokes
    P(S) = 237/600 = .395

12. There were 59,377,306 + 82,796 + 1,664 = 59,461,766 total potential passengers.
    let I = a passenger is involuntarily denied boarding
    P(I) = 1,664/59,461,766 = .0000280

13. let A = a selected 20-24 year old has an accident
P(A) = 136/400 = .34
NOTE: This is properly an estimate of the probability that such a person <u>had</u> an accident last year. Using this to estimate the probability that such a person <u>will have</u> an accident next year is frequently done but requires the additional assumption that relevant conditions (speed limits, weather, alcohol laws, the economy, etc., etc.) remain relatively unchanged.

14. let C = the question is answered correctly
P(C) = 1107/1733 = .639

15. There were 70 + 711 persons tested.
let D = a selected user becomes drowsy
P(D) = 70/781 = .0896

16. There were 76,000 + 25,000 + 2,000 = 103,000 convicted burglars in the sample.
let J = the convicted burglar serves jail time
P(J) = 76,000/103,000 = .738

17. There were 90 + 80 + 20 + 10 = 200 people surveyed.
let E = a selected person has group AB blood
P(E) = 10/200 = .05

18. There were 243 + 85 + 52 + 46 = 426 cases of fraud studied.
let C = the fraud results from a counterfeit card
P(C) = 85/426 = .200

19. There were 10,589 + 636 + 22,551 + 963 = 34,739 convictions.
There were 636 + 963 = 1599 DWI convictions.
let D = a selected conviction is for DWI
P(D) = 1599/34739 = .0460

20. There were 228 + 672 + 240 = 1140 respondents.
let T = a respondent brushes three times a day
P(T) = 240/1140 = .211

21. a. The 4 equally likely outcomes are: BB BG GB GG.
b. P(2 girls) = 1/4 = .25
c. P(1 of each sex) = 2/4 = .5

22. a. The 16 different possible outcomes are listed at the right.
b. P(GGGG) = 1/16
c. P(at least one of each) = 1 - P(BBBB or GGGG)
            = 1 - 2/16
            = 14/16 [or 7/8]
d. P(exactly 2 of each sex)
   = P(BBGG or BGBG or BGGB or GGBB or GBGB or GBBG)
   = 6/16 [or 3/8]

BBBB	GGGG
BBBG	GGGB
BBGB	GGBG
BGBB	GBGG
GBBB	BGGG
BBGG	GGBB
BGBG	GBGB
BGGB	GBBG

23. a. The 8 equally likely outcomes are at the right.
let C = correct
let N = not correct
b. P(3 correct) = 1/8 = .125   [CCC outcome]
c. P(0 correct) = 1/8 = .125   [NNN outcome]
d. P(at least 2 correct) = 4/8 = .5 [column 1 outcomes]

CCC	CNN
CCN	NCN
CNC	NNC
NCC	NNN

24. let B = the brown gene
  b = the blue gene
  Bb = the father's gene pair
  Bb = the mother's gene pair
  a. possible outcomes (father's gene listed first): BB, Bb, bB, bb
  b. P(bb) = 1/4
  c. P(brown eyes) = P(BB or Bb or bB) = 3/4

25. There were 20 managers surveyed.
  There were 12 managers with values of 2.1 or higher.
  let M = a selected manager so spends more than 2 hours
  P(M) = 12/20 = .6

26. let B = a person's birthday is October 18
  a. In one repeating 4 year cycle there are $3 \cdot (365) + 1 \cdot (366) = 1461$ days.
    In one repeating 4 year cycle there are 4 October 18's.
    P(B) = 4/1461
  b. In one repeating 400 year cycle there are $303 \cdot (365) + 97 \cdot (366) = 146,097$ days.
    In one repeating 400 year cycle there are 400 October 18's.
    P(B) = 400/146,097
  NOTE: comparing the answers as decimals -- #6, .00273972
                    #26a, .00273785 (a slight decrease from #6)
                    #26b, .00273791 (a slight increase from #26a)

27. No matter where the two flies land it is possible to cut the orange in half to include both flies on the same half. Since this is a certainty, the probability is 1. [Compare the orange to a globe. Turn the orange so the spot where fly A lands corresponds to, say, New York City. Consider all the circles of longitude. Wherever fly B lands it is possible to slice the globe in half along some circle of longitude that places fly A and fly B on the same half.] NOTE: If the orange is marked into two predesignated halves before the flies land, the problem is different; once fly A lands, fly B has 1/2 a chance of landing on the same half. If both flies are to land on a specified one of the two predesignated halves, the problem is different still; fly A has 1/2 a chance of landing on the specified half, and only 1/2 the time will fly B pick the same half -- the final answer would then be 1/2 of 1/2, or 1/4.

28. This is a difficult problem that will be broken into two events and a conclusion. Let L denote the length of the stick. Label the midpoint of the stick M, and label the first and second breaking points X and Y.

  Event A: X and Y must be on opposite sides of M. If X and Y are on the same side of M, the side without X and Y will be longer than .5L and no triangle will be possible. Once X is determined, P(Y is on the same side as X) = P(Y is on the opposite side from X) = .5.

  Event B: The distance $|XY|$ must be less than .5L. Assuming X and Y are on opposite sides of M, let Q denote the end of the stick closest to X and let R denote the end of the stick closest to Y so that the following sketch applies.

```
|---|--------|--|---------|
Q   X        M  Y         R
```

In order form a triangle, it must be true that $|XY| = |XM| + |MY| < .5L$. This happens only when $|MY| < |QX|$ so that $|XM| + |MY| = .5L - |QX| + |MY| < .5L - |QX| + |QX| = .5L$. With all choices random, there is no reason for $|QX|$ to be larger than $|MY|$ or vice-versa. This means $P(|MY| < |QX|) = .5$.

  Conclusion: For a triangle, we need both events A and B. Since P(A) = .5 and P(B occurs assuming that A has occurred) = .5, the probability of a triangle is (.5)(.5) = .25. In the notation of section 3-4 for the joint probability of dependent events,
  $P(A \text{ and } B) = P(A) \cdot P(B|A) = (.5)(.5) = .25$.

29. let W = winning by selecting the correct slot on the wheel
    P(W) = 1/38
    P($\overline{W}$) = 37/38
    a. odds against = P($\overline{W}$)/P(W) = (37/38)(1/38) = 37/1 or 37:1
    b. odds in favor = 1:37
    NOTE: If the odds against are 37:1, then a win gives $37 in winnings to a person putting up $1. Since P(W) = 1/38, someone who plays 38 times expects to win one time in 38. That one win will provide the $37 to cover the 37 losses, and the game is fair. In practice, a casino will offer odds of, say, 35:1 so that it will be likely to earn a profit.

3-3 Addition Rule

1. a. No, a student may regularly attend statistics class and own a computer.
   b. No, a student may have blond hair and brown eyes.
   c. Yes, a course cannot be required and be an elective at the same time (i.e., in the same context).

2. a. No, a person may color his hair and read *The Greening of America*.
   b. Yes, a viewer cannot be unmarried and have a spouse.
   c. Yes, a Car cannot be free of defects and have inoperative headlights.

3. a. P($\overline{A}$) = 1 - P(A)
     = 1 - .45
     = .55

   b. P(girl) = P($\overline{boy}$)
     = 1 - P(boy)
     = 1 - .513
     = .487

4. a. P($\overline{A}$) = 1 - P(A)
     = 1 - 3/8
     = 5/8

   b. P(pass) = P($\overline{fail}$)
     = 1 - P(fail)
     = 1 = .43
     = .57

5. It may be helpful to make a chart like the one given at the right.

   let A = a person is 18-21
   N = a person does not respond

		RESPOND?		
		Yes	No	
AGE	18-21	73	11	84
	22-29	255	20	275
		328	31	359

   There are two correct approaches.
   * Use broad categories and allow for double-counting.
     P(A or N) = P(A) + P(N) - P(A and N)
       = 84/359 + 31/359 - 11/359
       = 104/359
       = .290
   * Use individual mutually exclusive categories that involve no double-counting. NOTE: For simplicity in this problem, we use B for "a person is 22-29" and AY for "A and Y."
     P(A or N) = P(AY or AN or BN)
       = P(AY) + P(AN) + P(BN)
       = 73/359 + 11/359 + 20/359
       = 104/359
       = .290
   NOTE: In general, using broad categories and allowing for double-counting is a "more powerful" technique that "lets the formula do the work" and requires less detailed analysis by the solver. Except when such detailed analysis is instructive, the solutions in this manual utilize the first approach.

6. Refer to the chart and notation of exercise #5.

$$P(A \text{ or } Y) = P(A) + P(Y) - P(A \text{ and } Y)$$
$$= 84/359 + 328/359 - 73/359$$
$$= 339/359$$
$$= .944$$

7. It may be helpful to make a chart like the one given at the right.

		HARASSMENT?		
		Yes	No	
SEX	Male	240	380	620
	Female	180	200	380
		420	580	1000

let N = a person says no harassment

$$P(N) = 580/1000$$
$$= .58$$

8. Refer to the chart and notation of exercise #7.

$$P(M \text{ or } N) = P(M) + P(N) - P(M \text{ and } N)$$
$$= 620/1000 + 580/1000 - 380/1000$$
$$= 820/1000$$
$$= .820$$

9. It may be helpful to make a chart like the one given at the right.

		SMOKING?		
		Yes	No	
LOCATION	Aisle	16	64	80
	Off-A	24	96	120
		40	160	200

let A = an aisle seat
　　Y = a seat with smoking permitted

$$P(A \text{ or } Y) = P(A) + P(Y) - P(A \text{ and } Y)$$
$$= 80/200 + 40/200 - 16/200$$
$$= 104/200$$
$$= .52$$

10. It may be helpful to make a chart like the one given at the right.

		CRIME			
		Hom	Rob	Asl	
REL	Strngr	12	379	727	1118
	Friend	39	106	642	787
	Unknwn	18	20	57	95
		69	505	1426	2000

let H,R,A = the crimes
　　S,F,U = the relationships

$$P(H \text{ and } S) = 12/2000 = .00600$$

11. It may be helpful to make a chart like the one given at the right.

		SMOKE?		
		Yes	No	
STATUS	Married	54	146	200
	Divorced	38	62	100
	Alw Single	11	39	50
		103	247	350

let D = a person is divorced
　　Y = a person smokes

$$P(D \text{ or } Y) = P(D) + P(Y) - P(D \text{ and } Y)$$
$$= 100/350 + 103/350 - 38/350$$
$$= 165/350$$
$$= .471$$

12. Refer to the chart and notation of exercise #11.

$$P(A \text{ or } N) = P(A) + P(N(- P(A \text{ and } N)$$
$$= 50/350 + 247/350 - 39/350$$
$$= 258/350$$
$$= .737$$

13. The total number in the survey is 95,277.
    let x = the age of a person in the survey
    $P(x < 5 \text{ or } x > 74) = P(x < 5) + P(x > 74)$ [mutually exclusive]
    $= 3843/95277 + 16800/95277$
    $= 20643/95277$
    $= .217$

14. The total number in the survey is 95,277.
    There are 61,909 between 15 and 64 inclusive.
    let x = the age of a person in the survey
    $P(15 \le x \le 64) = 61909/95277 = .650$

15. The total number in the survey is 95,277.
    There are 55,245 younger than 45.
    There are 50,433 between 25 and 74 inclusive.
    let x = the age of a person in the survey
    $P(x < 45 \text{ or } 25 \le x \le 74) = P(x < 45) + P(25 \le x \le 74) - P(25 \le x < 45)$
    $= 55245/95277 + 50433/95277 - 27201/95277$
    $= 78477/95277$
    $= .824$
    NOTE: This is of the form $P(A \text{ or } B) = P(A) + P(B) - P(A \text{ and } B)$
    where, $A = x < 45$
    $B = 25 \le x \le 74$
    $A \text{ and } B = x < 45 \text{ and } 25 \le x \le 74$
    $= 25 \le x < 45$

16. The total number in the survey is 95,277.
    There are 28.044 younger than 25.
    There are 47,176 between 15 and 44 inclusive.
    let x = the age of a person in the survey
    $P(x < 25 \text{ or } 15 \le x \le 44) = P(x < 25) + P(15 \le x \le 44) - P(15 \le x < 25)$
    $= 28044/95277 + 47176/95277 - 19975/95277$
    $= 55245/95277$
    $= .580$

17. It may be helpful to make a chart like
    the one given at the right.

    $P(A \text{ or } B) = P(A) + P(B)$
    $= 40/100 + 10/100$
    $= 50/100$
    $= .5$

		RH FACTOR		
		+	–	
	A	35	5	40
	B	8	2	10
GROUP	AB	4	1	5
	O	39	6	45
		86	14	100

18. Refer to exercise #17.
    $P(R+) = 86/100 = .86$

19. Refer to exercise #17.
    $P(A \text{ or } R\text{-}) = P(A) + P(R\text{-}) - P(A \text{ and } R\text{-})$
    $= 40/100 + 14/100 - 5/100$
    $= 49/100$
    $= .49$

20. Refer to exercise #17.
$$P(O \text{ or } R+) = P(O) + P(R+) - P(O \text{ and } R+)$$
$$= 45/100 + 86/100 - 39/100$$
$$= 92/100$$
$$= .92$$

21. Refer to exercise #17.
$$P(\text{not } AB) = 1 - P(AB) = 1 - 5/100 = 1 - .05 = .95$$

22. Refer to exercise #17.
$$P(\overline{R\text{-}}) = 1 - P(R\text{-}) = 1 - 14/100 = 86/100 = .86$$

23. Refer to exercise #17.
$$P(O \text{ or } R\text{-}) = P(O) + P(R\text{-}) - P(O \text{ and } R\text{-})$$
$$= 45/100 + 14/100 - 6/100$$
$$= 53/100$$
$$= .53$$

24. $P[A \text{ or } O \text{ or } R\text{-}] = P[(A \text{ or } O) \text{ or } R\text{-}]$
$$= P(A \text{ or } O) + P(R\text{-}) - P[(A \text{ or } O) \text{ or } R\text{-}]$$
$$= P(A \text{ or } O) + P(R\text{-}) - P[(A \text{ or } R\text{-}) \text{ or } (O \text{ or } R\text{-})]$$
$$= [40/100 + 45/100]^* + 14/100 - [5/100 + 6/100]^*$$
$$= 85/100 + 14/100 - 11/100$$
$$= 88/100$$
$$= .88 \qquad \textbf{NOTE: } ^*\text{since the 2 events are mutually exclusive}$$

25. a. $P(A \text{ or } B) = P(A) + P(B) - P(A \text{ and } B)$
$$1/3 = P(A) + 1/4 - 1/5$$
$$.333 = P(A) + .250 - .200$$
$$.333 = P(A) + .050$$
$$.283 = P(A)$$
   b. for mutually exclusive events, $P(A \text{ or } B) = P(A) + P(B) = .4 + .5 = .9$
   c. if A and B are not mutually exclusive, $P(A \text{ and } B) > 0$
   and so, $P(A \text{ or } B) = P(A) + P(B) - P(A \text{ and } B)$
$$= .4 + .5 - P(A \text{ and } B)$$
$$= .9 - P(A \text{ and } B)$$
$$= \text{something less than } .9$$
   NOTE: More precisely, whenever A and B are not mutually exclusive it must be true that
$$0 < P(A \text{ and } B) \leq \min[P(A), P(B)]$$
   here, $0 < P(A \text{ and } B) \leq .4$
   and so, $.5 \leq P(A \text{ or } B) < .9$

26. No. Consider tossing a die and defining A,B,C as follows.
   let A = getting an odd number
       B = getting an even number
       C = getting a 3

27. If the exclusive *or* is used instead of the inclusive *or*, then the double-counted probability must be completely removed (i.e., must be subtracted twice) and the formula becomes
$$P(A \text{ or } B) = P(A) + P(B) - 2 \cdot P(A \text{ and } B)$$

28. P(A or B or C)

$$= P[(A \text{ or } B) \text{ or } C]$$
$$= P(A \text{ or } B) + P(C) - P[(A \text{ or } B) \text{ and } C]$$
$$= [P(A) + P(B) - P(A \text{ and } B)] + P(C) - P[(A \text{ and } C) \text{ or } (B \text{ and } C)]$$
$$= [P(A) + P(B) - P(A \text{ and } B)] + P(C) - [P(A \text{ and } C) + P(B \text{ and } C) - P(A \text{ and } B \text{ and } C)]$$
$$= P(A) + P(B) - P(A \text{ and } B) + P(C) - P(A \text{ and } C) - P(B \text{ and } C) + P(A \text{ and } B \text{ and } C)$$
$$= P(A) + P(B) + P(C) - P(A \text{ and } B) - P(A \text{ and } C) - P(A \text{ and } B) + P(A \text{ and } B \text{ and } C)$$

3-4 Multiplication Rule

1. a. Dependent, since removing the first watch changes the probabilities of getting a defective watch on the second selection.
   b. Independent, since getting the first question right or wrong does not affect the chance of guessing correctly on the second question.

2. a. Independent, since the operability of the electric microwave is in no way related to the operability of the battery-operated smoke detector.
   b. Dependent, since both the light and the microwave receive from the same source the electric power that makes them operable.

3. let N = the reservation is for a non-smoking table
   P(N) = .65, for all selections
   P(N$_1$ and N$_2$ and N$_3$ and N$_4$) = P(N$_1$)·P(N$_2$)·P(N$_3$)·P(N$_4$) = (.65)(.65)(.65)(.65) = .179

4. let V = a person considers vulgar language to be the worst turnoff
   P(V) = .74, for all selections
   P(V$_1$ and V$_2$) = P(V$_1$)·P(V$_2$) = (.74)(.74) = .548

5. let H = the derailment was caused by human error
   P(H) = .24, for all selections
   P(D$_1$ and D$_2$ and D$_3$ and D$_4$ and D$_5$) = P(D$_1$)·P(D$_2$)·P(D$_3$)·P(D$_4$)·P(D$_5$)
   $$= (.24)(.24)(.24)(.24)(.24)$$
   $$= .000796$$

6. let H = hypnosis successfully helps a person quit smoking
   P(H) = .60, for all selections
   P(H$_1$ and H$_2$ and ... and H$_8$) = P(H$_1$)·P(H$_2$)·...·P(H$_8$) = (.60)^8 = .0168

7. let D = the disk selected is defective
   P(D) = 10/30 for the first selection only
   P(D$_1$ and D$_2$ and D$_3$ and D$_4$) = P(D$_1$)·P(D$_2$ | D$_1$)...
   $$= (10/30)(9/29)(8/28)(7/27)$$
   $$= .00766$$

8. let R = a resistor works successfully
   P(R) = .992, for each resistor
   P(R$_1$ and R$_2$ and R$_3$ and R$_4$) = P(R$_1$)·P(R$_2$)·P(R$_3$)·P(R$_4$) = (.992)(.992)(.992)(.992) = .968

9. let W = a new car buyer is a woman
   a. P(W) = .5, for each sale
      P(W$_1$ and W$_2$ and W$_3$ and W$_4$) = P(W$_1$)·P(W$_2$)·P(W$_3$)·P(W$_4$) = (.5)(.5)(.5)(.5) = .0625
   b. P(W) = .6, for each sale
      P(W$_1$ and W$_2$ and W$_3$ and W$_4$) = P(W$_1$)·P(W$_2$)·P(W$_3$)·P(W$_4$) = (.6)(.6)(.6)(.6) =. 130

10. let M = an audited taxpayer owes more money
$P(M) = .70$, for all selections
$P(M_1 \text{ and } M_2 \text{ and } \ldots \text{ and } M_8) = P(M_1) \cdot P(M_2) \cdot \ldots \cdot P(M_8) = (.70)^8 = .0576$

11. let J = a person is born on July 4
$P(J) = 1/365$, for all random selections
$P(J_1 \text{ and } J_2 \text{ and } J_3) = P(J_1) \cdot P(J_2) \cdot P(J_3) = (1/365)(1/365)(1/365) = .0000000206$
NOTE: This is, as the problem requests, the probability of 3 randomly chosen people being born on July 4. There might be special circumstances that change that probability for the family in question -- suppose, for example, the husband is a migrant worker who is away all summer and returns to his wife each year around October 4 (i.e., 9 months before July 4).

12. let C = a question is answered correctly
a. $P(C) = .50$, for each question
$P(C_1 \text{ and } C_2) = P(C_1) \cdot P(C_2) = (.50)(.50) = .250$
b. $P(C) = .20$, for each question
$P(C_1 \text{ and } C_2) = P(C_1) \cdot P(C_2) = (.20)(.20) = .040$

13. let A = a death is accidental
$P(A) = .0478$, for all random selections
$P(A_1 \text{ and } A_2 \text{ and } A_3 \text{ and } A_4 \text{ and } A_5) = P(A_1) \cdot P(A_2) \cdot P(A_3) \cdot P(A_4) \cdot P(A_5)$
$$= (.0478)^5$$
$$= .000000250$$
NOTE: This is, as the problem requests, the probability of 4 randomly chosen deaths being accidental. The deaths investigated by the Baltimore detective, however, involved special circumstances and were not chosen at random.

14. let G = a selected fuse is good
a. $P(G) = 74/80$, for each selection
$P(G_1 \text{ and } G_2 \text{ and } G_3) = P(G_1) \cdot P(G_2) \cdot P(G_3)$
$$= (74/80)(74/80)(74/80)$$
$$= .791$$
b. $P(G) = 74/80$ for the first selection only
$P(G_1 \text{ and } G_2 \text{ and } G_3) = P(G_1) \cdot P(G_2 \mid G_1) \cdot P(G_3 \mid G_1 \text{ and } G_2)$
$$= (74/80)(73/79)(72/78)$$
$$= .791$$
c. Since $.789 < .791$, the procedure in part (b) is more likely to reveal a defective fuse. If, however, 6 defectives in 80 meets the contract specifications, the lot should be accepted and the procedure in part (b) makes the proper decision less likely. At any rate, since the procedure in part (b) guarantees 3 different fuses, it should be preferred because it guarantees the largest possible portion of the sample will be examined.

15. let V = an eligible voter actually votes
$P(V) = .6$, for each selection
$P(V_1 \text{ and } V_2 \text{ and} \ldots \text{and } V_9 \text{ and } V_{10}) = (.6)^{10} = .00605$

16. let G = a selected token is good
a. $P(G) = 1,995,000/2,000,000$ for each selection
$P(G_1 \text{ and } G_2) = P(G_1) \cdot P(G_2) = (1995000/2000000)(1995000/2000000)$
$$= .955 \quad [\text{actually, } .995006250]$$
b. $P(G) = 1,995,000/2,000,000$ for the first selection only
$P(G_1 \text{ and } G_2) = P(G_1) \cdot P(G_2 \mid G_1) = (1995000/2000000)(1994999/1999999)$
$$= .995 \quad [\text{actually, } .995006248]$$

17. let N = the flight from New York is on time
    S = the flight from St. Louis is on time
    P(N) = .80
    P(S) = .60
    P(N and S) = P(N) · P(S)   [assuming independence]
                = (.8)(.6)
                = .48

18. let H = the selected employee is honest
    P(H) = 34/36 for the first selection only
        P(H_1 and H_2 and H_3 and H_4) = P(H_1) · P(H_2 | H_1) · ...
                = (34/36)(33/35)(32/34)(31/33)
                = .787

19. let F = a selection has faulty brakes
    P(F) = 3/12 for the first selection (if the owner is correct)
    P(F_1 and F_2 and F_3) = P(F_1) · P(F_2 | F_1) · P(F_3 | F_1 and F_2)
                = (3/12) · (2/11) · (1/10)
                = .00455

20. let S = a person identifies the same voice
    P(S_1) = 5/5   [the first person can pick any voice to be "the same"]
    P(S_i) = 1/5, for each person after the first person
    P(S_1 and S_2 and ... and S_9) = P(S_1) · P(S_2) · ... · P(S_9) = (5/5)(1/5)^8 = .00000256

21. let L = losing when betting on #7
    P(L) = 37/38, each time
    P(L_1 and L_2 and...and L_{37} and L_{38}) = [P(L)]^{38} = [37/38]^{38} = .363

22. let S = a such a person survives for the year
    P(S) = .9982, for each such person
    P(S_1 and S_2 and ... and S_{12}) = [P(S)]^{12} = [.9982]^{12} = .979

23. let N = no reply is received from the coupon
    P(N) = .8, for each coupon
    P(N_1 and N_2 and N_3 and N_4 and N_5 and N_6) = [P(N)]^6 = [.8]^6 = .262

24. let M = the selected person is a male
    P(M) = 20/40 for the first selection only
    P(M_1 and M_2 and ... and M_{12})
        = P(M_1) · P(M_2 | M_1) · ...
        = (20/40)(19/39)(18/38)(17/37)(16/36)(15/35)(14/34)(13/33)(12/34)(11/31)(10/30)(9/29)
        = .0000225

25. let N = an individual tests negative
    P(N) = .985, for each person
    P(group is positive) = 1 - P(the group is negative)
                = 1 - P(N_1 and N_2 and N_3 and N_4 and N_5)
                = 1 - (.985)(.985)(.985)(.985)(.985)
                = 1 - .9272
                = .0728

26. let L = a person has a listed telephone number
    P(L) = .72, for each person
    P(at least one listed) = 1 - P(all not listed) = 1 - [P($\bar{L}$)]^5 = 1 - (.28)^5 = 1 - .002 = .998

27. let F = a component fails
    P(F) = .181, for each component
    P(circuit works) = 1 - P(circuit fails)
    $$= 1 - P(F_1 \text{ and } F_2 \text{ and } F_3 \text{ and } F_4)$$
    $$= 1 - (.181)(.181)(.181)(.181)$$
    $$= 1 - .001$$
    $$= .999$$

28. let W = the selected voter is a woman
    P(W) = .52, for each selection
    P(at least one woman) = 1 - P(all not women) = $1 - [P(\overline{W})]^4 = 1 - (.48)^4 = 1 - .053 = .947$

29. Refer to the table reproduced from the text and given at the right.

		CRIME			
		Hom	Rob	Asl	
	stranger	12	379	727	1118
CRIMINAL	frnd/rel	39	106	642	787
	unknown	18	20	57	95
		69	505	1426	2000

let H,R A stand for homicide robbery, assault
let S,F,U stand for stranger, friend/relative, unknown
a. This problem may be done either of two ways. NOTE: This is a conditional probability, given that the crime was a robbery. When using the first method below, the robbery column becomes the relevant sample space and is the only column used to answer the question.
   * reading directly from the table
   P(S | R) = 379/505 = .750
   * using the formula
   P(S | R) = P(S and R)/P(R)
   $$= (379/2000)/(505/2000)$$
   $$= 379/505$$
   $$= .750$$
b. P(R and S) = 379/2000 = .1895
c. P(R or U) = P(R) + P(U) - P(R and U)
   $$= (505/2000) + (95/2000) - (20/2000)$$
   $$= 580/2000$$
   $$= .290$$

30. Refer to the table and notation of exercise #29.
   a. P(F | H) = 39/69 = .565
   b. P(F and H) = 39/2000 = .0195

31. Refer to the table and notation of exercise #29. NOTE: This is a good exercise to verify that one can distinguish whether a word problem is asking an "or" or "and" or "given" question.
   a. P(A or S) = P(A) + P(S) - P(A and S)
   $$= (1426/2000) + (1118/2000) - (727/2000)$$
   $$= 1817/2000$$
   $$= .9085$$
   b. P(A and S) = 727/2000 = .3635
   c. * reading directly from the table
   P(A | S) = 727/1118 = .650
   * using the formula
   P(A | S) = P(A and S)/P(S)
   $$= (727/2000)/(1118/2000)$$
   $$= 727/1118$$
   $$= .650$$

32. Refer to the table and notation of exercise #29.

a. **NOTE:** The wording of the exercise is misleading. While the phrase "someone unknown to the victim" could be interpreted as "a stranger" [denoted S], it is assumed in this context to mean "someone whose relationship to the victim is unknown" [denoted U].

$P(H \text{ or } U) = P(H) + P(U) - P(H \text{ and } U)$
$= 69/2000 + 95/2000 - 18/2000 = 146/2000 = .0730$

b. $P(H \mid S) = 12/1118 = .0107$

c. $P(S \mid H) = 12/69 = .174$

33. a. let D = a birthday is different from any yet selected

$P(D_1) = 366/366$ NOTE: With nothing to match, it <u>must</u> be different.
$P(D_2 \mid D_1) = 365/366$
$P(D_3 \mid D_1 \text{ and } D_2) = 364/366$
...
$P(D_{25} \mid D_1 \text{ and } D_2 \text{ and}...\text{and } D_{24}) = 342/366$
$P(\text{no match}) = P(D_1 \text{ and } D_2...D_{25}) = (366/366) \cdot (365/366) \cdot (364/366) \cdots (342/366) = .432$

NOTE: Programs to perform this calculation can be constructed as follows.

*** in BASIC	*** in Minitab
10 LET P=1	*MTB > LET K1=1*
20 FOR K=1 TO 24	*MTB > LET K2=1*
30 LET P=P*(366-K)/366	*MTB > STORE*
40 NEXT K	*STOR> LET K1=K1*(366-K2)/366*
50 PRINT P	*STOR> LET K2=K2+1*
60 END	*STOR> END*
	MTB > NOECHO
	MTB > EXECUTE 24
	MTB > PRINT K1

The STORE command in Minitab allows the creation of a file of commands. The EXECUTE N command instructs Minitab to execute the stored file N times. The NOECHO command suppresses intermediate printout. When February 29 is ignored and a 365-day year is used, the answer is .431.

b. $P(\text{at least one match}) = 1 - P(\text{no match}) = 1 - .432 = .568$

34. a. $P(A \text{ and } B) = P(A) \cdot P(B) = (3/4)(5/6) = 15/24$ [or 5/8]

b. $P(A \text{ and } B) = P(A) \cdot P(B \mid A) = (3/4)(1/2) = 3/8$

c. $P(A \text{ and } B) = P(B \text{ and } A) = P(B) \cdot P(A \mid B) = (5/6)(1/3) = 5/18$

d. $P(A \text{ and } B) = P(B \text{ and } A) = P(B) \cdot P(B \mid A) = (5/6)(3/4) = 15/24$ [or 5/8]

NOTE: If $P(A \mid B) = P(A)$, A and B are independent and the answer agrees with part (a).

35. The following English/logic facts are used in this exercise.

* not (A or B) = not A and not B
   Is your sister either artistic or bright?
   No? Then she is not artistic and she is not bright.

* not (A and B) = not A or not B
   Is your brother artistic and bright?
   No? Then either he is not artistic or he is not bright.

a. $P(\overline{A \text{ or } B}) = P(\overline{A} \text{ and } \overline{B})$ [from the first fact above]
    or
   $P(\overline{A \text{ or } B}) = 1 - P(A \text{ or } B)$ [rule of complementary events]

b. $P(\overline{A} \text{ or } \overline{B}) = 1 - P(A \text{ and } B)$ [from the second fact above]

c. They are different: (a) gives the complement of "A or B"
                          while (b) gives the complement of "A and B."

36. It may be helpful to make a chart like
the one given at the right.

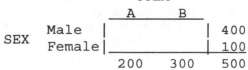

a. $P(M) = 400/500 = .800$
b. $P(A) = 200/500 = .400$
c. [assuming independence] $P(M$ and $A) = P(M) \cdot P(A) = (.800)(.400) = .320$
d. $(.320)(500) = 160$

37. This is problem can be done by two different methods. In either case,
let T = getting a 10
  C = getting a club
* consider the sample space
  The first card could be any of 52 cards; for each first card, there are 51 possible second
  cards. This makes a total of $52 \cdot 51 = 2652$ equally likely outcomes in the sample space.
  How many of them are T_1C_2?
  The 10's of hearts, diamonds and spades can be paired with any of the 13 clubs for a total
  of $3 \cdot 13 = 39$ favorable possibilities. The 10 of clubs can only be paired with any of the
  remaining 12 members of that suit for a total of 12 favorable possibilities. Since there are
  $39 + 12 = 51$ favorable possibilities among the equally likely outcomes,
$$P(T_1C_2) = 51/2652 = .0192$$
* use the formulas
  Let Tc and To represent the 10 of clubs and the 10 of any other suit respectively. Break
  T_1C_2 into mutually exclusive parts so the probability can be found by adding and without
  having to consider double-counting. $P(T_1C_2) = P(Tc_1C_2$ or $To_1C_2)$
$$= P(Tc_1C_2) + P(To_1C_2)$$
$$= (1/52)(12/51) + (3/52)(13/51)$$
$$= 12/2652 + 39/2652$$
$$= 51/2652$$
$$= .0192$$

3-5 Probabilities Through Simulations

1. Following the example in the text and using even digits to represent boys and odd digits to
   represent girls produces the following list of $50 \cdot 3 = 150$ simulated births. A slash follows
   the birth of each girl, and the family size necessary to produce a girl is the number of children
   since the last the slash. NOTE: The preceding system is only one possible method for
   converting the random numbers to sex of children. One could have, for example, used the
   digits 0-4 to represent boys and 5-9 to represent girls to produce a different list. The end
   result, however, should be similar for any valid simulation.
   BBBG/G/G/G/BBBBBBBG/G/BG/BBG/BG/G/BG/G/G/G/G/BBBG/G/G/
   G/BG/BG/BG/G/G/G/BG/BBG/BBBBG/BG/G/G/G/BG/BBBG/BG/G/BB
   BG/BG/BBG/BBG/G/G/BG/G/G/BBG/BBG/BBBG/G/G/G/G/BG/BG/BB
   G/G/BG/G/G/BBBG/G/BG/BG/G/G/G/BBG/BBG/G/G/BG/G/G/G/G/
   G/G/G/BG/G/BG/B (the final B is not used)

   The 82 x values (necessary family sizes) generated are:
   4 1 1 1 8 1 2 3 2 1 2 1 1 1 1 4 1 1 1 2 2 2 1 1 1 2 3 5
   2 1 1 1 2 4 2 1 4 2 3 3 1 1 2 1 1 3 3 4 1 1 1 1 2 2 3 1
   2 1 1 4 1 2 2 1 1 1 3 3 1 1 2 1 1 1 1 1 1 1 2 1 2
   Using the sample mean from the simulation to estimate the true average number of births
   necessary to get a girl, $\bar{x} = \Sigma x/n = 149/82 = 1.8$.

NOTE: Some problems are so complicated that the true answer will probably never be known and the simulated answer (or the average of several simulated answers) must be assumed to be close enough to allow the research to proceed. In this instance, the true average number may be found from probability theory to be 2.0. It is a weighted average of the possible necessary family sizes 1,2,3... where the probabilities are the weights and

$$P(x=1) = P(G) = .5$$
$$P(x=2) = P(BG) = P(B_1) \cdot P(G_2 \mid B_1) = (.5)(.5) = (.5)^2$$
$$P(x=3) = P(BBG) = (.5)(.5)(.5) = (.5)^3$$
...

in general, $P(x) = (.5)^x$ and $\Sigma P(x) = 1$

The true average is the weighted mean $\mu = (\Sigma x \cdot w)/(\Sigma w)$
$$= (\Sigma x \cdot P(x))/(\Sigma P(x))$$
$$= \Sigma x \cdot P(x)$$
$$= \Sigma x \cdot (.5)^x \quad \text{for the summation } x=1,2,3...$$

The more mathematically inclined may wish to follow the proof given below that $\mu = 2$.

Replacing x with y+1,

$$\mu = \Sigma(y+1) \cdot (.5)^{y+1} \qquad \text{for the summation } y=0,1,2,3...$$
$$= 1 \cdot (.5)^1 + \Sigma(y+1) \cdot (.5)^{y+1} \qquad \text{for the summation } y=1,1,2,3...$$
$$= .5 \qquad + \Sigma y \cdot (.5)^{y+1} \quad + \Sigma 1 \cdot (.5)^{y+1} \qquad \text{for the summation } y=1,2,3...$$
$$= .5 \qquad + (.5) \cdot \Sigma y \cdot (.5)^y + (.5) \cdot \Sigma (.5)^y \qquad \text{for the summation } y=1,2,3...$$
$$= .5 \qquad + (.5) \cdot \mu \qquad + (.5) \cdot 1$$

Which produces the result
$$\mu = .5 + .5\mu + .5$$
$$.5\mu = 1$$
$$\mu = 2$$

2. Refer to the method and notation of exercise #1.

BBBGG/GGBB/BBBBBGG/BGBBG/BGGGB/GGGGGBB/BGGG
GB/GBGB/GGGGBGB/BGBBBBG/BGGGGB/GBBBG/BGGB/B
BGBG/BBGBBG/GGBGGGB/BGBBG/BBBGG/GGGBGB/GBB
G/GBGGGB/BBGG/BGBG/GGGBB/GBBG/GGBGGGGGG
GGGB/GGBGB/

The 27 x values (necessary family sizes) generated are:

5 4 7 5 4 7 6 4 7 7 6 5 4 5 6 7 5 5 6 4 6 4 4 5 4 13 5

Using the sample mean from the simulation to estimate the true average number of births necessary to get at least 2 of each sex, $\bar{x} = \Sigma x/n = 150/27 = 5.6$. **NOTE:** The true average number may be found from probability theory to be 5.5.

3. Using 0 to represent a good chip and 1,2,...,9 to represent a defective one produces the following list of 13 simulated numbers of chips necessary to get a good one:

1 1 1 44 6 1 3 15 1 12 10 20 20

NOTE: The last 15 digits could not be used.

Using the sample mean from the simulation to estimate the true average number of chips necessary to get a good one, $\bar{x} = \Sigma x/n = 135/13 = 10.4$. NOTE: The true average number may be found from probability theory to be 10.0.

4. Using 0,1 to represent a good chip and 2,3,...,9 to represent a defective one produces the following list of 29 simulated numbers of chips necessary to get a good one:

1 1 1 2 21 1 7 10 2 1 6 1 3 3 7 2 3 1 12 10 7 7 1 2 2 1 14 6 14

NOTE: The last digit could not be used.

Using the sample mean from the simulation to estimate the true average number of chips necessary to get a good one, $\bar{x} = \Sigma x/n = 149/29 = 5.1$. NOTE: The true average number may be found from probability theory to be 5.0.

5. Using 0 to represent a good chip and 1,2,...,9 to represent a defective one produces the following list of 6 simulated numbers of chips necessary to get 2 good ones:

    2 45 7 18 13 30

    NOTE: The last 35 digits could not be used.

    Using the sample mean from the simulation to estimate the true average number of chips necessary to get a good one, $\bar{x} = \Sigma x/n = 115/6 = 19.2$. NOTE: The true average number may be found from probability theory to be 20.0

6. Using 1,2,3,4,5,6 to represent those die results and ignoring 7,8,9,0 produces 18 5's in 94 simulated tosses and an estimated probability for tossing a 6 of $18/94 = .191$.

7. Using 1,2,3,4,5,6 to represent those die results and ignoring 7,8,9,0 produces the following list of 18 simulated numbers of tosses necessary to get a 6:

    4 2 1 3 1 3 1 9 6 3 7 6 2 4 22 1 3 16

    Using the sample mean from the simulation to estimate the true average number of tosses necessary to get a 6, $\bar{x} = \Sigma x/n = 94/18 = 5.2$. NOTE: The true average number may be found from probability theory to be 6.0.

8. A student guessing on T-F questions has P(correct) = .5. Using odd digits to represent correct answers and even digits to represent incorrect ones produces the following list of 50 correct answers per 3-question test:

    0 3 1 0 1 2 1 2 2 3 0 3 2 1 3 2 1 0 1 3 2 0 2 1 1
    1 1 3 2 1 1 1 2 3 1 1 2 3 0 2 2 3 1 1 2 3 3 3 2 1

    Using the sample mean from the simulation to estimate the true average number of correct answers in a 3-question test, $\bar{x} = \Sigma x/n = 82/50 = 1.64$. NOTE: The true average number may be found from probability theory to be 1.5.

9. Using even digits to represent boys and odd digits to represent girls produces the following list of 50 simulated numbers of girls per 3-child family:

    0 3 1 0 1 2 1 2 2 3 0 3 2 1 3 2 1 0 1 3 2 0 2 1 1
    1 1 3 2 1 1 1 2 3 1 1 2 3 0 2 2 3 1 1 2 3 3 3 2 1

    Using the sample mean from the simulation to estimate the true average number of girls in a 3-child family, $\bar{x} = \Sigma x/n = 82/50 = 1.64$. NOTE: The true average number may be found from probability theory to be 1.5.

    Using the sample standard deviation mean from the simulation to estimate the true standard deviation among the numbers of girls in a 3-child family,

    $s^2 = [n(\Sigma x^2) - (\Sigma x)^2]/[n(n-1)]$
    $\quad = [50(182) - (82)^2]/[50(49)] = 2376/2450 = .970$
    $s = .98$

    NOTE: The true standard deviation among the numbers may be found from probability theory to be .866

10. a. The following commands place 25 random integers from 1 to 365 inclusive into column 1.
    MTB > RANDOM 25 C2;
    SUBC> INTEGER 1 365.

    b. After entering the above commands, the following output was obtained for this manual. Every execution of the commands, of course, produces a different sequence.
    MTB > PRINT C1

    C1
    | 125 | 248 | 332 | 160 | 290 | 20 | 248 | 312 | 284 | 70 | 281 |
    | 6 | 343 | 162 | 171 | 82 | 23 | 253 | 181 | 255 | 104 | 146 |
    | 272 | 32 | 283 |

    There are two 248's, indicating a shared birthday on that day of the year.

c. The following commands place 25 random integers from 1 to 365 inclusive into columns 2 through 10. See exercise #33 of the previous section for comments on the STORE, EXECUTE and NOECHO commands. The commands at the right place the numbers in ascending order and make it easier to identify repeats.

MTB > LET K1=2	*MTB > LET K1=1*
MTB > STORE	*MTB > STORE*
STOR> RANDOM 25 CK1;	*STOR> SORT CK1 CK1*
STOR> INTEGER 1 365.	*STOR> LET K1=K1+1*
STOR> LET K1=K1+1	*STOR> END*
STOR> END	*MTB > EXECUTE 10*
MTB > NOECHO	*MTB > PRINT C1-C10*
MTB > EXECUTE 9	

Entering the above commands to obtained for this manual produced repetitions in 5 of the 10 columns for an estimated P(at least 2 people in 25 share a birthday) = 5/10 = .500. The true value may be found from probability theory to be .568.

3-6 Counting

1. $7! = 7 \cdot 6 \cdot 5 \cdot 4 \cdot 3 \cdot 2 \cdot 1 = 5040$

2. $9! = 9 \cdot 8 \cdot 7 \cdot 6 \cdot 5 \cdot 4 \cdot 3 \cdot 2 \cdot 1 = 362,880$

3. $(70!)/(68!) = (70 \cdot 69 \cdot 68 \cdot 67 \cdot 66 \cdots)/(68 \cdot 67 \cdot 66 \cdots) = 70 \cdot 69 = 4830$
   NOTE: This technique of "cancelling out" or "reducing" the problem by removing the factors $68 \cdot 67 \cdot 66 \cdots 1$ from both the numerator and the denominator should be preferred over actually evaluating 70!, actually evaluating 68!, and then dividing those two very large numbers. In general, a smaller factorial in the denominator can be completely divided into a larger factorial in the numerator to leave only the "excess" factors not appearing the in the denominator. This is the technique employed in this manual -- e.g., see #9 below, where the 7! is cancelled from both the numerator and the denominator. In addition, the, the answer to a counting problem (but not a probability problem) must always be a whole number; a fractional number indicates that a mistake has been made.

4. $(92!)/(89!) = (92 \cdot 91 \cdot 90 \cdot 89!)/(89!) = 92 \cdot 91 \cdot 90 = 753,480$

5. $(9-3)! = (6)! = 6 \cdot 5 \cdot 4 \cdot 3 \cdot 2 \cdot 1 = 720$

6. $(20-12)! = (8)! = 8 \cdot 7 \cdot 6 \cdot 5 \cdot 4 \cdot 3 \cdot 2 \cdot 1 = 40,320$

7. $_6P_2 = 6!/(6-2)! = 6!/4! = 6 \cdot 5 = 30$

8. $_6C_2 = 6!/(2!4!) = 6 \cdot 5/2! = 30/2 = 15$

9. $_{10}C_3 = 10!/(7!3!) = (10 \cdot 9 \cdot 8)/3! = 720/6 = 120$ [see the NOTE in #3 above]

10. $_{10}P_3 = 10!/(10-3)! = 10!/7! = 10 \cdot 9 \cdot 8 = 720$

11. $_{52}C_2 = 52!/(50!2!) = (52 \cdot 51)/2! = 2652/2 = 1326$ [see the NOTE in #3 above]

12. $_{52}P_2 = 52!/(52-2)! = 52!/50! = 52 \cdot 51 = 2652$

13. $_nP_n = n!/(n-n)! = n!/0! = n!/1 = n!$

14. $_nC_n = n!/(n!0!) = n!/(n! \cdot 1) = n!/n! = 1$

15. $_nC_0 = n!/(n!0!) = n!/(n! \cdot 1) = n!/n! = 1$
NOTE: In words, this represents the number of ways to choose zero objects from among n objects. There is only 1 way -- to leave them all unselected.

16. $_nP_0 = n!/(n-0)! = n!/n! = 1$

17. $(3)(2) = 6$

18. $_5P_5 = 5!/(5-5)! = 5!/0! = 5!/1 = 5! = 120$

19. $_{10}P_4 = 10!/(10-4)! = 10!/6! = 10 \cdot 9 \cdot 8 \cdot 7 = 5040$

20. $_{10}C_4 = 10!/(4!6!) = (10 \cdot 9 \cdot 8 \cdot 7)/4! = 210$

21. $_{22}C_{12} = 22!/(10!12!) = (22 \cdot 21 \cdot 20 \cdot 19 \cdot 18 \cdot 17 \cdot 16 \cdot 15 \cdot 14 \cdot 13)/10! = 646,646$

22. $_{20}P_7 = 20!/(20-7)! = 20!/13! = 20 \cdot 19 \cdot 18 \cdot 17 \cdot 16 \cdot 15 \cdot 14 = 390,700,800$

23. $10! = 10 \cdot 9 \cdot 8 \cdot 7 \cdot 6 \cdot 5 \cdot 4 \cdot 3 \cdot 2 \cdot 1 = 3,628,800$

24. $8 \cdot 10 \cdot 10 \cdot 10 \cdot 10 \cdot 10 = 8,000,000$

25. a. $7! = 5040$
    b. Since only 1 of the 5040 possibilities can be the correct alphabetic order, the probability of that arrangement occurring is 1/5040.

26. a. $_{30}C_6 = 30!/(6!24!) = (30 \cdot 29 \cdot 28 \cdot 27 \cdot 26 \cdot 25)/6! = 593,775$
    b. Since only 1 of the 593,775 possibilities represents the 6 with the most time in service, the probability of that arrangement occurring is 1/593,775.

27. Because each of the 9 digits in the SS number could be any one of the 10 possibilities 0,1,2,3,4,5,6,7,8,9, there are $10 \cdot 10 \cdot 10 \cdot 10 \cdot 10 \cdot 10 \cdot 10 \cdot 10 \cdot 10 = 10^9 = 1,000,000,000$ (i.e., one billion) possible SS numbers.

28. $_{12}C_3 = 12!/(3!9!) = (12 \cdot 11 \cdot 10)/3! = 220$

29. $_{16}P_4 = 16!/(16-4)! = 16!/12! = 16 \cdot 15 \cdot 14 \cdot 13 = 43,680$

30. a. $10 \cdot 10 \cdot 10 \cdot 10 \cdot 10 = 100,000$
    b. The probability of randomly producing any one particular zip code is 1/100,000.

31. There are $50 \cdot 50 \cdot 50 = 50^3 = 125,000$ sequences possible. While such sequences are popularly called "combinations," they are really (i.e., in correct mathematical terminology) permutations.

32. $8! = 40,320$

33. The selections within each category can be thought of as a combination problem (i.e., choosing so many from a given group when the order is not important). The final menu can then be thought of as a sequence of four category selections to which the fundamental counting rule may be applied.
$_{10}C_1 \cdot {}_8C_1 \cdot {}_{13}C_2 \cdot {}_3C_1 = 10 \cdot 8 \cdot 78 \cdot 3 = 18,720$

34. Since the first symbol can be chosen only 1 way (as an asterisk) and the remaining 15 symbols can be chosen in any of 3 different ways, the total number of patterns is $1 \cdot 3^{15} = 14,348,907$.

35. a. Assuming the order of the shows is irrelevant in this context, use combinations.
$_{14}C_5 = 14!/(9!5!) = (14 \cdot 13 \cdot 12 \cdot 11 \cdot 10)/5! = 2002$
b. There are 2002 - 650 = 1352 compatible combinations; the probability of selecting a compatible combination by random selection is therefore 1352/2002 = .675.

36. a. $40! = 8.16 \cdot 10^{47}$
b. $2/40! = 2.45 \cdot 10^{-48}$
NOTE: There are $\underline{2}$ shortest ways because that route can be done "frontwards" or "backwards."

37. The number of possible combinations is $_{49}C_6 = 49!/(43!/6!) = (49 \cdot 48 \cdot 47 \cdot 46 \cdot 45 \cdot 44)/6!$
$$= 13,983,816$$
The probability of winning with one random selection is therefore 1/13,983,816 -- about 1/20 of the 1/700,000 probability of getting hit with lightning during the year. This means a person is 20 times more likely to be hit by lightning during the year than to win such a lottery. NOTE: This does not necessarily mean that there should be about 20 times as many people in the state that are hit with lightning than that win the lottery. The <u>number</u> of lottery winners depends on the number that actually play, not only on the probability of winning. If few people play the lottery, there are few winners; if many people play, there are many winners -- we expect about 1 winner for every 13,983,816 tickets purchased. The <u>probability</u> of winning the lottery with one ticket, the probability of being hit by lightning in one year, and the number of people hit by lightning in one year are not so subject to human manipulation.

38. a. Since each of the 10 questions could be answered in any of 5 ways, there are $5^{10} =$ 9,765,625 possible arrangements.
b. Since only 1 of the 9,765,625 possible arrangements answers each question correctly, the probability of obtaining a perfect paper by random guessing is 1/9,7675,635 = .000000102.

39. $24! = 24 \cdot 23 \cdot 22 \cdots 3 \cdot 2 \cdot 1 = 6.20 \times 10^{23}$

40. a. $_{12}P_5 = 12!/(12-5)! = 12!/7! = (12 \cdot 11 \cdot 10 \cdot 9 \cdot 8) = 95,040$
b. Since only 1 of the 95,040 orderings gives the 5 youngest secretaries in order of age (assumed to be from youngest to oldest), the probability that random selections produce such a list is 1/95,040. **NOTE:** The problem doesn't specify ascending or descending order. If the 5 youngest secretaries could be given from oldest to youngest, the probability is 2/95,040.

41. $10!/(7!3!) = (10 \cdot 9 \cdot 8)/3! = 120$
NOTE: This is the same as $_{10}C_3$, the number of combinations of 3 objects chosen from among 10 objects. Picking a committee of 3 from 10 people, for example, can be thought of lining up the 10 people (e.g, in alphabetical order) and marking 3 as successes (committee members) and 7 as failures (nonmembers). Each different arrangement of successes and failures corresponds to a particular committee of 3.

42. See the NOTE for exercise #41. The expression n!/x!(n-x)! equals $_nC_x$ and gives for n children (lined up, for example, in order of birth) the number of ways x of them can be "chosen" to be boys. The solution to this exercise is the fraction

$\dfrac{N}{D} = \dfrac{\text{\# of B-G arrangements of 12 children that are 4 boys and 8 girls}}{\text{total \# of B-G arrangements of 12 children}}$

By exercise #41, N = 12!/4!8! = 495.
By the fundamental counting rule, since each child can be either B or G, $D = 2^{12} = 4096$.
The desired fraction is 495/4096 = .121.

43. The numbers of possible names with 1,2,3,...,8 letters must be calculated separately using the fundamental counting rule and then added together to determine the total number of possible names. A chart will help to organize the work.

# of letters	# of possible names		
1	26	$= 26$	$= 26$
2	$26 \cdot 36$	$= 26 \cdot 36^1$	$= 936$
3	$26 \cdot 36 \cdot 36$	$= 26 \cdot 36^2$	$= 33,696$
4	$26 \cdot 36 \cdot 36 \cdot 36$	$= 26 \cdot 36^3$	$= 1,213,056$
5	$26 \cdot 36 \cdots 36$	$= 26 \cdot 36^4$	$= 43,670,016$
6	$26 \cdot 36 \cdots 36$	$= 26 \cdot 36^5$	$= 1,572,120,576$
7	$26 \cdot 36 \cdots 36$	$= 26 \cdot 36^6$	$= 56,596,340,736$
8	$26 \cdot 36 \cdots 36$	$= 26 \cdot 36^7$	$= \underline{2,037,468,266,496}$
			$2,095,681,645,538$

44. a. The number of handshakes is the number of ways 2 people can be chosen from 5 (i.e., the number of possible pairs), $_5C_2 = 5!/(2!3!) = 5 \cdot 4/2! = 10$.
   b. The number of handshakes is the number of ways 2 people can be chosen from n (i.e., the number of possible pairs), $_nC_2 = n!/[2!(n-2)!] = n \cdot (n-1)/2! = n(n-1)/2$.

45. Visualize the people entering one at a time, each person sitting to the right of the person who entered before him. Where the first person sits is irrelevant -- i.e., it merely establishes a reference point but does not affect the number of seating arrangements.
   a. Once the first person sits, there are 2 possibilities for the person at his right. Once the second person sits, there is only one possibility for the person at his right. Therefore, there are $2 \cdot 1 = 2$ possible arrangements.
   b. Once the first person sits, there are n-1 possibilities for the person at his right. Once the second person sits, there are n-2 possibilities for the person at his right...[and so on]. Once the next to last person sits, there is only one possibility for the person at his right. Therefore, there are $(n-1) \cdot (n-2) \cdots 1 = (n-1)!$ possible arrangements.

46. a. The total number of such samples possible is $_{60}C_5 = 60!/(5!55!) = (60 \cdot 59 \cdot 58 \cdot 57 \cdot 56)/5! = 5,461,512$. Since all such samples are equally likely, the probability of selecting any one of them is $1/5,461,512$.
   b. In general, the probability is $1/[N!/n!(N-n)!] = n!(N-n)!/N!$.

47. The calculator factorial key gives $50! = 3.04140932 \times 10^{64}$.
   Using the approximation, $K = (50.5) \cdot \log(50) + .39908993 - .43429448(50)$
   $$= 85.79798522 + .39930993 - 21.71472400$$
   $$= 64.48257115$$
   And then $n! = 10^K$
   $$= 10^{64.48257115}$$
   $$= 3.037883739 \times 10^{64}$$
   NOTE: The two answers differ by 3.5×10^{61} (i.e., by 35 followed by 60 zeros -- "zillions and zillions"). While that error may seem large, the numbers being dealt with are very large and that difference is only a $(3.5 \times 10^{61})/(3.04 \times 10^{64}) = 1.2\%$ error.

48. $n = 300$
   $K = (n+.5)(\log n) + .39908993 - .43429448n$
   $= (300.5)(\log 300) + .39908993 - .43429448(300)$
   $= 614.4$
   $300! = 10^{614.4}$
   Since the number of digits in 10^x is the next whole number above x, the number of digits in $300!$ is 615.

49. a. Assuming the judge knows there are to be 4 computers and 4 humans and frames his guess accordingly, the number of possible ways he could guess is the number of ways 4 of the 8 could be labeled as "computer" and is given by $_8C_4 = 8!/(4!4!)$
$$= (8 \cdot 7 \cdot 6 \cdot 5)/4!$$
$$= 70$$

The probability he guesses the right combination by chance alone is therefore 1/70.
NOTE: If the judge merely guesses on each one individually, either not knowing or not considering that there should be 4 computers and 4 humans, then the probability he guesses all 8 correctly is $(1/2)^8 = 1/256$.

   b. Under the assumptions of part a, the probability that all 10 judges make all correct guesses is $(1/70)^8 = 3.54 \times 10^{-19}$.
NOTE: In the statement of the problem, the author states the Turing criterion to be "the person believes he or she is communicating with another person instead of a computer." In the comment in parentheses in part (b), he appears to take the criterion to be "the person cannot tell whether he or she is communicating with another person or with a computer." These are not the same criteria. Part of the problem is the ambiguous directive that the judges "cannot distinguish between computers and people" -- which could mean either "the judges recognize people, but also think the computers are people" or "the judges aren't sure whether any one of the communicators is a person or a computer." This, coupled with the NOTE in part (a), indicates some of the complications that often arise in real research and the necessity for precision in terminology before attempting any statistical analysis.

Review Exercises

1. let M = the driver is a male
   a. P(all males) = $P(M_1$ and M_2 and M_3 and M_4 and $M_5)$
   $$= P(M_1) \cdot P(M_2) \cdot P(M_3) \cdot P(M_4) \cdot P(M_5)$$
   $$= (.90)(.90)(.90)(.90)(.90)$$
   $$= .590$$
   b. P(at least one female) = $P(\overline{\text{all males}})$
   $$= 1 - P(\text{all males})$$
   $$= 1 - .590$$
   $$= .410$$

2. let W = a component works properly
   $P(W_1) = 1 - .010 = .990$; $P(W_2) = 1 - .005 = .995$; $P(W_3) = 1 - .012 = .988$
   P(at least one fails) = 1 - P(all work properly)
   $$= 1 - P(W_1 \text{ and } W_2 \text{ and } W_3)$$
   $$= 1 - P(W_1) \cdot P(W_2) \cdot P(W_3) \quad \text{[assuming independence]}$$
   $$= 1 - (.990)(.995)(.988)$$
   $$= 1 - .973$$
   $$= .027$$

3. let S = a hang glider survives for the year
   $P(S) = 1 - .008$
   $$= .992$$
   a. P(all survive) = $P(S_1$ and S_2 and...and $S_{10})$
   $$= P(S_1) \cdot P(S_2) \cdots P(S_{10})$$
   $$= (.992)(.992)...(.992) = (.992)^{10} = .923$$
   b. P(at least does not survive) = $P(\overline{\text{all survive}}) = 1 - P(\text{all survive}) = 1 - .923 = .077$

4. Refer to the chart and notation of exercise #5.
   $P(Y) = 98/150 = .653$

5. It may be helpful to make a chart like the one given at the right.

		SEAT BELTS?		
		Yes	No	
SEX	Male	29	21	50
	Female	69	31	100
		98	52	150

let F = a person is female

Y = a person uses seat belts

P(F or Y) = P(F) + P(Y) - P(F and Y)

= 100/150 + 98/150 - 69/150 = 129/150 = .86

6. $10 \cdot 10 \cdot 10 \cdot 10 = 10,000$

7. Since 2945 of the 6665 films have an R rating, the probability that a single randomly selected film is one with an R rating is 2945/6665 = .442.

8. Since 873 + 2505 = 3378 of the 6665 films have a G or PG rating, the probability that a single randomly selected film is one with such a rating is 3378/6665 = .507.

9. The number of different possible combinations of 6 such numbers is

$_{40}C_6 = 40!/(34!6!) = (40 \cdot 39 \cdot 38 \cdot 37 \cdot 36 \cdot 35)/6! = 3,838,380$

The probability of winning with one random selection is therefore 1/3,838,380.

10. Ignoring order yields $_{54}C_6 = 54!/(6!48!) = 54 \cdot 53 \cdot 52 \cdot 51 \cdot 50 \cdot 49/6! = 25,827,165$ possibilities.
Counting order yields $_{54}P_6 = 54!/48! = 54 \cdot 53 \cdot 52 \cdot 51 \cdot 50 \cdot 49 = 18,595,558,800$ possibilities.
The desired probability is 1/18,595,558,800.

11. let D = the selected circuit is defective

a. $P(D_1 \text{ and } D_2) = P(D_1) \cdot P(D_2 \mid D_1) = (18/120) \cdot (18/120) = .0225$

NOTE: The "and" formula used above <u>always</u> works. When the events are independent, $P(D_2 \mid D_1) = P(D_2) = P(D_1) = P(D)$ is always 18/120 and the formula for independent events (really a special case of the above formula) can be used.

$P(D_1 \text{ and } D_2) = P(D_1) \cdot P(D_2) = (18/120) \cdot (18/120) = .0225$

b. $P(D_1 \text{ and } D_2) = P(D_1) \cdot P(D_2 \mid D_1) = (18/120) \cdot (17/119) = .0214$

12. let A = a divorced woman is awarded alimony

P(A) = 1 - .85 = .15, for each selection

P(at least one not awarded alimony) = 1 - P(all awarded alimony)

$= 1 - (.15)_4 = 1 - .000506 = .999494$

13. let R = a Republican is selected

P(R) = .30 for each selection

$P(R_1 \text{ and } R_2 \text{ and...and } R_{12}) = P(R_1) \cdot P(R_2) \cdots P(R_{12})$

$= (.30) \cdot (.30) \cdots (.30) = (.30)^{12} = .000000531$

NOTE: P(R) = .30 for each selection assumes independence, which technically is not so. If selecting a voter means that voter cannot be selected again during subsequent selections,

$P(R_1) = 60,000/200,000 = .3000000$

$P(R_2) = 59,999/199,999 = .2999965$

$P(R_3) = 59,998/199,998 = .2999930$

and so on

The guideline in the text suggests ignoring the lack of independence whenever the total sample is less than 5% of the population. In this case, the sample is 12/200,000 = .006% of the population, well within the suggested limits for proceeding under the assumption of independence.

14. let G = the child is a girl

P(G) = .5, for each birth

$P(G_1 \text{ and } G_2 \text{ and } G_3 \text{ and } G_4 \text{ and } G_5) = [P(G)]^5 = [.5]^5 = .03125$

15. a. for independent events, $P(A \text{ and } B) = P(A) \cdot P(B) = (.2) \cdot (.4) = .08$

b. for mutually exclusive events, $P(A \text{ or } B) = P(A) + P(B) = .2 + .4 = .6$

c. in all cases, $P(\bar{A}) = 1 - P(A) = 1 - .2 = .8$

16. a. $0! = 1$ [by definition]

b. $8! = 8 \cdot 7 \cdot 6 \cdot 5 \cdot 4 \cdot 3 \cdot 2 \cdot 1 = 40,320$

c. $_8P_6 = 8!/(8-6)! = 8!/2! = 8 \cdot 7 \cdot 6 \cdot 5 \cdot 4 \cdot 3 = 20,160$

d. $_{10}C_8 = 10!/(8!2!) = 10 \cdot 9/2! = 45$

e. $_{80}C_{78} = 80!/(78!2!) = 80 \cdot 79/2! = 3160$

17. There are $5! = 5 \cdot 4 \cdot 3 \cdot 2 \cdot 1 = 120$ possible orderings, only one of which is correct. The probability of obtaining the correct ordering by a single random guess, therefore, is 1/120.

18. let N = a woman aged 18 to 24 does not live at home with her parents

P(N) = .52, for each selection

$P(N_1 \text{ and } N_2 \text{ and } N_3 \text{ and } N_4 \text{ and } N_5) = [P(N)]^5 = [.52]^5 = .0380$

19. let F = the applicant selected is female

    M = the applicant is male

P(F) = P(M) = .5, for each selection

a. P(all women) = $P(F_1 \text{ and } F_2 \text{ and } F_3 \text{ and } F_4)$

$= P(F_1) \cdot P(F_2) \cdot P(F_3) \cdot P(F_4)$

$= (.5) \cdot (.5) \cdot (.5) \cdot (.5)$

$= .0625$

b. P(all men) = $P(M_1 \text{ and } M_2 \text{ and } M_3 \text{ and } M_4)$

$= P(M_1) \cdot P(M_2) \cdot P(M_3) \cdot P(M_4)$

$= (.5) \cdot (.5) \cdot (.5) \cdot (.5)$

$= .0625$

c. P(same sex) = P(all women or all men)   [NOTE: these events are mutually exclusive]

$= $ P(all women) + P(all men)

$= .0625 + .0625$

$= .1250$

20. let N = a person has not witnessed a robbery or mugging

P(N) = 1 - .19 = .81, for each selection

$P(N_1 \text{ and } N_2) = P(N_1) \cdot P(N_2) = (.81)(.81) = .656$

Chapter 4

Probability Distributions

4-2 Random Variables

1. This is a probability distribution since $\Sigma P(x)=1$ is true and $0 \leq P(x) \leq 1$ is true for each x.

x	P(x)	x·P(x)	x^2	x^2·P(x)	
4	.120	.480	16	1.920	$\mu = \Sigma x \cdot P(x)$
5	.253	1.265	25	6.325	$= 5.9$
6	.217	1.302	36	7.812	$\sigma^2 = \Sigma x^2 \cdot P(x) - \mu^2$
7	.410	2.870	49	20.090	$= 36.147 - (5.917)^2$
	1.000	5.917		36.147	$= 1.136$, rounded to 1.1
					$\sigma = 1.1$

NOTE: Several important statements should be kept in mind when working with probability distributions and the above formulas.
* If one of the conditions for a probability distribution does not hold, the formulas do not apply and produce numbers that have no meaning.
* $\Sigma x \cdot P(x)$ gives the mean of the x values and must be a number between the highest and lowest x values.
* $\Sigma x^2 \cdot P(x)$ gives the mean of the x^2 values and must be a number between the highest and lowest x^2 values.
* $\Sigma P(x)$ must always equal 1.000.
* Σx and Σx^2 have no meaning and should not be calculated.
* The quantity $[\Sigma x^2 \cdot P(x) - \mu^2]$ cannot possibly be negative; if it is, then there is a mistake.
* Always be careful to use the <u>unrounded</u> mean in the calculation of the variance and to take the square root of the <u>unrounded</u> variance to find the standard deviation.

2. This is <u>not</u> a probability distribution since $\Sigma P(x) = .96 \neq 1.00$.

3. This is <u>not</u> a probability distribution since $\Sigma P(x) = .78 \neq 1.00$.

4. This is a probability distribution since $\Sigma P(x)=1$ is true and $0 \leq P(x) \leq 1$ is true for each x.

x	P(x)	x·P(x)	x^2	x^2·P(x)	
0	.2	0	0	0	$\mu = \Sigma x \cdot P(x)$
1	.2	.2	1	.2	$= 2.0$
2	.2	.4	4	.8	$\sigma^2 = \Sigma x^2 \cdot P(x) - \mu^2$
3	.2	.6	9	1.8	$= 6.0 - (2.0)^2$
4	.2	.8	16	3.2	$= 2.0$
	1.0	2.0		6.0	$\sigma = 1.4$

5. This is a probability distribution since $\Sigma P(x)=1$ is true and $0 \leq P(x) \leq 1$ is true for each x.

x	P(x)	x·P(x)	x^2	x^2·P(x)	
0	.125	0	0	0	$\mu = \Sigma x \cdot P(x)$
1	.375	.375	1	.375	$= 1.5$
2	.375	.750	4	1.500	$\sigma^2 = \Sigma x^2 \cdot P(x) - \mu^2$
3	.125	.375	9	1.125	$= 3.000 - (1.500)^2$
	1.000	1.500		3.000	$= .75$
					$\sigma = .9$

6. This is a probability distribution since $\Sigma P(x)=1$ is true and $0 \le P(x) \le 1$ is true for each x.

x	P(x)	x·P(x)	x²	x²·P(x)	$\mu = \Sigma x \cdot P(x)$
0	.512	0	0	0	$= .73$, rounded to .7
1	.301	.301	1	.301	$\sigma^2 = \Sigma x^2 \cdot P(x) - \mu^2$
2	.132	.264	4	.528	$= 1.324 - (.73)^2$
3	.055	.165	9	.495	$= .791$, rounded to .8
	1.000	.730		1.324	$\sigma = .9$

7. This is a probability distribution since $\Sigma P(x)=1$ is true and $0 \le P(x) \le 1$ is true for each x.

x	P(x)	x·P(x)	x²	x²·P(x)	$\mu = \Sigma x \cdot P(x)$
0	.26	0	0	0	$= 2.8$
1	.16	.16	1	.16	$\sigma^2 = \Sigma x^2 \cdot P(x) - \mu^2$
2	.12	.24	4	.48	$= 14.20 - (2.80)^2$
3	.09	.27	9	.81	$= 6.36$, rounded to 6.4
4	.07	.28	16	1.12	$\sigma = 2.5$
5	.09	.45	25	2.25	
6	.07	.42	36	2.52	
7	.14	.98	49	6.86	
	1.00	2.80		14.20	

8. This is <u>not</u> a probability distribution since $\Sigma P(x) = .765 \ne 1.00$.

9. This is a probability distribution since $\Sigma P(x)=1$ is true and $0 \le P(x) \le 1$ is true for each x.

x	P(x)	x·P(x)	x²	x²·P(x)	$\mu = \Sigma x \cdot P(x)$
0	.36	0	0	0	$= .8$
1	.48	.48	1	.48	$\sigma^2 = \Sigma x^2 \cdot P(x) - \mu^2$
2	.16	.32	4	.64	$= 1.12 - (.80)^2$
	1.00	.80		1.12	$= .48$, rounded to .5
					$\sigma = .7$

10. This is a probability distribution since $\Sigma P(x)=1$ is true and $0 \le P(x) \le 1$ is true for each x.

x	P(x)	x·P(x)	x²	x²·P(x)	$\mu = \Sigma x \cdot P(x)$
0	.960	0	0	0	$= .044$
1	.036	.036	1	.036	$\sigma^2 = \Sigma x^2 \cdot P(x) - \mu^2$
2	.004	.008	4	.016	$= .052 - (.044)^2$
	1.000	.044		.052	$= .050$
					$\sigma = .224$

NOTE: Since the relevant numbers are so small (following the usual guideline and rounding to one more place than the original data would produce $\mu = 0.0$), the requested values are given with three decimal accuracy.

11. This is a probability distribution since $\Sigma P(x)=1$ is true and $0 \le P(x) \le 1$ is true for each x.

x	P(x)	x·P(x)	x²	x²·P(x)	$\mu = \Sigma x \cdot P(x)$
0	.026	0	0	0	$= 2.011$, rounded to 2.0
1	.345	.345	1	.345	$\sigma^2 = \Sigma x^2 \cdot P(x) - \mu^2$
2	.346	.692	4	1.384	$= 5.179 - (2.011)^2$
3	.154	.462	9	1.386	$= 1.139$, rounded to 1.1
4	.129	.516	16	2.064	$\sigma = 1.1$
	1.000	2.011		5.179	

12. This is a probability distribution since $\Sigma P(x)=1$ is true and $0\le P(x)\le 1$ is true for each x.

x	P(x)	x·P(x)	x^2	x^2·P(x)	
0	.245	0	0	0	$\mu = \Sigma x \cdot P(x)$
1	.370	.370	1	.370	$= 1.395$, rounded to 1.4
2	.210	.420	4	.840	$\sigma^2 = \Sigma x^2 \cdot P(x) - \mu^2$
3	.095	.285	9	.855	$= 3.345 - (1.395)^2$
4	.080	.320	16	1.280	$= 1.399$, rounded to 1.4
	1.000	1.395		3.345	$\sigma = 1.2$

13.

x	P(x)	x·P(x)	
500	.25	125	$E = \Sigma x \cdot P(x)$
200	.75	150	$= \$275$
	1.00	275	

14.

x	P(x)	x·P(x)	
200	.1	20	$E = \Sigma x \cdot P(x)$
0	.6	0	$= -\$70$ (i.e., a <u>loss</u> of $70)
-300	.3	-90	
	1.00	-70	

15.

x	P(x)	x·P(x)	
-30	.15	-4.50	$E = \Sigma x \cdot P(x)$
120	.85	102.00	$= \$97.50$
	1.00	97.50	

NOTE: Giving precedence to the standard notation for reporting monetary values over the rounding guideline stated in the text, we give two decimal place accuracy in the answer.

16.

x	P(x)	x·P(x)	
175,000	.7	122,500	$E = \Sigma x \cdot P(x)$
0	.3	0	$= \$122,500$
	1.0	122,500	

Yes, the bid should be submitted. Not only is E much greater than zero, but it is also true that he cannot lose money; the worst he can do is pay all the bills, pay all the help, and break even.

17. This problem can be worked in two different ways.

* Considering the $156 she paid up front, the woman loses 156 if she lives and gains 100,000 - 156 = 99844 if she dies.

x	P(x)	x·P(x)	
-156	.9995	-155.92	$E = \Sigma x \cdot P(x)$
99844	.0005	44.92	$= -\$106.00$ (i.e., a loss of $106)
	1.0000	-106.00	

* Ignoring the cost of the insurance, one may calculate the expected value of the policy itself.

x	P(x)	x·P(x)	
0	.9995	0	$E = \Sigma x \cdot P(x)$
100000	.0005	50.00	$= \$50.00$
	1.0000	50.00	

If the policy is worth $50 in expected returns and the woman pays $156 to purchase it, her net expectation is 50 - 156 = -$106.

18.

x	P(x)	x·P(x)
5,000,000	.000000004	.024875621
150,000	.000000004	.000746268
100,000	.000000004	.000497512
25,000	.000000009	.000248756
10,000	.000000019	.000199004
5,000	.000000039	.000199004
200	.000000124	.000024875
125	.000000995	.000124378
89	.000264970	.023582405
0	.999733832	0
	1.000000000	.050497823

a. E = Σx·P(x)
    = .05 (i.e., 5 cents)

b. mean loss
    = price of stamp - 5¢
    = 29¢ - 5¢
    = 24¢
    (for first class stamps
    as of fall 1994)

19. The 8 equally like outcomes in the sample
space are given at the right. If x
represents the number of girls, counting
the numbers of favorable outcomes indicates
    P(x = 0) = 1/8 = .125
    P(x = 1) = 3/8 = .375
    P(x = 2) = 3/8 = .375
    P(x = 3) = 1/8 = .125

outcome	x
BBB	0
BBG	1
BGB	1
GBB	1
GGB	2
GBG	2
BGG	2
GGG	3

x	P(x)	x·P(x)	x^2	x^2·P(x)
0	.125	0	0	0
1	.375	.375	1	.375
2	.375	.750	4	1.500
3	.125	.375	9	1.125
	1.000	1.500		3.000

μ = Σx·P(x)
    = 1.5
σ^2 = Σx^2·P(x) - μ^2
    = 3.000 - $(1.500)^2$
    = .75
σ = .9

20. The 16 equally like outcomes in the sample
space are given at the right. If x
represents the number of boys, counting
the numbers of favorable outcomes indicates
    P(x = 0) = 1/16 = .0625
    P(x = 1) = 4/16 = .2500
    P(x = 2) = 6/16 = .3750
    P(x = 3) = 4/16 = .2500
    P(x = 4) = 1/16 = .0625

outcome	x	outcome	x
BBBB	4	GBGB	2
BBBG	3	GBBG	2
BBGB	3	GGBB	2
BGBB	3	BGGG	1
GBBB	3	GBGG	1
BBGG	2	GGBG	1
BGBG	2	GGGB	1
BGGB	2	GGGG	0

x	P(x)	x·P(x)	x^2	x^2·P(x)
0	.0625	0	0	0
1	.2500	.2500	1	.2500
2	.3750	.7500	4	1.5000
3	.2500	.7500	9	2.2500
4	.0625	.2500	16	1.0000
	1.0000	2.0000		5.0000

μ = Σx·P(x)
    = 2.0
σ^2 = Σx^2·P(x) - μ^2
    = 5.0000 - $(2.0000)^2$
    = 1.0
σ = 1.0

21. There are 4 possible outcomes, but they are not equally likely. If G is a good (i.e., "made")
shot, B is a bad (i.e., "missed") shot, and x represents the number of shots made, then
    P(G) = .85
    P(B) = .15
and the following summary table applies.

outcome	probability						x
GG	$P(G_1$ and $G_2)$	$= P(G_1) \cdot P(G_2)$	$= (.85)(.85)$	$= .7225$	2
GB	$P(G_1$ and $B_2)$	$= P(G_1) \cdot B(G_2)$	$= (.85)(.15)$	$= .1275$	1
BG	$P(B_1$ and $G_2)$	$= P(B_1) \cdot P(G_2)$	$= (.15)(.85)$	$= .1275$	1
BB	$P(B_1$ and $B_2)$	$= P(B_1) \cdot P(B_2)$	$= (.15)(.15)$	$=\underline{~.0225~}$	0
				1.0000

x	P(x)	$x \cdot P(x)$	x^2	$x^2 \cdot P(x)$
0	.0225	0	0	0
1	.2550	.2550	1	.2550
2	.7225	1.4450	4	2.8900
	1.0000	1.7000		3.1450

$\mu = \Sigma x \cdot P(x)$
$= 1.7$
$\sigma^2 = \Sigma x^2 \cdot P(x) - \mu^2$
$= 3.1450 - (1.7000)^2$
$= .2550$, rounded to .3
$\sigma = .5$

22. Let x be the number of defective chips.
$P(x = 0) = P(G_1$ and $G_2) = P(G_1) \cdot P(G_2 \mid G_1) = (14/20) \cdot (13/19) = .479$
$P(x = 2) = P(D_1$ and $D_2) = P(D_1) \cdot P(D_2 \mid G_1) = (6/20) \cdot (5/19) = .079$
$P(x = 1) = 1 - [.479 + .079] = .442$

x	P(x)	$x \cdot P(x)$	x^2	$x^2 \cdot P(x)$
0	.479	0	0	0
1	.442	.442	1	.442
2	.079	.158	4	.316
	1.000	.600		.758

$\mu = \Sigma x \cdot P(x)$
$= .6$
$\sigma^2 = \Sigma x^2 \cdot P(x) - \mu^2$
$= .758 - (.600)^2$
$= .398$, rounded to .4
$\sigma = .6$

23. In order for P(x) to be a probability distribution, it must be true that
(1) $\Sigma P(x) = 1$
(2) $0 \le P(x) \le 1$ for every permissible value of x.
NOTE: Parts (a) and (b) below use the fact from algebra that
$1 + r + r^2 + r^3 + r^4 + ... = 1/(1-r)$ for any r such that $-1 < r < 1$

a. P(x) is a probability distribution because for x = 1,2,3...
(1) $\Sigma P(x) = \Sigma(.4)(.6)^{x-1}$
$= (.4) \cdot \Sigma(.6)^{x-1}$
$= (.4) \cdot [1 + .6 + .6^2 + .6^3 + ...]$
$= (.4) \cdot [1/(1-.6)]$
$= (.4) \cdot [1/.4]$
$= 1$
(2) $0 \le P(x) \le 1$, since P(x) > 0 for all x
and P(x) = .4 for x = 1
and P(x) keeps decreasing as x grows larger

b. P(x) is a probability distribution because for x = 1,2,3...
(1) $\Sigma P(x) = \Sigma(.5)^x$
$= (.5) \cdot \Sigma(.5)^{x-1}$
$= (.5) \cdot [1 + .5 + .5^2 + .5^3 + ...]$
$= (.5) \cdot [1/(1-.5)]$
$= (.5) \cdot [1/.5]$
$= 1$
(2) $0 \le P(x) \le 1$, since P(x) > 0 for all x
and P(x) = .5 for x = 1
and P(x) keeps decreasing as x grows larger

c. P(x) is not a probability distribution since for x = 1,2,3,...

(1) $\Sigma P(x) = \Sigma[1/(2x)]$

$= 1/2 + 1/4 + 1/6 + 1/8 + ...$

$= .5000 + .2500 + .1666 + .1250 + ...$

$= 1.0416 + ...$

> 1

(2) [Condition (2) happens to be satisfied, but that is irrelevant since condition (1) is not satisfied and both conditions must be met to have a probability distribution.]

24. Let x be the number of acceptable chips.

$P(x = 0) = (.65)^4 = .179$

$P(x = 4) = (.35)^4 = .015$

$P(x = 3) = 1 - [.179 + .384 + .311 + .015] = .111$

25. a. Exercise 21(c) of section 2-5 notes that "multiplying each score by a constant also multiplies the distances between scores by that constant, and hence the measures of dispersion in the original units are multiplies by that constant (and the variance is multiplied by the square of that constant)." Multiplying each score by 5, therefore, multiplies the variance by $5^2 = 25$ and the new variance is $25 \cdot (1.25) = 31.25$.

b. Refer to part (a). Dividing by 5 is the same as multiplying by 1/5. Multiplying each score by 1/5, therefore, multiplies the variance by $(1/5)^2 = 1/25 = .04$ and the new variance is $.04 \cdot (1.25) = .05$.

c. Exercise 21(a) of section 2-5 notes that "adding a constant to each score does not affect the spread of the scores of the measures of dispersion." Adding 5 to each score, therefore, does not change the variance and it remains 1.25.

d. Exercise 21(b) of section 2-5 notes that "subtracting a constant from each score does not affect the spread of the scores of the measures of dispersion." Subtracting 5 from each score, therefore, does not change the variance and it remains 1.25. NOTE: An alternative answer is "Refer to part (c). Subtracting 5 from each score is the same as adding -5. Adding -5 to each score, therefore, does not change the variance and it remains 1.25."

26.

x	P(x)	x·P(x)	x^2	x^2·P(x)
1	1/n	1/n	1	1/n
2	1/n	2/n	4	4/n
3	1/n	3/n	9	9/n
.	.	.	.	.
.	.	.	.	.
.	.	.	.	.
n	1/n	n/n	n^2	n^2/n
	1	[Σx]/n		[Σx²]/n

$\mu = \Sigma x \cdot P(x)$

$= (n+1)/2$

$\sigma^2 = \Sigma x^2 \cdot P(x) - \mu^2$

$= (n+1)(2n+1)/6 - (n+1)^2/4$

$= 2(n+1)(2n+1)/12 - 3(n+1)^2/12$

$= [(n+1)/12] \cdot [2(2n+1) - 3(n+1)]$

$= [(n+1)/12] \cdot [n-1]$

$= (n+1)(n-1)/12$

$= (n^2-1)/12$

$[\Sigma x]/n = [n(n+1)/2]/n = (n+1)/2$

$[\Sigma x^2]/n = [n(n+1)(2n+1)/6]/n = (n+1)(2n+1)/6$

27. The equivalence between the two expressions may be demonstrated in 7 steps as follows.

(1) $\Sigma(x-\mu)^2 \cdot P(x) = \Sigma(x^2 - 2\mu x + \mu^2) \cdot P(x)$

(2) $= \Sigma(x^2 \cdot P(x) - 2\mu x \cdot P(x) + \mu^2 \cdot P(x))$

(3) $= \Sigma x^2 \cdot P(x) - \Sigma 2\mu x \cdot P(x) + \Sigma \mu^2 \cdot P(x)$

(4) $= \Sigma x^2 \cdot P(x) - 2\mu \cdot \Sigma x \cdot P(x) + \mu^2 \cdot \Sigma P(x)$

(5) $= \Sigma x^2 \cdot P(x) - 2\mu \cdot \mu + \mu^2 \cdot \Sigma P(x)$

(6) $= \Sigma x^2 \cdot P(x) - 2\mu^2 + \mu^2$

(7) $= \Sigma x^2 \cdot P(x) - \mu^2$

The algebraic justification for each of the preceding steps is as follows.
(1) $(a-b)^2 = a^2 - 2ab + b^2$
(2) $(a-b-c)d = ad - bd - cd$
(3) $\Sigma(A - B + C) = \Sigma A - \Sigma B + \Sigma C$
(4) $\Sigma cX = c\Sigma X$, for any constant c [and μ is a constant]
(5) $\Sigma x \cdot P(x) = \mu$
(6) $\Sigma P(x) = 1$, for any probability distribution
(7) $-2a + a = -a$

4-3 Binomial Experiments

NOTE: To use the binomial formula, one must identify 3 quantities: n,x,p. Table A-1, for example, requires only these 3 values to supply a probability. Since what the text calls "q" always equals 1-p, it can be so designated without introducing unnecessary notation [just as no special notation is utilized for the quantity n-x, even though it appears twice in the binomial formula]. Accordingly, in the interest of simplicity, this manual uses the formula

$$P(x) = [n!/x!(n-x)!] \cdot p^x \cdot (1-p)^{n-x}$$

without introducing a special name for the quantity 1-p. This has the additional advantage of ensuring that the probabilities p and 1-p sum to 1.00 and protecting against an error in the separate calculation and/or identification of "q." Interestingly, this 1-p notation is employed in the text when presenting the geometric distribution in exercise #33 of this section.

1. The 4 requirements are:
    #1 There are a fixed number of trials.
    #2 The trials are independent.
    #3 Each trial has two possible named outcomes.
    #4 The probabilities remain constant for each trial.
   a. Yes, all 4 requirements are met.
   b. No, requirements #2 and #4 are not met. The probability the first lens is defective is 8/20. The probability the second lens is defective is either 8/19 or 7/19, depending on the first selection.
   c. Yes, all 4 requirements are met.
   d. Yes, all 4 requirements are met.
   e. Yes, all 4 requirements are met. NOTE: Assuming the selection process and/or the actual conducting of the experiment would not all ow the same person to be selected twice, the scenario is very technically not a binomial experiment. Assuming the population under consideration is so large that removing the people already sampled from the population does not appreciably affect the population or the probabilities, however, the binomial formulas may be applied.

2. The 4 requirements are restated in exercise #1 above.
   a. No, requirement #3 is not met. There are more than two possible responses.
   b. Yes, all 4 requirements are met.
   c. Yes, all 4 requirements are met.
   d. No, requirement #3 is not met. There are more than two possible responses.
   e. Yes, all 4 requirements are met.

3. From table A-1, .101.

4. From table A-1, .735.

5. From table A-1, .349.

6. From table A-1, .172.

7. $P(x) = [n!/x!(n-x)!] \cdot p^x \cdot (1-p)^{n-x}$
$= [5!/3!2!] \cdot (1/4)^3 \cdot (3/4)^2$
$= [10] \cdot (1/64) \cdot (9/16)$
$= 90/1024$ or $45/512$

8. $P(x) = [n!/x!(n-x)!] \cdot p^x \cdot (1-p)^{n-x}$
$= [4!/2!2!] \cdot (1/3)^2 \cdot (2/3)^2$
$= [6] \cdot (1/9) \cdot (4/9)$
$= 24/81$ or $8/27$

9. $P(x) = [n!/x!(n-x)!] \cdot p^x \cdot (1-p)^{n-x}$
$= [6!/2!4!] \cdot (1/2)^2 \cdot (1/2)^4$
$= [15] \cdot (1/4) \cdot (1/16)$
$= 15/64$

10. $P(x) = [n!/x!(n-x)!] \cdot p^x \cdot (1-p)^{n-x}$
$= [4!/4!0!] \cdot (2/3)^4 \cdot (1/3)^0$
$= [1] \cdot (16/81) \cdot (1)$
$= 16/81$

11. let x = the number of girls; n = 10, x = 4, p = .5
$P(x) = [n!/x!(n-x)!] \cdot p^x \cdot (1-p)^{n-x}$
$= [10!/4!6!] \cdot (.5)^4 \cdot (.5)^6$
$= [210] \cdot (.0625) \cdot (.015625)$
$= .205$

IMPORTANT NOTE: The intermediate values of 210, .0625 and .015625 are given above to help students with an incorrect answer to identify the portion of the problem in which the mistake was made. This practice will be followed in all problems (i.e., not just binomial problems) throughout the manual. In practice, all calculations can be done in one step on the calculator. You may choose to (or be asked to) write down such intermediate values for your own (or the instructor's) benefit, but never round off in the middle of a problem. Do not write the values down on paper and then re-enter them in the calculator -- use the memory to let the calculator remember with complete accuracy any intermediate values that will be used in subsequent calculations. In addition, always make certain that the quantity [n!/x!(n-x)!] is a whole number and that the final answer is between 0 and 1.

12. n = 10, x = 6, p = .6
$P(x) = [n!/x!(n-x)!] \cdot p^x \cdot (1-p)^{n-x}$
$= [10!/6!4!] \cdot (.6)^6 \cdot (.4)^4$
$= .251$

13. n = 3, x = 2, p = .01
$P(x) = [n!/x!(n-x)!] \cdot p^x \cdot (1-p)^{n-x}$
$= [3!/2!1!] \cdot (.01)^2 \cdot (.99)^1$
$= [3] \cdot (.0001) \cdot (.99)$
$= .000297$

14. n = 3, x = 2, p = .01
$P(x) = [n!/x!(n-x)!] \cdot p^x \cdot (1-p)^{n-x}$
$= [3!/2!1!] \cdot (.01)^2 \cdot (.99)^1$
$= [3] \cdot (.0001) \cdot (.99)$
$= .000297$

15. let x = the number that are late; n = 10, x = 2 or more, p = .10
NOTE: Since x represents the number that are late (i.e., being late is considered a "success"), then p must represent the probability of being late. Beginning each problem with a statement identifying what x represents helps to avoid errors caused by inconsistent identifications.
$$P(x \geq 2) = 1 - P(x \leq 1)$$
$$= 1 - [P(x=0) + P(x=1)]$$
$$= 1 - [.349 + .387] \quad \text{[from Table A-1]}$$
$$= 1 - .736$$
$$= .264$$

16. let x = the number that must pay more money; n = 8, x = at least 6, p = .70
$$P(x \geq 6) = P(x=6) + P(x=7) + P(x=8)$$
$$= .296 + .198 + .058 \quad \text{[from Table A-1]}$$
$$= .552$$

17. let x = the number that arrive for the flight; n = 14, x = 13 or more, p = .90
$$P(x \geq 13) = P(x=13) + P(x=14)$$
$$= .356 + .229 \quad \text{[from Table A-1]}$$
$$= .585$$

18. let x = the number that recognize Coke; n = 12, x = 11, p = .95
$$P(x) = [n!/x!(n-x)!] \cdot p^x \cdot (1-p)^{n-x}$$
$$= [12!/11!1!] \cdot (.95)^{11} \cdot (.05)^1$$
$$= .341$$

19. let x = the number that survive; n = 20, x = 18, p = .95
$$P(x) = [n!/x!(n-x)!] \cdot p^x \cdot (1-p)^{n-x}$$
$$= [20!/18!2!] \cdot (.95)^{18} \cdot (.05)^2$$
$$= [190] \cdot (.3972) \cdot (.0025)$$
$$= .189$$

20. let x = the number experiencing at least one burglary; n = 30, x = fewer than 2, p = .05
$$P(x<2) = P(x=0) + P(x=1)$$
$$= [30!/0!30!] \cdot (.05)^0 \cdot (.95)^{30} + [30!/1!29!] \cdot (.05)^1 \cdot (.95)^{29}$$
$$= .215 + .339$$
$$= .554$$

21. let x = the number with HS diploma but no college; n = 15, x = at least 10, p = .40
$$P(x \geq 10) = P(x=10) + P(x=11) + P(x=12) + P(x=13) + P(x=14) + P(x=15)$$
$$= .024 + .007 + .002 + .000^+ + .000^+ + .000^+ \quad \text{[from Table A-1]}$$
$$= .033$$

22. let x = the number opposed; n = 15, x = fewer than 3, p = .20
$$P(x<3) = P(x=0) + P(x=1) + P(x=2)$$
$$= .035 + .132 + .231 \quad \text{[from Table A-1]}$$
$$= .398$$

23. let x = the number living within 50 miles of a coastal shoreline; n = 20, x = 12, p = .53
$$P(x) = [n!/x!(n-x)!] \cdot p^x \cdot (1-p)^{n-x}$$
$$= [20!/12!8!] \cdot (.53)^{12} \cdot (.47)^8$$
$$= [125970] \cdot (.000491) \cdot (.00238)$$
$$= .147$$

24. let x = the number of correct answers; n = 10, x ≥ 6 [i.e., 60% of 10], p = .2 [i.e., 1 of 5]
P(x≥6) = P(x=6) + P(x=7) + P(x=8) + P(x=9) + P(x=10)
$\qquad$ = .006 + .001 + 0^+ + 0^+ + 0^+ $\qquad$ [from Table A-1]
$\qquad$ = .007

25. a. let x = the number that contain errors; n = 50, x = 50, p = .98
P(x) = [n!/x!(n-x)!]·p^x·$(1-p)^{n-x}$
$\qquad$ = [50!/50!0!]·$(.98)^{50}$·$(.02)^0$
$\qquad$ = [1]·(.3642)·(1.0000)
$\qquad$ = .364
b. n = 50, x = 49, p = .98
P(x) = [n!/x!(n-x)!]·p^x·$(1-p)^{n-x}$
$\qquad$ = [50!/49!1!]·$(.98)^{49}$·$(.02)^1$
$\qquad$ = [50]·(.3716)·(.0200)
$\qquad$ = .372

26. let x = the number that use computers; n = 9, x = more than 7, p = .80
P(x>7) = P(x=8) + P(x=9)
$\qquad$ = .302 + .134 $\qquad$ [from Table A-1]
$\qquad$ = .436

27. a. let x = the number that arrive on time; n = 12, x = at least 9, p = .60
P(x≥9) = P(x=9) + P(x=10) + P(x=11) + P(x=12)
$\qquad$ = .142 + .064 + .017 + .002 $\qquad$ [from Table A-1]
$\qquad$ = .225
b. n = 30, x = 20, p = .60
P(x) = [n!/x!(n-x)!]·p^x·$(1-p)^{n-x}$
$\qquad$ = [30!/20!10!]·$(.60)^{20}$·$(.5)^{10}$
$\qquad$ = [30,045,015]·(.0000366)·(.000105)
$\qquad$ = .115

28. a. let x = the number of girls; n = 5, x = at least 4, p = .5
P(x≥4) = P(x=4) + P(x=5)
$\qquad$ = .156 + .031 $\qquad$ [from Table A-1]
$\qquad$ = .187
b. n = 5, x = at least 4, p = 1 - .513 = .487
P(x≥4) = P(x=4) + P(x=5)
$\qquad$ = [5!/4!1!]·$(.487)^4$·$(.513)^1$ + [5!/5!0!]·$(.487)^5$·$(.513)^0$
$\qquad$ = .144 + .027
$\qquad$ = .171

NOTE: In mathematics and statistics, including the Poisson formula, the letter e is commonly used for Euler's constant and has the value (accurate to five decimal places) e = 2.71828. Most calculators have an "e^x" key to find e raised to a power -- or one could also use the y^x key to raise 2.71828 to the desired power. Since any non-zero constant raised to the zero power is 1, e^0 = 1. In the Poisson formula, e will always be raised to a non-positive power and $e^{-\mu}$ will always be between 0 and 1.

29. let x = the number of births per day
μ = 2.25
a. x = 0
P(x) = μ^x·$e^{-\mu}$/x!
$\qquad$ = $(2.25)^0$·$e^{-2.25}$/0!
$\qquad$ = (1)·(.1054)/1
$\qquad$ = .105

b. $x = 1$
  $P(x) = \mu^x \cdot e^{-\mu}/x!$
    $= (2.25)^1 \cdot e^{-2.25}/1!$
    $= (2.25) \cdot (.1054)/1$
    $= .237$
c. $x = 4$
  $P(x) = \mu^x \cdot e^{-\mu}/x!$
    $= (2.25)^4 \cdot e^{-2.25}/4!$
    $= (25.629) \cdot (.1054)/24$
    $= .113$

30. $\mu = .5$; $P(x) = \mu^x \cdot e^{-\mu}/x!$
  a. $P(x=0) = (.5)^0 \cdot e^{-.5}/0! = .607$
  b. $P(x=1) = (.5)^1 \cdot e^{-.5}/1! = .303$
  c. $P(x=2) = (.5)^2 \cdot e^{-.5}/2! = .076$

31. let x = the number of defects
  $\mu = 2.0$
  x = more than 1
  $P(x \geq 1) = 1 - P(x \leq 1)$
    $= 1 - [P(x=0) + P(x=1)]$
    $= 1 - [\mu^0 \cdot e^{-\mu}/0! + \mu^1 \cdot e^{-\mu}/1!]$
    $= 1 - [(2.0)^0 \cdot e^{-2.0}/0! + (2.0)^1 \cdot e^{-2.0}/1!]$
    $= 1 - [(1) \cdot (.1353)/1 + (2.0) \cdot (.1353)/1]$
    $= 1 - [.1353 + .2707]$
    $= 1 - .4060$
    $= .594$

32. $\mu = .2$; $P(x) = \mu^x \cdot e^{-\mu}/x!$
  a. $P(x=0) = (.2)^0 \cdot e^{-.2}/0! = .819$
  b. $P(x=1) = (.2)^1 \cdot e^{-.2}/1! = .164$
  c. $P(x=2) = (.2)^2 \cdot e^{-.2}/2! = .016$

33. let x = the number of components tested to find 1st defect; $x = 7$, $p = .2$
  $P(x) = p \cdot (1-p)^{x-1}$
    $= (.2) \cdot (.8)^6$
    $= (.2) \cdot (.2621)$
    $= .0524$

34. let x = the number of winning selections
  $A = 6$ (winning numbers), $B = 48$ (losing numbers), $n = 6$ (selections)
  $P(x) = [A!/(A-x)!x!] \cdot [B!/(B-n+x)!(n-x)!] \div [(A+B)!/(A+B-n)!n!]$
  a. $P(x=6) = [6!/0!6!] \cdot [48!/48!0!] \div [54!/48!6!]$
        $= [1] \cdot [1] \div [25,827,165]$
        $= .0000000387$
  b. $P(x=5) = [6!/1!5!] \cdot [48!/47!1!] \div [54!/48!6!]$
        $= [6] \cdot [48] \div [25,827,165]$
        $= .0000112$
  c. $P(x=3) = [6!/3!3!] \cdot [48!/45!3!] \div [54!/48!6!]$
        $= [20] \cdot [17296] \div [25,827,165]$
        $= .0134$
  d. $P(x=0) = [6!/6!0!] \cdot [48!/42!6!] \div [54!/48!6!]$
        $= [1] \cdot [12,271,512] \div [25,827,165]$
        $= .475$
  NOTE: The formula is really $[_AC_x] \cdot [_BC_{n-x}] \div [_{A+B}C_n]$ and follows from the methods of Chapter 3.

35. a. Extending the pattern to cover 6 types of outcomes, where $\Sigma x = n$ and $\Sigma p = 1$,

$$P(x_1,x_2,x_3,x_4,x_5,x_6) = [n!/(x_1!x_2!x_3!x_4!x_5!x_6!)] \cdot p_1^{x_1} \cdot p_2^{x_2} \cdot p_3^{x_3} \cdot p_4^{x_4} \cdot p_5^{x_5} \cdot p_6^{x_6}$$

  b. $n = 20$

$x_1 = 5$, $x_2 = 4$, $x_3 = 3$, $x_4 = 2$, $x_5 = 3$, $x_6 = 3$

$p_1 = p_2 = p_3 = p_4 = p_5 = p_6 = 1/6$

$$\begin{aligned}
P(x_1,x_2,x_3,x_4,x_5,x_6) &= [n!/(x_1!x_2!x_3!x_4!x_5!x_6!)] \cdot p_1^{x_1} \cdot p_2^{x_2} \cdot p_3^{x_3} \cdot p_4^{x_4} \cdot p_5^{x_5} \cdot p_6^{x_6} \\
&= [20!/(5!4!3!2!3!3!)] \cdot (1/6)^5 \cdot (1/6)^4 \cdot (1/6)^3 \cdot (1/6)^2 \cdot (1/6)^3 \cdot (1/6)^3 \\
&= [20!/(5!4!3!2!3!3!)] \cdot (1/6)^{20} \\
&= [1.955 \cdot 10^{12}] \cdot (2.735 \cdot 10^{-16}) \\
&= .000535
\end{aligned}$$

36. binomial; let x = the number of defectives in the sample

$n = 25$, $x \leq 1$

  a. $p = .01$

$$\begin{aligned}
P(x \leq 1) &= P(x=0) + P(x=1) \\
&= [25!/0!25!] \cdot (.01)^0 \cdot (.99)^{25} + [25!/1!24!] \cdot (.01)^1 \cdot (.99)^{24} \\
&= .778 + .196 \\
&= .974
\end{aligned}$$

  b. $p = .20$

$$\begin{aligned}
P(x \leq 1) &= P(x=0) + P(x=1) \\
&= [25!/0!25!] \cdot (.20)^0 \cdot (.80)^{25} + [25!/1!24!] \cdot (.20)^1 \cdot (.80)^{24} \\
&= .0038 + .0236 \\
&= .0274
\end{aligned}$$

37. a. binomial; let x = the number of wins in 20 spins

$n = 20$, $x = 1$, $p = 1/38$

$$\begin{aligned}
P(x) &= [n!/x!(n-x)!] \cdot p^x \cdot (1-p)^{n-x} \\
&= [20!/1!19!] \cdot (1/38)^1 \cdot (19/38)^{19} \\
&= [20] \cdot (.0263) \cdot (.6025) \\
&= .317
\end{aligned}$$

  b. Poisson; let x = the number of wins in 20 spins

$\mu = 20/28$

NOTE: If the probability of a win on any one spin is 1/38, then the expected number of wins is n/38 for n spins (i.e., 1/38 for 1 spin, 38/38 = 1 for 38 spins, 20/38 for 30 spins, etc.). Keep in mind that this is the <u>expected number</u> of wins in n spins, not the <u>probability</u> of winning in n spins.

$x = 1$

$$\begin{aligned}
P(x) &= \mu^x \cdot e^{-\mu}/x! \\
&= (20/30)^1 \cdot e^{-20/38}/1! \\
&= (.5263)^1 \cdot e^{-.5263}/1! \\
&= (.5263) \cdot (.5908)/1 \\
&= .311
\end{aligned}$$

4.4 Mean, Variance and Standard Deviation for the Binomial Distribution

1. $\mu = n \cdot p = (49) \cdot (.5) = 24.5$

$\sigma^2 = n \cdot p \cdot (1-p) = (49) \cdot (.5) \cdot (.5) = 12.25$

$\sigma = 3.5$

2. $\mu = n \cdot p = (64) \cdot (.25) = 16.0$

$\sigma^2 = n \cdot p \cdot (1-p) = (64) \cdot (.25) \cdot (.75) = 12.0$

$\sigma = 3.5$

3. $\mu = n \cdot p = (473) \cdot (.855) = 404.4$
   $\sigma^2 = n \cdot p \cdot (1-p) = (473) \cdot (.855) \cdot (.145) = 58.6$
   $\sigma = 7.7$

4. $\mu = n \cdot p = (1067) \cdot (.375) = 400.1$
   $\sigma^2 = n \cdot p \cdot (1-p) = (1067) \cdot (.375) \cdot (.625) = 250.1$
   $\sigma = 15.8$

5. let x = the number of correct answers; n = 50, p = .5
   $\mu = n \cdot p = (50) \cdot (.5) = 25.0$
   $\sigma^2 = n \cdot p \cdot (1-p) = (50) \cdot (.5) \cdot (.5) = 12.5$
   $\sigma = 3.5$

6. let x = the number of correct answers; n = 30, p = .25
   $\mu = n \cdot p = (30) \cdot (.25) = 7.5$
   $\sigma^2 = n \cdot p \cdot (1-p) = (30) \cdot (.25) \cdot (.75) = 5.6$
   $\sigma = 2.4$

7. let x = the number of R-rated films; n = 20, p = 2945/6665 = .4419
   NOTE: Technically, the binomial formulas do not apply. Once a film is selected and removed from the population from which further selections are made, the probability of selecting an R-rated film is no longer 2945/6665. Since removing a sample of 20 from such a large population does not appreciably change the population or the probabilities, however, the binomial formulas may still be applied.
   $\mu = n \cdot p = (20) \cdot (.4419) = 8.8$
   $\sigma^2 = n \cdot p \cdot (1-p) = (20) \cdot (.4419) \cdot (.5581) = 4.9$
   $\sigma = 2.2$

8. let x = the number with regular savings accounts; n = 50, p = .64
   $\mu = n \cdot p = (50) \cdot (.64) = 32.0$
   $\sigma^2 = n \cdot p \cdot (1-p) = (50) \cdot (.64) \cdot (.36) = 11.5$
   $\sigma = 3.4$

9. let x = the number of injuries caused by seat failure; n = 200, p = .47
   $\mu = n \cdot p = (200) \cdot (.47) = 94.0$
   $\sigma^2 = n \cdot p \cdot (1-p) = (200) \cdot (.47) \cdot (.53) = 49.8$
   $\sigma = 7.1$

10. let x = the number sleeping at least 9 hours; n = 36, p = .125
    $\mu = n \cdot p = (36) \cdot (.125) = 4.5$
    $\sigma^2 = n \cdot p \cdot (1-p) = (36) \cdot (.125) \cdot (.875) = 3.9$
    $\sigma = 2.0$

11. let x = the number of returns with taxpayer errors; n = 12, p = .37
    $\mu = n \cdot p = (12) \cdot (.37) = 4.4$
    $\sigma^2 = n \cdot p \cdot (1-p) = (12) \cdot (.37) \cdot (.63) = 2.8$
    $\sigma = 1.7$

12. let x = the number of letter e's; n = 2600, p = .130
    $\mu = n \cdot p = (2600) \cdot (.130) = 338.0$
    $\sigma^2 = n \cdot p \cdot (1-p) = (2600) \cdot (.130) \cdot (.870) = 294.1$
    $\sigma = 17.1$

13. a. let x = the number of pizzas delivered on time; n = 300, p = .90
μ = n·p = (300)·(.90) = 270.0
σ^2 = n·p·(1-p) = (300)·(.90)·(.10) = 27.0
σ = 5.2
b. Unusual values are those outside $\mu \pm 2\cdot\sigma$
270.0 $\pm$ 2·(5.2)
270.0 $\pm$ 10.4
259.6 to 280.4
As 244 on-time deliveries is not within these limits, it would be considered an unusual event.

14. a. let x = the number of rotary phones; n = 250, p = .30
μ = n·p = (250)·(.30) = 75.0
σ^2 = n·p·(1-p) = (250)·(.30)·(.70) = 52.5
σ = 7.2
b. Unusual values are those outside $\mu \pm 2\cdot\sigma$
75.0 $\pm$ 2·(7.25)
75.0 $\pm$ 14.5
60.5 to 89.5
c. As 50 rotary phones is not within these limits, it would be considered an unusual event.

15. a. let x = the number of colors correctly guessed; n = 50, p = 1/6
μ = n·p = (50)·(1/6) = 8.3
σ^2 = n·p·(1-p) = (50)·(1/6)·(5/6) = 6.9
σ = 2.6
b. Unusual values are those outside $\mu \pm 2\cdot\sigma$
8.3 $\pm$ 2·(2.6)
8.3 $\pm$ 5.2
3.1 to 13.5
As 12 correct guesses is within these limits, it would not be considered an unusual event.

16. a. let x = the number of myocardial infarction deaths; n = 5000, p = .149
μ = n·p = (5000)·(.149) = 745.0
σ^2 = n·p·(1-p) = (5000)·(.149)·(.851) = 634.0
σ = 25.2
b. Unusual values are those outside $\mu \pm 2\cdot\sigma$
745.0 $\pm$ 2·(25.2)
745.0 $\pm$ 50.4
694.6 to 795.4
As 896 such deaths is not within these limits, it would be considered an unusual event and there is cause for concern and/or further investigation.

17. let x = the number of accidents involving injuries; n = 80, p = .612
μ = n·p = (80)·(.612) = 49.0
σ^2 = n·p·(1-p) = (80)·(.612)·(.388) = 19.0
σ = 4.4
Unusual values are those outside $\mu \pm 2\cdot\sigma$
49.0 $\pm$ 2·(4.4)
49.0 $\pm$ 8.8
40.2 to 57.8
As 72 accidents with injuries is not within these limits, it would be considered an unusual event.

18. a. let x = the number of babies that are girls; n = 64, p = .50
$\mu = n \cdot p = (64) \cdot (.50) = 32.0$
$\sigma^2 = n \cdot p \cdot (1-p) = (64) \cdot (.50) \cdot (.50) = 16.0$
$\sigma = 4.0$

b. Unusual values are those outside $\mu \pm 2 \cdot \sigma$
$$32.0 \pm 2 \cdot (4.0)$$
$$32.0 \pm 8.0$$
$$24.0 \text{ to } 40.0$$

As 36 female babies is within these limits, it would not be considered an unusual event. While in harmony with the expectations of the method, the results do not provide evidence that the method is effective.

19. a. let x = the number of consumers recognizing the Coke brand name; n = 200, p = .95
$\mu = n \cdot p = (200) \cdot (.95) = 191.0$
$\sigma^2 = n \cdot p \cdot (1-p) = (200) \cdot (.95) \cdot (.05) = 9.5$
$\sigma = 3.1$

b. Unusual values are those outside $\mu \pm 2 \cdot \sigma$
$$191.0 \pm 2 \cdot (3.1)$$
$$191.0 \pm 6.2$$
$$184.8 \text{ to } 197.2$$

As all 200 consumers recognizing the Coke brand name is not within these limits, it would be considered an unusual event.

20. a. let x = the number of households experiencing a burglary; n = 300, p = .90
$\mu = n \cdot p = (150) \cdot (.05) = 7.5$
$\sigma^2 = n \cdot p \cdot (1-p) = (150) \cdot (.05) \cdot (.95) = 7.1$
$\sigma = 2.7$

b. Unusual values are those outside $\mu \pm 2 \cdot \sigma$
$$7.5 \pm 2 \cdot (2.67)$$
$$7.5 \pm 5.3$$
$$2.2 \text{ to } 12.8$$

As 12 households experiencing burglaries is within these limits, it would not be considered an unusual event.

21. a. let x = the number of correct responses; n = 100, p = .20
$\mu = n \cdot p = (100) \cdot (.20) = 20.0$
$\sigma^2 = n \cdot p \cdot (1-p) = (100) \cdot (.20) \cdot (.80) = 16.0$
$\sigma = 4.0$

b. $z_{28} = (x-\mu)/\sigma$ $z_{12} = (x-\mu)/\sigma$
$\quad = (28-20)/4$ $= (12-20)/4$
$\quad = 2.00$ $= -2.00$

The z scores indicate that the range from 20 to 28 includes scores within 2 standard deviations of the mean. Since Chebyshev's Theorem (section 2-5) states that there must be at least $1 - 1/K^2$ of the scores within K standard deviations of the mean, there must be at least $1 - 1/2^2 = 1 - 1/4 = 3/4$ of the scores within K=2 standard deviations of the mean. In other words, we expect to find that at least 75% of the students who guess will score between 12 and 28. NOTE: Chebyshev's Theorem makes a statement about what <u>must</u> be true in a set of scores with a specific mean and standard deviation. While it can be used to make statements about what we may <u>expect</u> in a sample, it cannot make definitive predictions about samples using population and/or theoretical means and standard deviations.

c. $z_{30} = (x-\mu)/\sigma$
$\quad = (30-20)/4$
$\quad = 2.50$

22. a. Use Table A-1 to find $P(x \geq 5)$ for $p=.80$ and various values of n.

```
n= 5: P(x≥5) = P(x=5)
             =    .328
n= 6: P(x≥5) = P(x=5) + P(x=6)
             =    .393  +   .262  =   .655
n= 7: P(x≥5) = P(x=5) + P(x=6) + P(x=7)
             =    .275  +   .367  +   .210  =   .852
n= 8: P(x≥5) = P(x=5) + P(x=6) + P(x=7) + P(x=8)
             =    .147  +   .294  +   .336  +   .168  =   .945
n= 9: P(x≥5) = P(x=5) + P(x=6) + P(x=7) + P(x=8) + P(x=9)
             =    .066  +   .176  +   .302  +   .302  +   .134  =   .980
n=10: P(x≥5) = P(x=5) + P(x=6) + P(x=7) + P(x=8) + P(x=9) + P(x=10)
             =    .026  +   .088  +   .201  +   .302  +   .268  +   .107  =  .992
```

The minimum n for which $P(x \geq 5) \geq .99$ is $n = 10$.

b. $\mu = n \cdot p = (10) \cdot (.80) = 8.0$
$\sigma^2 = n \cdot p \cdot (1-p) = (10) \cdot (.80) \cdot (.20) = 1.6$
$\sigma = 1.3$

Review Exercises

1. binomial problem; let x = the number of students owning videocassette recorders
$n = 10$, $p = .3$
a. $x = 5$
    $P(x=5) = .103$ [from Table A-1]
b. x = at least 5
    $P(X \geq 5) = P(x=5) + P(x=6) + P(x=7) + P(x=8) + P(x=9) + P(x=10)$
        $= .103 + .037 + .009 + .001 + .000^+ + .000^+$ [from Table A-1]
        $= .150$
c. $\mu = n \cdot p = (10) \cdot (.30) = 3.0$
$\sigma^2 = n \cdot p \cdot (1-p) = (10) \cdot (.30) \cdot (.70) = 2.1$
$\sigma = 1.4$

2. a. let x = the randomly selected number

x	P(x)	x·P(x)	x^2	$x^2 \cdot P(x)$	
1	.2	.2	1	.2	$\mu = \Sigma x \cdot P(x)$
2	.2	.4	4	.8	$\quad = 3.0$
3	.2	.6	9	1.8	$\sigma^2 = \Sigma x^2 \cdot P(x) - \mu^2$
4	.2	.8	16	3.2	$\quad = 11.0 - (3.0)^2$
5	.2	1.0	25	5.0	$\quad = 2.0$
	1.0	3.0		11.0	$\sigma = 1.4$

b. binomial; let x = number of correct guesses
$n = 50$, $p = .2$
$\mu = n \cdot p = (50) \cdot (.2) = 10.0$

c. $\sigma^2 = n \cdot p \cdot (1-p) = (50) \cdot (.2) \cdot (.8) = 8.0$
$\sigma = 2.8$
Unusual values are those outside $\mu \pm 2 \cdot \sigma$
$10.0 \pm 2 \cdot (2.83)$
10.0 ± 5.7
4.3 to 15.7
No; if 15 correct responses occurred, that would not be considered an unusual event.

3. $n = 10$, $x = 6$, $p = .05$
$$P(x) = [n!/x!(n-x)!] \cdot p^x \cdot (1-p)^{n-x}$$
$$= [10!/6!4!] \cdot (.05)^6 \cdot (.95)^4$$
$$= [210] \cdot (.0000000156) \cdot (.8145)$$
$$= .00000267$$

4. This is <u>not</u> a probability distribution since $\Sigma P(x) = 0/5 + 1/5 + 2/5 + 3/5 = 6/5 \neq 1$.

5. binomial problem; let x = the number of Florida passengers wearing seat belts

$n = 5$, $x = 4$, $p = .57$ $n = 5$, $x = 5$, $p = .57$
$$P(x) = [n!/x!(n-x)!] \cdot p^x \cdot (1-p)^{n-x} \qquad P(x) = [n!/x!(n-x)!] \cdot p^x \cdot (1-p)^{n-x}$$
$$= [5!/4!1!] \cdot (.57)^4 \cdot (.43)^1 \qquad\qquad = [5!/5!0!] \cdot (.57)^5 \cdot (.43)^0$$
$$= [5] \cdot (.1056) \cdot (.43) \qquad\qquad\qquad = [1] \cdot (.0229) \cdot (1.00)$$
$$= .227 \qquad\qquad\qquad\qquad\qquad\qquad = .060$$

x	P(x)	x·P(x)	x²	x²·P(x)	
0	.015	0	0	0	$\mu = \Sigma x \cdot P(x)$
1	.097	.097	1	.097	$= 2.85$
2	.258	.516	4	1.032	$\sigma^2 = \Sigma x^2 \cdot P(x) - \mu^2$
3	.342	1.026	9	3.078	$= 9.339 - (2.847)^2$
4	.227	.908	16	3.632	$= 1.23$
5	.060	.300	25	1.500	$\sigma = 1.11$
	1.000	2.847		9.339	

NOTE: Once $P(x=4) = .227$ was calculated, $P(x=5)$ could have been found by subtracting the other probabilities from 1.000. Calculating $P(x=5)$ separately and verifying the probabilities sum to 1.000, however, provides a check. In this instance, the sum of the probabilities is actually .999 due to rounding off. The values for μ and σ are reported to one more decimal point than usual to see how the answers check with the true values obtained below (without round-off error) using the special formulas that apply only to the binomial.

$$\mu = n \cdot p = (5) \cdot (.57) = 2.85$$
$$\sigma^2 = n \cdot p \cdot (1-p) = (5) \cdot (.57) \cdot (.43) = 1.23$$
$$\sigma = 1.11$$

And, to two decimal accuracy, the answers agree exactly. Of course, the answers always agree exactly when all decimal places are carried throughout to eliminate all round-off error.

6. binomial problem

let x = the number of CA passengers wearing seat belts
$n = 5$, $p = .70$
use Table A-1

x	P(x)
0	.002
1	.028
2	.132
3	.309
4	.360
5	.168
	1.000

7. binomial problem

$n = 500$, $p = .70$
let x = the number of CA passengers wearing seat belts

a. $\mu = n \cdot p = (500) \cdot (.70) = 350.0$
$\sigma^2 = n \cdot p \cdot (1-p) = (500) \cdot (.70) \cdot (.30) = 105.0$
$\sigma = 10.2$

b. Unusual values are those outside $\mu \pm 2 \cdot \sigma$
$$350.0 \pm 2 \cdot (10.2)$$
$$350.0 \pm 20.4$$
$$329.6 \text{ to } 370.4$$

c. The result suggests that the campaign was successful and that the proportion is now greater than 70%, since 375 California drivers out of 500 is an unusual result if the true proportion is still 70%.

8.

x	P(x)	x·P(x)	x^2	x^2·P(x)
0	.0004	0	0	0
1	.0094	.0094	1	.0094
2	.0870	.1740	4	.3480
3	.3562	1.0686	9	3.2058
4	.5470	2.1880	16	8.7520
	1.0000	3.4400		12.3152

$\mu = \Sigma x \cdot P(x)$
$= 3.4$
$\sigma^2 = \Sigma x^2 \cdot P(x) - \mu^2$
$= 12.3152 - (3.44)^2$
$= .5$
$\sigma = .7$

9. binomial problem; let x = the number of worker firings citing incompatibility
   n = 5, p = .17
   a. for x = 0, $P(x) = [n!/x!(n-x)!] \cdot p^x \cdot (1-p)^{n-x}$
      $= [5!/0!5!] \cdot (.17)^0 \cdot (.83)^5$
      $= [1] \cdot (1.00) \cdot (.3939)$
      $= .394$
   b. for x = 4, $P(x) = [n!/x!(n-x)!] \cdot p^x \cdot (1-p)^{n-x}$
      $= [5!/4!1!] \cdot (.17)^4 \cdot (.83)^1$
      $= [5] \cdot (.0008352) \cdot (.83)$
      $= .00347$
   c. for x = 5, $P(x) = [n!/x!(n-x)!] \cdot p^x \cdot (1-p)^{n-x}$
      $= [5!/5!0!] \cdot (.17)^5 \cdot (.83)^0$
      $= [1] \cdot (.0001420) \cdot (1.00)$
      $= .000142$
   d. $P(x \geq 3) = ?$
      for x = 3, $P(x) = [n!/x!(n-x)!] \cdot p^x \cdot (1-p)^{n-x}$
      $= [5!/3!2!] \cdot (.17)^3 \cdot (.83)^2$
      $= [10] \cdot (.004913) \cdot (.6889)$
      $= .03385$
      $P(x \geq 3) = P(x=3) + P(x=4) + P(x=5)$
      $= .03385 + .00347 + .00014$
      $= .0375$

10. binomial problem; let x = the number of worker firings citing incompatibility
    n = 5, p = .17   [see exercise #9 for P(x=0), P(x=3), P(x=4), P(x=5)]
    $P(x=1) = [5!/1!4!] \cdot (.17)^1 \cdot (.83)^4 = .4034$
    $P(x=2) = [5!/2!3!] \cdot (.17)^2 \cdot (.83)^3 = .1652$

x	P(x)	x·P(x)	x^2	x^2·P(x)
0	.3939	0	0	0
1	.4034	.4034	1	.4034
2	.1652	.3304	4	.6608
3	.0338	.1014	9	.3042
4	.0035	.0140	16	.0560
5	.0001	.0005	25	.0025
	1.0000	.8497		1.4269

$\mu = \Sigma x \cdot P(x)$
$= .8497$
$\sigma^2 = \Sigma x^2 \cdot P(x) - \mu^2$
$= 1.4269 - (.8497)^2$
$= .7049$
$\sigma = .8396$

a. $\mu = .85$ [use two decimal accuracy since the mean is less than 1]
b. $\sigma = .84$
c. Unusual values are those outside $\mu \pm 2 \cdot \sigma$
   $.85 \pm 2 \cdot (.84)$
   $.85 \pm 1.68$
   0 to 2.53

As having 4 employees fired for this reason is not within these limits, it would be considered an unusual event.

Chapter 5

Normal Probability Distributions

5-2 The Standard Normal Distribution

1. The height of the rectangle is .2. Probability corresponds to area, and the area of a rectangle is (width)·(height).
$$P(3 < x < 5) = (width)·(height)$$
$$= (5\text{-}3)·(.2)$$
$$= (2)·(.2)$$
$$= .4$$

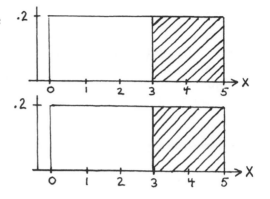

2. $P(x > 3) = (width)·(height)$
$$= (5\text{-}3)·(.2)$$
$$= (2)·(.2)$$
$$= .4$$

3. The height of the rectangle is .2. Probability corresponds to area, and the area of a rectangle is (width)·(height).
$$P(x < 4.5) = (width)·(height)$$
$$= (4.5\text{-}0)·(.2)$$
$$= (4.5)·(.2)$$
$$= .9$$

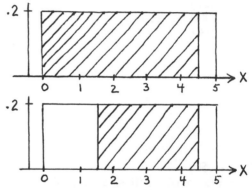

4. $P(1.5 < x < 4.5) = (width)·(height)$
$$= (4.5\text{-}1.5)·(.2)$$
$$= (3)·(.2)$$
$$= .6$$

NOTE: The sketch is the key to exercises 5-28. It tells whether to add two Table A-2 probabilities, to subtract two Table A-2 probabilities, to subtract a Table A-2 probability from .5000, to add a Table A-2 probability to .5000, etc., etc. It also often provides a check against gross errors by indicating at a glance whether the final probability is less than or greater than .5000. Remember that the symmetry of the normal curve implies two important facts:
    * There is always .5000 above and below the middle (i.e., at z = 0).
    * $P(\text{-}a < z < 0) = P(0 < z < a)$ for all values of "a."

5. $P(0 < z < 2.00)$
$$= .4772$$

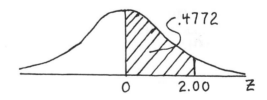

6. P(-2.34 < z < 0)
  = .4904

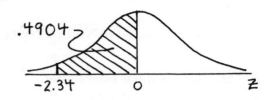

7. P(z > 1.05)
  = P(z > 0) - P(0 < z < 1.05)
  = .5000 - .3531
  = .1469

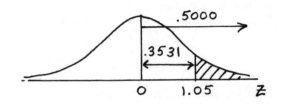

8. P(z < .82)
  = .5000 + .2939
  = .7939

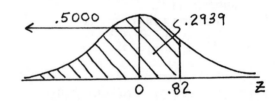

9. P(z > .50)
  = P(z > 0) - P(0 < z < .50)
  = .5000 - .1915
  = .3085

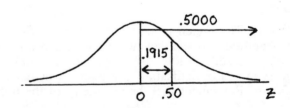

10. P(z < -3.00)
  = .5000 - .4987
  = .0013

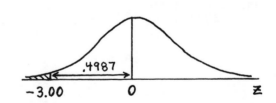

11. P(z > -1.09)
   = P(-1.09 < z < 0) + P(z > 0)
   = .3621 + .5000
   = .8621

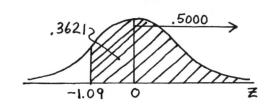

12. P(z > -2.37)
   = .4911 + .5000
   = .9911

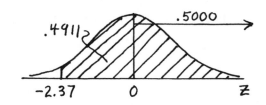

13. P(-1.00 < z < 1.00)
   = P(-1.00 < z < 0) + P(0 < z < 1.00)
   = .3413 + .3413
   = .6826

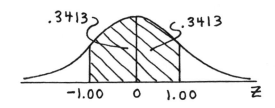

14. P(-2.00 < z < 2.00)
   = .4722 + .4722
   = .9544

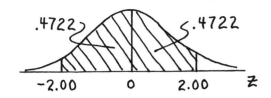

15. P(-1.15 < z < 2.60)
   = (-1.15 < z < 0) + P(0 < z < 2.60)
   = .3749 + .4953
   = .8702

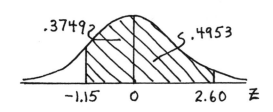

16. P(-.09 < z < 1.02)
   = .0359 + .3461
   = .3820

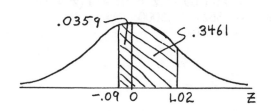

17. P(1.20 < z < 1.80)
   = P(0 < z < 1.80) - P(0 < z < 1.20)
   = .4641 - .3849
   = .0792

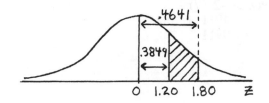

18. P(.25 < z < 2.25)
   = .4878 - .0987
   = .3891

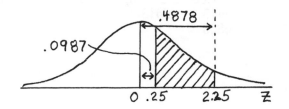

19. P(-2.30 < z < -1.05)
   = P(-2.30 < z < 0) - P(-1.05 < z < 0)
   = .4893 - .3531
   = .1362

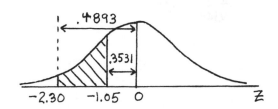

20. P(-2.88 < z < -1.44)
   = .4980 - .4251
   = .0729

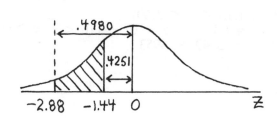

21. P(z > 0)
      = .5000

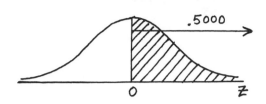

22. P(z < 0.00)
      = .5000

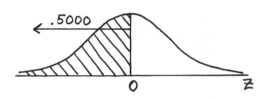

23. P(z < -1.96 or z > 1.96)
      = 1 - P(-1.96 < z < 1.96)
      = 1 - [P(-1.96 < z < 0) + P(0 < z < 1.96)]
      = 1 - [.4750 + .4750]
      = 1 - .9500
      = .0500

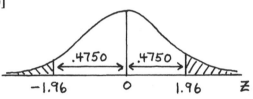

24. P(z < -2.00 or z > 1.00)
      = (.5000 - .4772) + (.5000 - .3413)
      = .0228 + .1587
      = .1815

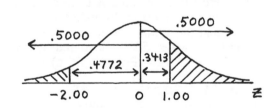

25. P(z > 1.96)
      = P(z > 0) - P(0 < z < 1.96)
      = .5000 - .4750
      = .0250

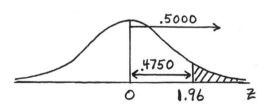

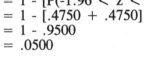

26. P(-.77 < z < .77)
   = .2794 + .2794
   = .5588

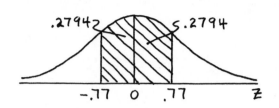

27. P(z < 2.33)
   = P(z < 0) + P(0 < z < 2.33)
   = .5000 + .4901
   = .9901

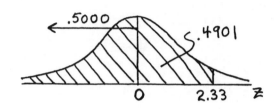

28. P(1.11 < z < 2.22)
   = .4868 - .3665
   = .1203

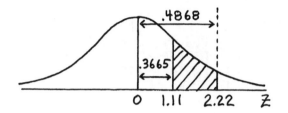

NOTE: The sketch is the key to exercises 29-36. It tells what probability is between 0 and the z score of interest (i.e. the area A to look up when reading Table A-2 "backwards." It also provides a check against gross errors by indicating at a glance whether a z score is above or below 0. Remember that the symmetry of the normal curve implies two important facts:
   * There is always .5000 above and below the middle (i.e., at z = 0).
   * P(-a < z < 0) = P(0 < z < a) for all values of "a."

29. For P₉₀, A = .4000.
   The closest entry is A = .3997,
   for which z = 1.28.

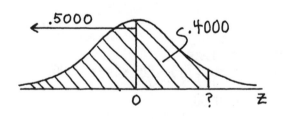

30. A = .1000 [.0987]
   z = -.25

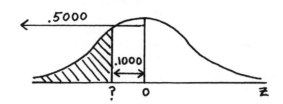

31. For the top 15%, A = .3500.
   The closest entry is A = .3508,
   for which z = 1.04 [positive
   since it is above the middle,
   where z = 0].

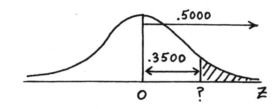

32. A = .3300 [.3289]
   z = -.95

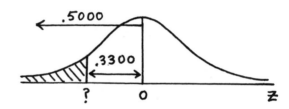

33. For the bottom 1%, A = .4900.
   The closest entry is A = .4901,
   for which z = -2.33 [negative
   since it is below the middle,
   where z = 0].

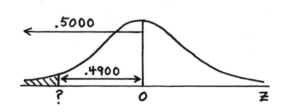

34. A = .4750
   z = 1.96

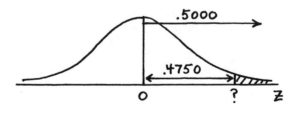

35. For the top 5%, A = .4500.
The entry A = .4500 is at the top,
for which z = 1.645.
For the bottom 5%, A = .4500.
The entry A = .4500 is at the top,
for which z = -1.645 [negative
since it is below the middle,
where z = 0].

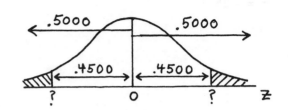

36. A_L = .3000 [.2995] A_U = .3500 [.3508]
z_L = -.84 z_U = 1.04

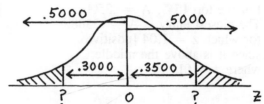

37. a. P(0 < z < a) = .4778
A = .4778
a = 2.01

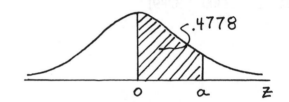

b. P(-b < z < b) = .7814
P(0 < z < b) = .7814/2
= .3907
A = .3907
b = 1.23

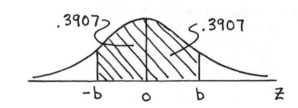

c.   P(z > c) = .0329
P(0 < z < c) = .5000 - .0329
= .4671
A = .4671
c = 1.84

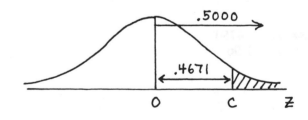

d. $P(z > d) = .8508$
  $P(d < z < 0) = .8508 - .5000$
                $= .3508$
  $A = .3508$
  $d = -1.04$ [negative since
    it falls below the
    middle, where z=0]

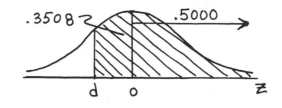

e. $P(z < e) = .0062$
  $P(e < z < 0) = .5000 - .0062$
                $= .4938$
  $A = .4938$
  $e = -2.50$ [negative since
    it falls below the
    middle, where z=0]

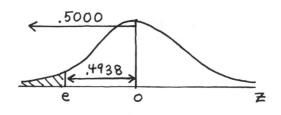

38. a. $P(-1.00 < z < 1.00)$
    $= .3413 + .3413$
    $= .6826$

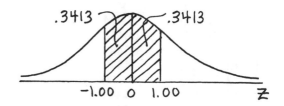

b. $P(-1.96 < z < 1.96)$
   $= .4750 + .4750$
   $= .9500$

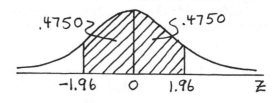

c. $P(-3.00 < z < 3.00)$
   $= .4987 + .4987$
   $= .9974$

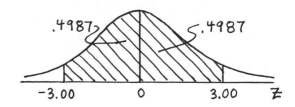

d. $P(-1.00 < z < 2.00)$
$\quad = .3413 + .4772$
$\quad = .8185$

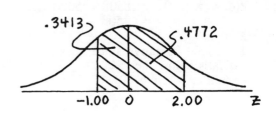

e. $P(z < -2.00 \text{ or } z > 2.00)$
$\quad = (.5000 - .4772) + (.5000 - .4772)$
$\quad = .0228 + .0228$
$\quad = .0456$

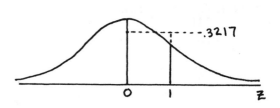

39.

x	$[2.7^{-x \cdot x/2}]/2.5$	y
-4	$(2.7)^{-8}/2.5$	.00014
-3	$(2.7)^{-4.5}/2.5$	.00458
-2	$(2.7)^{-2}/2.5$	.05487
-1	$(2.7)^{-.5}/2.5$	.24343
0	$(2.7)^{-0}/2.5$	.40000
1	$(2.7)^{-.5}/2.5$	.24343
2	$(2.7)^{-2}/2.5$	.05487
3	$(2.7)^{-4.5}/2.5$	.00458
4	$(2.7)^{-8}/2.5$	.00014

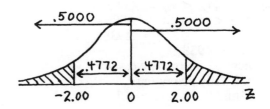

Approximate the area between $z=0$ and $z=1$ using a rectangle with width 1.0 and height $(.40000 + .24343)/2 = .3217$, the average of the heights of the curve at $z=0$ and $z=1$. The approximate area of $(1.0) \cdot (.3217) = .3217$ compares well with the true area from Table A-2 of .3413.

40. a. $\sqrt{3} = 1.732$. Since (total area) $= 1.00$
$\qquad\qquad$ (width) $\cdot$ (height) $= 1.00$
$\qquad\qquad\qquad$ $(3.464) \cdot h = 1.00$
$\qquad\qquad\qquad\qquad h = 1/3.464$
$\qquad\qquad\qquad\qquad\quad = .2887$
$\quad P(1 < x < 1.732) = $ (width) $\cdot$ (height)
$\qquad\qquad\qquad\qquad = (1.732 - 1.000) \cdot (.2887)$
$\qquad\qquad\qquad\qquad = .2113$

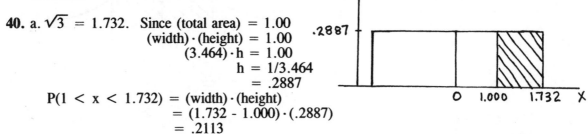

b. $P(z > 1) = .5000 - .3413$
$\qquad\qquad = .1587$

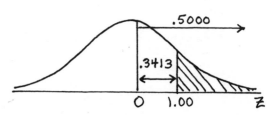

c. Yes; the .0526 error is approximately 25% of the correct answer.

5-3 Nonstandard Normal Distributions

NOTE: In each nonstandard normal distribution, x scores are converted to z scores using the formula $z = (x-\mu)/\sigma$ and rounded to two decimal places. The preceding formula may also be solved for x in terms of z to produce $x = \mu + z\sigma$. As in the previous section, drawing and labeling the sketch is the key to successful completion of the exercises.

1. $\mu = 100$
   $\sigma = 15$
   $P(100 < x < 127) = P(0 < z < 1.80)$
   $= .4641$

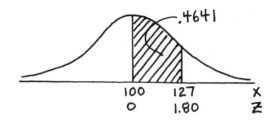

2. $\mu = 100$
   $\sigma = 15$
   $P(x < 118) = P(z < 1.20)$
   $= .5000 + .3849$
   $= .8849$

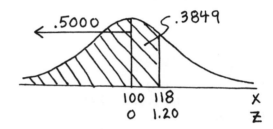

3. $\mu = 100$
   $\sigma = 15$
   $P(x > 76) = P(z > -1.60)$
   $= .4452 + .5000$
   $= .9452$

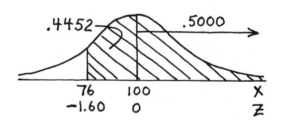

4. $\mu = 100$
   $\sigma = 15$
   $P(85 < x < 112) = P(-1.00 < z < .80)$
   $= .3413 + .2881$
   $= .6294$

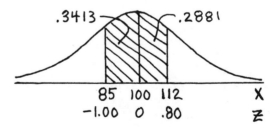

5. $\mu = 100$
   $\sigma = 15$
   For P_{90}, A = .4000.
   The closest entry is A = .3997,
   for which z = 1.28.
   $x = \mu + z\sigma$
   $= 100 + (1.28) \cdot (15)$
   $= 100 + 19.2$
   $= 119.2$

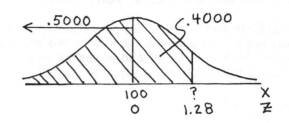

6. A = .4000 [.3997]
   z = -1.28
   $x = \mu + z\sigma$
   $= 100 + (-1.28) \cdot (15)$
   $= 80.8$

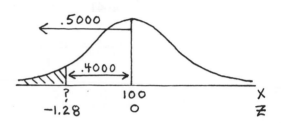

7. $\mu = 100$
   $\sigma = 15$
   For P_{20}, A = .3000.
   The closest entry is A = .2995,
   for which z = -.84.
   $x = \mu + z\sigma$
   $= 100 + (-.84) \cdot (15)$
   $= 100 - 12.6$
   $= 87.4$

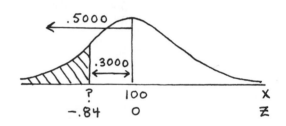

8. A = .4500
   z = 1.645 [see note above Table A-2]
   $x = \mu + z\sigma$
   $= 100 + (1.645) \cdot (15)$
   $= 124.7$

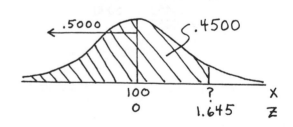

9. $\mu = 69.0$
   $\sigma = 2.8$
   $P(x < 64) = P(z < -1.79)$
   $= .5000 - .4633$
   $= .0367$

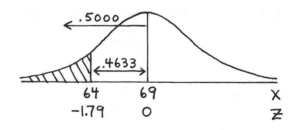

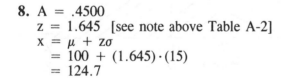
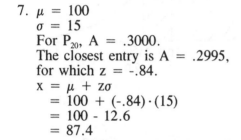
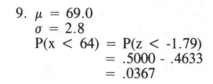

10. $\mu = 63.6$
$\sigma = 2.5$
$P(58 < x < 73) = P(-2.24 < z < 3.76)$
$= .4875 + .4999$ [see note above Table A-2]
$= .9874$

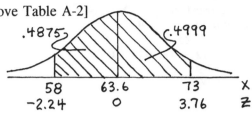

11. $\mu = 9.4$
$\sigma = 4.2$
For P_5, $A = .4500$ and $z = -1.645$.
$x = \mu + z\sigma$
$= 9.4 + (-1.645) \cdot (4.2)$
$= 9.4 - 6.9$
$= 2.5$
For P_{95}, $A = .4500$ and $z = 1.645$.
$x = \mu + z\sigma$
$= 9.4 + (1.645) \cdot (4.2)$
$= 9.4 + 6.9$
$= 16.3$

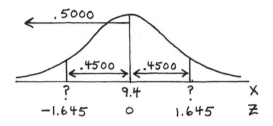

12. $\mu = 9.4$
$\sigma = 4.2$
$P(x \geq 5.0) = P(z > -1.05)$
$= .3531 + .5000$
$= .8531$

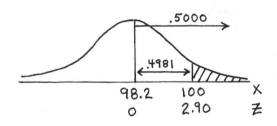

13. $\mu = 98.2$
$\sigma = .62$
$P(x > 100) = P(z > 2.90)$
$= .5000 - .4981$
$= .0019$

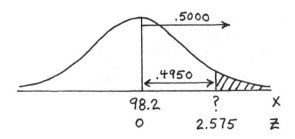

14. $\mu = 98.2$
$\sigma = .62$
$A = .4950$
$z = 2.575$ [see note above Table A-2]
$x = \mu + z\sigma$
$= 98.2 + (2.575) \cdot (.62)$
$= 99.8$

15. μ = 4.89
    σ = .63
    For the bottom 8%, A = .4200.
    The closest entry is A = .4207,
    for which z = -1.41.
    x = μ + zσ
      = 4.89 + (-1.41)·(.63)
      = 4.89 - .89
      = 4.00

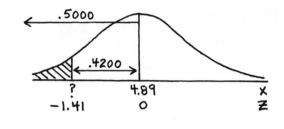

16. μ = 3.93
    σ = .75
    P(x > 2.75) = P(z > -1.57)
               = .4418 + .5000
               = .9418

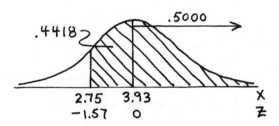

17. μ = 3.80
    σ = .95
    P(x > 4.00) = P(z > .21)
               = .5000 - .0832
               = .4168

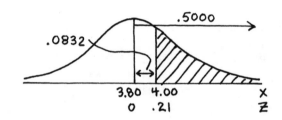

18. μ = 178.1
    σ = 40.7
    P(100 < x < 200) = P(-1.92 < z < .54)
                     = .4726 + .2054
                     = .6780

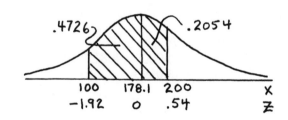

19. μ = 99.56
    σ = 25.84
    P(110 < x < 150) = P(.40 < z < 1.95)
                    = .4744 - .1554
                    = .3190

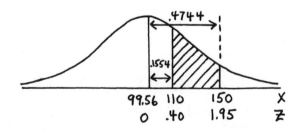

20. $\mu = 615$
$\sigma = 107$
A $= .2000$ [.1985]
z $= .52$
x $= \mu + z\sigma$
  $= 615 + (.52) \cdot (107)$
  $= 670.6$

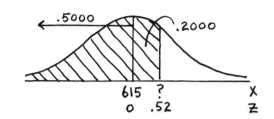

21. $\mu = 58.84$
$\sigma = 15.94$
$P(x < 27.00) = P(z < -2.00)$
  $= .5000 - .4772$
  $= .0228$

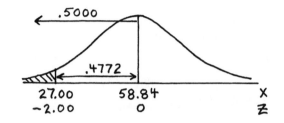

22. $\mu = 119.3$
$\sigma = 32.4$
$P(x \le 172) = P(z < 1.63)$
  $= .5000 + .4484$
  $= .9484$

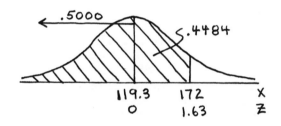

23. $\mu = 22.83$
$\sigma = 8.55$
$P(4.02 < x < 22.83) = P(-2.20 < z < 0)$
  $= .4861$

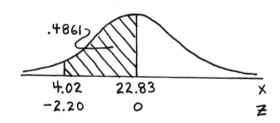

24. $\mu = 18.4$
$\sigma = 5.1$
$P(x > 16.0) = P(z > -.47)$
  $= .1808 + .5000$
  $= .6808$

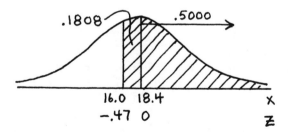

25. $\mu = 8.0$
    $\sigma = 2.6$
    $P(6.0 < x < 7.0) = P(-.77 < z < -.38)$
    $\qquad\qquad\qquad = .2794 - .1480$
    $\qquad\qquad\qquad = .1314$
    $(.1314) \cdot (600) = 78.8$

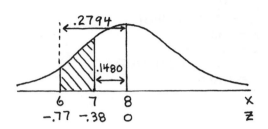

26. $\mu = 35,600$
    $\sigma = 4275$
    $A = .4700 \ [.4699]$
    $z = -1.88$
    $x = \mu + z\sigma$
    $\quad = 35,600 + (-1.88) \cdot (4275)$
    $\quad = 27,563$

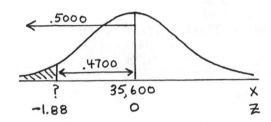

In round numbers, a guarantee of 27,500 miles would meet the target of replacing 3% or less.

27. $\mu = 268$
    $\sigma = 15$
    $P(x > 308) = P(z > 2.67)$
    $\qquad\qquad\quad = .5000 - .4962$
    $\qquad\qquad\quad = .0038$
    Such an occurrence would be
    rare, occurring less than 1%
    of the time, suggesting her
    husband might not be the father.
    On the other hand, in a small city
    of 100,000 people there should be about
    $(.0038) \cdot (100,000) = 380$ people who were born under just such circumstances.

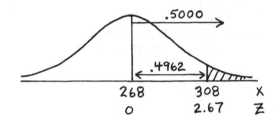

28. $\mu = 69.0$
    $\sigma = 2.8$

    $A_L = .4500 \qquad\qquad A_U = .4500$
    $z_L = -1.645 \qquad\qquad z_U = 1.645$
    $x_L = \mu + z\sigma \qquad\quad x_U = \mu + z\sigma$
    $\quad = 69.0 + (-1.645) \cdot (2.8) \quad = 69.0 + (1.645) \cdot (2.8)$
    $\quad = 64.4 \qquad\qquad\qquad = 73.6$

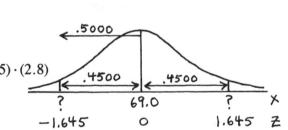

29. $\mu = 5.67$
    $\sigma = .0700$
    $P(\text{acceptance}) = P(5.50 < x < 5.80)$
    $\qquad\qquad\qquad\; = P(-.2.43 < z < 1.86)$
    $\qquad\qquad\qquad\; = .4925 + .4686$
    $\qquad\qquad\qquad\; = .9611$
    $P(\text{rejection}) = 1 - P(\text{acceptance})$
    $\qquad\qquad\quad = 1.0000 - .9611$
    $\qquad\qquad\quad = .0389$

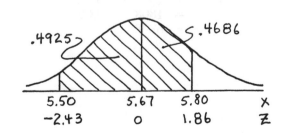

30. $\mu = 5.67$
$\sigma = .0700$

$A_L = .3500$ [.3508]	$A_U = .4200$ [.4207]
$z_L = -1.04$	$z_U = 1.41$
$x_L = \mu + z\sigma$	$x_U = \mu + z\sigma$
$\quad = 5.67 + (-1.04)\cdot(.07)$	$\quad = 5.67 + (1.41)\cdot(.07)$
$\quad = 5.60$	$\quad = 5.77$

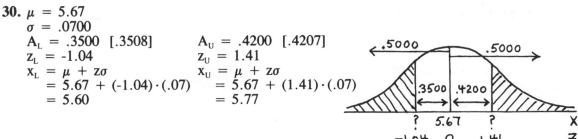

31. preliminary values: $n = 40$, $\Sigma x = 8018.4$, $\Sigma x^2 = 1,607,371.66$

$\begin{aligned}
\bar{x} &= \Sigma x/n \\
&= 8018.4/40 \\
&= 200.46 \\
s^2 &= [n(\Sigma x^2) - (\Sigma x)^2]/[n(n-1)] \\
&= [40(1,607,371.66) - (8018.4)^2]/[40(39)] \\
&= (127.84)/1560 \\
&= .0819 \\
s &= .286
\end{aligned}$

a. $3/40 = .075$ or 7.5%
b. $\bar{x} = 200.46$
c. $s = .29$
d. $\mu = 200.46$
$\sigma = .286$

$\begin{aligned}
P(x > 201.0) &= P(z > 1.89) \\
&= .5000 - .4706 \\
&= .0294
\end{aligned}$

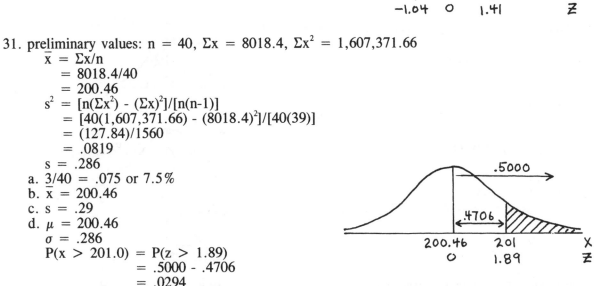

NOTE: The observed proportion in part (a) of .0750 differs from this value predicted assuming a normal distribution. This may suggest that the times do not follow a normal distribution.

e. No observations occurred below 198.0 or above 202.0.
Furthermore, $z_{198.0} = (198.0-200.46)/.286$
$\qquad\qquad\quad = -8.60$
$\quad z_{202.0} = (202.0-200.46)/.286$
$\qquad\qquad\quad = 5.38$

It appears that given the sample data, 198.0 and 202.0 would be extremely rare events with virtually zero probability of occurring. The specifications seem to be met.

32. uniform: $\mu = 100$, $\sigma = 15$.
Since (total area) $= 1.00$
(width)$\cdot$(height) $= 1.00$
$\qquad\qquad 52\cdot h = 1.00$
$\qquad\qquad\quad h = 1/52$
$\qquad\qquad\qquad = .01923$

$\begin{aligned}
P(80 < x < 110) &= (\text{width})\cdot(\text{height}) \\
&= (30)\cdot(.01923) \\
&= .5769
\end{aligned}$

normal: $\mu = 100$, $\sigma = 15$

$\begin{aligned}
P(80 < x < 110) &= P(-1.33 < z < .67) \\
&= .4082 + .2486 \\
&= .6568
\end{aligned}$

33. The cut-off points are P_{90}, P_{70}, P_{30}, P_{10}.

$\mu = 50$
$\sigma = 10$

For P_{90}, A = .4000.

The closest entry is A = .3997, for which z = 1.28

$x = \mu + z\sigma$
$= 50 + (1.28) \cdot (10)$
$= 50 + 12.8$
$= 62.8$

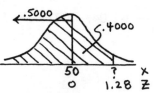

For P_{70}, A = .2000.

The closest entry is A = .1985, for which z = .52

$x = \mu + z\sigma$
$= 50 + (.52) \cdot (10)$
$= 50 + 5.2$
$= 55.2$

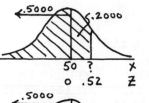

For P_{30}, A = .2000.

The closest entry is A = .1985, for which z = -.52

$x = \mu + z\sigma$
$= 50 + (-.52) \cdot (10)$
$= 50 - 5.2$
$= 44.8$

For P_{10}, A = .4000.

The closest entry is A = .3997, for which z = -1.28

$x = \mu + z\sigma$
$= 50 + (-1.28) \cdot (10)$
$= 50 - 12.8$
$= 37.2$

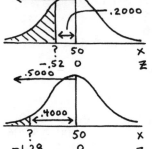

The grades are assigned as follows.

A: at least 62.8
B: at least 55.2 and less than 62.8
C: at least 44.8 and less than 55.2
D: at least 37.2 and less than 44.8
F: less than 37.2

34. $\mu = 475$
$\sigma = ?$

A = .3300 [.3289]   A = .4900 [.4901]
z = .95   z = 2.33
$z = (x-\mu)/\sigma$ $x = \mu + z\sigma$
$\sigma = (x-\mu)/z$ $= 475 + (2.33) \cdot (131.6)$
$= 125/.95$ $= 781.6$
$= 131.6$

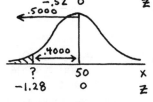

35. $\mu = 36.2$
$\sigma = 3.8$

For the top 200/4830 = .0414, A = .4586.

The closest entry is A = .4582, for which z = -1.73.

NOTE: The top runners have the <u>lower</u> times (i.e., they occur below the mean and have negative z scores.

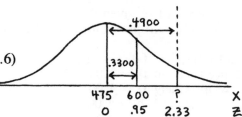

$x = \mu + z\sigma$
$= 36.2 + (-1.73) \cdot (3.8)$
$= 36.2 - 6.6$
$= 29.6$

36. $\mu = 650$
$\sigma = ?$
$A_L = .2500\ [.2486]$ <u>OR</u> $A_U = .2500\ [.2486]$
$z_L = -.67$ $z_U = .67$
$z = (x-\mu)/\sigma$ $z = (x-\mu)/\sigma$
$\sigma = (x-\mu)/z$ $\sigma = (x-\mu)/z$
     $= (-78)/(-.67)$ $= (78)/(.67)$
     $= 116.4$ $= 116.4$

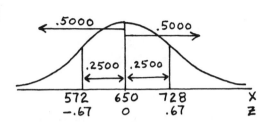

37. First, find the z score for which $P(-a < z < a) = 2/3$ -- i.e., for which
$$P(0 < z < a) = 1/3$$
$$= .3333$$
The closest entry is $A = .3340$, for which the score is .97.
Now use the above z score and the condition that the difference between x and μ is 30 to solve
$z = (x-\mu)/\sigma$ for σ to get $\sigma = (x-\mu)/z$
$$= 30/.97$$
$$= 30.9$$

5-4 The Central Limit Theorem

NOTE: When working with individual scores (i.e., making a statement about a single x scores from the original distribution), convert x to z using the mean and standard deviation of the x's and $z = (x-\mu)/\sigma$.

When working with a sample of n scores (i.e., making a statement about $\bar{x}$), convert $\bar{x}$ to z using the mean and standard deviation of the $\bar{x}$'s and $z = (\bar{x}-\mu_{\bar{x}})/\sigma_{\bar{x}}$.

1. a. normal distribution
$\mu = 100$
$\sigma = 15$
$P(100 < x < 103) = P(0 < z < .20)$
                   $= .0793$

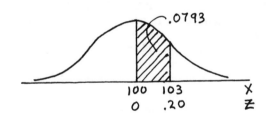

b. normal distribution, since the original distribution is so
$\mu_{\bar{x}} = \mu = 100$
$\sigma_{\bar{x}} = \sigma/\sqrt{n} = 15/\sqrt{25} = 3$
$P(100 < \bar{x} < 103) = P(0 < z < 1.00)$
                   $= .3413$

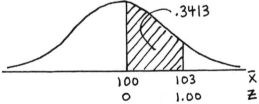

2. a. normal distribution

$\mu = 114.8$

$\sigma = 13.1$

$P(x > 120) = P(z > .40)$

$\qquad\qquad = .5000 - .1554$

$\qquad\qquad = .3446$

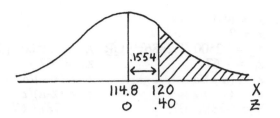

b. normal distribution, since the original distribution is so

$\mu_{\bar{x}} = \mu = 114.8$

$\sigma_{\bar{x}} = \sigma/\sqrt{n} = 13.1/\sqrt{30} = 2.392$

$P(\bar{x} > 120) = P(z > 2.17)$

$\qquad\qquad = .5000 - .4850$

$\qquad\qquad = .0150$

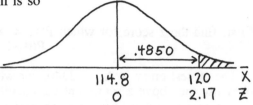

3. normal distribution, since the original distribution is so

$\mu_{\bar{x}} = \mu = 9.43$

$\sigma_{\bar{x}} = \sigma/\sqrt{n} = 4.17/\sqrt{12} = 1.204$

$P(10.0 < \bar{x} < 12.0) = P(.47 < z < 2.13)$

$\qquad\qquad\qquad\qquad = .4834 - .1808$

$\qquad\qquad\qquad\qquad = .3026$

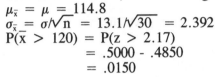

IMPORTANT NOTE: After calculating $\sigma_{\bar{x}}$, <u>STORE IT</u> in the calculator to recall it with total accuracy whenever it is needed in subsequent calculations. <u>DO NOT</u> write it down on paper rounded off (even to several decimal places) and then re-enter it in the calculator whenever it is needed. This avoids both round off errors and recopying errors.

4. normal distribution, by the Central Limit Theorem

$\mu_{\bar{x}} = \mu = 23.08$

$\sigma_{\bar{x}} = \sigma/\sqrt{n} = 15.58/\sqrt{60} = 2.011$

$P(20.00 < \bar{x} < 24.00) = P(-1.53 < z < .46)$

$\qquad\qquad\qquad\qquad\qquad = .4370 + .1772$

$\qquad\qquad\qquad\qquad\qquad = .6142$

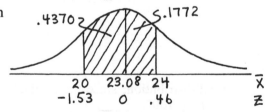

5. normal distribution, since the original distribution is so

$\mu_{\bar{x}} = \mu = 34.8$

$\sigma_{\bar{x}} = \sigma/\sqrt{n} = 7.02/\sqrt{36} = 1.17$

$P(34.8 < \bar{x} < 37.0) = P(0 < z < 1.88)$

$\qquad\qquad\qquad\qquad\qquad = .4699$

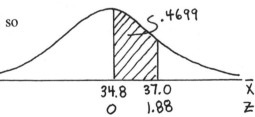

6. normal distribution, by the Central Limit Theorem

$\mu_{\bar{x}} = \mu = 7.06$

$\sigma_{\bar{x}} = \sigma/\sqrt{n} = 5.32/\sqrt{55} = .7173$

$P(\bar{x} > 7.00) = P(z > -.08)$
$\qquad = .0319 + .5000$
$\qquad = .5319$

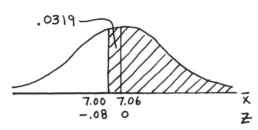

7. normal distribution, by the Central Limit Theorem

$\mu_{\bar{x}} = \mu = 10.7$

$\sigma_{\bar{x}} = \sigma/\sqrt{n} = 11.2/\sqrt{42} = 1.728$

$P(\bar{x} < 12.0) = P(z < .75)$
$\qquad = .5000 + .2734$
$\qquad = .7734$

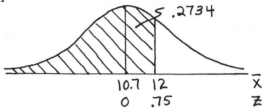

8. normal distribution, by the Central Limit Theorem

$\mu_{\bar{x}} = \mu = .500$

$\sigma_{\bar{x}} = \sigma/\sqrt{n} = .289/\sqrt{45} = .0431$

$P(\bar{x} < .565) = P(z < 1.51)$
$\qquad = .5000 + .4345$
$\qquad = .9345$

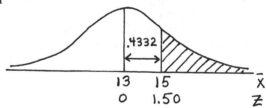

9. normal distribution, by the Central Limit Theorem

$\mu_{\bar{x}} = \mu = 13.0$

$\sigma_{\bar{x}} = \sigma/\sqrt{n} = 7.9/\sqrt{35} = 1.335$

$P(\bar{x} > 15) = P(z > 1.50)$
$\qquad = .5000 - .4332$
$\qquad = .0668$

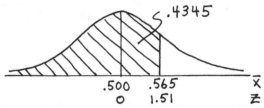

10. normal distribution, by the Central Limit Theorem

$\mu_{\bar{x}} = \mu = 2.7$

$\sigma_{\bar{x}} = \sigma/\sqrt{n} = 1.4/\sqrt{75} = .1617$

$P(\bar{x} < 2.5) = P(z < -1.24)$
$\qquad = .5000 - .3925$
$\qquad = .1075$

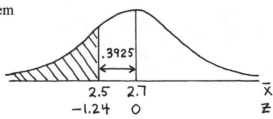

11. normal distribution, by the Central Limit Theorem
[since $31/630 < .05$, the finite population correction factor is not needed]
$\mu_{\bar{x}} = \mu = 2.97$
$\sigma_{\bar{x}} = \sigma/\sqrt{n} = .60/\sqrt{31} = .108$
$P(3.00 < \bar{x} < 3.10) = P(.28 < z < 1.21)$
$\qquad = .3869 - .1103$
$\qquad = .2766$

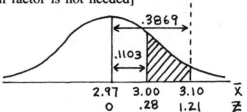

12. normal distribution, by the Central Limit Theorem
$\mu_{\bar{x}} = \mu = 462.1$
$\sigma_{\bar{x}} = \sigma/\sqrt{n} = 76.3/\sqrt{32} = 13.49$
$P(447.0 < \bar{x} < 456.8) = P(-1.12 < z < -.39)$
$\qquad = .3686 - .1517$
$\qquad = .2169$

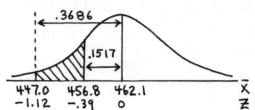

13. normal distribution, since the original distribution is so
$\mu_{\bar{x}} = \mu = 8.5$
$\sigma_{\bar{x}} = \sigma/\sqrt{n} = 3.96/\sqrt{36} = .66$
$P(7.0 < \bar{x} < 10.0) = P(-2.27 < z < 2.27)$
$\qquad = .4884 + .4884$
$\qquad = .9768$

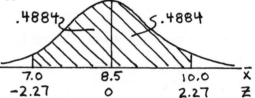

14. originally, $\sigma_{\bar{x}} = \sigma/\sqrt{n} = 3.96/\sqrt{36} = .66$
now, $\sigma_{\bar{x}} = \sigma/\sqrt{n} = 3.96/\sqrt{144} = .33$
Increasing the sample size by a factor of 4 reduces the standard error by a factor of $\sqrt{4} = 2$.

15. normal distribution, by the Central Limit Theorem
$\mu_{\bar{x}} = \mu = 3.54$
$\sigma_{\bar{x}} = \sigma/\sqrt{n} = .96/\sqrt{80} = .107$
$P(3.50 < \bar{x} < 3.60) = P(-.37 < z < .56)$
$\qquad = .1443 + .2123$
$\qquad = .3566$

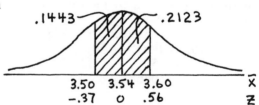

16. normal distribution, by the Central Limit Theorem
$\mu_{\bar{x}} = \mu = 7.37$
$\sigma_{\bar{x}} = \sigma/\sqrt{n} = .79/\sqrt{40} = .1249$
$P(\bar{x} > 7.00) = P(z > -2.96)$
$\qquad = .4985 + .5000$
$\qquad = .9985$

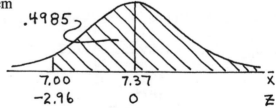

17. a. normal distribution

$\mu = 430$

$\sigma = 120$

$P(x > 440) = P(z > .08)$

$= .5000 - .0319$

$= .4681$

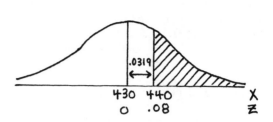

b. normal distribution, since the original distribution is so

$\mu_{\bar{x}} = \mu = 430$

$\sigma_{\bar{x}} = \sigma/\sqrt{n} = 120/\sqrt{100} = 12$

$P(\bar{x} > 440) = P(z > .83)$

$= .5000 - .2967$

$= .2033$

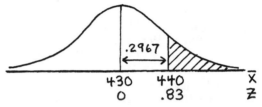

c. Not necessarily; 20% of the time 100 students with <u>no</u> special training can be expected to achieve a mean score of 440 or higher.

18. a. normal distribution

$\mu = 268$

$\sigma = 15$

$P(x < 260) = P(z < -.53)$

$= .5000 - .2019$

$= .2981$

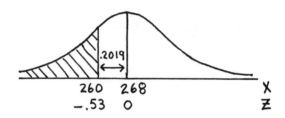

b. normal distribution, since the original distribution is so

$\mu_{\bar{x}} = \mu = 268$

$\sigma_{\bar{x}} = \sigma/\sqrt{n} = 15/\sqrt{25} = 3$

$P(\bar{x} < 260) = P(z < -2.67)$

$= .5000 - .4962$

$= .0038$

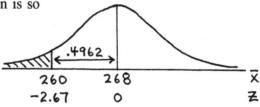

c. This is evidence that the special diet shortens the length of pregnancy. The supervisors should be concerned if babies are now being born at less than full term -- i.e., if the decreased length of pregnancy was not accompanied by a corresponding increase in the rate of fetal development.

19. normal distribution, by the Central Limit Theorem
   $\mu_{\bar{x}} = \mu = 38.9$
   $\sigma_{\bar{x}} = \sigma/\sqrt{n} = 12.4/\sqrt{150} = 1.012$
   a. $P(\bar{x} > 42.0) = P(z > 3.06)$
   $= .5000 - .4989$
   $= .0011$
   b. Yes; there is only a .1%
   chance of getting such a
   high sample mean if the
   true population mean is 38.9.

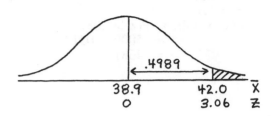

20. normal distribution, since the original distribution is so
   $\mu_{\bar{x}} = \mu = 173$
   $\sigma_{\bar{x}} = \sigma/\sqrt{n} = 30/\sqrt{32} = 5.303$
   $P(\bar{x} > 186) = P(z > 2.45)$
   $= .5000 - .4929$
   $= .0071$
   The club may expect about 7 overloads
   for every 1000 trips of 32 passengers.
   If the elevator routinely carries 32
   passengers at a time, there is reason
   concern. If the elevator typically
   carries far fewer passengers, and 32-person loads occur only a few times a year, then there is
   not cause for concern. **NOTE:** Limiting the occupancy to 30 passengers reduces P(overload)
   $= P(\Sigma x > 5952)$ to $P(\bar{x} > 198.4) = P(z > 4.64)$, which is beyond the limits of Table A-2.

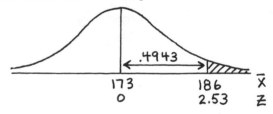

21. normal distribution, since the original distribution is so
   [since 32/500 > .05, the finite population correction factor must be used]
   $\mu_{\bar{x}} = \mu = 173$
   $\sigma_{\bar{x}} = [\sigma/\sqrt{n}] \cdot \sqrt{(N-n)/(N-1)}$
   $= [30/\sqrt{32}] \cdot \sqrt{(500-32)/(500-1)}$
   $= [30/\sqrt{32}] \cdot \sqrt{(468)/(499)}$
   $= 5.136$
   $P(\bar{x} > 186) = P(z > 2.53)$
   $= .5000 - .4943$
   $= .0057$

22. a.

x	x−μ	$(x-\mu)^2$
2	−6	36
3	−5	25
6	−2	4
8	0	0
11	3	9
18	10	100
48	0	174

$\mu = (\Sigma x)/N = 48/6 = 8$

$\sigma^2 = \Sigma(x-\mu)^2/N = 174/6 = 29$

$\sigma = 5.385$

b. and c.

sample	$\bar{x}$	$\bar{x}-\mu_{\bar{x}}$	$(\bar{x}-\mu_{\bar{x}})^2$
2,3	2.5	−5.5	30.25
2,6	4.0	−4.0	16.00
2,8	5.0	−3.0	9.00
2,11	6.5	−1.5	2.25
2,18	10.0	2.0	4.00
3,6	4.5	−3.5	12.25
3,8	5.5	−2.5	6.25
3,11	7.0	−1.0	1.00
3,18	10.5	2.5	6.25
6,8	7.0	−1.0	1.00
6,11	8.5	.5	.25
6,18	12.0	4.0	16.00
8,11	9.5	1.5	2.25
8,18	13.0	5.0	25.00
11,18	14.5	6.5	42.25
	120.0	0.0	174.00

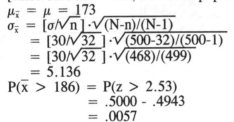

d. Refer to the 15 $\bar{x}$ values and the two "extra" columns in the table for parts (b) and (c).

$\mu_{\bar{x}} = \Sigma\bar{x}/15 = 120/15 = 8$

$\sigma_{\bar{x}}^2 = \Sigma(\bar{x}-\mu_{\bar{x}})^2/15 = 174/15 = 11.6$

$\sigma_{\bar{x}} = \sqrt{11.6} = 3.406$

e. $\mu = 8 = \mu_{\bar{x}}$

$[\sigma/\sqrt{n}]\cdot\sqrt{(N-n)/(N-1)} = [5.385/\sqrt{2}]\cdot\sqrt{(6-2)/(6-1)} = 3.406 = \sigma_{\bar{x}}$

23. normal distribution, since the original distribution appears so
    [since 15/350 < .05, the finite population correction factor is not needed]
    NOTE: From the boxplot we determine,
    $\bar{x} = 50$
    $s = 22.4$ In a normal distribution, P(-.67 < z < .67) = .5000 from Table A-2.
    The hinges indicate that P(35 < x < 65) = .5000.
    Solving z = (x-$\bar{x}$)/s for s yields s = (x-$\bar{x}$)/z
    = (65-50)/.67
    = 15/.67
    = 22.4

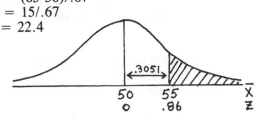

Using $\bar{x}$ and s to estimate μ and σ
$\mu_{\bar{x}} = \mu = 50$
$\sigma_{\bar{x}} = \sigma/\sqrt{n} = 22.4/\sqrt{15} = 5.78$
$P(\bar{x} > 55) = P(z > .86)$
= .5000 - .3051
= .1949

24. The appropriate summary statistics are: n = 62; $\Sigma x = 118.47$; $\Sigma x^2 = 295.6233$
    Using the sample values $\bar{x} = 1.911$ and s = 1.065 as suggested produces the following.
    normal distribution, by the Central Limit Theorem
    $\mu_{\bar{x}} = \mu = 1.911$
    $\sigma_{\bar{x}} = \sigma/\sqrt{n} = 1.065/\sqrt{100} = .1065$
    $P(\bar{x} > 1.75) = P(z > -1.51)$
    = .4345 + .5000
    = .9345

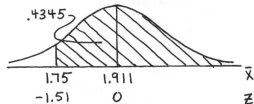

5-5 Normal Distribution as Approximation to Binomial Distribution

1. a. Table A-1 with n = 12 and p = .50
      P(x = 8) = .121

   b. normal approximation appropriate since
      np = 12(.50) = 6 ≥ 5
      n(1-p) = 12(.50) = 6 ≥ 5
      $\mu = np = 12(.50) = 6$
      $\sigma = \sqrt{np(1-p)} = \sqrt{12(.50)(.50)} = 1.732$
      $P(x = 8) = P_c(7.5 < x < 8.5)$
      = P(.87 < z < 1.44)
      = .4251 - .3078
      = .1173

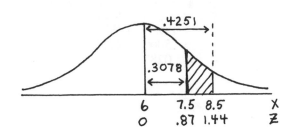

IMPORTANT NOTE: As in the previous section, store σ in the calculator so that it may be recalled with complete accuracy whenever it is needed in subsequent calculations. As before, P(E) represents the probability of an event E; this manual uses $P_c(E)$ to represent the probability of an event E with the continuity correction applied.

2. a. $P(x = 7) = .177$

b. normal approximation appropriate since

$$np = 15(.40) = 6 \geq 5$$
$$n(1-p) = 15(.60) = 9 \geq 5$$
$$\mu = np = 15(.40) = 6$$
$$\sigma = \sqrt{np(1-p)} = \sqrt{15(.40)(.60)} = 1.897$$
$$P(x = 7) = P_c(6.5 < x < 7.5)$$
$$= P(.26 < z < .79)$$
$$= .2852 - .1026$$
$$= .1826$$

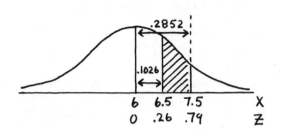

3. a. Table A-1 with $n = 10$ and $p = .80$

$$P(x \geq 8) = P(8) + P(9) + P(10)$$
$$= .302 + .268 + .107$$
$$= .677$$

b. normal approximation not appropriate since

$$np = 10(.80) = 8 \geq 5$$
$$n(1-p) = 10(.20) = 2 < 5$$

4. a. $P(x < 9) = P(x = 0) + P(x = 1) + P(x = 2) + ... + P(x = 8)$
$$= 0^+ + 0^+ + .001 + ... + .207$$
$$= .515$$

b. normal approximation appropriate since

$$np = 14(.60) = 8.4 \geq 5$$
$$n(1-p) = 14(.40) = 5.6 \geq 5$$
$$\mu = np = 14(.60) = 8.4$$
$$\sigma = \sqrt{np(1-p)} = \sqrt{14(.60)(.40)} = 1.833$$
$$P(x < 9) = P_c(x < 8.5)$$
$$= P(z < .05)$$
$$= .5000 + .0199$$
$$= .5199$$

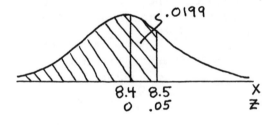

5. binomial with $n = 100$ and $p = .50$
normal approximation appropriate since

$$np = 100(.50) = 50 \geq 5$$
$$n(1-p) = 100(.50) = 50 \geq 5$$
$$\mu = np = 100(.50) = 50$$
$$\sigma = \sqrt{np(1-p)} = \sqrt{100(.50)(.50)} = 5$$
$$P(x \geq 60) = P_c(x > 59.5)$$
$$= P(z > 1.90)$$
$$= .5000 - .4713$$
$$= .0287$$

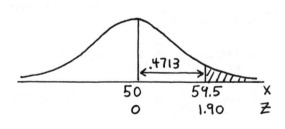

6. binomial with n = 80 and p = .50
normal approximation appropriate since
$np = 80(.50) = 40 \geq 5$
$n(1-p) = 80(.50) = 40 \geq 5$
$\mu = np = 80(.50) = 40$
$\sigma = \sqrt{np(1-p)} = \sqrt{80(.50)(.50)} = 4.472$
$P(x = 40) = P_c(39.5 < x < 40.5)$
$= P(-.11 < z < .11)$
$= .0438 + .0438$
$= .0876$

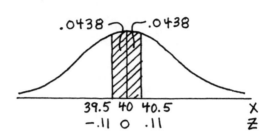

7. binomial with n = 50 and p = .50
normal approximation appropriate since
$np = 50(.50) = 25 \geq 5$
$n(1-p) = 50(.50) = 25 \geq 5$
$\mu = np = 50(.50) = 25$
$\sigma = \sqrt{np(1-p)} = \sqrt{50(.50)(.50)} = 3.536$
$P(x \geq 30) = P_c(x > 29.5)$
$= P(z > 1.27)$
$= .5000 - .3980$
$= .1020$

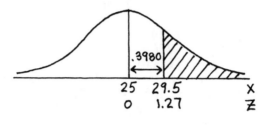

8. binomial with n = 40 and p = .20   ["at most 30% correct" is the same as "x ≤ 12"]
normal approximation appropriate since
$np = 40(.20) = 8 \geq 5$
$n(1-p) = 40(.80) = 32 \geq 5$
$\mu = np = 40(.20) = 8$
$\sigma = \sqrt{np(1-p)} = \sqrt{40(.20)(.80)} = 2.530$
$P(x \leq 12) = P_c(x < 12.5)$
$= P(z < 1.78)$
$= .5000 + .4625$
$= .9625$

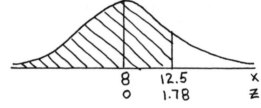

9. binomial with n = 1000 and p = .66
normal approximation appropriate since
$np = 1000(.66) = 660 \geq 5$
$n(1-p) = 1000(.34) = 340 \geq 5$
$\mu = np = 1000(.66) = 660$
$\sigma = \sqrt{np(1-p)} = \sqrt{1000(.66)(.34)} = 14.980$
$P(x \geq 700) = P_c(x > 699.5)$
$= P(z > 2.64)$
$= .5000 - .4959$
$= .0041$

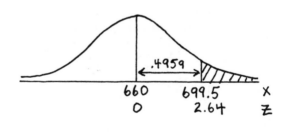

10. binomial with n = 150 and p = .75
normal approximation appropriate since
$np = 150(.75) = 112.5 \geq 5$
$n(1-p) = 150(.25) = 37.5 \geq 5$
$\mu = np = 150(.75) = 112.5$
$\sigma = \sqrt{np(1-p)} = \sqrt{150(.75)(.25)} = 5.303$
$P(x < 100) = P_c(x < 99.5)$
$= P(z < -2.45)$
$= .5000 - .4929$
$= .0071$

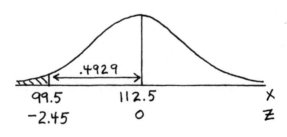

11. binomial with n = 500 and p = .26
   normal approximation appropriate since
   np = 500(.26) = 130 $\geq$ 5
   n(1-p) = 500(.74) = 370 $\geq$ 5
   μ = np = 500(.26) = 130
   $\sigma = \sqrt{np(1-p)} = \sqrt{500(.26)(.74)}$ = 9.808
   P(125 $\leq$ x $\leq$ 150) = P$_c$(124.5 < x < 150.5)
   $\qquad\qquad\qquad$ = P(-.56 < z < 2.09)
   $\qquad\qquad\qquad$ = .2132 + .4817
   $\qquad\qquad\qquad$ = .6940

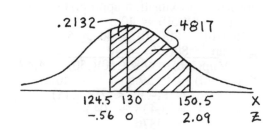

12. binomial with n = 220 and p = .80
   normal approximation appropriate since
   np = 220(.80) = 176 $\geq$ 5
   n(1-p) = 220(.20) = 44 $\geq$ 5
   μ = np = 220(.80) = 176
   $\sigma = \sqrt{np(1-p)} = \sqrt{220(.80)(.20)}$ = 5.933
   P(180 $\leq$ x $\leq$ 185) = P$_c$(179.5 < x < 185.5)
   $\qquad\qquad\qquad$ = P(.59 < z < 1.60)
   $\qquad\qquad\qquad$ = .4452 - .2224
   $\qquad\qquad\qquad$ = .2228

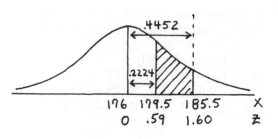

13. binomial with n = 2500 and p = .24
   normal approximation appropriate since
   np = 2500(.24) = 600 $\geq$ 5
   n(1-p) = 2500(.76) = 1900 $\geq$ 5
   μ = np = 2500(.24) = 600
   $\sigma = \sqrt{np(1-p)} = \sqrt{2500(.24)(.76)}$ = 21.354
   P(x > 650) = P$_c$(x > 650.5)
   $\qquad\qquad$ = P(z > 2.36)
   $\qquad\qquad$ = .5000 - .4909
   $\qquad\qquad$ = .0091

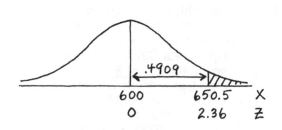

14. binomial with n = 140 and p = .12
   normal approximation appropriate since
   np = 140(.12) = 16.8 $\geq$ 5
   n(1-p) = 140(.88) = 123.2 $\geq$ 5
   μ = np = 140(.12) = 16.8
   $\sigma = \sqrt{np(1-p)} = \sqrt{140(.12)(.88)}$ = 3.845
   P(x $\geq$ 20) = P$_c$(x > 19.5)
   $\qquad\qquad$ = P(z > .70)
   $\qquad\qquad$ = .5000 - .2580
   $\qquad\qquad$ = .2420

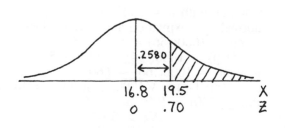

15. binomial with n = 600 and p = .35
   normal approximation appropriate since
   np = 600(.35) = 210 $\geq$ 5
   n(1-p) = 600(.65) = 390 $\geq$ 5
   μ = np = 600(.35) = 210
   $\sigma = \sqrt{np(1-p)} = \sqrt{600(.35)(.65)}$ = 11.683
   P(x $\geq$ 210) = P$_c$(x > 209.5)
   $\qquad\qquad$ = P(z > -.04)
   $\qquad\qquad$ = .0160 + .5000
   $\qquad\qquad$ = .5160

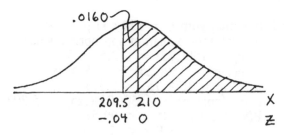

16. binomial with n = 500 and p = .60
  normal approximation appropriate since
  $np = 500(.60) = 300 \geq 5$
  $n(1-p) = 500(.40) = 200 \geq 5$
  $\mu = np = 500(.60) = 300$
  $\sigma = \sqrt{np(1-p)} = \sqrt{500(.60)(.40)} = 10.954$
  $P(x > 325) = P_c(x > 325.5)$
  $\qquad\qquad = P(z > 2.33)$
  $\qquad\qquad = .5000 - .4901$
  $\qquad\qquad = .0099$

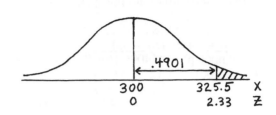

17. binomial with n = 320 and p = .75
  normal approximation appropriate since
  $np = 320(.75) = 240 \geq 5$
  $n(1-p) = 320(.25) = 80 \geq 5$
  $\mu = np = 320(.75) = 240$
  $\sigma = \sqrt{np(1-p)} = \sqrt{320(.75)(.25)} = 7.746$
  $P(x > 250) = P_c(x > 250.5)$
  $\qquad\qquad = P(z > 1.36)$
  $\qquad\qquad = .5000 - .4131$
  $\qquad\qquad = .0869$

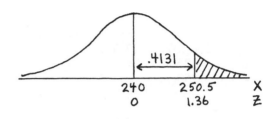

18. binomial with n = 350 and p = .26
  normal approximation appropriate since
  $np = 350(.26) = 91 \geq 5$
  $n(1-p) = 350(.74) = 259 \geq 5$
  $\mu = np = 350(.26) = 91$
  $\sigma = \sqrt{np(1-p)} = \sqrt{350(.26)(.74)} = 8.206$
  $P(80 \leq x \leq 90) = P_c(79.5 < x < 90.5)$
  $\qquad\qquad = P(-1.40 < z < -.06)$
  $\qquad\qquad = .4192 - .0239$
  $\qquad\qquad = .3953$

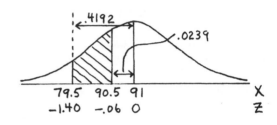

19. binomial with n = 100 and p = .20
  normal approximation appropriate since
  $np = 100(.20) = 20 \geq 5$
  $n(1-p) = 100(.80) = 80 \geq 5$
  $\mu = np = 100(.20) = 20$
  $\sigma = \sqrt{np(1-p)} = \sqrt{100(.20)(.80)} = 4$
  $P(x \leq 17) = P_c(x < 17.5)$
  $\qquad\quad = P(z < -.625)$
  $\qquad\quad = P(z < -.62)$ or $P(z < -.63)$
  $\qquad\quad = .5000 - .2324$ or $.5000 - .2357$
  $\qquad\quad = .2676$ or $.2643$

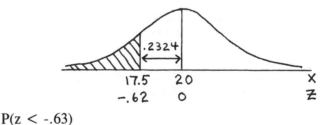

20. binomial with n = 200 and p = 1/38
  normal approximation appropriate since
  $np = 200(1/38) = 5.26 \geq 5$
  $n(1-p) = 200(37/38) = 194.74 \geq 5$
  $\mu = np = 200(1/38) = 5.263$
  $\sigma = \sqrt{np(1-p)} = \sqrt{200(1/38)(37/38)} = 2.263$
  $P(x \geq 6) = P_c(x > 5.5)$
  $\qquad\quad = P(z > .10)$
  $\qquad\quad = .5000 - .0398$
  $\qquad\quad = .4602$

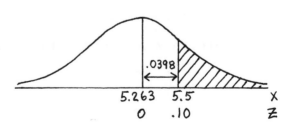

21. binomial with n = 400 and p = .16
    normal approximation appropriate since
    np = 400(.16) = 64 $\geq$ 5
    n(1-p) = 400(.84) = 336 $\geq$ 5
    μ = np = 400(.16) = 64
    σ = $\sqrt{np(1-p)}$ = $\sqrt{400(.16)(.84)}$ = 7.332
    P(x $\geq$ 100) = P$_c$(x > 99.5)
    = P(z > 4.84)
    = .5000 - .4999 [see note at top of Table A-2]
    = .0001

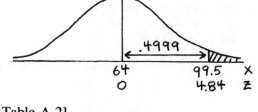

    No; it does not seem plausible that the state arrests women at the 16% rate.

22. binomial with n = 250 and p = .07
    normal approximation appropriate since
    np = 250(.07) = 17.5 $\geq$ 5
    n(1-p) = 250(.93) = 232.5 $\geq$ 5
    μ = np = 250(.07) = 17.5
    σ = $\sqrt{np(1-p)}$ = $\sqrt{250(.07)(.93)}$ = 4.034
    P(x $\leq$ 4) = P$_c$(x < 4.5)
    = P(z < -3.22)
    = .5000 - .4999
    = .0001

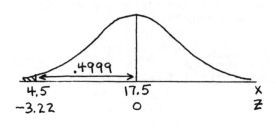

    Yes; if p = .07 were still true, then the above result would be extremely unlikely.

23. binomial with n = 40 and p = .25
    normal approximation appropriate since
    np = 40(.25) = 10 $\geq$ 5
    n(1-p) = 40(.75) = 30 $\geq$ 5
    μ = np = 40(.25) = 10
    σ = $\sqrt{np(1-p)}$ = $\sqrt{40(.25)(.75)}$ = 2.739
    P(x $\leq$ 8) = P$_c$(x < 8.5)
    = P(z < -.55)
    = .5000 - .2088
    = .2912

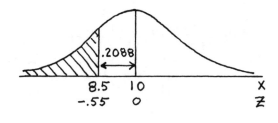

    The result is well within the reasonable expectation for a 25% rate does not give any indication that the 25% rate is incorrect.

24. binomial with n = 500 and p = .70
    normal approximation appropriate since
    np = 500(.70) = 350 $\geq$ 5
    n(1-p) = 500(.30) = 150 $\geq$ 5
    μ = np = 500(.70) = 350
    σ = $\sqrt{np(1-p)}$ = $\sqrt{500(.70)(.30)}$ = 10.247
    P(x $\leq$ 340) = P$_c$(x < 340.5)
    = P(z < -.93)
    = .5000 - .3238
    = .1762

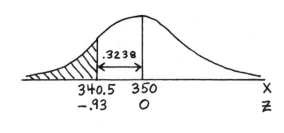

    Tom's results are within the reasonable expectations for p = .70.

25. binomial with n = 300 and p = .184
normal approximation appropriate since
$np = 300(.184) = 55.2 \geq 5$
$n(1-p) = 300(.816) = 244.80 \geq 5$
$\mu = np = 300(.184) = 55.2$
$\sigma = \sqrt{np(1-p)} = \sqrt{300(.184)(.816)} = 6.711$
$P(x \geq 72) = P_c(x > 71.5)$
$\quad\quad\quad = P(z > 2.43)$
$\quad\quad\quad = .5000 - .4925$
$\quad\quad\quad = .0075$

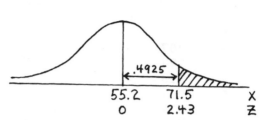

Under the conditions of the problem, getting 72 or more smokers by chance alone is very
unlikely. It appears either that the 18.4% figure is incorrect or that there is something wrong
with the sample.

26. binomial with n = 40 and p = .25
normal approximation appropriate since
$np = 40(.25) = 10 \geq 5$
$n(1-p) = 40(.75) = 30 \geq 5$
$\mu = np = 40(.25) = 10$
$\sigma = \sqrt{np(1-p)} = \sqrt{40(.25)(.75)} = 2.739$
$P(x < 5) = P_c(x < 4.5)$
$\quad\quad\quad = P(z < -2.01)$
$\quad\quad\quad = .5000 - .4778$
$\quad\quad\quad = .0222$

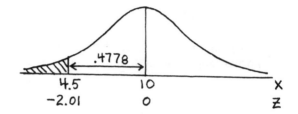

27. binomial with n = 400 and p = .45
normal approximation appropriate since
$np = 400(.45) = 180 \geq 5$
$n(1-p) = 400(.55) = 220 \geq 5$
$\mu = np = 400(.45) = 180$
$\sigma = \sqrt{np(1-p)} = \sqrt{400(.45)(.55)} = 9.950$
$P(x \geq 177) = P_c(x > 176.5)$
$\quad\quad\quad = P(z > -.35)$
$\quad\quad\quad = .1368 + .5000$
$\quad\quad\quad = .6368$

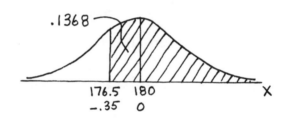

28. binomial with n = 50 and p = .10
normal approximation appropriate since
$np = 50(.10) = 5 \geq 5$
$n(1-p) = 50(.90) = 45 \geq 5$
$\mu = np = 50(.10) = 5$
$\sigma = \sqrt{np(1-p)} = \sqrt{50(.10)(.90)} = 2.121$
$P(x \geq 2) = P_c(x > 1.5)$
$\quad\quad\quad = P(z > -1.65)$
$\quad\quad\quad = .4505 + .5000$
$\quad\quad\quad = .9505$

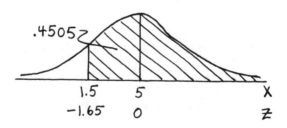

29. binomial with $n = 15$ and $p = .4$

a. using Table A-1

x	P(x)
5	.186
6	.207
7	.177
8	.118
9	.061
10	.024
11	.007
12	.002
13	0+
14	0+
15	0+
	.782

b. using the binomial formula $P(x) = [n!/x!(n-x)!]p^x(1-p)^{n-x}$

x	P(x)	
5	$3003(.4)^5(.6)^{10}$	$= .1859378$
6	$5005(.4)^6(.6)^9$	$= .2065976$
7	$6435(.4)^7(.6)^8$	$= .1770837$
8	$6435(.4)^8(.6)^7$	$= .1880558$
9	$5005(.4)^9(.6)^6$	$= .0612142$
10	$3003(.4)^{10}(.6)^5$	$= .0244856$
11	$1365(.4)^{11}(.6)^4$	$= .0074190$
12	$455(.4)^{12}(.6)^3$	$= .0016489$
13	$105(.4)^{13}(.6)^2$	$= .0002537$
14	$15(.4)^{14}(.6)^1$	$= .0000242$
15	$1(.4)^{15}(.6)^0$	$= .0000011$
		$.7827215$

c. binomial with $n = 15$ and $p = .4$

normal approximation appropriate since

$np = 15(.4) = 6 \geq 5$

$n(1-p) = 15(.6) = 9 \geq 5$

$\mu = np = 15(.4) = 6$

$\sigma = \sqrt{np(1-p)} = \sqrt{15(.4)(.6)} = 1.897$

$P(x \geq 5) = P_c(x > 4.5)$

$\qquad = P(z > -.79)$

$\qquad = .2852 + .5000$

$\qquad = .7852$

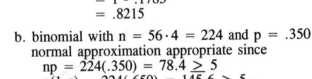

Of the three answers (a) $P(x \geq 5) = .782$

(b) $P(x \geq 5) = .7827215$

(c) $P(x \geq 5) = .7852$,

answer (b) is the one closest to the correct answer.

30. a. binomial with $n = 4$ and $p = .350$

$P(x \geq 1) = 1 - P(x = 0)$

$\qquad = 1 - [4!/0!4!] \cdot (.350)^0(.650)^4$

$\qquad = 1 - .1785$

$\qquad = .8215$

b. binomial with $n = 56 \cdot 4 = 224$ and $p = .350$

normal approximation appropriate since

$np = 224(.350) = 78.4 \geq 5$

$n(1-p) = 224(.650) = 145.6 \geq 5$

$\mu = np = 224(.350) = 78.4$

$\sigma = \sqrt{np(1-p)} = \sqrt{224(.350)(.650)} = 7.139$

$P(x \geq 56) = P_c(x > 55.5)$

$\qquad = P(z > -3.20)$

$\qquad = .4999 + .5000$

$\qquad = .9999$

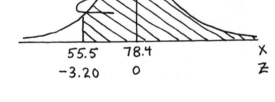

c. let H = getting at least one hit in 4 times at bat

$P(H) = .8215$ [from part (a) above]

for 56 consecutive games, $[P(H)]^{56} = [.8215]^{56} = .0000165$

d. The solution below employs the methods and notation of parts (a) and (c).
for $[P(H)]^{56} > .10$, it is required that
$$P(H) > (.10)^{1/56}$$
$$P(H) > .9597$$
for $P(H) = P(x \geq 1) > .9507$, it is required that
$$1 - P(x = 0) > .9597$$
$$.0403 > P(x = 0)$$
$$.0403 > [4!/0!4!] \cdot p^0(1-p)^4$$
$$.0403 > (1-p)^4$$
$$(.0403)^{1/4} > 1-p$$
$$p > 1 - (.0403)^{1/4}$$
$$p > 1 - .448$$
$$p > .552$$

31. Letting x represent the number of persons with advanced reservations who <u>do</u> show up for a flight, the problem is a binomial one with n unknown and $p = .93$. The airline wants to find the largest n for which $P(x \leq 250) = .95$. The problem may be solved in two different manners -- (a) increasing n and calculating $P(x \leq 250)$ until $P(x \leq 250)$ drops below .95, (b) solving directly for the value of n at which $P(x \leq 250)$ equal .95.

(a) The following table summarizes the procedure for finding $P(x \leq 250) = P_c(x < 250.5)$, where $p = .93$.

n	$\mu = np$	$\sigma = \sqrt{np(1-p)}$	$z = (250.5-\mu)/\sigma$	$P(x<250.5)$
260	241.8	4.11	2.11	.9826
261	242.7	4.12	1.89	.9706
262	243.7	4.13	1.66	.9515
263	244.6	4.14	1.43	.9236
264	245.5	4.15	1.20	.8849
265	246.5	4.15	.98	.8365
266	247.4	4.16	.75	.7734
267	248.3	4.17	.53	.7019
268	249.2	4.18	.30	.6179
269	250.2	4.18	.08	.5319
270	251.1	4.19	-.14	.4443
271	252.0	4.20	-.36	.3594
272	253.0	4.21	-.58	.2810
273	253.9	4.22	-.80	.2119
274	254.8	4.22	-1.02	.1539
275	255.8	4.23	-1.24	.1075
276	256.7	4.24	-1.46	.0721
277	257.6	4.25	-1.67	.0475

And so for $n = 262$ there is a 95% probability that everyone who shows up with advanced reservations will have a seat. For $n > 262$ the probability falls below that acceptable level.

(b) The z value below which 95% of the probability occurs is 1.645. Solving to find the n for which 250.5 or fewer persons with advanced reservations will show up 95% of the time,
$$(x-\mu)/\sigma = z$$
$$(250.5 - .93n)/\sqrt{(.93)(.07)n} = 1.645$$
$$(250.5) - .93n = 1.645 \cdot \sqrt{(.93)(.07)n}$$
$$62750.25 - 465.93n + .8649n^2 = .1762n$$
$$.8649n^2 - 466.1062n + 62750.25 = 0$$

Solving for n using the quadratic formula,
$$n = [466.1062 \pm \sqrt{(-466.1062)^2 - 4(.8649)(62750.25)}]/2(.8649)$$
$$= [466.1062 \pm 12.8136]/1.7298$$
$$= 453.2925/1.7298 \quad or \quad 478.9198/1.7298$$
$$= 262.05 \quad or \quad 276.86$$
And so z = 1.645 when n = 262.05. If n > 262.05 then z < 1.645 and the probability is less than 95%, so we round down to n = 262. Note that z = -1.645 when n = 276.86 [and that extraneous root was introduced when both sides of the equation were squared].

32. binomial with n = ? and p = .77   [let getting a good chip be a success]
normal approximation appropriate since n ≥ 5000
np ≥ 5000(.77) = 3850 ≥ 5
n(1-p) ≥ 5000(.23) = 1150 ≥ 5
$\mu = np = n(.77) = .77n$
$\sigma = \sqrt{np(1-p)} = \sqrt{n(.77)(.23)} = \sqrt{.1771n}$
given the condition: $P(x \geq 5000) = .90 \quad \rightarrow \quad P_c(x > 4999.5) = .90$
$A = .4000 \ [.3997] \quad \rightarrow \quad z = -1.28$
solve the following equation for n: $x = \mu + z\sigma$
$$4999.5 = .77n - 1.28\sqrt{.1771n}$$
$$4999.5 - .77n = -1.28\sqrt{.1771n}$$
$$24,885,999.25 - 7699.23n + .5929n^2 = .29n$$
$$.5929n^2 - 7699.52n + 24,995,000.25 = 0$$
using the quadratic formula,
$$n = [7699.52 \pm \sqrt{(-7699.52)^2 - 4(.5929)(24,995,000.25)}]/[2(.5929)]$$
$$= [.7699.52 \pm \sqrt{4468.1112}]/1.1858$$
$$= 6549.47 \ or \ 6436.73$$
If n ≥ 6549.47, then P(x ≥ 5000) ≥ .90.   [corresponding to z = -1.28]
If n ≥ 6436.73, then P(x ≥ 5000) ≥ .10.   [corresponding to z = +1.28]
The desired solution is n = 6550.   [rounded up]

33. binomial with n = 200 and p = P(a woman is taller than 66 inches)
= [not given directly, must be calculated]
adult female heights follow a normal distribution with $\mu = 63.6$ and $\sigma = 2.5$

$P(x > 66) = P(z > .96)$
$= .5000 - .3315$
$= .1685$

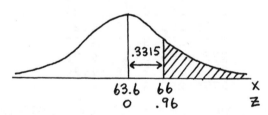

binomial with n = 200 and p = .1685
normal approximation appropriate since
np = 200(.1685) = 33.7 ≥ 5
n(1-p) = 200(.8315) = 166.3 ≥ 5
$\mu = np = 200(.1685) = 33.7$
$\sigma = \sqrt{np(1-p)} = \sqrt{200(.1685)(.8315)} = 5.294$
$P(x \geq 40) = P_c(x > 39.5)$
$= P(z > 1.10)$
$= .5000 - .3643$
$= .1357$

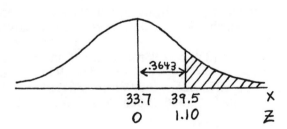

Review Exercises

1. normal distribution with $\mu = 106$ and $\sigma = 3$

a. $P(x < 100) = P(z < -2.00)$
$= .5000 - .4772$
$= .0228$

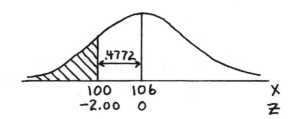

b. $P(x > 110) = P(z > 1.33)$
$= .5000 - .4082$
$= .0918$

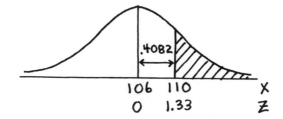

c. $P(x < 112) = P(z < 2.00)$
$= .5000 + .4772$
$= .9772$

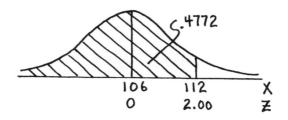

d. $P(100 < x < 109) = P(-2.00 < z < 1.00)$
$= .4772 + .3413$
$= .8185$

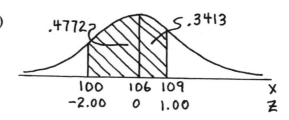

e. $P(110 < x < 115) = P(1.33 < z < 3.00)$
$= .4987 - .4082$
$= .0905$

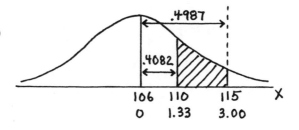

2. a. A = .4000 [.3997]   b. A = .3000 [.2995]
   z = 1.28   z = -.84
   x = μ + zσ x = μ + zσ
     = 106 + 1.28(3)   = 106 - .84(3)
     = 109.84   = 103.48

3. normal distribution with μ = 0 and σ = 1
   [Because this is a standard normal, the variable is already expressed in terms of z scores.]
   a. P(0 < z < 1.25) = .3994

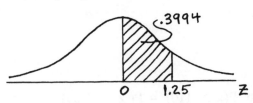

   b. P(z > .50) = .5000 - .1915
                 = .3085

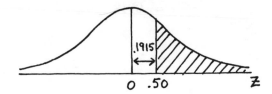

   c. P(z > -1.08) = .3599 + .5000
                   = .8599

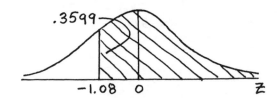

   d. P(-.50 < z < 1.50) = .1915 + .4332
                         = .6247

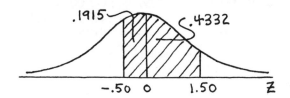

   e. P(-1.00 < z < -.25) = .3413 - .0987
                          = .2426

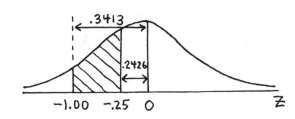

4. binomial with n = 850 and p = .04
normal approximation appropriate since
$np = 850(.04) = 34 \geq 5$
$n(1-p) = 850(.96) = 816 \geq 5$
$\mu = np = 850(.04) = 34$
$\sigma = \sqrt{np(1-p)} = \sqrt{850(.04)(.96)} = 5.713$
$P(x \geq 50) = P_c(x > 49.5)$
$= P(z > 2.71)$
$= .5000 - .4966$
$= .0034$

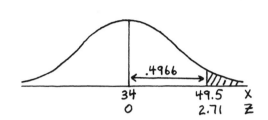

5. normal distribution, by the Central Limit Theorem
$\mu_{\bar{x}} = \mu = 41182$
$\sigma_{\bar{x}} = \sigma/\sqrt{n} = 19990/\sqrt{125} = 1787.96$
$P(\bar{x} > 40000) = P(z > -.66)$
$= .2454 + .5000$
$= .7454$

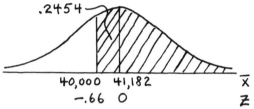

No additional assumptions about the data are necessary -- the Central Limit Theorem applies to <u>any</u> distribution with a given μ and σ.

6. a. normal with $\mu = 26.9$ and $\sigma = 8.4$.
$P(x < 25) = P(z < -.23)$
$= .5000 - .0910$
$= .4090$

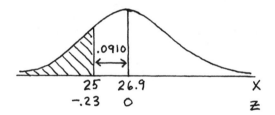

b. normal, since the original distribution is so
$\mu_{\bar{x}} = \mu = 26.9$
$\sigma_{\bar{x}} = \sigma/\sqrt{n} = 8.4/\sqrt{40} = 1.328$
$P(\bar{x} < 25) = P(z < -1.43)$
$= .5000 - .4236$
$= .0764$

c. $A = .1000 \quad [.0987]$
$z = -.25$
$x = \mu + z\sigma$
$= 26.9 - .25(8.4)$
$= 24.8$

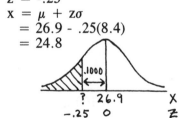

7. binomial with n = 225 and p = .08
normal approximation appropriate since
$np = 225(.08) = 18 \geq 5$
$n(1-p) = 225(.92) = 207 \geq 5$
$\mu = np = 225(.08) = 18$
$\sigma = \sqrt{np(1-p)} = \sqrt{225(.08)(.92)} = 4.07$
$P(x \geq 20) = P_c(x > 19.5)$
$= P(z > .37)$
$= .5000 - .1443$
$= .3557$

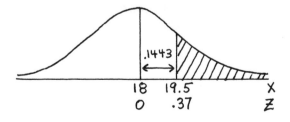

8. a. normal with $\mu = 2.22$ and $\sigma = 1.09$.
$$P(x > 2.00) = P(z > -.20)$$
$$= .0793 + .5000$$
$$= .5793$$

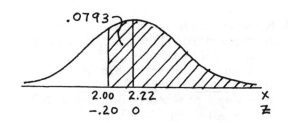

b. normal, since the original distribution is so
$$\mu_{\bar{x}} = \mu = 2.22$$
$$\sigma_{\bar{x}} = \sigma/\sqrt{n} = 1.09/\sqrt{25} = .218$$
$$P(\bar{x} > 2.50) = P(z > 1.28)$$
$$= .5000 - .3997$$
$$= .1003$$

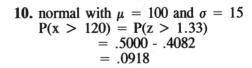

c. $A = .2000$ [.1985]
$$z = -.52$$
$$x = \mu + z\sigma$$
$$= 2.22 - .52(1.09)$$
$$= 1.65$$

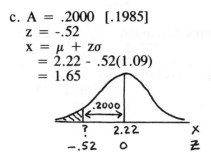

9. binomial with $n = 200$ and $p = .10$
normal approximation appropriate since
$$np = 200(.10) = 20 \geq 5$$
$$n(1-p) = 200(.90) = 180 \geq 5$$
$$\mu = np = 200(.10) = 20$$
$$\sigma = \sqrt{np(1-p)} = \sqrt{200(.10)(.90)} = 4.24$$
a. $P(x = 18) = P_c(17.5 < x < 18.5)$
$$= P(-.59 < z < -.35)$$
$$= .2224 - .1368$$
$$= .0856$$

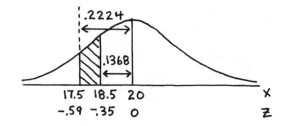

b. $P(x < 25) = P_c(x < 24.5)$
$$= P(z < 1.06)$$
$$= .5000 + .3554$$
$$= .8554$$

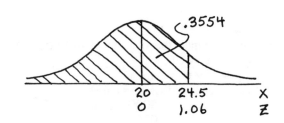

10. normal with $\mu = 100$ and $\sigma = 15$
$$P(x > 120) = P(z > 1.33)$$
$$= .5000 - .4082$$
$$= .0918$$

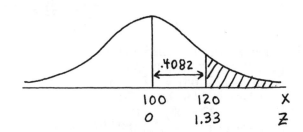

Chapter 6

Estimates and Sample Sizes

6-2 Estimating a Population Mean

IMPORTANT NOTE: This manual uses the following conventions.
(1) The designation "df" stands for "degrees of freedom."
(2) Since the t value depends on both the degrees of freedom and the probability lying beyond it, double subscripts are used to identify points on t distributions. The t distribution with 15 degrees of freedom and .025 beyond it, for example, is designated $t_{15,.025} = 2.132$.
(3) When df $\geq$ 30 the difference between the t and z distributions is negligible and Table A-3 uses one final row of z values to cover all such cases. Consequently, the z scores for certain "popular" α and $\alpha/2$ values may be found by reading Table A-3 "frontwards" instead of reading Table A-2 "backwards." This is not only easier but also more accurate, since Table A-3 includes one more decimal place.

1. a. $\alpha = .05$; $\alpha/2 = .025$
    $z_{.025} = 1.960$ [from Table A-3]
    b. $\alpha = .02$; $\alpha/2 = .01$
    $z_{.01} = 2.327$ [from Table A-3]
    c. $\alpha = .04$; $\alpha/2 = .02$
    $z_{.02} = 2.05$ [from Table A-2 with A = .4800 (or closest entry)]
    d. df = 19
    $\alpha = .05$; $\alpha/2 = .025$
    $t_{19,.025} = 2.093$ [from Table A-3]
    e. df = 14
    $\alpha = .01$; $\alpha/2 = .005$
    $t_{14,.005} = 2.971$ [from Table A-3]

2. a. $z_{.05} = 1.645$
    b. $z_{.025} = 1.960$
    c. $z_{.10} = 1.282$
    d. $t_{9,.05} = 1.833$
    e. $t_{24,.01}\ 2.492$

3. σ known, use z
    $E = z_{.005} \cdot \sigma/\sqrt{n} = 2.575 \cdot 40/\sqrt{25} = 20.6$

4. n $\leq$ 30 and σ unknown, use t
    $E = t_{24,.025} \cdot s/\sqrt{n} = 2.064 \cdot 30/\sqrt{25} = 12.384$

5. n > 30, use z (with s for σ)
    $E = z_{.005} \cdot \sigma/\sqrt{n} = 2.575 \cdot 15/\sqrt{100} = 3.8625$

6. n > 30, use z (with s for σ)
    $E = z_{.005} \cdot \sigma/\sqrt{n} = 2.575 \cdot 30/\sqrt{64} = 9.656$

7. σ known, use z
$\bar{x} \pm z_{.025} \cdot \sigma/\sqrt{n}$
$70.4 \pm 1.960 \cdot 5/\sqrt{36}$
70.4 ± 1.6
$68.8 < \mu < 72.0$

8. σ known, use z
$\bar{x} \pm z_{.005} \cdot \sigma/\sqrt{n}$
$84.2 \pm 2.575 \cdot 7.3/\sqrt{40}$
84.2 ± 3.0
$81.2 < \mu < 87.2$

9. $n > 30$, use z (with s for σ)
$\bar{x} \pm z_{.005} \cdot \sigma/\sqrt{n}$
$5.15 \pm 2.575 \cdot 1.68/\sqrt{4400}$
$5.15 \pm .07$
$5.08 < \mu < 5.22$

10. $n > 30$, use z (with s for σ)
$\bar{x} \pm z_{.005} \cdot \sigma/\sqrt{n}$
$69.7 \pm 2.575 \cdot 2.8/\sqrt{772}$
$69.7 \pm .3$
$69.4 < \mu < 70.0$

11. $n \leq 30$ and σ unknown, use t
$\bar{x} \pm t_{19, .025} \cdot s/\sqrt{n}$
$2.40 \pm 2.093 \cdot 1.30/\sqrt{20}$
$2.40 \pm .61$
$1.79 < \mu < 3.01$

12. $n \leq 30$ and σ unknown, use t
$\bar{x} \pm t_{16, .05} \cdot s/\sqrt{n}$
$.9043 \pm 1.746 \cdot .0414/\sqrt{17}$
$.9043 \pm .0175$
$.8868 < \mu < .9218$

13. $E = 1.5$
$\alpha = .02$
$n = [z_{.01} \cdot \sigma/E]^2 = [2.327 \cdot 15/1.5]^2 = 541.49$ rounded up to 542

14. $n = [z_{.005} \cdot \sigma/E]^2 = [2.575 \cdot .068/.025]^2 = 49.06$ rounded up to 50

15. $n > 30$, use z (with s for σ)
$\bar{x} \pm z_{.005} \cdot \sigma/\sqrt{n}$
$5.622 \pm 2.575 \cdot 0.68/\sqrt{50}$
$5.622 \pm .025$
$5.597 < \mu < 5.647$
While this confidence interval does not include the value 5.670, it does not contradict the mint's claim that quarters are produced at that weight. The mint is making a claim about the weight of new (i.e., uncirculated) quarters, but the quarters in the sample were taken from general circulation and had experienced wear. This problem reinforces the importance of making certain that the sample selected for observation is representative of the population of interest.

16. $\underline{n} \leq 30$ and σ unknown, use t
$\overline{x} \pm t_{19,.05} \cdot s/\sqrt{n}$
$24.2 \pm 1.729 \cdot 8.7/\sqrt{20}$
24.2 ± 3.4
$20.8 < \mu < 27.6$

17. $E = 2.0$
$\alpha = .05$
$n = [z_{.025} \cdot \sigma/E]^2 = [1.960 \cdot 21.2/2.0]^2 = 431.64$ rounded up to 432

18. $n = [z_{.01} \cdot \sigma/E]^2 = [2.327 \cdot 12.46/2]^2 = 210.17$ rounded up to 211

19. $\underline{n} > 30$, use $\underline{z}$ (with s for σ)
$\overline{x} \pm z_{.005} \cdot \sigma/\sqrt{n}$
$27.44 \pm 2.575 \cdot 12.46/\sqrt{62}$
27.44 ± 4.07
$23.37 < \mu < 31.51$
No; since we are 99% certain that the true mean is less than 35, we are 99% certain that the facility will not be overburdened.

20. $\underline{n} > 30$, use $\underline{z}$ (with s for σ)
$\overline{x} \pm z_{.015} \cdot \sigma/\sqrt{n}$
$191.7 \pm 2.17 \cdot 41.0/\sqrt{1525}$
191.7 ± 2.3
$189.4 < \mu < 194.0$
No; since we are 97% certain that the true mean is less than 200, we are 97% certain that the doctor is not correct.

21. a. $\underline{n} > 30$, use $\underline{z}$ (with s for σ)
$\overline{x} \pm z_{.005} \cdot \sigma/\sqrt{n}$
$41.8 \pm 2.575 \cdot 16.7/\sqrt{570}$
41.8 ± 1.8
$40.0 < \mu < 43.6$
   b. $E = .5$
$\alpha = .01$
$n = [z_{.005} \cdot \sigma/E]^2 = [2.757 \cdot 16.7/.5]^2 = 7396.86$ rounded up to 7397

22. a. $\underline{n} > 30$, use $\underline{z}$ (with s for σ)
$\overline{x} \pm z_{.025} \cdot \sigma/\sqrt{n}$
$3.94 \pm 1.960 \cdot .75/\sqrt{231}$
$3.94 \pm .10$
$3.84 < \mu < 3.94$
   b. $n = [z_{.025} \cdot \sigma/E]^2 = [1.960 \cdot .75/.05]^2 = 864.36$ rounded up to 865

23. $\underline{n} = 16 \qquad \Sigma x = 133.3 \qquad \Sigma x^2 = 1167.99$
$\overline{x} = 8.33$
$s = 1.957$ [do not round, store <u>all</u> the digits in the calculator]
$\underline{n} \leq 30$ and σ unknown, use t
$\overline{x} \pm t_{15,.025} \cdot \sigma/\sqrt{n}$
$8.33 \pm 2.132 \cdot 1.957/\sqrt{16}$
8.33 ± 1.04
$7.29 < \mu < 9.37$

24. $\underline{n} = 36$ $\Sigma x = 1889.7$ $\Sigma x^2 = 128840.56$ [for ease of calculations, work in $1000's]
$\bar{x} = 52.49$
$s = 29.104$ [do not round, store all the digits in the calculator]
$\underline{n} > 30$, use $\underline{z}$ (with s for σ)
$\bar{x} \pm z_{.05} \cdot \sigma / \sqrt{n}$
$52.49 \pm 1.645 \cdot 29.104 / \sqrt{36}$
52.49 ± 7.98
$44.51 < \mu < 60.47$ [or $44,510 < \mu < $60,470]

25. $\underline{n} = 62$ $\Sigma x = 584.54$ $\Sigma x^2 = 6570.8216$
$\bar{x} = 9.428$
$s = 4.168$ [do not round, store all the digits in the calculator]
$\underline{n} > 30$, use $\underline{z}$ (with s for σ)
$\bar{x} \pm z_{.025} \cdot \sigma / \sqrt{n}$
$9.428 \pm 1.960 \cdot 4.168 / \sqrt{62}$
9.428 ± 1.038
$8.390 < \mu < 10.466$

26. $\underline{n} = 40$ $\Sigma x = 536.5$ $\Sigma x^2 = 9872.67$
$\bar{x} = 13.41$
$s = 8.285$ [do not round, store all the digits in the calculator]
$\underline{n} > 30$, use $\underline{z}$ (with s for σ)
$\bar{x} \pm z_{.005} \cdot \sigma / \sqrt{n}$
$13.41 \pm 2.575 \cdot 8.285 / \sqrt{40}$
13.41 ± 3.37
$10.04 < \mu < 16.79$

27. a. Use only films with the R rating.
$\underline{n} = 35$ $\Sigma x = 3890$ $\Sigma x^2 = 446620$
$\bar{x} = 111.1$
$s = 20.49$ [do not round, store all the digits in the calculator]
$\underline{n} > 30$, use $\underline{z}$ (with s for σ)
$\bar{x} \pm z_{.025} \cdot \sigma / \sqrt{n}$
$111.1 \pm 1.960 \cdot 20.49 / \sqrt{35}$
111.1 ± 6.8
$104.3 < \mu < 117.9$

b. Use only films with the PG or PG-13 rating.
$\underline{n} = 23$ $\Sigma x = 2552$ $\Sigma x^2 = 297902$
$\bar{x} = 111.0$
$s = 25.89$ [do not round, store all the digits in the calculator]
$\underline{n} \leq 30$ and σ unknown, use t
$\bar{x} \pm t_{22,.025} \cdot \sigma / \sqrt{n}$
$111.0 \pm 2.074 \cdot 25.89 / \sqrt{23}$
111.0 ± 11.2
$99.8 < \mu < 122.2$

c. Since there were 35 (i.e., more than 30) films with the R rating, part (a) used z scores; since there were only 23 (i.e., less than or equal to 30) films with the PG or PG-13 rating, part (b) used t scores. The confidence interval in part (b) is wider because (1) the PG and PG-13 films exhibited more variability (as measured by the standard deviation) than the R films and (2) the larger number of R films produced a larger denominator in the standard error and a smaller table value (i.e., t or z value) by which to multiply the standard error to determine E.

28. An individual woman with a serum cholesterol level of 200 should not be concerned about being unusual. If the levels are normally distributed with $\mu = 191.7$ and $\sigma = 41.0$, then $P(x \geq 200) = P(z > .20) = .5000 - .0793 = .4207$ -- i.e., over 42% of the population have levels of 200 or higher. Be careful; the above statistics indicate that the woman is not unusual, but they say nothing about whether she is in good health.

29. $n > 30$, use z (with s for σ)   NOTE: For the original 23 data values
$$E = z_{.025} \cdot \sigma / \sqrt{n}$$
$$= 1.960 \cdot 7.26 / \sqrt{850}$$
$$= .5$$

$$\Sigma x = 19962.2$$
$$\Sigma x^2 = 13284510.72$$
$$\overline{x} = 665.41$$
$$s = 7.26$$

30. For the 95% confidence interval $430 < \mu < 470$, $\overline{x} = (430 + 470)/2 = 450$ and $E_{95\%} = 20$.
  a. Since $E_{95\%} = 1.960 \cdot \sigma / \sqrt{n}$ and $E_{99\%} = 2.575 \cdot \sigma / \sqrt{n}$,
$$E_{99\%} = (2.575/1.960) \cdot E_{95\%} = (2.575/1.960) \cdot 20 = 26$$
    and the 99% confidence interval is $\overline{x} \pm E_{99\%}$
$$450 \pm 26$$
$$424 < \mu < 476$$
  b. Since the given interval is symmetric about the sample mean, $\overline{x} = (430 + 470)/2 = 450$.
  c. $E_{95\%} = 1.960 \cdot \sigma / \sqrt{n}$
$$20 = 1.960 \cdot \sigma / \sqrt{100}$$
$$\sigma = 200/1.960 = 102$$
  d. For the interval $432 < \mu < 468$, $\overline{x} = (432 + 468)/2 = 450$ and $E = 18$.
$$E_? = z \cdot \sigma / \sqrt{n}$$
$$18 = z \cdot 102 / \sqrt{100}$$
$$z = 180/102 = 1.76$$
    For $z = 1.76$, Table A-2 indicates $A = .4608$ or about 46%.
    The confidence level is $2 \cdot A = 2 \cdot 46\% = 92\%$.

31. $E = 1.5$
  $\alpha = .02$
  $N = 200$
$$n = [N\sigma^2(z_{.01})^2]/[(N-1)E^2 + \sigma^2(z_{.01})^2]$$
$$= [(200)(15)^2(2.327)^2]/[(199)(1.5)^2 + (15)^2(2.327)^2]$$
$$= [243671.8]/[1666.109]$$
$$= 146.25 \text{ rounded up to } 147$$

32. $n \leq 30$ and σ unknown, use t
  $n > .05 \cdot N$, use the finite population correction factor
$$\overline{x} \pm t_{29,.025} \cdot s / \sqrt{n} \cdot \sqrt{(N-n)/(N-1)}$$
$$132 \pm 2.045 \cdot 10 / \sqrt{30} \cdot \sqrt{70/99}$$
$$132 \pm 3$$
$$129 < \mu < 135 \text{ [Since } \overline{x} \text{ is an integer, decimal accuracy is not possible in the endpoints.]}$$

6-3 Estimating a Population Proportion

IMPORTANT NOTE: When calculating confidence intervals using the formula
$$\hat{p} \pm E$$
$$\hat{p} \pm z_{\alpha/2}\sqrt{\hat{p}\hat{q}/n}$$
do not round off in the middle of the problem. This may be accomplished conveniently on most calculators having a memory as follows.
  (1) Calculate $\hat{p} = x/n$ and STORE the value
  (2) Calculate E as 1 - RECALL = * RECALL = ÷ n = $\sqrt{}$ * $z_{\alpha/2}$ =

(3) With the value of E showing on the display, the upper confidence limit is calculated by + RECALL.

(4) With the value of the upper confidence limit showing on the display, the lower confidence limit is calculated by - RECALL $\pm$ + RECALL

You must become familiar with your own calculator. [Do your homework using the same type of calculator you will be using for the exams.] The above procedure works on most calculators; make certain you understand why it works and verify whether it works on your calculator. If it does not seem to work on your calculator, or if your calculator has more than one memory so that you can STORE both $\hat{p}$ and E at the same time, ask your instructor for assistance.

NOTE: It should be true that $0 \leq \hat{p} \leq 1$ and that $E \leq .5$ [usually, <u>much</u> less than .5]. If such is not the case, an error has been made.

1. $\hat{p} = x/n = 100/500 = .20$
$\alpha = .05$
$E = z_{.025}\sqrt{\hat{p}\hat{q}/n} = 1.960\sqrt{(.20)(.80)/500} = .0351$

2. $\hat{p} = x/n = 300/2000 = .15$
$E = z_{.025}\sqrt{\hat{p}\hat{q}/n} = 1.960\sqrt{(.15)(.85)/2000} = .0156$

3. $\hat{p} = x/n = 325/1068 = .304$
$\alpha = .05$
$E = z_{.025}\sqrt{\hat{p}\hat{q}/n} = 1.960\sqrt{(.304)(.696)/1068} = .0276$

4. $\hat{p} = x/n = 50/1776 = .028$
$E = z_{.025}\sqrt{\hat{p}\hat{q}/n} = 1.960\sqrt{(.028)(.912)/1776} = .00769$

5. $\hat{p} = x/n = 100/400 = .250$
$\alpha = .05$
$\hat{p} \pm z_{.025}\sqrt{\hat{p}\hat{q}/n}$
$.250 \pm 1.960\sqrt{(.250)(.750)/400}$
$.250 \pm .042$
$.208 < p < .292$

6. $\hat{p} = x/n = 400/900 = .444$
$\hat{p} \pm z_{.025}\sqrt{\hat{p}\hat{q}/n}$
$.444 \pm 1.960\sqrt{(.444)(.556)/900}$
$.444 \pm .032$
$.412 < p < .477$

7. $\hat{p} = x/n = 309/512 = .604$
$\alpha = .02$
$\hat{p} \pm z_{.01}\sqrt{\hat{p}\hat{q}/n}$
$.604 \pm 2.327\sqrt{(.604)(.396)/512}$
$.604 \pm .050$
$.553 < p < .654$

8. $\hat{p} = x/n = 3456/12485 = .277$
$\hat{p} \pm z_{.005}\sqrt{\hat{p}\hat{q}/n}$
$.277 \pm 2.575\sqrt{(.277)(.723)/12485}$
$.277 \pm .010$
$.267 < p < .287$

9. E = .02
α = .03
$\hat{p}$ unknown, use $\hat{p}$ = .5
n = $[(z_{.015})^2\hat{p}\hat{q}]/E^2$
  = $[(2.17)^2(.5)(.5)]/(.02)^2$
  = 2943.06 rounded up to 2944

NOTE: To find $z_{.015}$ in Table A-3,
find A = .4850 (or closest entry)
read z from the margins of the table

10. $\hat{p}$ = x/n = 79/1180 = .067
$\hat{p} \pm z_{.025}\sqrt{\hat{p}\hat{q}/n}$
.067 $\pm$ 1.960$\sqrt{(.067)(.933)/1180}$
.067 $\pm$ .014
.0527 < p < .0812

11. $\hat{p}$ = .63
α = .02
$\hat{p} \pm z_{.01}\sqrt{\hat{p}\hat{q}/n}$
.63 $\pm$ 2.327$\sqrt{(.63)(.37)/1500}$
.63 $\pm$ .03
.60 < p < .66
NOTE: Since $\hat{p}$ was given with only two decimal accuracy and the actual value of x was not given, the final answer is limited to two decimal accuracy. Any x from 938 to 952 rounds to 63% with varying digits for the next decimal point.

12. $\hat{p}$ = .24
n = $[(z_{.005})^2\hat{p}\hat{q}]/E^2$ = $[(2.575)^2(.24)(.76)]/(.04)^2$ = 755.89 rounded up to 756

13. $\hat{p}$ = x/n = 1280/2000 = .640
α = .05
$\hat{p} \pm z_{.025}\sqrt{\hat{p}\hat{q}/n}$
.640 $\pm$ 1.960$\sqrt{(.640)(.360)/2000}$
.640 $\pm$ .021
.553 < p < .654

14. $\hat{p}$ unknown, use $\hat{p}$ = .5
n = $[(z_{.04})^2\hat{p}\hat{q}]/E^2$
  = $[(1.75)^2(.5)(.5)]/(.03)^2$
  = 850.7 rounded up to 851

NOTE: To find $z_{.04}$ in Table A-3,
find A = .4600 (or closest entry)
read z from the margins of the table

15. E = .02
α = .10
$\hat{p}$ = .29
n = $[(z_{.05})^2\hat{p}\hat{q}]/E^2$ = $[(1.645)^2(.29)(.71)]/(.02)^2$ = 1392.93 rounded up to 1393

16. $\hat{p}$ = .67
$\hat{p} \pm z_{.01}\sqrt{\hat{p}\hat{q}/n}$
.67 $\pm$ 2.327$\sqrt{(.67)(.33)/24350}$
.67 $\pm$ .01
.66 < p < .68
NOTE: Since $\hat{p}$ was given with only two decimal accuracy and the actual value of x was not given, the final answer is limited to two decimal accuracy. Any x from 16,193 to 16,436 rounds to 67% with varying digits for the next decimal point.

17. $\hat{p} = .75$
$\alpha = .10$
$\hat{p} \pm z_{.05}\sqrt{\hat{p}\hat{q}/n}$
$.75 \pm 1.645\sqrt{(.75)(.25)/600}$
$.75 \pm .03$
$.72 < p < .78$
NOTE: Since $\hat{p}$ was given with only two decimal accuracy and the actual value of x was not given, the final answer is limited to two decimal accuracy. Any x from 447 to 453 rounds to 75% with varying digits for the next decimal point.

18. $\hat{p} = .183$
$\hat{p} \pm z_{.01}\sqrt{\hat{p}\hat{q}/n}$
$.183 \pm 2.327\sqrt{(.183)(.817)/785}$
$.183 \pm .032$
$.151 < p < .215$

19. $\hat{p} = x/n = 312/650 = .480$
$\alpha = .05$
$\hat{p} \pm z_{.025}\sqrt{\hat{p}\hat{q}/n}$
$.480 \pm 1.960\sqrt{(.480)(.520)/650}$
$.480 \pm .038$
$.442 < p < .518$

20. $\hat{p} = .85$
$n = [(z_{.05})^2\hat{p}\hat{q}]/E^2 = [(1.645)^2(.85)(.15)]/(.025)^2 = 552.03$ rounded up to 553

21. $E = .03$ NOTE: To find $z_{.03}$ in Table A-3,
$\alpha = .06$ find A = .4700 (or closest entry)
$\hat{p} = .27$ read z from the margins of the table
$n = [(z_{.03})^2\hat{p}\hat{q}]/E^2$
$= [(1.88)^2(.27)(.73)]/(.03)^2$
$= 774.03$ rounded up to 775

22. $\hat{p}$ unknown, use $\hat{p} = .5$
$n = [(z_{.025})^2\hat{p}\hat{q}]/E^2 = [(1.960)^2(.5)(.5)]/(.04)^2 = 600.25$ rounded up to 601

23. $\hat{p} = x/n = 180/650 = .277$
$\alpha = .05$
$\hat{p} \pm z_{.025}\sqrt{\hat{p}\hat{q}/n}$
$.277 \pm 1.960\sqrt{(.277)(.723)/650}$
$.277 \pm .034$
$.243 < p < .311$

24. $\hat{p}$ unknown, use $\hat{p} = .5$
$n = [(z_{.025})^2\hat{p}\hat{q}]/E^2 = [(1.960)^2(.5)(.5)]/(.015)^2 = 4268.4$ rounded up to 4269

25. $\hat{p} = .72$
$\alpha = .01$
$\hat{p} \pm z_{.005}\sqrt{\hat{p}\hat{q}/n}$
$.72 \pm 2.575\sqrt{(.72)(.28)/4664}$
$.72 \pm .02$
$.70 < p < .74$
NOTE: Since $\hat{p}$ was given with only two decimal accuracy and the actual value of x was not given, the final answer is limited to two decimal accuracy. Any x from 3335 to 3381 rounds to 72% with varying digits for the next decimal point.

26. $\hat{p} = x/n = 5463/6503 = .840$
$\hat{p} \pm z_{.005}\sqrt{\hat{p}\hat{q}/n}$
$.840 \pm 2.575\sqrt{(.840)(.160)/6503}$
$.840 \pm .012$
$.828 < p < .852$
Yes; since the confidence interval is entirely above 80%, we are 99% certain that the question meets the stated criterion for being an easy question.

27. There are 36 of the 125 individuals that studied beyond high school.
$\hat{p} = x/n = 36/125 = .288$
$\alpha = .05$
$\hat{p} \pm z_{.025}\sqrt{\hat{p}\hat{q}/n}$
$.288 \pm 1.960\sqrt{(.288)(.712)/125}$
$.288 \pm .079$
$.209 < p < .367$

28. There are 17 of the 100 M & M's that were red.
$\hat{p} = x/n = 17/100 = .170$
$\hat{p} \pm z_{.025}\sqrt{\hat{p}\hat{q}/n}$
$.170 \pm 1.960\sqrt{(.170)(.830)/100}$
$.170 \pm .074$
$.096 < p < .244$
Yes; since .2 is within the interval, the result is consistent with the manufacturer's claim.

29. There are 35 of the 60 movies that received the R rating.
$\hat{p} = x/n = 35/60 = .583$
$\alpha = .05$
$\hat{p} \pm z_{.025}\sqrt{\hat{p}\hat{q}/n}$
$.583 \pm 1.960\sqrt{(.583)(.417)/60}$
$.583 \pm .125$
$.459 < p < .708$

30. There are 23 of the 62 households that discarded <u>more than</u> 10 pounds of paper.
$\hat{p} = x/n = 23/62 = .371$
$\hat{p} \pm z_{.025}\sqrt{\hat{p}\hat{q}/n}$
$.371 \pm 1.960\sqrt{(.371)(.629)/62}$
$.371 \pm .120$
$.251 < p < .491$

31. $E = .01$
$\alpha = .05$ [since 19/20 implies 95% confidence]
$\hat{p}$ unknown, use $\hat{p} = .5$
$n = [(z_{.025})^2\hat{p}\hat{q}]/E^2 = [(1.960)^2(.5)(.5)]/(.01)^2 = 9604$

32. $\hat{p} = x/n = 3/8 = .375$
$\hat{p} \pm z_{.025}\sqrt{\hat{p}\hat{q}/n}$
$.375 \pm 1.960\sqrt{(.375)(.625)/8}$
$.375 \pm .335$
$.040 < p < .710$
While the endpoints above fall in the same general areas as the correct endpoints, it is clear that using the normal approximation when n = 8 does not give an accurate answer.

33. $E = .002$
$\hat{p} = .08$
$n = 47000$
Solve $E = z_{\alpha/2}\sqrt{\hat{p}\hat{q}/n}$ for $z_{\alpha/2}$ to get
$z_{\alpha/2} = E/\sqrt{\hat{p}\hat{q}/n}$
$= .002/\sqrt{(.08)(.92)/47000}$
$= .002/.00125$
$= 1.60$
Since $P(-1.60 < z < 1.60) = .4452 + .4452$
$= .8904$,
the level of confidence is $.8904$, or about 89%.

34. a. normal distribution with $\mu = 100$ and $\sigma = 15$
$P(x > 130) = P(z > 2.00) = .5000 - .4772 = .0228$
b. $\hat{p} = .0228$
$n = [(z_{.01})^2\hat{p}\hat{q}]/E^2 = [(2.327)^2(.0228)(.9772)]/(.025)^2 = 193.03$ rounded up to 194

35. a.
$$E = z_{\alpha/2}\sqrt{\hat{p}\hat{q}/n}\sqrt{(N-n)/(N-1)}$$
$$E^2 = (z_{\alpha/2})^2 \cdot [\hat{p}\hat{q}/n] \cdot [(N-n)/(N-1)] \quad \text{squaring both sides}$$
$$n(N-1)E^2 = (z_{\alpha/2})^2\hat{p}\hat{q}(N-n) \quad \text{multiplying by } n(N-1)$$
$$n(N-1)E^2 = (z_{\alpha/2})^2\hat{p}\hat{q}N - (z_{\alpha/2})^2\hat{p}\hat{q}n \quad \text{distributing } (z_{\alpha/2})^2\hat{p}\hat{q}$$
$$(z_{\alpha/2})^2\hat{p}\hat{q}n + n(N-1)E^2 = (z_{\alpha/2})^2\hat{p}\hat{q}N \quad \text{adding } (z_{\alpha/2})^2\hat{p}\hat{q}$$
$$n[(z_{\alpha/2})^2\hat{p}\hat{q} + (N-1)E^2] = (z_{\alpha/2})^2\hat{p}\hat{q}N \quad \text{factoring out n}$$
$$n = (z_{\alpha/2})^2\hat{p}\hat{q}N/[(z_{\alpha/2})^2\hat{p}\hat{q} + (N-1)E^2]$$

b. $E = .03$
$\alpha = .06$
$\hat{p} = .27$
$n = (z_{\alpha/2})^2\hat{p}\hat{q}N/[(z_{\alpha/2})^2\hat{p}\hat{q} + (N-1)E^2]$
$= (1.88)^2(.27)(.73)(500)/[(1.88)^2(.27)(.73) + (499)(.03)^2]$
$= 348.315/1.14573$
$= 304.01$ rounded up to 305

36. $\hat{p} = x/n = 630/750 = .840$
$\hat{p} - z_{.05}\sqrt{\hat{p}\hat{q}/n}$
$.840 - 1.645\sqrt{(.840)(.160)/750}$
$.840 - .022$
$.818 < p$ [i.e., we are 95% confident that $p > 81.8\%$]

6-4 Estimating a Population Variance

1. a. From the 95% and σ^2 section of Table 6-2, n=97.
b. The best point estimate for σ^2 is $s^2 = 144.0$.
c. $\chi_L^2 = \chi_{26,.975}^2 = 13.844$
$\chi_R^2 = \chi_{26,.025}^2 = 41.923$
d. $(n-1)s^2/\chi_R^2 < \sigma^2 < (n-1)s^2/\chi_L^2$
$(26)(144.0)/41.923 < \sigma^2 < (26)(144.0)/13.844$
$89.3 < \sigma^2 < 270.4$
e. Taking the square roots, $9.5 < \sigma < 16.4$

2. a. From the 99% and σ section of Table 6-2, n=336.
b. The best point estimate for σ^2 is $s^2 = (1.50)^2 = 2.25$
c. $\chi_L^2 = \chi_{17,.995}^2 = 5.697$
$\chi_R^2 = \chi_{17,.005}^2 = 35.718$

d. $(n-1)s^2/\chi_R^2 < \sigma^2 < (n-1)s^2/\chi_L^2$
$(17)(2.25)/35.718 < \sigma^2 < (17)(2.25)/5.697$
$1.07 < \sigma^2 < 6.71$
e. Taking the square roots, $1.03 < \sigma < 2.59$

3. summary information
   n = 10
   $\Sigma x = 1018$ $\bar{x} = 101.8$
   $\Sigma x^2 = 104908$ $s^2 = 141.7$
   a. $s^2 = 141.7$
   b. $(n-1)s^2/\chi_{9,.025}^2 < \sigma^2 < (n-1)s^2/\chi_{9,.975}^2$
   $(9)(141.7)/19.023 < \sigma^2 < (9)(141.7)/2.700$
   $67.1 < \sigma^2 < 472.4$
   $8.2 < \sigma < 21.7$
   c. Yes, $8.2 < 15 < 21.7$

2. summary information
   n = 8
   $\Sigma x = 508.4$ $\bar{x} = 63.55$
   $\Sigma x^2 = 32329.84$ $s^2 = 3.00$
   a. $s^2 = 3.00$
   b. $(n-1)s^2/\chi_{7,.005}^2 < \sigma^2 < (n-1)s^2/\chi_{7,.995}^2$
   $(7)(3.00)/20.278 < \sigma^2 < (7)(3.00)/.989$
   $1.04 < \sigma^2 < 21.25$
   $1.02 < \sigma < 4.61$
   c. Yes, $1.02 < 2.5 < 4.61$

5. $(n-1)s^2/\chi_{100,.005}^2 < \sigma^2 < (n-1)s^2/\chi_{100,.995}^2$
$(100)(1.68)^2/140.169 < \sigma^2 < (100)(1.68)^2/67.328$
$2.01 < \sigma^2 < 4.19$
$1.42 < \sigma < 2.05$

6. $(n-1)s^2/\chi_{80,.005}^2 < \sigma^2 < (n-1)s^2/\chi_{80,.995}^2$
$(80)(2.8)^2/116.321 < \sigma^2 < (80)(2.8)^2/51.172$
$5.39 < \sigma^2 < 12.26$
$2.3 < \sigma < 3.5$

7. $(n-1)s^2/\chi_{19,.025}^2 < \sigma^2 < (n-1)s^2/\chi_{19,.975}^2$
$(19)(1.30)^2/32.852 < \sigma^2 < (19)(1.30)^2/8.907$
$.98 < \sigma^2 < 3.61$
$.99 < \sigma < 1.90$

8. $(n-1)s^2/\chi_{16,.05}^2 < \sigma^2 < (n-1)s^2/\chi_{16,.95}^2$
$(16)(.0414)^2/26.296 < \sigma^2 < (16)(.0414)^2/7.962$
$.001043 < \sigma^2 < .003444$
$.0323 < \sigma < .0587$

9. $(n-1)s^2/\chi_{49,.005}^2 < \sigma^2 < (n-1)s^2/\chi_{49,.995}^2$
$(49)(.068)^2/79.490 < \sigma^2 < (49)(.068)^2/27.991$
$.003 < \sigma^2 < .008$

10. $(n-1)s^2/\chi^2_{61,.005} < \sigma^2 < (n-1)s^2/\chi^2_{61,.995}$
$(61)(12.46)^2/91.952 < \sigma^2 < (61)(12.46)^2/35.534$
$102.99 < \sigma^2 < 266.52$
$10.15 < \sigma < 16.33$

11. $(n-1)s^2/\chi^2_{100,.01} < \sigma^2 < (n-1)s^2/\chi^2_{100,.99}$
$(100)(41.0)^2/135.807 < \sigma^2 < (100)(41.0)^2/70.065$
$1237.8 < \sigma^2 < 2399.2$
$35.2 < \sigma < 49.0$

12. $(n-1)s^2/\chi^2_{90,.025} < \sigma^2 < (n-1)s^2/\chi^2_{90,.975}$
$(90)(.75)^2/118.136 < \sigma^2 < (90)(.75)^2/65.647$
$.43 < \sigma^2 < .77$

13. summary information
n = 16
$\Sigma x = 133.3$ $\bar{x} = 8.33$
$\Sigma x^2 = 1167.99$ $s^2 = 3.829$
$(n-1)s^2/\chi^2_{15,.025} < \sigma^2 < (n-1)s^2/\chi^2_{15,.975}$
$(15)(3.829)/27.488 < \sigma^2 < (15)(3.829)/6.262$
$2.09 < \sigma^2 < 9.17$
$1.45 < \sigma < 3.03$

14. summary information [for ease of calculations, work in $1000's]
n = 36
$\Sigma x = 1889.7$ $\bar{x} = 52.49$
$\Sigma x^2 = 128840.56$ $s^2 = 847.06$
$(n-1)s^2/\chi^2_{35,.05} < \sigma^2 < (n-1)s^2/\chi^2_{35,.95}$
$(35)(847.06)/43.773 < \sigma^2 < (35)(847.06)/18.493$
$677.29 < \sigma^2 < 1603.15$
$26.02 < \sigma < 40.04$
$\$26,020 < \sigma < \$40,040$

NOTE: The desired χ^2 degrees of freedom of 35 falls exactly half way between the tabled values for 30 and 40. The above solution adopts the conservative approach of choosing the values for df = 30. Choosing the values for df = 40 [i.e., $\chi^2_{.05} = 26.509$ and $\chi^2_{.95} = 55.758$], produces the interval $\$23,060 < \sigma < \$33,440$. Other options would be to estimate the desired tabled value using the mean of the values for 30 and 40 degrees of freedom or to estimate the exact interval endpoints using the means of the given endpoints. At any rate, the guideline in the text produces endpoints rounded to $10's since the data was rounded to $100's.

15. summary information
n = 62
$\Sigma x = 584.54$ $\bar{x} = 9.428$
$\Sigma x^2 = 6570.8216$ $s^2 = 17.373$
$(n-1)s^2/\chi^2_{61,.025} < \sigma^2 < (n-1)s^2/\chi^2_{61,.975}$
$(61)(17.373)/83.298 < \sigma^2 < (61)(17.373)/40.482$
$12.722 < \sigma^2 < 26.178$
$3.567 < \sigma < 5.116$

16. summary information
$n = 40$
$\Sigma x = 536.5$ $\bar{x} = 13.41$
$\Sigma x^2 = 9872.67$ $s^2 = 68.64$
$$(n-1)s^2/\chi^2_{39,.005} < \sigma^2 < (n-1)s^2/\chi^2_{39,.995}$$
$$(39)(68.64)/66.766 < \sigma^2 < (39)(68.64)/20.707$$
$$40.09 < \sigma^2 < 129.27$$
$$6.33 < \sigma < 11.37$$

17. The given interval $2.8 < \sigma < 6.0$
$$7.84 < \sigma^2 < 36.00$$
and the usual calculations $(n-1)s^2/\chi^2_{19,\alpha/2} < \sigma^2 < (n-1)s^2/\chi^2_{19,1-\alpha/2}$
$$(19)(3.8)^2/\chi^2_{19,\alpha/2} < \sigma^2 < (19)(3.8)^2/\chi^2_{19,1-\alpha/2}$$
$$274.36/\chi^2_{19,\alpha/2} < \sigma^2 < 274.36/\chi^2_{19,1-\alpha/2}$$
imply that $7.84 = 274.37/\chi^2_{19,\alpha/2}$ and $36.00 = 274.36/\chi^2_{19,1-\alpha/2}$
$\chi^2_{19,\alpha/2} = 274.36/7.84$ $\chi^2_{19,1-\alpha/2} = 274.36/36.00$
$= 34.99$ $= 7.62$
The closest entries in Table A-4 are $\chi^2_{19,\alpha/2} = 34.805$ and $\chi^2_{19,1-\alpha/2} = 7.633$
which imply $\alpha/2 = .01$ $1 - \alpha/2 = .99$
$\alpha = .01$ $\alpha/2 = .01$
$\alpha = .02$
The level of confidence is therefore is $1-\alpha = 98\%$.

18. $(n-1)s^2/\chi^2_{11,.025} < \sigma^2 < (n-1)s^2/\chi^2_{11,.975}$
using the lower endpoint   OR
$(11)s^2/21.920 = (19.1)^2$
$s^2 = 726.97$
$s = 27.0$

using the upper endpoint
$(11)s^2/3.816 = (45.8)^2$
$s^2 = 727.69$
$s = 27.0$

19. $\chi^2 = \frac{1}{2}[\pm z_{.025} + \sqrt{2 \cdot df - 1}]^2$
$= \frac{1}{2}[\pm 1.960 + \sqrt{2 \cdot (771) - 1}]^2$
$= \frac{1}{2}[\pm 1.960 + 39.256]^2$
$= \frac{1}{2}[37.296]^2$ or $\frac{1}{2}[41.216]^2$
$= 695.48$ or 849.36
$$(n-1)s^2/\chi^2_{771,.025} < \sigma^2 < (n-1)s^2/\chi^2_{771,.975}$$
$$(771)(2.8)^2/849.36 < \sigma^2 < (771)(2.8)^2/695.48$$
$$7.117 < \sigma^2 < 8.691$$
$$2.7 < \sigma < 2.9$$

20. $(n-1)s^2/\chi^2_R < \sigma^2 < (n-1)s^2/\chi^2_L$
using the lower endpoint   OR
$499 \cdot (4.8)^2/\chi^2_R = (4.5459)^2$
$\chi^2_R = 556.344$
$\chi^2_R = .5 \cdot [+z + \sqrt{2k-1}]^2$
$556.344 = .5 \cdot [+z + \sqrt{997}]^2$
$33.357 = +z + \sqrt{997}$
$1.78 = +z$
$z = 1.78$

using the upper endpoint
$499 \cdot (4.8)^2/\chi^2_L = (5.0788)^2$
$\chi^2_L = 445.719$
$\chi^2_L = .5 \cdot [-z + \sqrt{2k-1}]^2$
$445.719 = .5 \cdot [-z + \sqrt{997}]^2$
$29.857 = -z + \sqrt{997}$
$-1.78 = -z$
$z = 1.72$

NOTE: This problem is very sensitive numerically. Lower endpoint 4.5 (rounded to the accuracy for s), for example, produces z = 2.12. In theory, either endpoint produces z -- and the two results should agree. Proceed using the mean $z = (1.78 + 1.72)/2 = 1.75$. Table A-2 area of .4599 corresponds to a confidence level of $2(.4599) = .9198$, rounded to 92%.

Review Exercises

1. a. The best point estimate for μ is $\bar{x} = 17.6$
   b. $n > 30$, use z (with s for σ)
   $\bar{x} \pm z_{.025} \cdot \sigma/\sqrt{n}$
   $17.6 \pm 1.960 \cdot 9.3/\sqrt{50}$
   17.6 ± 2.6
   $15.0 < \mu < 20.2$

2. a. The best point estimate for the population variance σ^2 is $s^2 = (9.3)^2 = 86.5$
   b. $(n-1)s^2/\chi^2_{49,.025} < \sigma^2 < (n-1)s^2/\chi^2_{49,.975}$
   $(49)(86.49)^2/71.420 < \sigma^2 < (49)(86.49)^2/32.357$
   $59.34 < \sigma^2 < 130.98$
   $7.7 < \sigma < 11.4$

3. $(n-1)s^2/\chi^2_{24,.025} < \sigma^2 < (n-1)s^2/\chi^2_{24,.975}$
   $(24)(3.74)^2/39.364 < \sigma^2 < (24)(3.74)^2/12.401$
   $8.528 < \sigma^2 < 27.071$
   $2.92 < \sigma < 5.20$

4. $n \leq 30$ and σ unknown, use t
   $\bar{x} \pm t_{24,.025} \cdot s/\sqrt{n}$
   $7.01 \pm 2.064 \cdot 3.74/\sqrt{25}$
   7.01 ± 1.54
   $5.47 < \mu < 8.55$

5. $\hat{p} = x/n = 70/781 = .0896$ NOTE: $n = 70 + 711 = 781$
   $\alpha = .05$
   $\hat{p} \pm z_{.025}\sqrt{\hat{p}\hat{q}/n}$
   $.0896 \pm 1.960\sqrt{(.0896)(.9104)/781}$
   $.0896 \pm .0200$
   $.0696 < p < .1097$

6. $\hat{p}$ unknown, use $\hat{p} = .5$
   $n = [(z_{.02})^2\hat{p}\hat{q}]/E^2 = [(2.05)^2(.5)(.5)]/(.06)^2 = 291.8$ rounded up to 292

7. $E = 250$
   $\alpha = .04$
   $n = [z_{.02} \cdot \sigma/E]^2 = [2.05 \cdot 3050/250]^2 = 625.50$ rounded up to 626

8. $\hat{p} = .93$
   $n = [(z_{.01})^2\hat{p}\hat{q}]/E^2 = [(2.327)^2(.93)(.07)]/(.04)^2 = 220.3$ rounded up to 221

9. $n > 30$, use z (with s for σ)
   $\bar{x} \pm z_{.005} \cdot \sigma/\sqrt{n}$
   $40.7 \pm 2.575 \cdot 10.2/\sqrt{40}$
   40.7 ± 4.2
   $36.5 < \mu < 44.9$

10. $(n-1)s^2/\chi^2_{39,.005} < \sigma^2 < (n-1)s^2/\chi^2_{39,.995}$
   $(39)(10.2)^2/66.766 < \sigma^2 < (39)(10.2)^2/20.707$
   $60.77 < \sigma^2 < 195.95$
   $7.8 < \sigma < 14.0$

11. E = .04
    α = .10
    $\hat{p}$ unknown, use $\hat{p}$ = .5
    n = $[(z_{.05})^2\hat{p}\hat{q}]/E^2$ = $[(1.645)^2(.5)(.5)]/(.04)^2$ = 422.82 rounded up to 423

12. $\hat{p}$ = x/n = x/1475 = .320 [and so x = 472]
    $\hat{p} \pm z_{.025}\sqrt{\hat{p}\hat{q}/n}$
    .320 $\pm$ 1.960$\sqrt{(.320)(.680)/1475}$
    .320 $\pm$ .024
    .296 < p < .344

13. E = 4
    α = .05
    n = $[z_{.025} \cdot \sigma/E]^2$ = $[1.960 \cdot 41.0/4]^2$ = 403.61 rounded up to 404

14. $\hat{p}$ = .135
    n = $[(z_{.02})^2\hat{p}\hat{q}]/E^2$ = $[(2.05)^2(.135)(.865)]/(.015)^2$ = 2181.1 rounded up to 2182

15. $\hat{p}$ = .24
    α = .01
    $\hat{p} \pm z_{.005}\sqrt{\hat{p}\hat{q}/n}$
    .24 $\pm$ 2.575$\sqrt{(.24)(.76)/1998}$
    .24 $\pm$ .02
    .22 < p < .26
    NOTE: Since $\hat{p}$ was given with only two decimal accuracy and the actual value of x was not given, the final answer is limited to two decimal accuracy. Any x from 470 to 489 rounds to 24% with varying digits for the next decimal point.

16. n $\leq$ 30 and σ unknown, use t
    $\bar{x} \pm t_{27,.05} \cdot s/\sqrt{n}$
    37.9 $\pm$ 1.703 $\cdot$ 7.3/$\sqrt{28}$
    37.9 $\pm$ 2.3
    35.6 < μ < 40.2

17. n = 16 Σx = 1162.6 Σx^2 = 84577.34
    $\bar{x}$ = 72.66 [do not round, store all the digits in the calculator]
    s = 2.581 [do not round, store all the digits in the calculator]
    n $\leq$ 30 and σ unknown, use t
    $\bar{x} \pm t_{15,.025} \cdot \sigma/\sqrt{n}$
    72.66 $\pm$ 2.132 $\cdot$ 2.581/$\sqrt{16}$
    72.66 $\pm$ 1.38
    71.29 < μ < 74.04

18. n = 16 Σx = 1162.6 Σx^2 = 84577.34
    $\bar{x}$ = 72.66
    s = 6.661
    $(n-1)s^2/\chi^2_{15,.025} < \sigma^2 < (n-1)s^2/\chi^2_{15,.975}$
    $(15)(6.661)^2/27.488 < \sigma^2 < (15)(6.661)^2/6.262$
    3.63 < σ^2 < 15.96
    1.91 < σ < 3.99

Chapter 7

Hypothesis Testing

7-2 Fundamentals of Hypothesis testing

1. a. $\mu = 120$
   b. $H_o:\mu = 120$
   c. $H_1:\mu \neq 120$
   d. two-tailed
   e. rejecting the hypothesis that the mean IQ is 120 when it really is 120
   f. failing to reject the hypothesis that the mean IQ is 120 when it really is not 120
   g. there is sufficient evidence to reject the claim that the mean IQ is 120
   h. there is not sufficient evidence to reject the claim that the mean IQ is 120

2. a. $\mu < 10$
   b. $H_o:\mu \geq 10$
   c. $H_1:\mu < 10$
   d. left-tailed
   e. rejecting the hypothesis that the mean wgt. is at least 10 when it really is at least 10
   f. failing to reject the hypothesis that the mean wgt. is at least 10 when it really is less than 10
   g. there is sufficient evidence to support the claim that the mean wgt. is less than 10
   h. there is not sufficient evidence to support the claim that the mean wgt. is less than 10

3. a. $\mu > 5$
   b. $H_o:\mu \leq 5$
   c. $H_1:\mu > 5$
   d. right-tailed
   e. rejecting the hypothesis that the mean time is at most 5 years when it really is
   f. failing to reject the hypothesis that the mean time is at most 5 years when it really is not
   g. there is sufficient evidence to support the claim that the mean time is more than 5 years
   h. there is not sufficient evidence to support the claim that the mean time is more than 5 years

4. a. $\mu = 191,000$
   b. $H_o:\mu = 191,000$
   c. $H_1:\mu \neq 191,000$
   d. two-tailed
   e. rejecting the hypothesis that the mean income is 191,000 when it really is 191,000
   f. failing to reject the hypothesis that the mean income is 191,000 when it really is not 191,000
   g. there is sufficient evidence to reject the claim that the mean income is 191,000
   h. there is not sufficient evidence to reject the claim that the mean income is 191,000

5. a. $\mu \geq 10$
   b. $H_o:\mu \geq 10$
   c. $H_1:\mu < 10$
   d. left-tailed
   e. rejecting the hypothesis that the mean age is at least 10 years when it really is
   f. failing to reject the hypothesis that the mean age is at least 10 years when it really is not
   g. there is sufficient evidence to reject the claim that the mean age is at least 10 years
   h. there is not sufficient evidence to reject the claim that the mean age is at least 10 years

6. a. $\mu \leq 17$
b. $H_o: \mu \leq 17$
c. $H_1: \mu > 17$
d. right-tailed
e. rejecting the hypothesis that the mean mpg is no more than 17 when it really is
f. failing to reject the hypothesis that the mean mpg is no more than 17 when it really is not
g. there is sufficient evidence to reject the claim that the mean mpg is no more than 17
h. there is not sufficient evidence to reject the claim that the mean mpg is no more than 17

7. the critical z values are $\pm z_{.005} = \pm 2.575$
   the critical region is z < -2.575
   $\qquad$ z > 2.575

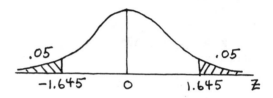

8. the critical z values are $\pm z_{.05} = \pm 1.645$
   the critical region is z < -1.645
   $\qquad$ z > 1.645

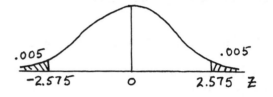

9. the critical z value is $z_{.01} = 2.327$ [from the last row of Table A-3]
   the critical region is z > 2.327

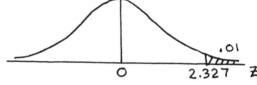

10. the critical z value is $-z_{.025} = -1.960$
   the critical region is z < -1.960

11. the critical z value is $-z_{.05} = -1.645$
    the critical region is z < -1.645

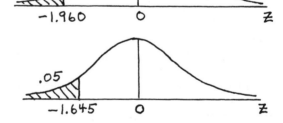

12. the critical z value is $z_{.02} = 2.05$
    the critical region is z > 2.05

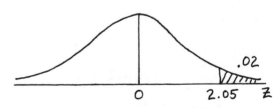

13. a. The conclusion indicates H_o was rejected; the null hypothesis must have been $\mu \leq 30$.
    b. The conclusion indicated H_o was not rejected; the null hypothesis must have been $\mu \geq 5$.
    c. The conclusion indicates H_o was rejected; the null hypothesis must have been $\mu = 70$.

14. a. Underestimating the mean would be the more serious error.
   b. H_o: $\mu \geq 27.44$
      H_1: $\mu < 27.44$
   c. Choose $\alpha = .01$ since we want the probability of incorrectly concluding H_o to be low.

15. Mathematically, in order for α to equal 0 the magnitude of the critical value would have to be infinite. Practically, the only way never to make a type I error is to always fail to reject H_o. From either perspective, the only way to achieve $\alpha = 0$ is to never reject H_o no matter how extreme the sample data might be.

7-3 Testing a Claim about a Mean: Large Samples

1. original claim: $\mu = 100$ [n > 30, use z (with s for σ)]
   H_o: $\mu = 100$
   H_1: $\mu \neq 100$
   $\alpha = .01$
   C.R. $z < -z_{.005} = -2.575$
      $z > z_{.005} = 2.575$
   calculations:
      $z_{\bar{x}} = (\bar{x} - \mu)/\sigma_{\bar{x}}$
         $= (100.8 - 100)/(5/\sqrt{81}\,)$
         $= .8/.556$
         $= 1.44$
   conclusion:
      Do not reject H_o; there is not sufficient evidence to conclude that $\mu \neq 100$.

2. original claim: $\mu \geq 20$ [n > 30, use z (with s for σ)]
   H_o: $\mu \geq 20$
   H_1: $\mu < 20$
   $\alpha = .05$
   C.R. $z < -z_{.05} = -1.645$
   calculations:
      $z_{\bar{x}} = (\bar{x} - \mu)/\sigma_{\bar{x}}$
         $= (18.7 - 20)/(3/\sqrt{100})$
         $= -1.3/.3$
         $= -4.333$
   conclusion:
      Reject H_o; there is sufficient evidence to conclude that $\mu < 20$.

3. original claim: $\mu = 500$ [n > 30, use z (with s for σ)]
   H_o: $\mu = 500$
   H_1: $\mu \neq 500$
   $\alpha = .10$
   C.R. $z < -z_{.05} = -1.645$
      $z > z_{.05} = 1.645$
   calculations:
      $z_{\bar{x}} = (\bar{x} - \mu)/\sigma_{\bar{x}}$
         $= (510 - 500)/(50/\sqrt{300})$
         $= 10/2.887$
         $= 3.464$
   conclusion:
      Reject H_o; there is sufficient evidence to conclude that $\mu \neq 500$ (in fact, $\mu > 500$).

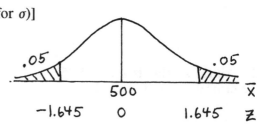

4. original claim: $\mu > 40$ [n > 30, use z (with s for σ)]
 $H_o: \mu \leq 40$
 $H_1: \mu > 40$
 $\alpha = .05$
 C.R. $z > z_{.05} = 1.645$
 calculations:
 $\quad z_{\bar{x}} = (\bar{x} - \mu)/\sigma_{\bar{x}}$
 $\quad\quad = (42 - 40)/(8/\sqrt{50}\,)$
 $\quad\quad = 2/1.131$
 $\quad\quad = 1.768$
 conclusion:
 $\quad$ Reject H_o; there is sufficient evidence to conclude that $\mu > 40$.

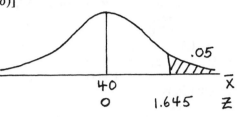

5. original claim: $\mu < 12$ [n > 30, use z (with s for σ)]
 $H_o: \mu \geq 12$
 $H_1: \mu < 12$
 $\alpha = .01$
 C.R. $z < -z_{.01} = -2.327$
 calculations:
 $\quad z_{\bar{x}} = (\bar{x} - \mu)/\sigma_{\bar{x}}$
 $\quad\quad = (11.82 - 12)/(.38/\sqrt{36}\,)$
 $\quad\quad = -.18/.0633$
 $\quad\quad = -2.842$
 conclusion:
 $\quad$ Reject H_o; there is sufficient evidence to conclude that $\mu < 12$.

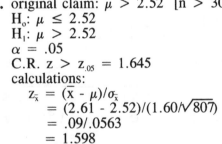

6. original claim: $\mu > 2.52$ [n > 30, use z (with s for σ)]
 $H_o: \mu \leq 2.52$
 $H_1: \mu > 2.52$
 $\alpha = .05$
 C.R. $z > z_{.05} = 1.645$
 calculations:
 $\quad z_{\bar{x}} = (\bar{x} - \mu)/\sigma_{\bar{x}}$
 $\quad\quad = (2.61 - 2.52)/(1.60/\sqrt{807})$
 $\quad\quad = .09/.0563$
 $\quad\quad = 1.598$
 conclusion:
 $\quad$ Do not reject H_o; there is not sufficient evidence to conclude that $\mu > 2.52$.

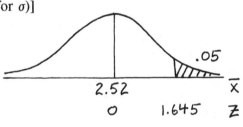

7. original claim: $\mu > .21$ [n > 30, use z (with s for σ)]
 $H_o: \mu \leq .21$
 $H_1: \mu > .21$
 $\alpha = .01$
 C.R. $z > z_{.01} = 2.327$
 calculations:
 $\quad z_{\bar{x}} = (\bar{x} - \mu)/\sigma_{\bar{x}}$
 $\quad\quad = (.83 - .21)/(.24/\sqrt{32}\,)$
 $\quad\quad = .62/.0424$
 $\quad\quad = 14.614$
 conclusion:
 $\quad$ Reject H_o; there is sufficient evidence to conclude that $\mu > .21$.

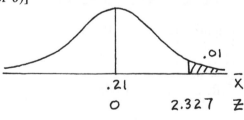

8. original claim: $\mu = 47.0$ [n > 30, use z (with s for σ)]
H_o: $\mu = 47.0$
H_1: $\mu \neq 47.0$
$\alpha = .05$
C.R. $z < -z_{.025} = -1.960$
$\quad z > z_{.025} = 1.960$
calculations:
$\quad z_{\bar{x}} = (\bar{x} - \mu)/\sigma_{\bar{x}}$
$\quad = (39.213 - 47.0)/(7.5675/\sqrt{244})$
$\quad = -7.787/.484$
$\quad = -16.075$

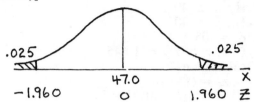

conclusion:
$\quad$ Reject H_o; there is sufficient evidence to conclude that $\mu \neq 47.0$ (in fact, $\mu < 47.0$).

9. original claim: $\mu < 1.39$ [n > 30, use z (with s for σ)]
H_o: $\mu \geq 1.39$
H_1: $\mu < 1.39$
$\alpha = .01$
C.R. $z < -z_{.01} = -2.327$
calculations:
$\quad z_{\bar{x}} = (\bar{x} - \mu)/\sigma_{\bar{x}}$
$\quad = (.83 - 1.39)/(.16/\sqrt{123})$
$\quad = -.56/.0144$
$\quad = -38.817$

conclusion:
$\quad$ Reject H_o; there is sufficient evidence to conclude that $\mu < 1.39$.

10. original claim: $\mu = 13,901$ [n > 30, use z (with s for σ)]
H_o: $\mu = 13,901$
H_1: $\mu \neq 13,901$
$\alpha = .05$
C.R. $z < -z_{.025} = -1.960$
$\quad z > z_{.025} = 1.960$
calculations:
$\quad z_{\bar{x}} = (\bar{x} - \mu)/\sigma_{\bar{x}}$
$\quad = (13,447 - 13,901)/(4883/\sqrt{200})$
$\quad = -454/345.28$
$\quad = -1.315$

conclusion:
$\quad$ Do not reject H_o; there is not sufficient evidence to conclude that $\mu \neq 13,901$.

11. original claim: $\mu > 0$ [n > 30, use z (with s for σ)]
H_o: $\mu \leq 0$
H_1: $\mu > 0$
$\alpha = .05$
C.R. $z > z_{.05} = 1.645$
calculations:
$\quad z_{\bar{x}} = (\bar{x} - \mu)/\sigma_{\bar{x}}$
$\quad = (.6 - 0)/(3.8/\sqrt{75})$
$\quad = .6/.439$
$\quad = 1.367$

conclusion:
$\quad$ Do not reject H_o; there is not sufficient evidence to conclude that $\mu > 0$.

12. original claim: $\mu > 90{,}000$ [n > 30, use z (with s for σ)]
H_o: $\mu \le 90{,}000$
H_1: $\mu > 90{,}000$
$\alpha = .05$
C.R. $z > z_{.05} = 1.645$
calculations:

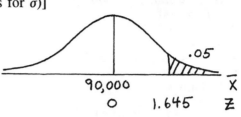

$z_{\bar{x}} = (\bar{x} - \mu)/\sigma_{\bar{x}}$
$= (96{,}700 - 90{,}000)/(37{,}500/\sqrt{191})$
$= 6700/2713$
$= 2.469$
conclusion:
  Reject H_o; there is sufficient evidence to conclude that $\mu > 90{,}000$.

13. original claim: $\mu > 51.0$ [n > 30, use z (with s for σ)]
H_o: $\mu \le 51.0$
H_1: $\mu > 51.0$
$\alpha = .05$
C.R. $z > z_{.05} = 1.645$
calculations:

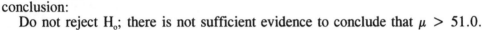

$z_{\bar{x}} = (\bar{x} - \mu)/\sigma_{\bar{x}}$
$= (52.4 - 51.0)/(13.14/\sqrt{150})$
$= 1.4/1.073$
$= 1.305$
conclusion:
  Do not reject H_o; there is not sufficient evidence to conclude that $\mu > 51.0$.

14. original claim: $\mu < 7.5$ [n > 30, use z (with s for σ)]
H_o: $\mu \ge 7.5$
H_1: $\mu < 7.5$
$\alpha = .05$
C.R. $z < -z_{.05} = -1.645$
calculations:

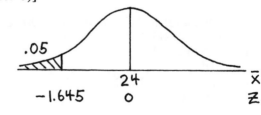

$z_{\bar{x}} = (\bar{x} - \mu)/\sigma_{\bar{x}}$
$= (7.01 - 7.5)/(3.74/\sqrt{100})$
$= -.49/.374$
$= -1.310$
conclusion:
  Do not reject H_o; there is not sufficient evidence to conclude that $\mu < 7.5$.

15. original claim: $\mu < 24$ [n > 30, use z (with s for σ)]
H_o: $\mu \ge 24$
H_1: $\mu < 24$
$\alpha = .05$
C.R. $z < -z_{.05} = -1.645$
calculations:

$z_{\bar{x}} = (\bar{x} - \mu)/\sigma_{\bar{x}}$
$= (22.1 - 24)/(8.6/\sqrt{200})$
$= -1.9/.608$
$= -3.124$
conclusion:
  Reject H_o; there is sufficient evidence to conclude that $\mu < 24$.

16. original claim: $\mu = 600$ [$n > 30$, use z (with s for σ)]

$H_o: \mu = 600$

$H_1: \mu \neq 600$

$\alpha = .01$

C.R. $z < -z_{.005} = -2.575$

   $z > z_{.005} = 2.575$

calculations:

   $z_{\bar{x}} = (\bar{x} - \mu)/\sigma_{\bar{x}}$

      $= (589 - 600)/(21/\sqrt{65}\,)$

      $= -11/2.605$

      $= -4.223$

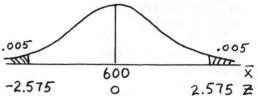

conclusion:

   Reject H_o; there is sufficient evidence to conclude that $\mu \neq 600$ (in fact, $\mu < 600$).

17. original claim: $\mu = 7124$ [$n > 30$, use z (with s for σ)]

$H_o: \mu \geq 7124$

$H_1: \mu < 7124$

$\alpha = .05$

C.R. $z < -z_{.05} = -1.645$

calculations:

   $z_{\bar{x}} = (\bar{x} - \mu)/\sigma_{\bar{x}}$

      $= (6047 - 7124)/(2944/\sqrt{750})$

      $= -1077/107.500$

      $= -10.019$

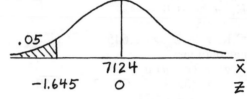

conclusion:

   Reject H_o; there is sufficient evidence to conclude that $\mu < 7124$.

18. original claim: $\mu = 180$ [$n > 30$, use z (with s for σ)]

$H_o: \mu = 180$

$H_1: \mu \neq 180$

$\alpha = .05$

C.R. $z < -z_{.025} = -1.960$

   $z > z_{.025} = 1.960$

calculations:

   $z_{\bar{x}} = (\bar{x} - \mu)/\sigma_{\bar{x}}$

      $= (182 - 180)/(52/\sqrt{45}\,)$

      $= 2/7.752$

      $= .258$

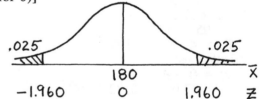

conclusion:

   Do not reject H_o; there is not sufficient evidence to conclude that $\mu \neq 180$.

19. original claim: $\mu > 40,000$ [$n > 30$, use z (with s for σ)]

$H_o: \mu \leq 40,000$

$H_1: \mu > 40,000$

$\alpha = .05$

C.R. $z > z_{.05} = 1.645$

calculations:

   $z_{\bar{x}} = (\bar{x} - \mu)/\sigma_{\bar{x}}$

      $= (41.182 - 40,000)/(19,990/\sqrt{1700})$

      $= 1182/484.829$

      $= 2.438$

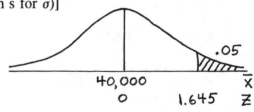

conclusion:

   Reject H_o; there is sufficient evidence to conclude that $\mu > 40,000$.

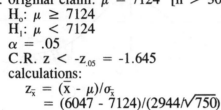

20. original claim: $\mu > 420$ [$n > 30$, use z (with s for σ)]
   H_o: $\mu \leq 420$
   H_1: $\mu > 420$
   $\alpha = .05$
   C.R. $z > z_{.05} = 1.645$
   calculations:

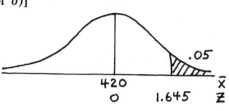

   $z_{\bar{x}} = (\bar{x} - \mu)/\sigma_{\bar{x}}$
   $= (385 - 420)/(24/\sqrt{35})$
   $= -35/4.057 = -8.628$
   conclusion:
   Do not reject H_o; there is not sufficient evidence to conclude that $\mu > 420$.

21. original claim: $\mu = 14$ [$n > 30$, use z (with s for σ)]
   H_o: $\mu = 14$
   H_1: $\mu \neq 14$
   $\alpha = .05$
   C.R. $z < -z_{.025} = -1.960$
   $\quad\; z > z_{.025} = 1.960$
   calculations:

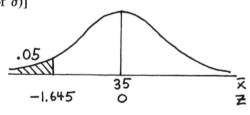

   $z_{\bar{x}} = (\bar{x} - \mu)/\sigma_{\bar{x}}$
   $= (13.41 - 14)/(8.28/\sqrt{40})$
   $= -.59/1.309$
   $= -.451$
   conclusion:
   Do not reject H_o; there is not sufficient evidence to conclude that $\mu \neq 14$.

22. original claim: $\mu < 35$ [$n > 30$, use z (with s for σ)]
   H_o: $\mu \geq 35$
   H_1: $\mu < 35$
   $\alpha = .05$
   C.R. $z < -z_{.05} = -1.645$
   calculations:

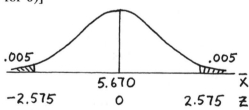

   $z_{\bar{x}} = (\bar{x} - \mu)/\sigma_{\bar{x}}$
   $= (27.44 - 35)/(12.46/\sqrt{62})$
   $= -7.56/1.582 = -4.777$
   conclusion:
   Reject H_o; there is sufficient evidence to conclude that $\mu < 35$.
   No; based on this result, there is no cause for concern that there might be too much garbage.

23. original claim: $\mu = 5.670$ [$n > 30$, use z (with s for σ)]
   H_o: $\mu = 5.670$
   H_1: $\mu \neq 5.670$
   $\alpha = .01$
   C.R. $z < -z_{.005} = -2.575$
   $\quad\; z > z_{.005} = 2.575$
   calculations:

   $z_{\bar{x}} = (\bar{x} - \mu)/\sigma_{\bar{x}}$
   $= (6.622 - 5.670)/(.068/\sqrt{50})$
   $= -.048/.00962$
   $= -4.991$
   conclusion:
   Reject H_o; there is sufficient evidence to conclude that $\mu \neq 5.670$ (in fact, $\mu < 5.670$).
   The fact that the mean weight of worn quarters found in circulation is less than 5.670 is not
      evidence that the mean weight of uncirculated quarters straight from the mint is different
      from 5.670.

24. original claim: $\mu = 453/496 = .9133$ [n > 30, use z (with s for σ)]
  H_o: $\mu = .9133$
  H_1: $\mu \neq .9133$
  $\alpha = .01$
  C.R. z < -$z_{.005}$ = -2.575
    z > $z_{.005}$ = 2.575
  calculations:

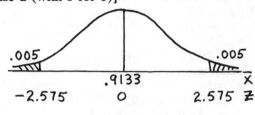

  $z_{\bar{x}} = (\bar{x} - \mu)/\sigma_{\bar{x}}$
    $= (.916 - .9133)/(.043/\sqrt{100})$
    $= .00269/.0043$
    $= .626$
  conclusion:
    Do not reject H_o; there is not sufficient evidence to conclude that $\mu \neq .9133$.

25. original claim: $\mu = 105$ [n > 30, use z (with s for σ)]
  NOTE: From the boxplot, $\bar{x} = 107$ and $P_{75} = 115$.
      Assuming a normal distribution, P_{75} corresponds to z = .67.
      Solving for s, z = $(x - \bar{x})/s$
          $.67 = (115 - 107)/s$
          $.67 = 8/s$
            $s = 8/.67$
              $= 11.940$

  H_o: $\mu = 105$
  H_1: $\mu \neq 105$
  $\alpha = .03$
  C.R. P-value < .03
  calculations:
    $z_{\bar{x}} = (\bar{x} - \mu)/\sigma_{\bar{x}}$
      $= (107 - 105)/(11.940/\sqrt{93})$
      $= 2/1.238$
      $= 1.62$
    P-value $= 2 \cdot P(z > 1.62)$
        $= 2 \cdot [.5000 - .4774]$
        $= 2 \cdot [.0526]$
        $= .1052$
  conclusion:
    Do not reject H_o; there is not sufficient evidence to conclude that $\mu \neq 105$.

26. original claim: $\mu = 100$ [n > 30, use z (with s for σ)]
  H_o: $\mu = 100$
  H_1: $\mu \neq 100$
  $\alpha = .01$
  C.R. z < -$z_{.005}$ = -2.575
    z > $z_{.005}$ = 2.575
  calculations:
    $z_{\bar{x}} = (\bar{x} - \mu)/\sigma_{\bar{x}}$
    $2.575 < (103.6 - 100)/(s/\sqrt{62})$
      $s < (103.6 - 100)/(2.575/\sqrt{62})$
      $s < 3.6/.327$
      $s < 11.008$
  NOTE: There is no conclusion because this exercise is not asking for the completion of a test of hypotheses.

27. original claim: $\mu = 23{,}460$ [$n > 30$, use z (with s for σ)]

    H_o: $\mu = 23{,}460$

    H_1: $\mu \neq 23{,}460$

    $\alpha = .02$

    C.R. $z < -z_{.01} = -2.327$

       $z > z_{.01} = 2.327$

    calculations:

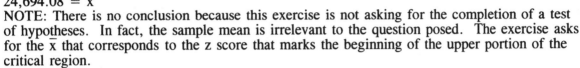

        $z_{\bar{x}} = (\bar{x} - \mu)/\sigma_{\bar{x}}$

      $2.327 = (\bar{x} - 23{,}460)/(3750/\sqrt{50}\,)$

      $2.327 = (\bar{x} - 23{,}460)/530.33$

    $1234.08 = \bar{x} - 23{,}460$

    $24{,}694.08 = \bar{x}$

NOTE: There is no conclusion because this exercise is not asking for the completion of a test of hypotheses. In fact, the sample mean is irrelevant to the question posed. The exercise asks for the $\bar{x}$ that corresponds to the z score that marks the beginning of the upper portion of the critical region.

28. original claim: $\mu = 98.6$ [$n > 30$, use z (with s for σ)]

    H_o: $\mu = 98.6$

    H_1: $\mu \neq 98.6$

    $\alpha = .05$

    C.R. $z < -z_{.025} = -1.960$

       $z > z_{.025} = 1.960$

    calculations:

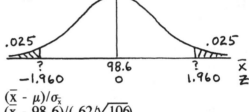

      $z_{\bar{x}} = (\bar{x} - \mu)/\sigma_{\bar{x}}$ $z_{\bar{x}} = (\bar{x} - \mu)/\sigma_{\bar{x}}$

  $-1.960 = (\bar{x} - 98.6)/(.62/\sqrt{106})$ $1.960 = (\bar{x} - 98.6)/(.62/\sqrt{106})$

        $\bar{x} = 98.482$ $\bar{x} = 98.718$

NOTE: There is no conclusion because this exercise is not asking for the completion of a test of hypotheses.

    a. $\beta = P(98.482 < \bar{x} < 98.718 \mid \mu = 98.7)$

       $= P(-3.62 < z < .30)$

       $= .4999 + .1179$

       $= .6178$

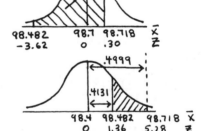

    b. $\beta = P(98.482 < \bar{x} < 98.718 \mid \mu = 98.4)$

       $= P(1.36 < z < 5.28)$

       $= .4999 - .4131$

       $= .0868$

29. original claim: $\mu < 32$ [$n > 30$, use z (with s for σ)]

    H_o: $\mu \geq 32$

    H_1: $\mu < 32$

    $\alpha = .01$

    C.R. $z < -z_{.01} = -2.327$

    calculations:

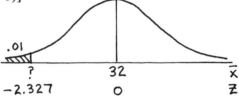

        $z_{\bar{x}} = (\bar{x} - \mu)/\sigma_{\bar{x}}$

    $-2.327 = (\bar{x} - 32)/(.75/\sqrt{50}\,)$

    $-2.327 = (\bar{x} - 32)/.106$

    $-.2468 = \bar{x} - 32$

    $31.7532 = \bar{x}$

NOTE: There is no conclusion because this exercise is not asking for the completion of a test of hypothesis. In fact, the sample mean is irrelevant to the question posed. The above calculations find the $\bar{x}$ that corresponds to the z score that marks the beginning of the critical region. In terms of $\bar{x}$, then, the critical region is $\bar{x} < 31.7532$.

The exercise asks for the value of μ for which P(rejecting H_o) = $P(\bar{x} < 31.7532)$ = .80.
NOTE: The z score with .80 below it [i.e., with A = .3000 in Table A-1] is z = .84.

$z_{\bar{x}} = (\bar{x} - \mu)/\sigma_{\bar{x}}$
$.84 = (31.7532 - \mu)/(.75/\sqrt{50}\,)$
$.84 = (31.7532 - \mu)/.106$
$.0891 = 31.7532 - \mu$
$\mu = 31.7532 - .0891$
$= 31.6641$

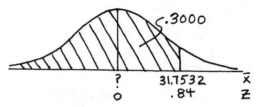

7-4 Testing a Claim about a Mean: Small Samples

1. a. $\pm t_{26,.025} = \pm 2.056$ b. $t_{16,.10} = 1.337$ c. $-t_{5,.01}\ -3.365$

2. a. $t_{26,.10} = 1.315$ b. $\pm t_{15,.025} = \pm 2.132$ c. $-t_{11,.05}\ -1.796$

3. original claim: $\mu \leq 10$ [n $\leq$ 30 and σ unknown, use t]
    $H_o: \mu \leq 10$
    $H_1: \mu > 10$
    $\alpha = .05$
    C.R. $t > t_{8,.05} = 1.860$
    calculations:
    $t_{\bar{x}} = (\bar{x} - \mu)/s_{\bar{x}}$
    $= (11 - 10)/(2/\sqrt{9}\,)$
    $= 1/.667 = 1.500$
    conclusion:
    Do not reject H_o; there is not sufficient evidence to conclude that $\mu > 10$.

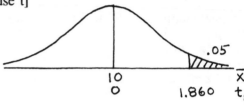

4. original claim: $\mu \geq 100$ [n $\leq$ 30 and σ unknown, use t]
    $H_o: \mu \geq 100$
    $H_1: \mu < 100$
    $\alpha = .05$
    C.R. $t < -t_{21,.05} = -1.721$
    calculations:
    $t_{\bar{x}} = (\bar{x} - \mu)/s_{\bar{x}}$
    $= (95 - 100)/(18/\sqrt{22}\,)$
    $= -5/3.838 = -1.303$
    conclusion:
    Do not reject H_o; there is not sufficient evidence to conclude that $\mu < 100$.

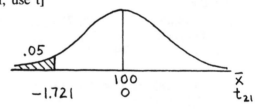

5. original claim: $\mu = 75$ [n $\leq$ 30 and σ unknown, use t]
    $H_o: \mu = 75$
    $H_1: \mu \neq 75$
    $\alpha = .05$
    C.R. $t < -t_{14,.025} = -2.145$
    $t > t_{14,.025} = 2.145$
    calculations:
    $t_{\bar{x}} = (\bar{x} - \mu)/s_{\bar{x}}$
    $= (77.6 - 75)/(5/\sqrt{15}\,)$
    $= 2.6/1.291 = 2.014$
    conclusion:
    Do not reject H_o; there is not sufficient evidence to conclude that $\mu \neq 75$.

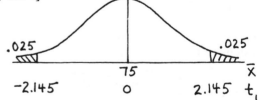

6. original claim: $\mu = 500$ [n $\leq$ 30 and σ unknown, use t]
H_o: $\mu = 500$
H_1: $\mu \neq 500$
$\alpha = .10$
C.R. t $< -t_{19,.05} = -1.729$
        t $> t_{19,.05} = 1.729$
calculations:
$t_{\bar{x}} = (\bar{x} - \mu)/s_{\bar{x}}$
        $= (541 - 500)/(115/\sqrt{20})$
        $= 41/25.715 = 1.594$
conclusion:
    Do not reject H_o; there is not sufficient evidence to conclude that $\mu \neq 500$.

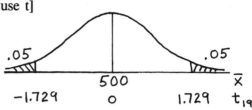

7. original claim: $\mu = 98.6$ [n $\leq$ 30 and σ unknown, use t]
H_o: $\mu = 98.6$
H_1: $\mu \neq 98.6$
$\alpha = .05$
C.R. t $< -t_{24,.025} = -2.064$
        t $> t_{24,.025} = 2.064$
calculations:
$t_{\bar{x}} = (\bar{x} - \mu)/s_{\bar{x}}$
        $= (98.24 - 98.6)/(.56/\sqrt{25})$
        $= -.36/.112 = -3.214$
conclusion:
    Reject H_o; there is sufficient evidence to conclude that $\mu \neq 98.6$ (in fact, $\mu < 98.6$).

8. original claim: $\mu = 98.6$ [n $>$ 30, use z (with s for σ)]
H_o: $\mu = 98.6$
H_1: $\mu \neq 98.6$
$\alpha = .05$
C.R. z $< -z_{.025} = -1.960$
        z $> z_{.025} = 1.960$
calculations:
$z_{\bar{x}} = (\bar{x} - \mu)/\sigma_{\bar{x}}$
        $= (98.27 - 98.6)/(.65/\sqrt{35})$
        $= -.33/.110 = -3.004$
conclusion:
    Reject H_o; there is sufficient evidence to conclude that $\mu \neq 98.6$ (in fact, $\mu < 98.6$).

9. original claim: $\mu > 32$ [n $\leq$ 30 and σ unknown, use t]
H_o: $\mu \leq 32$
H_1: $\mu > 32$
$\alpha = .10$
C.R. t $> t_{26,.10} = 1.315$
calculations:
$t_{\bar{x}} = (\bar{x} - \mu)/s_{\bar{x}}$
        $= (32.2 - 32)/(.4/\sqrt{27})$
        $= .2/.0770 = 2.598$
conclusion:
    Reject H_o; there is sufficient evidence to conclude that $\mu > 32$.

10. The calculated t_{26} for the above one-tailed test is 2.598.
    Since $t_{26,.005} = 2.779 > 2.598 > 2.479 = t_{26,.01}$, $.005 <$ P-value $< .01$.

11. original claim: $\mu > 1800$ [$n \leq 30$ and σ unknown, use t]
    H_o: $\mu \leq 1800$
    H_1: $\mu > 1800$
    $\alpha = .01$
    C.R. t $> t_{11,.01} = 2.718$
    calculations:

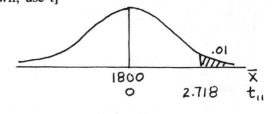

$$t_{\bar{x}} = (\bar{x} - \mu)/s_{\bar{x}}$$
$$= (2133 - 1800)/(345/\sqrt{12}\,)$$
$$= 333/99.593 = 3.344$$

conclusion:
    Reject H_o; there is sufficient evidence to conclude that $\mu > 1800$.

12. original claim: $\mu \geq 154$ [$n \leq 30$ and σ unknown, use t]
    H_o: $\mu \geq 154$
    H_1: $\mu < 154$
    $\alpha = .005$
    C.R. t $< -t_{19,.005} = -2.861$
    calculations:

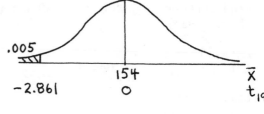

$$t_{\bar{x}} = (\bar{x} - \mu)/s_{\bar{x}}$$
$$= (141 - 154)/(12/\sqrt{20}\,)$$
$$= -13/2.683 = -4.845$$

conclusion:
    Reject H_o; there is sufficient evidence to conclude that $\mu < 154$.

13. The calculated t is larger in magnitude than the largest t (i.e., the t for the smallest α) in Table A-2 for its degrees of freedom. For this one-tailed test, therefore, conclude P-value $< .005$.

14. original claim: $\mu = 500$ [$n \leq 30$ and σ unknown, use t]
    H_o: $\mu = 500$
    H_1: $\mu \neq 500$
    $\alpha = .05$
    C.R. t $< -t_{11,.025} = -2.201$
        t $> t_{11,.025} = 2.201$
    calculations:

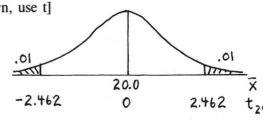

$$t_{\bar{x}} = (\bar{x} - \mu)/s_{\bar{x}}$$
$$= (524 - 500)/(23/\sqrt{12}\,)$$
$$= 24/6.640 = 3.615$$

conclusion:
    Reject H_o; there is sufficient evidence to conclude that $\mu \neq 500$ (in fact, $\mu > 500$).
    No; the manufacturer cannot justifiably claim a mean of 500 m.

15. original claim: $\mu = 20.0$ [$n \leq 30$ and σ unknown, use t]
    H_o: $\mu = 20.0$
    H_1: $\mu \neq 20.0$
    $\alpha = .02$
    C.R. t $< -t_{29,.01} = -2.462$
        t $> t_{29,.01} = 2.462$
    calculations:
        $t_{\bar{x}} = (\bar{x} - \mu)/s_{\bar{x}}$
        $= (20.5 - 20.0)/(1.5/\sqrt{30}\,)$
        $= .5/.274 = 1.826$
    conclusion:
        Do not reject H_o; there is not sufficient evidence to conclude that $\mu \neq 20.0$.

16. original claim: $\mu < 10.00$ [n $\leq$ 30 and σ unknown, use t]

$H_o: \mu \geq 10.00$
$H_1: \mu < 10.00$
$\alpha = .01$
C.R. t $< -t_{15,.01} = -2.602$
calculations:

$t_{\bar{x}} = (\bar{x} - \mu)/s_{\bar{x}}$
$\quad = (8.33 - 10.00)/(1.96/\sqrt{16})$
$\quad = -1.67/.49 = -3.408$

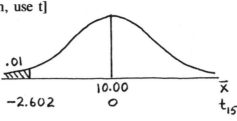

conclusion:
    Reject H_o; there is sufficient evidence to conclude that $\mu < 10.00$.

17. original claim: $\mu > 2000$ [n > 30, use z (with s for σ)]

$H_o: \mu \leq 2000$
$H_1: \mu > 2000$
$\alpha = .025$
C.R. z $> z_{.025} = 1.960$
calculations:

$z_{\bar{x}} = (\bar{x} - \mu)/\sigma_{\bar{x}}$
$\quad = (2177 - 2000)/(843/\sqrt{100})$
$\quad = 177/84.3 = 2.100$

conclusion:
    Reject H_o; there is sufficient evidence to conclude that $\mu > 2000$.

18. original claim: $\mu < .88$ [n $\leq$ 30 and σ unknown, use t]

$H_o: \mu \geq .88$
$H_1: \mu < .88$
$\alpha = .01$
C.R. t $< -t_{7,.01} = -2.998$
calculations:

$t_{\bar{x}} = (\bar{x} - \mu)/s_{\bar{x}}$
$\quad = (.76 - .88/(.04/\sqrt{8})$
$\quad = -.12/.0141 = -8.458$

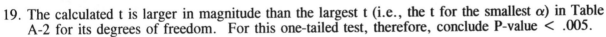

conclusion:
    Reject H_o; there is sufficient evidence to conclude that $\mu < .88$.

19. The calculated t is larger in magnitude than the largest t (i.e., the t for the smallest α) in Table A-2 for its degrees of freedom. For this one-tailed test, therefore, conclude P-value $< .005$.

20. original claim: $\mu > 10.7$ [n $\leq$ 30 and σ unknown, use t]

$H_o: \mu \leq 10.7$
$H_1: \mu > 10.7$
$\alpha = .05$
C.R. t $> t_{24,.05} = 1.711$
calculations:

$t_{\bar{x}} = (\bar{x} - \mu)/s_{\bar{x}}$
$\quad = (12.3 - 10.7)/(11.2/\sqrt{25})$
$\quad = 1.6/2.24 = .714$

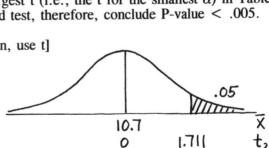

conclusion:
    Do not reject H_o; there is not sufficient evidence to conclude that $\mu > 10.7$.

21. original claim: $\mu = 243.5$ [n ≤ 30 and σ unknown, use t]
 H_0: $\mu = 243.5$
 H_1: $\mu \neq 243.5$
 $\alpha = .05$
 C.R. t < $-t_{21,.025} = -2.080$
 t > $t_{21,.025} = 2.080$
 calculations:

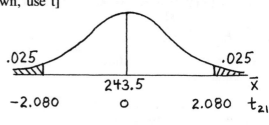

 $t_{\bar{x}} = (\bar{x} - \mu)/s_{\bar{x}}$
 $= (170.2 - 243.5)/(35.3/\sqrt{22})$
 $= -73.3/7.526 = -9.740$
 conclusion:
    Reject H_0; there is sufficient evidence to conclude that $\mu \neq 243.5$ (in fact, $\mu < 243.5$).

22. The calculated t_{21} for the above two-tailed test is -9.740.
    Since $-9.740 < -2.831 = t_{21,.005}$, P-value $< 2 \cdot [.005] = .01$

23. original claim: $\mu > 5$ [n > 30, use z (with s for σ)]
 H_0: $\mu \leq 5$
 H_1: $\mu > 5$
 $\alpha = .10$
 C.R. z > $z_{.10} = 1.282$
 calculations:

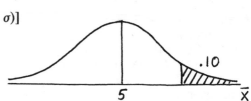

 $z_{\bar{x}} = (\bar{x} - \mu)/\sigma_{\bar{x}}$
 $= (5.15 - 5)/(1.68/\sqrt{80})$
 $= .15/.188 = .799$
 conclusion:
    Do not reject H_0; there is not sufficient evidence to conclude that $\mu > 5$.

24. original claim: $\mu = 41.9$ [n ≤ 30 and σ unknown, use t]
 H_0: $\mu = 41.9$
 H_1: $\mu \neq 41.9$
 $\alpha = .01$
 C.R. t < $-t_{14,.005} = -2.977$
 t > $t_{14,.005} = 2.977$
 calculations:

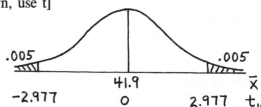

 $t_{\bar{x}} = (\bar{x} - \mu)/s_{\bar{x}}$
 $= (31.0 - 41.9)/(10.5/\sqrt{15})$
 $= -10.9/2.711 = -4.021$
 conclusion:
    Reject H_0; there is sufficient evidence to conclude that $\mu \neq 41.9$ (in fact, $\mu < 41.9$).

25. original claim: $\mu = 650$ [n ≤ 30 and σ unknown, use t]
 H_0: $\mu = 650$
 H_1: $\mu \neq 650$
 $\alpha = .05$
 C.R. t < $-t_{29,.025} = -2.045$
 t > $t_{29,.025} = 2.045$
 calculations:

 $t_{\bar{x}} = (\bar{x} - \mu)/s_{\bar{x}}$
 $= (665.41 - 650)/(7.26/\sqrt{30})$
 $= 15.41/1.325 = 11.626$
 conclusion:
    Reject H_0; there is sufficient evidence to conclude that $\mu \neq 650$ (in fact, $\mu > 650$).

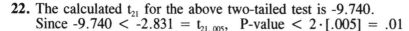
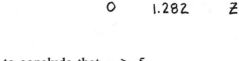
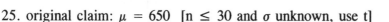

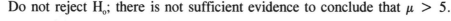

26. original claim: $\mu \geq .913$ [$n \leq 30$ and σ unknown, use t]

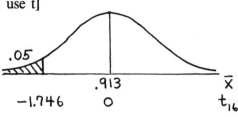

H_o: $\mu \geq .913$
H_1: $\mu < .913$
$\alpha = .05$
C.R. $t < -t_{16,.05} = -1.746$
calculations:
   $t_{\bar{x}} = (\bar{x} - \mu)/s_{\bar{x}}$
     $= (.9043 - .913)/(.0414/\sqrt{17}\,)$
     $= -.0087/.0100 = -.866$
conclusion:
   Do not reject H_o; there is not sufficient evidence to conclude that $\mu < .913$.
NOTE: Using $\mu = 453/496 = .9133$, so that the both μ and $\bar{x}$ are accurate to four decimal places, gives a calculated $t = -.897$.

27. original claim: $\mu > 75$ [$n \leq 30$ and σ unknown, use t]

   $n = 20$
   $\Sigma x = 1561$ $\bar{x} = 78.05$
   $\Sigma x_2 = 122431$ $s = 5.596$
H_o: $\mu \leq 75$
H_1: $\mu > 75$
$\alpha = .05$
C.R. $t > t_{19,.05} = 1.729$

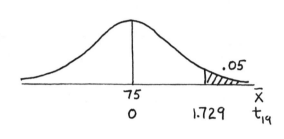

calculations:
   $t_{\bar{x}} = (\bar{x} - \mu)/s_{\bar{x}}$
     $= (78.05 - 75)/(5.596)/\sqrt{20}\,)$
     $= 3.05/1.251 = 2.438$
conclusion:
   Reject H_o; there is sufficient evidence to conclude that $\mu > 75$.

28. original claim: $\mu = 11,000$ [$n \leq 30$ and σ unknown, use t]

   $n = 7$
   $\Sigma x = 74,209$ $\bar{x} = 10,601.286$
   $\Sigma x_2 = 793,035,705$ $s = 1026.718$
H_o: $\mu = 11,000$
H_1: $\mu \neq 11,000$
$\alpha = .05$
C.R. $t < -t_{6,.025} = -2.447$
     $t > t_{6,.025} = 2.447$

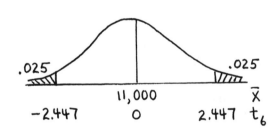

calculations:
   $t_{\bar{x}} = (\bar{x} - \mu)/s_{\bar{x}}$
     $= (10,601.286 - 11,000)/(1026.718/\sqrt{7}\,)$
     $= -398.714/388.063 = -1.027$
conclusion:
   Do not reject H_o; there is not sufficient evidence to conclude that $\mu \neq 11,000$.

29. Since the calculated t is between $t_{6,.10} = 1.440$ and $t_{6,.25} = .718$, the P-value for this two-tailed test is between $2 \cdot (.10)$ and $2 \cdot (.25)$ -- i.e., $.20 < \text{P-value} < .50$.

30. original claim: $\mu = 3.39$ [n $\leq$ 30 and σ unknown, use t]

  n = 16

  $\Sigma x = 58.80$ $\bar{x} = 3.675$

  $\Sigma x_2 = 222.5710$ s = .6573

  H_o: $\mu = 3.39$

  H_1: $\mu \neq 3.39$

  $\alpha = .05$

  C.R. t < $-t_{15,.025} = -2.132$

    t > $t_{15,.025} = 2.132$

  calculations:

  $t_{\bar{x}} = (\bar{x} - \mu)/s_{\bar{x}}$

    $= (3.675 - 3.39)/(.6573/\sqrt{16}\,)$

    $= .285/.164 = 1.734$

  conclusion:

  Do not reject H_o; there is not sufficient evidence to conclude that $\mu \neq 3.39$.

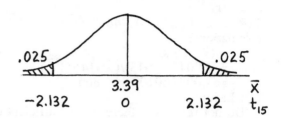

31. original claim: $\mu = 176$ [n $\leq$ 30 and σ unknown, use t]

  The following are the relevant entries from Data Set 3.

#	age	sex	married	weight
17	35	1	2	157
20	41	1	2	176
33	38	1	2	155
34	40	1	2	235
69	38	1	2	176
74	44	1	2	200
82	44	1	2	175
91	42	1	2	216
92	44	1	2	185
95	37	1	2	137
98	39	1	2	176
106	41	1	2	162
110	40	1	2	220
123	42	1	2	191
124	40	1	2	261

  n = 15

  $\Sigma x = 2822$

  $\Sigma x^2 = 546,348$

  $\bar{x} = 188.133$

  s = 33.205

  H_o: $\mu = 176$

  H_1: $\mu \neq 176$

  $\alpha = .05$

  C.R. t < $-t_{14,.025} = -2.145$

    t > $t_{14,.025} = 2.145$

  calculations:

  $t_{\bar{x}} = (\bar{x} - \mu)/s_{\bar{x}}$

    $= (188.133 - 176)/(33.205/\sqrt{15}\,)$

    $= 12.133/8.573 = 1.415$

  conclusion:

  Do not reject H_o; there is not sufficient evidence to conclude that $\mu \neq 176$.

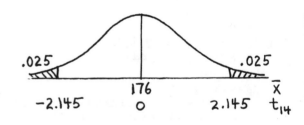

32. Because the z distribution has less spread than a t distribution, $z_\alpha < t_\alpha$ for any α. Since a smaller z score (in absolute value) is needed to reject H_o, rejection is more likely with z than t.

33. $A = z_{.05} \cdot [(8 \cdot df + 3)/(8 \cdot df + 1)]$
$= 1.645 \cdot [(8 \cdot 9 + 3)/(8 \cdot 9 + 1)]$
$= 1.645 \cdot [75/73] = 1.690$
$t = \sqrt{df \cdot (e^{A \cdot A/df} - 1)}$
$= \sqrt{9 \cdot (e^{(1.690)(1.690)/9} - 1)}$
$= \sqrt{9 \cdot (e^{.317} - 1)}$
$= \sqrt{9 \cdot (.3735)}$
$= 1.833$

This agrees exactly with $t_{9,.05} = 1.833$ given in Table A-3.

34. original claim: $\mu > 32$ [$n \leq 30$ and σ unknown, use t]
  H_o: $\mu \leq 32$
  H_1: $\mu > 32$
  $\alpha = .10$
  C.R. $t > t_{26,.10} = 1.315$
  calculations:
  $t_{\bar{x}} = (\bar{x} - \mu)/s_{\bar{x}}$
  $1.315 = (\bar{x} - 32)/(.4/\sqrt{27})$
  $\bar{x} = 32.101$

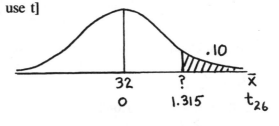

NOTE: There is no conclusion because this exercise is not asking for the completion of a test of hypotheses. The sample mean is irrelevant to the question posed, and the exercise requires finding the $\bar{x}$ that corresponds to the z score that marks the beginning of the critical region.

$\beta = P(\bar{x} < 32.101 \mid \mu = 32.3)$
$= P(t_{26} < -2.582)$
Since $-t_{25,.005} = -2.779 < -2.582 < -t_{26,.01} = -2.471$, $.005 < \beta < .01$.

7-5 Testing a Claim about a Proportion

NOTE: To be consistent with the notation of the previous sections, and thereby reinforcing the patterns and concepts presented in those sections, the manual uses the "usual" z formula written to apply to $\hat{p}$'s
$z_{\hat{p}} = (\hat{p} - \mu_{\hat{p}})/\sigma_{\hat{p}}$
When the normal approximation to the binomial applies, the $\hat{p}$'s are normally distributed
  with $\mu_{\hat{p}} = p$
  and $\sigma_{\hat{p}} = \sqrt{pq/n}$
And so the formula for the z statistic may also be written as
$z_{\hat{p}} = (\hat{p} - p)/\sqrt{pq/n}$

1. original claim: $p \leq .04$ [normal approximation to the binomial, use z]
  H_o: $p \leq .04$ $\hat{p} = x/n = 9/150 = .06$
  H_1: $p > .04$
  $\alpha = .05$
  C.R. $z > z_{.05} = 1.645$
  calculations:
  $z_{\hat{p}} = (\hat{p} - \mu_{\hat{p}})/\sigma_{\hat{p}}$
  $= (.06 - .04)/\sqrt{(.04)(.96)/150}$
  $= .02/.016 = 1.25$

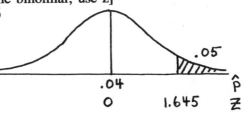

conclusion:
  Do not reject H_o; there is not sufficient evidence to conclude that $p > .04$.
  P-value $= P(z > 1.25) = .5000 - .3944 = .1056$
  We cannot be 95% certain that corrective action is needed.

2. original claim: p = .50 [normal approximation to the binomial, use z]
   H_o: p = .50 $\hat{p}$ = x/n = .43
   H_1: p ≠ .50
   α = .05
   C.R. z < - $z_{.025}$ = -1.960
        z > $z_{.025}$ = 1.960
   calculations:

   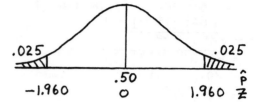

   $z_{\hat{p}} = (\hat{p} - \mu_p)/\sigma_{\hat{p}}$
   $= (.43 - .50)/\sqrt{(.50)(.50)/603}$
   $= -.07/.0204 = -3.438$
   conclusion:
      Reject H_o; there is sufficient evidence to conclude that p ≠ .50 (in fact, p < .50).
   P-value = 2·P(z < -3.44) = 2·[.5000 - .4999] = 2·[.0001] = .0002

3. original claim: p = .71 [normal approximation to the binomial, use z]
   H_o: p = .71 $\hat{p}$ = x/n = .74
   H_1: p ≠ .71
   α = .10
   C.R. z < - $z_{.05}$ = -1.645
        z > $z_{.05}$ = 1.645
   calculations:

   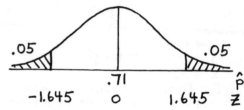

   $z_{\hat{p}} = (\hat{p} - \mu_p)/\sigma_{\hat{p}}$
   $= (.74 - .71)/\sqrt{(.71)(.29)/500}$
   $= .03/.0203 = 1.478$
   conclusion:
      Do not reject H_o; there is not sufficient evidence to conclude that p ≠ .71.
   P-value = 2·P(z > 1.48) = 2·[.5000 - .4306] = 2·[.0694] = .1388

4. original claim: p < .50 [normal approximation to the binomial, use z]
   H_o: p ≥ .50 $\hat{p}$ = x/n = .39
   H_1: p < .50
   α = .005
   C.R. z < - $z_{.005}$ = -2.575
   calculations:

   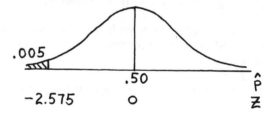

   $z_{\hat{p}} = (\hat{p} - \mu_p)/\sigma_{\hat{p}}$
   $= (.39 - .50)/\sqrt{(.50)(.50)/400}$
   $= -.11/.025 = -4.400$
   conclusion:
      Reject H_o; there is sufficient evidence to conclude that p < .50.
   P-value = P(z < -4.40) = .5000 - .4999 = .0001

5. original claim: p = .20 [normal approximation to the binomial, use z]
   H_o: p = .20 $\hat{p}$ = x/n = 288/(288 + 962) = 288/1250 = .2304
   H_1: p ≠ .20
   α = .02
   C.R. z < - $z_{.01}$ = -2.327
        z > $z_{.01}$ = 2.327
   calculations:

   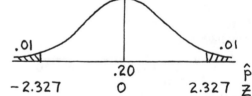

   $z_{\hat{p}} = (\hat{p} - \mu_p)/\sigma_{\hat{p}}$
   $= (.2304 - .20)/\sqrt{(.20)(.80)/1250}$
   $= .0304/.0113 = 2.687$
   conclusion:
      Reject H_o; there is sufficient evidence to conclude that p ≠ .20 (in fact, p > .20).
   P-value = 2·P(z > 2.69) = 2·[.5000 - .4964] = 2·[.0036] = .0072

6. original claim: $p > .784$ [normal approximation to the binomial, use z]
$H_0: p \le .784$ $\qquad\qquad$ $\hat{p} = x/n = 680/750 = .840$
$H_1: p > .784$
$\alpha = .01$
C.R. $z > z_{.01} = 2.327$
calculations:

$z_{\hat{p}} = (\hat{p} - \mu_{\hat{p}})/\sigma_{\hat{p}}$
$\quad = (.840 - .784)/\sqrt{(.784)(.216)/750}$
$\quad = .056/.0150 = 3.727$

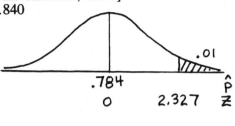

conclusion:
$\quad$ Reject H_0; there is sufficient evidence to conclude that $p > .784$.
P-value $= P(z > 3.73) = .5000 - .4999 = .0001$

7. original claim: $p > .25$ [normal approximation to the binomial, use z]
$H_0: p \le .25$ $\qquad\qquad$ $\hat{p} = x/n = .33$
$H_1: p > .25$
$\alpha = .02$
C.R. $z > z_{.02} = 2.05$
calculations:

$z_{\hat{p}} = (\hat{p} - \mu_{\hat{p}})/\sigma_{\hat{p}}$
$\quad = (.33 - .25)/\sqrt{(.25)(.75)/1400}$
$\quad = .08/.0116 = 6.913$

conclusion:
$\quad$ Reject H_0; there is sufficient evidence to conclude that $p > .25$.
P-value $= P(z > 6.91) = .5000 - .4999 = .0001$

8. original claim: $p > .70$ [normal approximation to the binomial, use z]
$H_0: p \le .70$ $\qquad\qquad$ $\hat{p} = x/n = 813/1084 = .75$
$H_1: p > .70$
$\alpha = .01$
C.R. $z > z_{.01} = 2.327$
calculations:

$z_{\hat{p}} = (\hat{p} - \mu_{\hat{p}})/\sigma_{\hat{p}}$
$\quad = (.75 - .70)/\sqrt{(.70)(.30)/1084}$
$\quad = .05/.0139 = 3.592$

conclusion:
$\quad$ Reject H_0; there is sufficient evidence to conclude that $p > .70$.
P-value $= P(z > 3.59) = .5000 - .4999 = .0001$

9. original claim: $p = .125$ [normal approximation to the binomial, use z]
$H_0: p = .125$ $\qquad\qquad$ $\hat{p} = x/n = 83/500 = .166$
$H_1: p \ne .125$
$\alpha = .02$
C.R. $z < -z_{.01} = -2.327$
$\quad\quad z > z_{.01} = 2.327$
calculations:

$z_{\hat{p}} = (\hat{p} - \mu_{\hat{p}})/\sigma_{\hat{p}}$
$\quad = (.166 - .125)/\sqrt{(.125)(.875)/500}$
$\quad = .041/.0148 = 2.772$

conclusion:
$\quad$ Reject H_0; there is sufficient evidence to conclude that $p \ne .125$ (in fact, $p > .125$).
P-value $= 2 \cdot P(z > 12.77 = 2 \cdot [.5000 - .4972] = 2 \cdot [.0028] = .0056$

10. original claim: $p < .65$ [normal approximation to the binomial, use z]

H_o: $p \geq .65$ $\hat{p} = x/n = 1280/2000 = .64$

H_1: $p < .65$

$\alpha = .05$

C.R. $z < -z_{.05} = -1.645$

calculations:

$z_{\hat{p}} = (\hat{p} - \mu_{\hat{p}})/\sigma_{\hat{p}}$

$= (.64 - .65)/\sqrt{(.65)(.35)/2000}$

$= -.01/.0107 = -.938$

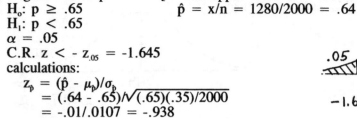 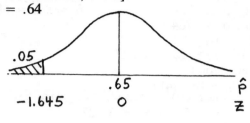

conclusion:

Do not reject H_o; there is not sufficient evidence to conclude that $p < .65$.

P-value $= P(z < -.94) = .5000 - .3264 = .1736$

11. original claim: $p > .08$ [normal approximation to the binomial, use z]

H_o: $p \leq .08$ $\hat{p} = x/n = 70/781 = .0896$

H_1: $p > .08$

$\alpha = .05$

C.R. $z > z_{.05} = 1.645$

calculations

$z_{\hat{p}} = (\hat{p} - \mu_{\hat{p}})/\sigma_{\hat{p}}$

$= (.0896 - .08)/\sqrt{(.08)(.92)/781}$

$= .00962/.00971 = .992$

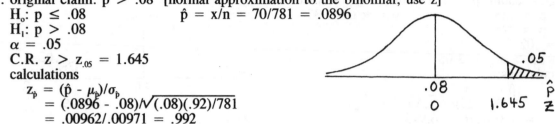

conclusion:

Do not reject H_o; there is not sufficient evidence to conclude that $p > .08$.

P-value $= P(z > .99) = .5000 - .3389 = .1611$

12. original claim: $p > .75$ [normal approximation to the binomial, use z]

H_o: $p \leq .75$ $\hat{p} = x/n = .91$

H_1: $p > .75$

$\alpha = .01$

C.R. $z > z_{.01} = 2.327$

calculations:

$z_{\hat{p}} = (\hat{p} - \mu_{\hat{p}})/\sigma_{\hat{p}}$

$= (.91 - .75)/\sqrt{(.75)(.25)/500}$

$= .16/.0194 = 8.262$

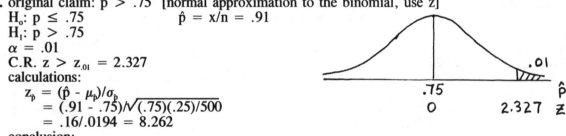

conclusion:

Reject H_o; there is sufficient evidence to conclude that $p > .75$.

P-value $= P(z > 8.26) = .5000 - .4999 = .0001$

Yes; based on these sample results, the funding should be approved.

13. original claim: $p < .10$ [normal approximation to the binomial, use z]

H_o: $p \geq .10$ $\hat{p} = x/n = 64/1068 = .0599$

H_1: $p < .10$

$\alpha = .01$

C.R. $z < -z_{.01} = -2.327$

calculations

$z_{\hat{p}} = (\hat{p} - \mu_{\hat{p}})/\sigma_{\hat{p}}$

$= (.0599 - .10)/\sqrt{(.10)(.90)/1068}$

$= -.0401/.00918 = -4.366$

conclusion:

Reject H_o; there is sufficient evidence to conclude that $p < .10$.

P-value $= P(z < -4.37) = .5000 - .43999 = .0001$

14. original claim: $p < .32$ [normal approximation to the binomial, use z]
   H_o: $p \geq .32$ $\hat{p} = x/n = .183$
   H_1: $p < .32$
   $\alpha = .04$
   C.R. $z < -z_{.04} = -1.75$
   calculations:

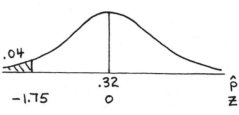

$$z_{\hat{p}} = (\hat{p} - \mu_{\hat{p}})/\sigma_{\hat{p}}$$
$$= (.183 - .32)/\sqrt{(.32)(.68)/785}$$
$$= -.137/.0166 = -8.229$$

conclusion:
   Reject H_o; there is sufficient evidence to conclude that $p < .32$.
P-value $= P(z < -8.23) = .5000 - .4999 = .0001$

15. original claim: $p < .07$ [normal approximation to the binomial, use z]
   H_o: $p \geq .07$ $\hat{p} = x/n = 333/5218 = .0638$
   H_1: $p < .07$
   $\alpha = .05$ [or any other appropriate value]
   C.R. $z < -z_{.05} = -1.645$
   calculations

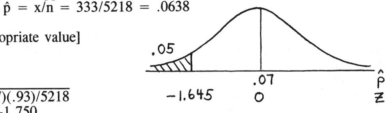

$$z_{\hat{p}} = (\hat{p} - \mu_{\hat{p}})/\sigma_{\hat{p}}$$
$$= (.0638 - .07)/\sqrt{(.07)(.93)/5218}$$
$$= -.00618/.00353 = -1.750$$

conclusion:
   Reject H_o; there is sufficient evidence to conclude that $p < .07$.
P-value $= P(z < -1.75) = .5000 - .4599 = .0401$
Statistically, we are 95% certain the new system reduces the percent of no shows; whether the estimated .6% reduction qualifies as "effective" must be decided by the airline executives.

16. original claim: $p < .50$ [normal approximation to the binomial, use z]
   H_o: $p \geq .50$ $\hat{p} = x/n = 400/850 = .471$
   H_1: $p < .50$
   $\alpha = .05$
   C.R. $z < -z_{.05} = -1.645$
   calculations:

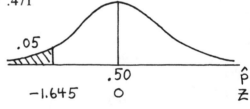

$$z_{\hat{p}} = (\hat{p} - \mu_{\hat{p}})/\sigma_{\hat{p}}$$
$$= (.471 - .50)/\sqrt{(.50)(.50)/850}$$
$$= -.0294/.0171 = -1.715$$

conclusion:
   Reject H_o; there is sufficient evidence to conclude that $p < .50$.
P-value $= P(z < -1.71) = .5000 - .4564 = .0436$
Yes; the results indicate we can be 95% sure that the usage rate is really less than 50%.

17. original claim: $p > .50$ [normal approximation to the binomial, use z] NOTE: A success is
   H_o: $p \leq .50$ $\hat{p} = x/n = 268/506 = .530$ not being aware of the Holocaust.
   H_1: $p > .50$
   $\alpha = .05$
   C.R. $z > z_{.05} = 1.645$
   calculations

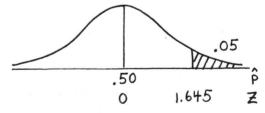

$$z_{\hat{p}} = (\hat{p} - \mu_{\hat{p}})/\sigma_{\hat{p}}$$
$$= (.530 - .50)/\sqrt{(.50)(.50)/506}$$
$$= .0296/.0222 = 1.334$$

conclusion:
   Do not reject H_o; there is not sufficient evidence to conclude that $p > .50$.
P-value $= P(z > 1.33) = .5000 - .4082 = .0918$
We cannot be 95% certain p is greater than 50% and the curriculum needs to be revised.

18. original claim: p > .50 [normal approximation to the binomial, use z]

H_o: p ≤ .50 $\hat{p} = x/n = 34,227,000/(34,227,000 + 34,108,000) = .5008707$

H_1: p > .50

α = .01

C.R. z > $z_{.01}$ = 2.327

calculations:

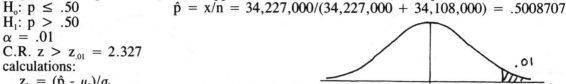

$z_{\hat{p}} = (\hat{p} - \mu_{\hat{p}})/\sigma_{\hat{p}}$

$= (.5008707 - .50)/\sqrt{(.50)(.50)/68,335,000}$

$= .0008707/.0000605 = 14.395$

conclusion:

Reject H_o; there is sufficient evidence to conclude that p > .50.

P-value = P(z > 14.40) = .5000 - .4999 = .0001

19. original claim: p ≥ .50 [normal approximation to the binomial, use z]

H_o: p ≥ .50 $\hat{p} = x/n = .28$

H_1: p < .50

α = .005

C.R. z < - $z_{.005}$ = -2.575

calculations

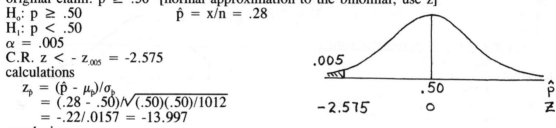

$z_{\hat{p}} = (\hat{p} - \mu_{\hat{p}})/\sigma_{\hat{p}}$

$= (.28 - .50)/\sqrt{(.50)(.50)/1012}$

$= -.22/.0157 = -13.997$

conclusion:

Reject H_o; there is sufficient evidence to conclude that p < .50.

P-value = P(z < -14.00) = .5000 - .4999 = .0001

The sample results are extremely unlikely to have occurred by chance if p ≥ .50. It appears that the fruitcake producers have a problem.

20. original claim: p = .50 [normal approximation to the binomial, use z]

H_o: p = .50 $\hat{p} = x/n = 48/100 = .48$

H_1: p ≠ .50

α = .05

C.R. z < - $z_{.025}$ = -1.960

$\quad$ z > $z_{.025}$ = 1.960

calculations:

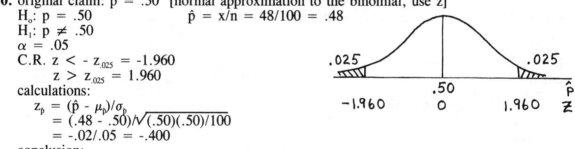

$z_{\hat{p}} = (\hat{p} - \mu_{\hat{p}})/\sigma_{\hat{p}}$

$= (.48 - .50)/\sqrt{(.50)(.50)/100}$

$= -.02/.05 = -.400$

conclusion:

Do not reject H_o; there is not sufficient evidence to conclude that p ≠ .50.

P-value = 2·P(z < -.40) = 2·[.5000 - .1554] = 2·[.3446] = .6892

21. a. original claim: p = .20 [normal approximation to the binomial, use z]

H_o: p = .20 $\hat{p} = x/n = 17/100 = .17$

H_1: p ≠ .20

α = .05

C.R. z < - $z_{.025}$ = -1.960

$\quad$ z > $z_{.025}$ = 1.960

calculations

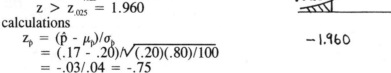

$z_{\hat{p}} = (\hat{p} - \mu_{\hat{p}})/\sigma_{\hat{p}}$

$= (.17 - .20)/\sqrt{(.20)(.80)/100}$

$= -.03/.04 = -.75$

conclusion:

Do not reject H_o; there is not sufficient evidence to conclude that p ≠ .20.

b. original claim: p = .71 [normal approximation to the binomial, use z]
   H_o: p = .20 $\hat{p}$ = x/n = 17/100 = .17
   H_1: p ≠ .20
   α = .05
   C.R. P-value < .05
   calculations

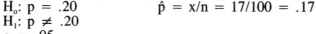

$z_{\hat{p}} = (\hat{p} - \mu_{\hat{p}})/\sigma_{\hat{p}}$
   = (.17 - .20)/$\sqrt{(.20)(.80)/100}$
   = -.03/.04 = -.75

   P-value = 2·P(z < -.75) = 2·[.5000 - .2734] = 2·[.2266] = .4532
   conclusion:
      Do not reject H_o; there is not sufficient evidence to conclude that p ≠ .20.

c. There are 17 of the 100 M&M's that were red.
   H_o: p = .20 $\hat{p}$ = x/n = 17/100 = .170
   H_1: p ≠ .20
   α = .05
   calculations:

   $\hat{p} \pm z_{.025}\sqrt{\hat{p}\hat{q}/n}$
   .170 ± 1.960$\sqrt{(.170)(.830)/100}$
   .170 ± .074
   .096 < p < .244
   conclusion:
      Since .20 is within the confidence interval, do not reject H_o; there is not sufficient
      evidence to conclude that p ≠ .20.

22. original claim: p ≤ d, a rate to be determined [normal approximation to the binomial, use z]
   H_o: p ≤ d $\hat{p}$ = x/n = .03
   H_1: p > d
   α = .05
   C.R. z > $z_{.05}$ = 1.645
   calculations:

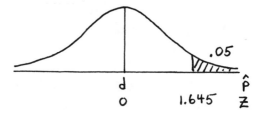

   $z_{\hat{p}} = (\hat{p} - \mu_{\hat{p}})/\sigma_{\hat{p}}$
      1.645 = (.03 - d)/$\sqrt{(d)(1-d)/500}$
   1.645$\sqrt{d(1-d)}$ = $\sqrt{500}$ (.03 - d)
      2.706(d-d^2) = 500(.0009 - .06d + d^2)
         d - d^2 = 184.7729(.0009 - .06d + d^2)
         d - d^2 = .1663 - 11.0864d + 184.7729d^2
   185.7729d^2 - 12.0864d + .1163 = 0
      d = [12.0864 ± $\sqrt{(-12.0864)^2 - 4(185.7729)(.1163)}$]/2(185.7729)
         = [12.0864 ± 4.7442]/371.5457
         = .01976 or .04530 [the d for which .03 has a z score of -1.645]
   The lowest defective rate he can claim and still have $\hat{p}$ = x/500 = .03 <u>not</u> cause the claim to
      be rejected is .01977 (rounded up).

23. original claim: $p = .10$ [np = (15)(.10) = 1.5 < 5; use binomial distribution]

x = 0

H_o: p = .10

H_1: p ≠ .10

α = .05

C.R. x < 0   NOTE: obtained by placing .025 (or

x > 4   much as possible) in each tail

calculations

x = 0

conclusion:

Do not reject H_o; there is not sufficient evidence
to conclude that p ≠ .10.

P-value = P(a value as extreme as our value)

= P(x=0 or x≥3)

= .206 + (.129 + .043 + .010 + .002 + 0⁺ +...)

= .206 + .184

= .390

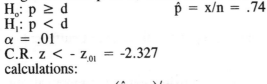

NOTE: Refer to this probability
distribution for n=15 and p=.10.

x	P(x)
0	.206
1	.343
2	.267
3	.129
4	.043
5	.010
6	.002
7	0⁺
·	·
·	·
·	·
	1.000

24. original claim: $p \geq d$, a rate to be determined [normal approximation to the binomial, use z]

H_o: p ≥ d $\hat{p}$ = x/n = .74

H_1: p < d

α = .01

C.R. z < $-z_{.01}$ = -2.327

calculations:

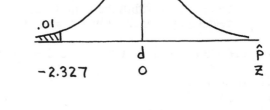

$$z_{\hat{p}} = (\hat{p} - \mu_{\hat{p}})/\sigma_{\hat{p}}$$

$$-2.327 = (.91 - d)/\sqrt{(d)(1-d)/500}$$

$$-2.327\sqrt{d(1-d)} = \sqrt{500}\ (.91 - d)$$

$$5.415(d-d^2) = 500(.8281 - 1.82d + d^2)$$

$$d - d^2 = 92.3373(.8281 - 1.82d + d^2)$$

$$d - d^2 = 76.4645 - 168.0539d + 92.3373d^2$$

$$13.3373d^2 - 169.0539d + 76.4645 = 0$$

$$d = [169.0539 \pm \sqrt{(-169.0539)^2 - 4(93.3373)(76.4645)}\ ]/2(93.3373)$$

$$= [169.0539 \pm 5.5901]/186.6746$$

= .8757 [the d for which .91 has a z score of +2.327] or .9356

The highest rate he can claim and still have $\hat{p}$ = x/500 = .91 <u>not</u> cause the claim to
be rejected is .9355 (rounded down).

25. original claim: p < 32 [normal approximation to the binomial, use z]

H_o: p ≥ .50

H_1: p < .50

α = .05

C.R. z < $-z_{.05}$ = -1.645

calculations:

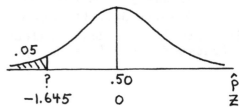

$$z_{\hat{p}} = (\hat{p} - \mu_{\hat{p}})/\sigma_{\hat{p}}$$

$$-1.645 = (\hat{p} - .50)/\sqrt{(.50)(.50)/1998}$$

$$-1.645 = (\hat{p} - .50)/.0112$$

$$-.0184 = \hat{p} - .50$$

$$.4816 = \hat{p}$$

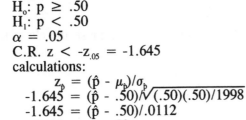

NOTE: There is no conclusion because this exercise is not asking for the completion of a test of hypothesis. In fact, the sample proportion is irrelevant to the question posed. The above calculations find the $\hat{p}$ that corresponds to the z score that marks the beginning of the critical region. In terms of $\hat{p}$, then, the critical region is $\hat{p} < .4816$.

The exercise asks for P(not rejecting H_o) = P($\hat{p} > .4816$) when p = .45.

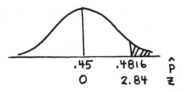

$z_{\hat{p}} = (\hat{p} - \mu_{\hat{p}})/\sigma_{\hat{p}}$
$= (.4816 - .45)/\sqrt{(.45)(.55)/1998}$
$= .0316/.0111$
$= 2.839$
P(z > 2.84) = .5000 - .4977 = .0023

7-6 Testing a Claim about a Standard Deviation or Variance

1. a. $\chi^2_{19,.975} = 8.907$
   $\chi^2_{19,.025} = 32.852$
   b. $\chi^2_{19,.950} = 10.117$
   c. $\chi^2_{74,.975} = 48.758$ [closest entry]
   or $48.758 + (4/10) \cdot (57.153 - 48.758) = 52.116$ [using interpolation]
   $\chi^2_{74,.025} = 95.023$ [closest entry]
   or $95.023 + (4/10) \cdot (106.629 - 95.023) = 99.665$ [using interpolation]

NOTE: Following the pattern used with the z and t distributions, this manual uses the closest entry from Table A-4 for χ^2 as if it were the precise value necessary and does not use interpolation. This procedure sacrifices very little accuracy -- and even interpolation does not yield precise values. When extreme accuracy is needed in practice, statisticians refer either to more accurate tables or to computer-produced values.

2. a. $\chi^2_{5,.90} = 1.160$
   b. $\chi^2_{40,.05} = 55.758$
   c. $\chi^2_{22,.01} = 40.289$

3. original claim: $\sigma = 43.7$
   H_o: $\sigma = 43.7$
   H_1: $\sigma \neq 43.7$
   $\alpha = .05$
   C.R. $\chi^2 < \chi^2_{80,.975} = 57.153$
   $\chi^2 > \chi^2_{80,.025} = 106.629$
   calculations:
   $\chi^2 = (n-1)s^2/\sigma^2 = (80)(52.3)^2/(43.7)^2 = 114.586$
   conclusion:
   Reject H_o; there is sufficient evidence to conclude that $\sigma \neq 43.7$ (in fact, $\sigma > 43.7$).

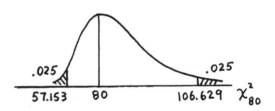

4. original claim: $\sigma > 4.00$
   H_o: $\sigma \leq 2$
   H_1: $\sigma > 2$
   $\alpha = .05$
   C.R. $\chi^2 > \chi^2_{69,.05} = 90.531$
   calculations:
   $\chi^2 = (n-1)s^2/\sigma^2 = (69)(5.33)^2/(4.00)^2 = 122.513$
   conclusion:
   Reject H_o; there is sufficient evidence to conclude that $\sigma > 4.00$.

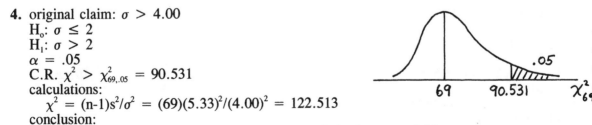

5. original claim: $\sigma < .75$
 $H_0: \sigma \geq .75$
 $H_1: \sigma < .75$
 $\alpha = .05$
 C.R. $\chi^2 < \chi^2_{60,.95} = 43.188$
 calculations:
   $\chi^2 = (n-1)s^2/\sigma^2 = (60)(.48)^2/(.75)^2 = 24.576$
 conclusion:
   Reject H_0; there is sufficient evidence to conclude that $\sigma < 43.7$.

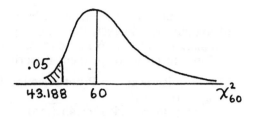

6. original claim: $\sigma > 2$
 $H_0: \sigma \leq 2$
 $H_1: \sigma > 2$
 $\alpha = .025$
 C.R. $\chi^2 > \chi^2_{11,.025} = 21.920$
 calculations:
   $\chi^2 = (n-1)s^2/\sigma^2 = (11)(2.85)^2/2^2 = 22.337$
 conclusion:
   Reject H_0; there is sufficient evidence to conclude that $\sigma > 2$.
 Yes; the variation does appear to be too high.

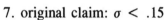

7. original claim: $\sigma < .15$
 $H_0: \sigma \geq .15$
 $H_1: \sigma < .15$
 $\alpha = .05$
 C.R. $\chi^2 < \chi^2_{70,.95} = 51.739$
 calculations:
   $\chi^2 = (n-1)s^2/\sigma^2 = (70)(.12)^2/(.15)^2 = 44.800$
 conclusion:
   Reject H_0; there is sufficient evidence to conclude that $\sigma < .15$.
 Yes, since Medassist can be 95% certain that the new machine fills bottles with less variation they should consider its purchase.

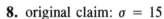

8. original claim: $\sigma = 15$
 $H_0: \sigma = 15$
 $H_1: \sigma \neq 15$
 $\alpha = .05$
 C.R. $\chi^2 < \chi^2_{23,.975} = 11.689$
   $\chi^2 > \chi^2_{23,.025} = 38.076$
 calculations:
   $\chi^2 = (n-1)s^2/\sigma^2 = (23)(10)^2/(15)^2 = 10.222$
 conclusion:
   Reject H_0; there is sufficient evidence to conclude that $\sigma \neq 15$ (in fact, $\sigma < 15$).

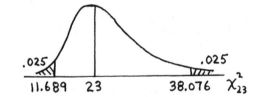

9. original claim: $\sigma > 19.7$
 $H_0: \sigma \leq 19.7$
 $H_1: \sigma > 19.7$
 $\alpha = .05$
 C.R. $\chi^2 > \chi^2_{49,.05} = 67.505$
 calculations:
   $\chi^2 = (n-1)s^2/\sigma^2 = (49)(23.4)^2/(19.7)^2 = 69.135$
 conclusion:
   Reject H_0; there is sufficient evidence to conclude that $\sigma > 19.7$.

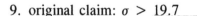

10. original claim: $\sigma > .05$
    H_o: $\sigma \leq .05$
    H_1: $\sigma > .05$
    $\alpha = .05$
    C.R. $\chi^2 > \chi^2_{49,.05} = 67.505$
    calculations:
       $\chi^2 = (n-1)s^2/\sigma^2 = (49)(.068)^2/(.05)^2 = 90.630$
    conclusion:
       Reject H_o; there is sufficient evidence to conclude that $\sigma > .05$.

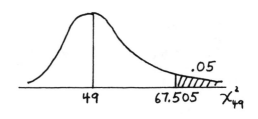

11. original claim: $\sigma = 2.4$
    H_o: $\sigma = 2.4$
    H_1: $\sigma \neq 2.4$
    $\alpha = .05$
    C.R. $\chi^2 < \chi^2_{49,.975} = 32.357$
       $\chi^2 > \chi^2_{49,.025} = 71.420$
    calculations:
       $\chi^2 = (n-1)s^2/\sigma^2 = (49)(2.7)^2/(2.4)^2 = 62.016$
    conclusion:
       Do not reject H_o; there is not sufficient evidence to conclude that $\sigma \neq 2.4$.

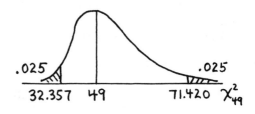

12. original claim: $\sigma < .05$
    H_o: $\sigma \geq .05$
    H_1: $\sigma < .05$
    $\alpha = .05$
    C.R. $\chi^2 < \chi^2_{16,.95} = 7.962$
    calculations:
       $\chi^2 = (n-1)s^2/\sigma^2 = (16)(.0414)^2/(.05)^2 = 10.969$
    conclusion:
       Do not reject H_o; there is not sufficient evidence to conclude that $\sigma < .05$.

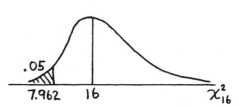

13. original claim: $\sigma = 3$
    H_o: $\sigma = 3$
    H_1: $\sigma \neq 3$
    $\alpha = .05$
    C.R. $\chi^2 < \chi^2_{61,.975} = 40.482$
       $\chi^2 > \chi^2_{61,.025} = 83.298$
    calculations:
       $\chi^2 = (n-1)s^2/\sigma^2 = (61)(3.297)^2/(3)^2 = 73.676$
    conclusion:
       Do not reject H_o; there is not sufficient evidence to conclude that $\sigma \neq 3$.

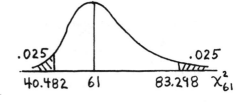

14. original claim: $\sigma < 10$
    H_o: $\sigma \geq 10$
    H_1: $\sigma < 10$
    $\alpha = .05$
    C.R. $\chi^2 < \chi^2_{39,.95} = 26.509$
    calculations:
       $\chi^2 = (n-1)s^2/\sigma^2 = (39)(8.28)^2/(10)^2 = 26.738$
    conclusion:
       Do not reject H_o; there is not sufficient evidence to conclude that $\sigma < 10$.

15. original claim: $\sigma < 2.0$

    $n = 12$

    $\Sigma x = 396.6$ $\bar{x} = 33.05$

    $\Sigma x^2 = 13121.66$ $s^2 = 1.275$

  $H_o: \sigma \geq 2.0$

  $H_1: \sigma < 2.0$

  $\alpha = .025$

  C.R. $\chi^2 < \chi^2_{11,.975} = 3.816$

  calculations:

    $\chi^2 = (n-1)s^2/\sigma^2 = (11)(1.275)/(2.0)^2 = 3.5075$

  conclusion:

    Reject H_o; there is sufficient evidence to conclude that $\sigma < 2.0$.

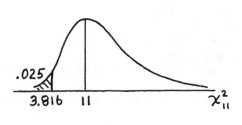

16. original claim: $\sigma < 2.9$

    $n = 25$

    $\Sigma x = 1721.07$ $\bar{x} = 68.843$

    $\Sigma x^2 = 118614.7057$ $s^2 = 5.476$

  $H_o: \sigma \geq 2.9$

  $H_1: \sigma < 2.9$

  $\alpha = .05$ [or any other appropriate value]

  C.R. $\chi^2 < \chi^2_{24,.95} = 13.848$

  calculations:

    $\chi^2 = (n-1)s^2/\sigma^2 = (24)(5.476)/(2.9)^2 = 15.628$

  conclusion:

    Do not reject H_o; there is not sufficient evidence to conclude that $\sigma < 2.0$.

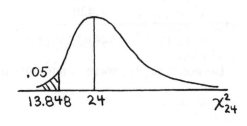

17. original claim: $\sigma = .470$

    $n = 16$

    $\Sigma x = 58.8$ $\bar{x} = 3.675$

    $\Sigma x^2 = 222.571$ $s^2 = .432$

  $H_o: \sigma = .470$

  $H_1: \sigma \neq .470$

  $\alpha = .05$ [or any other appropriate value]

  C.R. $\chi^2 < \chi^2_{15,.975} = 6.262$

       $\chi^2 > \chi^2_{15,.025} = 27.488$

  calculations:

    $\chi^2 = (n-1)s^2/\sigma^2 = (15)(.432)/(.470)^2 = 29.339$

  conclusion:

    Reject H_o; there is sufficient evidence to conclude that $\sigma \neq .470$ (in fact, $\sigma > .470$).

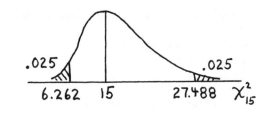

18. original claim: $\sigma = 1000$

    $n = 7$

    $\Sigma x = 74,209$ $\bar{x} = 10,601.3$

    $\Sigma x^2 = 793,035,705$ $s^2 = 1,054,148.9$

  $H_o: \sigma = 1000$

  $H_1: \sigma \neq 1000$

  $\alpha = .10$

  C.R. $\chi^2 < \chi^2_{6,.95} = 1.635$

      $\chi^2 > \chi^2_{6,.05} = 12.592$

  calculations:

    $\chi^2 = (n-1)s^2/\sigma^2 = (6)(1,054,148.9)/(1000)^2 = 6.325$

  conclusion:

    Do not reject H_o; there is not sufficient evidence to conclude that $\sigma \neq 1000$.

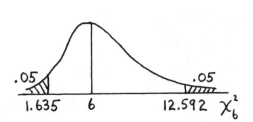

19. a. To find the P-value for $\chi_9^2 = 19.735$ in a one-tailed greater than test,
     note that $\chi_{9,.025}^2 = 19.023$ and $\chi_{9,.01}^2 = 21.666$.
     Therefore, $.01 < \text{P-value} < .025$.

   b. To find the P-value for $\chi_{19}^2 = 7.337$ in a one-tailed less than test,
     note that $\chi_{19,.995}^2 = 6.844$ and $\chi_{19,.99}^2 = 7.633$.
     Therefore, $.005 < \text{P-value} < .01$.

   c. To find the P-value for $\chi_{29}^2 = 54.603$ in a two-tailed test,
     note that $\chi_{29,.005}^2 = 52.336$ is the largest χ^2 value in the row.
     Therefore, P-value $< 2 \cdot (.005)$
                   P-value $< .01$

20. The 106 scores form frequency distribution given below at the left. The corresponding histogram given below at the right indicates that the distribution is approximately normal.

Total Weight	Freq.
0.00 - 9.99	2
10.00 - 19.99	16
20.00 - 29.99	22
30.00 - 39.99	11
40.00 - 49.99	8
50.00 - 59.99	3
	62

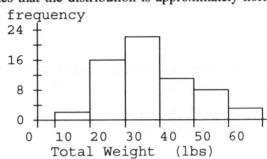

original claim $\sigma = 15.5$
   $n = 62$
   $\Sigma x = 1701.49$ $\bar{x} = 27.443$
   $\Sigma x^2 = 56161.88$ $s^2 = 155.201$

a. traditional method
   H_o: $\sigma = 15.5$
   H_1: $\sigma \neq 15.5$
   $\alpha = .05$
   C.R. $\chi^2 < \chi_{61,.975}^2 = 40.482$
       $\chi^2 > \chi_{61,.025}^2 = 83.298$
   calculations:
   $\chi^2 = (n-1)s^2/\sigma^2 = (61)(155.201)/(15.5)^2 = 39.406$
   conclusion:
   Reject H_o; there is sufficient evidence to conclude that $\sigma \neq 15.6$ (in fact, $\sigma < 15.5$).

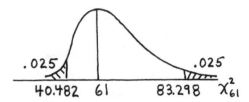

b. P-value method
   H_o: $\sigma = 15.5$
   H_1: $\sigma \neq 15.5$
   $\alpha = .05$
   C.R. P-value $< .05$
   calculations:
   $\chi^2 = (n-1)s^2/\sigma^2 = (61)(155.201)/(15.5)^2 = 39.406$
   Since $\chi_{61,.99}^2 = 37.485 < 39.406 < \chi_{61,.975}^2 = 40.482$ in a two-tailed test,
       $2(.01) < \text{P-value} < 2(.025)$
        $.02 < \text{P-value} < .05$
   conclusion:
   Reject H_o; there is sufficient evidence to conclude that $\sigma \neq 15.6$ (in fact, $\sigma < 15.5$ -- since the values used to obtain the P-value are from the lower tail of the distribution).

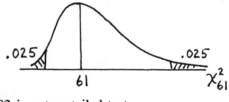

c. confidence interval method. **NOTE:** Because the confidence interval estimate for σ is not discussed in the text, the following derivation is given.

We have 95% confidence that $\chi^2_{.975} < \chi^2 < \chi^2_{.025}$

$$\chi^2_{.975} < (n-1)s^2/\sigma^2 < \chi^2_{.025}$$
$$1/\chi^2_{.975} > \sigma^2/(n-1)s^2 > 1/\chi^2_{.975}$$
$$(n-1)s^2/\chi^2_{.975} > \sigma^2 > (n-1)s^2/\chi^2_{.025}$$
$$(n-1)s^2/\chi^2_{.025} < \sigma^2 < (n-1)s^2/\chi^2_{.975}$$

H_o: $\sigma = 15.5$
H_1: $\sigma \neq 15.5$
$\alpha = .05$
C.R. 15.5 is <u>not</u> in the 95% confidence interval
calculations:

$$(n-1)s^2/\chi^2_{.025} < \sigma^2 < (n-1)s^2/\chi^2_{.975}$$
$$(61)(155.201)/83.298 < \sigma^2 < (61)(155.201)/40.482$$
$$113.655 < \sigma^2 < 233.863$$
$$10.661 < \sigma < 15.293$$

conclusion:
Reject H_o; there is sufficient evidence to conclude that $\sigma \neq 15.6$ (in fact, $\sigma < 15.5$ -- since the calculated interval lies entirely below 15.5).

21. a. upper $\chi^2 = \frac{1}{2}(z_{.025} + \sqrt{2 \cdot df - 1})^2$
$= \frac{1}{2}(1.960 + \sqrt{2 \cdot (100) - 1})^2$
$= \frac{1}{2}(1.960 + \sqrt{199})^2$
$= \frac{1}{2}(258.140)$
$= 129.070$ [compare to $\chi^2_{100,.025} = 129.561$ from Table A-4]
lower $\chi^2 = \frac{1}{2}(-z_{.025} + \sqrt{2 \cdot df - 1})^2$
$= \frac{1}{2}(-1.960 + \sqrt{2 \cdot (100) - 1})^2$
$= \frac{1}{2}(-1.960 + \sqrt{199})^2$
$= \frac{1}{2}(147.543)$
$= 73.772$ [compare to $\chi^2_{100,.975} = 74.222$ from Table A-4]

b. upper $\chi^2 = \frac{1}{2}(z_{.025} + \sqrt{2 \cdot df - 1})^2$
$= \frac{1}{2}(1.960 + \sqrt{2 \cdot (149) - 1})^2$
$= \frac{1}{2}(1.960 + \sqrt{297})^2$
$= \frac{1}{2}(368.398)$
$= 184.199$
lower $\chi^2 = \frac{1}{2}(-z_{.025} + \sqrt{2 \cdot df - 1})^2$
$= \frac{1}{2}(-1.960 + \sqrt{2 \cdot (149) - 1})^2$
$= \frac{1}{2}(-1.960 + \sqrt{297})^2$
$= \frac{1}{2}(233.286)$
$= 116.643$

22. a. upper $\chi^2 = df \cdot [1 - 2/(9 \cdot df) + z_{.025}\sqrt{2/(9 \cdot df)}]^3$
$= 100 \cdot [1 - 2/(9 \cdot 100) + 1.960\sqrt{2/(9 \cdot 100)}]^3$
$= 100 \cdot (1.0902)^3$
$= 129.565$ [compare to $\chi^2_{100,.025} = 129.561$ from Table A-4]
lower $\chi^2 = df \cdot [1 - 2/(9 \cdot df) - z_{.025}\sqrt{2/(9 \cdot df)}]^3$
$= 100 \cdot [1 - 2/(9 \cdot 100) - 1.960\sqrt{2/(9 \cdot 100)}]^3$
$= 100 \cdot (.9054)^3$
$= 74.216$ [compare to $\chi^2_{100,.975} = 74.222$ from Table A-4]

b. upper $\chi^2 = df \cdot [1 - 2/(9 \cdot df) + z_{.025}\sqrt{2/(9 \cdot df)}]^3$
$= 149 \cdot [1 - 2/(9 \cdot 149) + 1.960\sqrt{2/(9 \cdot 149)}]^3$
$= 149 \cdot (1.0742)^3$
$= 184.690$
lower $\chi^2 = df \cdot [1 - 2/(9 \cdot df) - z_{.025}\sqrt{2/(9 \cdot df)}]^3$
$= 149 \cdot [1 - 2/(9 \cdot 149) - 1.960\sqrt{2/(9 \cdot 149)}]^3$
$= 149 \cdot (.9228)^3$
$= 117.093$

23. original claim: $\sigma < 6.2$
    $H_o: \sigma \geq 6.2$
    $H_1: \sigma < 6.2$
    $\alpha = .05$
    C.R. $\chi^2 < \chi^2_{24,.95} = 13.848$
    calculations:

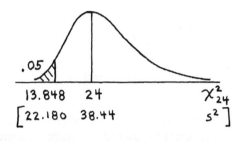

$\chi^2 = (n-1)s^2/\sigma^2$
$13.848 = (24)s^2/(6.2)^2$
$532.317 = (24)s^2$
$22.180 = s^2$
NOTE: There is no conclusion because this exercise is not asking for the completion of a test of hypothesis. In fact, the sample data is irrelevant to the question posed. The above calculations find the s^2 that corresponds to the χ^2 score that marks the beginning of the critical region. In terms of s^2, then, the critical region is $s^2 < 22.180$

The exercise asks for P(not rejecting H_o) = P($s^2 > 22.180$) when $\sigma = 4.0$.

$\chi^2 = (n-1)s^2/\sigma^2$
$= (24)(22.180)/(4.0)^2$
$= 33.270$
$P(\chi^2_{24} > 33.270) = .10$ [from Table A-4]

24. a. original claim: $\mu = 98.6$ [n > 30, use z (with s for σ)]
    $H_o: \mu = 98.6$
    $H_1: \mu \neq 98.6$
    $\alpha = .05$
    C.R. $z < -z_{.025} = -1.960$
    $\quad z > z_{.025} = 1.960$
    calculations:

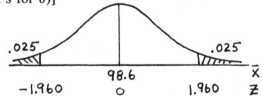

$z_{\bar{x}} = (\bar{x} - \mu)/\sigma_{\bar{x}}$
$-1.960 = (98.2 - 98.6)/(\sigma/\sqrt{106})$
$\sigma = 2.101$ (to the accuracy of this problem, rounded up to 2.11)

b. original claim: $\sigma \geq 2.11$
   $H_o: \sigma \geq 2.11$
   $H_1: \sigma < 2.11$
   $\alpha = .05$ [or any other appropriate value]
   C.R. $\chi^2 < \chi^2_{105,.95} = 77.929$
   calculations:

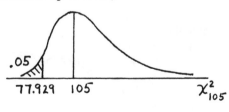

$\chi^2 = (n-1)s^2/\sigma^2 = (105)(.62)^2/(2.11)^2 = 9.066$
conclusion:
    Reject H_o; there is sufficient evidence to conclude that $\sigma < 2.11$.

c. No; based on the results of parts (a) and (b), we are confident that the true standard deviation is not so large that $\mu = 98.6$ should not be rejected.

Review Exercises

1. a. concerns p: normal approximation to the binomial, use z
   $-z_{.05} = -1.645$

   b. concerns μ: n > 30, use z (with s for σ)
   $z_{.01} = 2.327$

   c. concerns μ: n < 30 and σ unknown, use t
   $\pm t_{11,.005} = \pm 3.106$

   d. concerns σ, use χ^2
   $\chi^2_{24,.99} = 10.856$

   e. concerns σ, use χ^2
   $\chi^2_{29,.975} = 16.047$ and $\chi^2_{29,.025} = 45.772$

3. a. H_o: $\mu \geq 20.0$
   b. left-tailed [since the alternative hypothesis is H_1: $\mu < 20.0$]
   c. concluding that the average treatment time is less than 20.0 when it really is not
   d. failing to conclude that the average treatment time is less than 20.0 when it really is
   e. $\alpha = .01$

5. original claim: p > .50 [concerns p: normal approximation to the binomial, use z]
   $\hat{p} = x/n = .57$
   H_o: p $\leq$ .50
   H_1: p > .50
   $\alpha = .01$
   C.R. z > $z_{.01} = 2.327$
   calculations:
   $z_{\hat{p}} = (\hat{p} - \mu_{\hat{p}})/\sigma_{\hat{p}}$
   $= (.57 - .50)/\sqrt{(.50)(.50)/504}$
   $= .07/.0223$
   $= 3.143$
   conclusion:
   Reject H_o; there is sufficient evidence to conclude that p > .50.

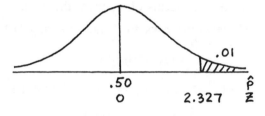

7. original claim: $\mu = 3393$ [concerns μ: n $\leq$ 30 and σ unknown, use t]
   H_o: $\mu = 3393$
   H_1: $\mu \neq 3393$
   $\alpha = .05$
   C.R. t < $-t_{29,.025} = -2.045$
   t > $t_{29,.025} = 2.045$
   calculations:
   $t_{\bar{x}} = (\bar{x} - \mu)/s_{\bar{x}}$
   $= (3264 - 3393)/(485/\sqrt{30})$
   $= -129/88.548$
   $= -1.457$
   conclusion:
   Do not reject H_o; there is not sufficient evidence to conclude that $\mu \neq 3393$.

9. original claim: $\mu < 5.00$ [concerns μ: $n > 30$, use z (with s for σ)]
   H_o: $\mu \geq 5.00$
   H_1: $\mu < 5.00$
   $\alpha = .01$
   C.R. z < $-z_{.01}$ = -2.327
   calculations:

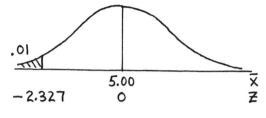

$$z_{\bar{x}} = (\bar{x} - \mu)/\sigma_{\bar{x}}$$
$$= (4.13 - 5.00)/(1.91/\sqrt{36})$$
$$= -.87/.3183$$
$$= -2.733$$

   conclusion:
   Reject H_o; there is sufficient evidence to conclude that $\mu < 5.00$.

11. original claim: $\mu > 55.0$ [concerns μ: $n > 30$, use z (with s for σ)]
   H_o: $\mu \leq 55.0$
   H_1: $\mu > 55.0$
   $\alpha = .025$
   C.R. z > $z_{.025}$ = 1.960
   calculations:

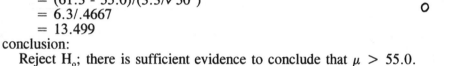

$$z_{\bar{x}} = (\bar{x} - \mu)/\sigma_{\bar{x}}$$
$$= (61.3 - 55.0)/(3.3/\sqrt{50})$$
$$= 6.3/.4667$$
$$= 13.499$$

   conclusion:
   Reject H_o; there is sufficient evidence to conclude that $\mu > 55.0$.

13. original claim: $\sigma^2 > 6410$ [concerns σ: use χ^2]
   H_o: $\sigma^2 \leq 6410$
   H_1: $\sigma^2 > 6410$
   $\alpha = .10$
   C.R. $\chi^2 > \chi^2_{59,.10} = 74.397$
   calculations:

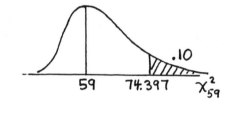

$$\chi^2 = (n-1)s^2/\sigma^2$$
$$= (59)(8464)/(6410)$$
$$= 77.906$$

   conclusion:
   Reject H_o; there is sufficient evidence to conclude that $\sigma^2 > 6410$.

15. original claim: p = .98 [concerns p: normal approximation to the binomial, use z]
   $\hat{p} = x/n = 831/849 = .9788$
   H_o: p = .98
   H_1: p ≠ .98
   $\alpha = .10$
   C.R. z < $-z_{.05}$ = -1.645
       z > $z_{.05}$ = 1.645
   calculations:
   $$z_{\hat{p}} = (\hat{p} - \mu_{\hat{p}})/\sigma_{\hat{p}}$$
   $$= (.9788 - .98)/\sqrt{(.98)(.02)/849}$$
   $$= -.0012/.00480$$
   $$= -.250$$

   conclusion:
   Do not reject H_o; there is not sufficient evidence to conclude that p ≠ .98.

17. original claim: $\sigma < 10.0$ [concerns σ: use χ^2]

    n = 15

    Σx = 1470 $\bar{x}$ = 98.0

    Σx^2 = 145050 s^2 = 70.714

    H_o: $\sigma \geq 10.0$

    H_1: $\sigma < 10.0$

    α = .05

    C.R. $\chi^2 < \chi^2_{14,.95}$ = 6.571

    calculations:

      $\chi^2 = (n-1)s^2/\sigma^2$

         = $(14)(70.714)/(10)^2$

         = 9.90

    conclusion:

    Do not reject H_o; there is not sufficient evidence to conclude that $\sigma < 10.0$.

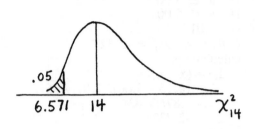

19. original claim: $p < .05$ [concerns p: normal approximation to the binomial, use z]

    $\hat{p}$ = x/n = 72/1500 = .048

    H_o: $p \geq .05$

    H_1: $p < .05$

    α = .10

    C.R. $z < - z_{.10}$ = -1.282

    calculations:

      $z_{\hat{p}} = (\hat{p} - \mu_{\hat{p}})/\sigma_{\hat{p}}$

         = $(.048 - .05)/\sqrt{(.05)(.95)/1500}$

         = -.002/.005627

         = -.355

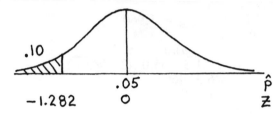

    conclusion:

    Do not reject H_o; there is not sufficient evidence to conclude that $p < .05$.

    P-value = $P(z < -.36)$ = .5000 - .1406 = .3594

Chapter 8

Inferences from Two Samples

8-2 Inferences about Two Means: Dependent Samples

NOTE: To be consistent with the notation of the previous sections, and thereby reinforcing the patterns and concepts presented in those sections, the manual uses the "usual" t formula written to apply to $\bar{d}$'s

$$t_{\bar{d}} = (\bar{d} - \mu_{\bar{d}})/s_{\bar{d}}$$

with $\mu_{\bar{d}} = \mu_d$
and $s_{\bar{d}} = s_d/\sqrt{n}$
And so the formula for the t statistic may also be written as

$$t_{\bar{d}} = (\bar{d} - \mu_d)/(s_d/\sqrt{n})$$

1. $d = x - y$: 1 -1 0 1 -1 2 3 3 2 2 3
   $n = 11$ $\Sigma d = 15$ $\Sigma d^2 = 43$
   a. $\bar{d} = (\Sigma d)/n = 15/11 = 1.363$
   b. $s_d^2 = [n \cdot \Sigma d^2 - (\Sigma d)^2]/[n(n-1)]$
      $= [11 \cdot 43 - (15)^2]/[11(10)] = 248/110 = 2.225$
      $s_d = 1.502$
   c. $t_{\bar{d}} = (\bar{d} - \mu_{\bar{d}})/s_{\bar{d}}$
      $= (1.364 - 0)/(1.502/\sqrt{11})$
      $= 1.363/.4527$
      $= 3.012$
   d. $\pm t_{10,.025} = \pm 2.228$

2. $d = x - y$: -1 1 0 0 -1 2 4 2 2 6 5 7
   $n = 12$ $\Sigma d = 27$ $\Sigma d^2 = 141$
   a. $\bar{d} = (\Sigma d)/n = 27/12 = 2.253$
   b. $s_d^2 = [n \cdot \Sigma d^2 - (\Sigma d)^2]/[n(n-1)]$
      $= [12 \cdot 141 - (27)^2]/[12(11)] = 963/132 = 7.295$
      $s_d = 2.701$
   c. $t_{\bar{d}} = (\bar{d} - \mu_{\bar{d}})/s_{\bar{d}}$
      $= (2.25 - 0)/(2.701/\sqrt{12})$
      $= 2.25/.780$
      $= 2.886$
   d. $\pm t_{11,.025} = \pm 2.201$

3. $\bar{d} \pm t_{10,.025} \cdot s_d/\sqrt{n}$
   $1.4 \pm 2.228 \cdot 1.502/\sqrt{11}$
   1.4 ± 1.0
   $.4 < \mu_d < 2.4$

4. $\bar{d} \pm t_{11,.005} \cdot s_d/\sqrt{n}$
   $2.25 \pm 3.106 \cdot 2.701/\sqrt{12}$
   2.25 ± 2.42
   $-.17 < \mu_d < 4.67$

5. original claim $\mu_d = 0$ [n ≤ 30 and σ_d unknown, use t]
   d = $x_H - x_S$: 2.1 -1.0 1.7 1.6 1.4 2.5 1.3 2.6
   n = 8
   $\Sigma d = 12.2$ $\quad\quad\quad\quad$ $\bar{d} = 1.525$
   $\Sigma d^2 = 27.52$ $\quad\quad\quad$ $s_d = 1.129$
   $H_o: \mu_d = 0$
   $H_1: \mu_d \neq 0$
   $\alpha = .05$
   C.R. $t < -t_{7,.025} = -2.365$
   $\quad\quad$ $t > t_{7,.025} = 2.365$
   calculations:
   $\quad$ $t_{\bar{d}} = (\bar{d} - \mu_{\bar{d}})/s_{\bar{d}}$
   $\quad\quad\quad = (1.525 - 0)/(1.129/\sqrt{8})$
   $\quad\quad\quad = 1.525/.399 = 3.822$
   conclusion:
   $\quad$ Reject H_o; there is sufficient evidence to conclude that $\mu_d \neq 0$ (in fact, $\mu_d > 0$).

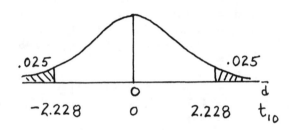

6. original claim $\mu_d = 0$ [n ≤ 30 and σ_d unknown, use t]
   d = $x_M - x_F$: 3 3 4 1 0 5 9 -1 3 7 -13
   n = 11
   $\Sigma d = 21$ $\quad\quad\quad\quad$ $\bar{d} = 1.909$
   $\Sigma d^2 = 369$ $\quad\quad\quad$ $s_d = 5.735$
   $H_o: \mu_d = 0$
   $H_1: \mu_d \neq 0$
   $\alpha = .05$
   C.R. $t < -t_{10,.025} = -2.228$
   $\quad\quad$ $t > t_{10,.025} = 2.228$
   calculations:
   $\quad$ $t_{\bar{d}} = (\bar{d} - \mu_{\bar{d}})/s_{\bar{d}}$
   $\quad\quad\quad = (1.909 - 0)/(5.735/\sqrt{11})$
   $\quad\quad\quad = 1.909/1.729 = 1.104$
   conclusion:
   $\quad$ Do not reject H_o; there is not sufficient evidence to conclude that $\mu_d \neq 0$.

7. d = $x_B - x_A$: 2 14 17 -7 17 12 0 15 33 16
   n = 10
   $\Sigma d = 119$ $\quad\quad\quad\quad$ $\bar{d} = 11.9$
   $\Sigma d^2 = 2541$ $\quad\quad\quad$ $s_d = 11.180$
   $\bar{d} \pm t_{9,.025} \cdot s_d/\sqrt{n}$
   $11.9 \pm 2.262 \cdot 11.180/\sqrt{10}$
   11.9 ± 8.0
   $3.9 < \mu_d < 19.9$

8. Refer to the notation and data summary for exercise #7. The original claim is $\mu_d = 0$.
   $H_o: \mu_d = 0$
   $H_1: \mu_d \neq 0$
   $\alpha = .05$
   C.R. $t < -t_{9,.025} = -2.262$
   $\quad\quad$ $t > t_{9,.025} = 2.262$
   calculations:
   $\quad$ $t_{\bar{d}} = (\bar{d} - \mu_{\bar{d}})/s_{\bar{d}}$
   $\quad\quad\quad = (11.9 - 0)/(11.180/\sqrt{10})$
   $\quad\quad\quad = 11.9/3.535 = 3.366$
   conclusion:
   $\quad$ Reject H_o; there is sufficient evidence to conclude that $\mu_d \neq 0$ (in fact, $\mu_d > 0$).

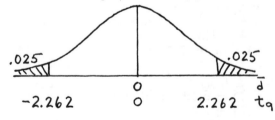

9. original claim $\mu_d > 0$ [$n \le 30$ and σ_d unknown, use t]

$d = x_B - x_A$: -.2 4.1 1.6 1.8 3.2 2.0 2.9 9.6

$n = 8$

$\Sigma d = 25.0$ $\bar{d} = 3.125$

$\Sigma d^2 = 137.46$ $s_d = 2.911$

H_o: $\mu_d \le 0$

H_1: $\mu_d > 0$

$\alpha = .05$

C.R. $t > t_{7,.05} = 1.895$

calculations:

$t_{\bar{d}} = (\bar{d} - \mu_{\bar{d}})/s_{\bar{d}}$

$= (3.125 - 0)/(2.911/\sqrt{8})$

$= 3.125/1.029 = 3.036$

conclusion:

Reject H_o; there is sufficient evidence to conclude that $\mu_d > 0$.

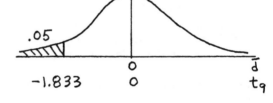

10. Refer to the notation and data summary for exercise #9.

$\bar{d} \pm t_{7,.025} \cdot s_d/\sqrt{n}$

$3.125 \pm 2.365 \cdot 2.911/\sqrt{8}$

3.125 ± 2.434

$.69 < \mu_d < 5.56$

11. $d = x_B - x_A$: -36 10 15 -2 0 -22 -36 -6 -26 7

$n = 10$

$\Sigma d = -96$ $\bar{d} = -9.6$

$\Sigma d^2 = 4106$ $s_d = 18.987$

$\bar{d} \pm t_{9,.025} \cdot s_d/\sqrt{n}$

$-9.6 \pm 2.262 \cdot 18.987/\sqrt{10}$

-9.6 ± 13.6

$-23.2 < \mu_d < 4.0$

12. Refer to the notation and data summary for exercise #11. The original claim is $\mu_d < 0$.

H_o: $\mu_d \ge 0$

H_1: $\mu_d < 0$

$\alpha = .05$

C.R. $t < -t_{9,.05} = -1.833$

calculations:

$t_{\bar{d}} = (\bar{d} - \mu_{\bar{d}})/s_{\bar{d}}$

$= (-9.6 - 0)/(18.987/\sqrt{10})$

$= -9.6/6.004 = -1.599$

conclusion:

Do not reject H_o; there is not sufficient evidence to conclude that $\mu_d < 0$.

Even though the sample changed in the desired direction, we cannot be 95% certain that the program is effective. If the program would not increase expenses and/or raise other concerns, however, it might be worth implementing -- or at least trying on a larger sample.

13. original claim $\mu_d = 0$ [$n \le 30$ and σ_d unknown, use t]

$d = x_{pre} - x_{post}$: 5 0 0 0 8 1 1 4 0 1

$n = 10$

$\Sigma d = 20$ $\bar{d} = 2.0$

$\Sigma d^2 = 108$ $s_d = 2.749$

H_o: $\mu_d = 0$
H_1: $\mu_d \neq 0$
$\alpha = .05$
C.R. t < $-t_{9,.025} = -2.262$
$\quad$ t > $t_{9,.025} = 2.262$
calculations:
$\quad t_{\bar{d}} = (\bar{d} - \mu_{\bar{d}})/s_{\bar{d}}$
$\quad\quad = (2.0 - 0)/(2.749/\sqrt{10})$
$\quad\quad = 2.0/.869 = 2.301$
conclusion:
$\quad$ Reject H_o; there is sufficient evidence to conclude that $\mu_d \neq 0$ (in fact, $\mu_d > 0$).
We can be 95% certain that physical training does affect weight (and that the weight is less after the training).

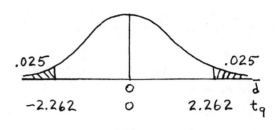

14. Refer to the notation and data summary for exercise #13.
$\bar{d} \pm t_{9,.025} \cdot s_d/\sqrt{n}$
$2.0 \pm 2.262 \cdot 2.749/\sqrt{10}$
2.0 ± 2.0
$0.0 < \mu_d < 4.0$

15. original claim $\mu_d = 0$ [n ≤ 30 and σ_d unknown, use t]
$d = x_8 - x_{12}$: -.7 -.8 -1.6 -.5 .8 .1 -.5 0 -1.3 -1.8 -2.3
$\quad$ n = 11
$\quad \Sigma d = -8.6 \quad\quad\quad \bar{d} = -.782$
$\quad \Sigma d^2 = 15.06 \quad\quad s_d = .913$
H_o: $\mu_d = 0$
H_1: $\mu_d \neq 0$
$\alpha = .05$
C.R. t < $-t_{10,.025} = -2.228$
$\quad$ t > $t_{10,.025} = 2.228$
calculations:
$\quad t_{\bar{d}} = (\bar{d} - \mu_{\bar{d}})/s_{\bar{d}}$
$\quad\quad = (-.782 - 0)/(.913/\sqrt{11})$
$\quad\quad = -.782/.275 = -2.840$
conclusion:
$\quad$ Reject H_o; there is sufficient evidence to conclude that $\mu_d \neq 0$ (in fact, $\mu_d < 0$).
No, it appears that a female's noon temperature is higher than her morning temperature.

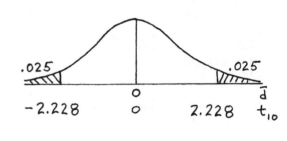

16. Refer to the notation and data summary for exercise #15.
$\bar{d} \pm t_{10,.025} \cdot s_d/\sqrt{n}$
$-.782 \pm 2.228 \cdot .913/\sqrt{11}$
$-.782 \pm .613$
$-1.40 < \mu_d < -.17$

17. a. To find the P-value for $t_7 = 3.822$ in a two-tailed test,
note that $t_{7,.005} = 3.500$ is the largest t value in the row.
$\quad$ Therefore, P-value < 2·(.005)
$\quad\quad\quad\quad$ P-value < .01

b. To find the P-value for $t_7 = 3.036$ in a one-tailed greater than test,
note that $t_{7,.01} = 2.998$ and $t_{7,.005} = 3.500$.
$\quad$ Therefore, .005 < P-value < .01.

18. $n > 30$, use z (with s for σ)
  a. $\pm z_{.05} = \pm 1.645$
  b. $-z_{.025} = -1.960$
  c. $z_{.02} = 2.05$

19. The $0 < \mu_d < 1.2$ confidence interval was obtained from $\bar{d} \pm t_{df,.025} \cdot s_d/\sqrt{n}$,
  which corresponds to a test of hypothesis critical region
      C.R. $t < -t_{df,.025}$
          $t > t_{df,.025}$.
  The one-tailed greater than test of hypothesis with the same relevant critical boundary has
      C.R. $t > t_{df,.025}$
  which corresponds to $\alpha = .025$.

8-3 Inferences about Two Means: Independent and Large Samples

NOTE: To be consistent with the notation of the previous sections, and thereby reinforcing the patterns and concepts presented in those sections, the manual uses the "usual" z formula written to apply to $\bar{x}_1 - \bar{x}_2$'s

$$z_{\bar{x}_1 - \bar{x}_2} = (\bar{x}_1 - \bar{x}_2 - \mu_{\bar{x}_1 - \bar{x}_2})/\sigma_{\bar{x}_1 - \bar{x}_2}$$

with $\mu_{\bar{x}_1 - \bar{x}_2} = \mu_1 - \mu_2$
and $\sigma_{\bar{x}_1 - \bar{x}_2} = \sqrt{\sigma_1^2/n_1 + \sigma_2^2/n_2}$
And so the formula for the z statistic may also be written as

$$z_{\bar{x}_1 - \bar{x}_2} = ((\bar{x}_1 - \bar{x}_2) - (\mu_1 - \mu_2))/\sqrt{\sigma_1^2/n_1 + \sigma_2^2/n_2}$$

1. original claim: $\mu_1 - \mu_2 = 0$ [$n_1 > 30$ and $n_2 > 30$, use z (with s's for σ's)]
  $\bar{x}_1 - \bar{x}_2 = 79.6 - 84.2 = -4.6$
  H_0: $\mu_1 - \mu_2 = 0$
  H_1: $\mu_1 - \mu_2 \neq 0$
  $\alpha = .05$
  C.R. $z < -z_{.025} = -1.960$
      $z > z_{.025} = 1.960$
  calculations:

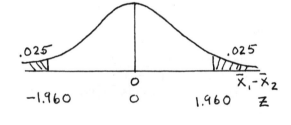

  $z_{\bar{x}_1 - \bar{x}_2} = (\bar{x}_1 - \bar{x}_2 - \mu_{\bar{x}_1 - \bar{x}_2})/\sigma_{\bar{x}_1 - \bar{x}_2}$
      $= (-4.6 - 0)/\sqrt{(12.4)^2/40 + (12.2)^2/40}$
      $= -4.6/2.750 = -1.672$
  conclusion:
    Do not reject H_0; there is not sufficient evidence to conclude that $\mu_1 - \mu_2 \neq 0$.

2. original claim: $\mu_1 - \mu_2 = 0$ [$n_1 > 30$ and $n_2 > 30$, use z (with s's for σ's)]
  $\bar{x}_1 - \bar{x}_2 = 8.49 - 8.41 = .08$
  H_0: $\mu_1 - \mu_2 = 0$
  H_1: $\mu_1 - \mu_2 \neq 0$
  $\alpha = .05$
  C.R. $z < -z_{.025} = -1.960$
      $z > z_{.025} = 1.960$
  calculations:
  $z_{\bar{x}_1 - \bar{x}_2} = (\bar{x}_1 - \bar{x}_2 - \mu_{\bar{x}_1 - \bar{x}_2})/\sigma_{\bar{x}_1 - \bar{x}_2}$
      $= (.08 - 0)/\sqrt{(.11)^2/32 + (.18)^2/60}$
      $= .08/.0303 = 2.640$
  conclusion:
    Reject H_0; there is sufficient evidence to conclude that $\mu_1 - \mu_2 \neq 0$ (in fact, $\mu_1 - \mu_2 > 0$).

3. $(\bar{x}_1 - \bar{x}_2) \pm z_{.025}\sqrt{\sigma_1^2/n_1 + \sigma_2^2/n_2}$
   $-4.6 \pm 1.960 \cdot \sqrt{(12.4)^2/40 + (12.2)^2/40}$
   -4.6 ± 5.4
   $-10.0 < \mu_1 - \mu_2 < .8$

4. $(\bar{x}_1 - \bar{x}_2) \pm z_{.025}\sqrt{\sigma_1^2/n_1 + \sigma_2^2/n_2}$
   $.08 \pm 1.960 \cdot \sqrt{(.11)^2/32 + (.18)^2/60}$
   $.08 \pm .06$
   $.02 < \mu_1 - \mu_2 < .14$

5. Let the males be group 1.
   original claim: $\mu_1 - \mu_2 > 0$ [$n_1 > 30$ and $n_2 > 30$, use z (with s's for σ's)]
   $\bar{x}_1 - \bar{x}_2 = 35,330 - 29,610 = 5720$
   $H_o: \mu_1 - \mu_2 \leq 0$
   $H_1: \mu_1 - \mu_2 > 0$
   $\alpha = .01$
   C.R. $z > z_{.01} = 2.327$
   calculations:

   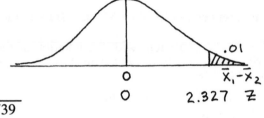

   $z_{\bar{x}_1 - \bar{x}_2} = (\bar{x}_1 - \bar{x}_2 - \mu_{\bar{x}_1 - \bar{x}_2})/\sigma_{\bar{x}_1 - \bar{x}_2}$
   $\quad = (5720 - 0)/\sqrt{(23,260)^2/86 + (16,480)^2/39}$
   $\quad = 5720/3640.72 = 1.571$
   conclusion:
   Do not reject H_o; there is not sufficient evidence to conclude that $\mu_1 - \mu_2 > 0$.

6. Refer to the notation and data summary for exercise #5.
   $(\bar{x}_1 - \bar{x}_2) \pm z_{.005}\sqrt{\sigma_1^2/n_1 + \sigma_2^2/n_2}$
   $5720 \pm 2.575 \cdot \sqrt{(23,260)^2/86 + (16,480)^2/39}$
   5720 ± 9375
   $-3655 < \mu_1 - \mu_2 < 15096$

7. Let the men aged 25-34 be group 1. [$n_1 > 30$ and $n_2 > 30$, use z (with s's for σ's)]
   $(\bar{x}_1 - \bar{x}_2) \pm z_{.005}\sqrt{\sigma_1^2/n_1 + \sigma_2^2/n_2}$
   $(176-164) \pm 2.575 \cdot \sqrt{(35.0)^2/804 + (27.0)^2/1657}$
   12 ± 4
   $8 < \mu_1 - \mu_2 < 16$
   NOTE: Since $\bar{x}_1 = 176$ and $\bar{x}_2 = 164$ are given only to the nearest pound, and the value of any decimal places is unknown, more accuracy than that cannot be given in the answer.

8. Refer to the notation and data summary for exercise #7. original claim: $\mu_1 - \mu_2 > 0$
   $H_o: \mu_1 - \mu_2 \leq 0$
   $H_1: \mu_1 - \mu_2 > 0$
   $\alpha = .01$
   C.R. $z > z_{.01} = 2.327$
   calculations:

   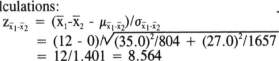

   $z_{\bar{x}_1 - \bar{x}_2} = (\bar{x}_1 - \bar{x}_2 - \mu_{\bar{x}_1 - \bar{x}_2})/\sigma_{\bar{x}_1 - \bar{x}_2}$
   $\quad = (12 - 0)/\sqrt{(35.0)^2/804 + (27.0)^2/1657}$
   $\quad = 12/1.401 = 8.564$
   conclusion:
   Reject H_o; there is sufficient evidence to conclude that $\mu_1 - \mu_2 > 0$.

9. Let American be group 1.
   original claim: $\mu_1-\mu_2 = 0$ [$n_1 > 30$ and $n_2 > 30$, use z (with s's for σ's)]
   $\bar{x}_1-\bar{x}_2 = 23{,}870 - 22{,}025 = 1845$
   H_o: $\mu_1-\mu_2 = 0$
   H_1: $\mu_1-\mu_2 \neq 0$
   $\alpha = .10$
   C.R. $z < -z_{.05} = -1.645$
   $z > z_{.05} = 1.645$
   calculations:

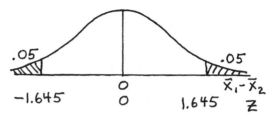

$$z_{\bar{x}_1-\bar{x}_2} = (\bar{x}_1-\bar{x}_2 - \mu_{\bar{x}_1-\bar{x}_2})/\sigma_{\bar{x}_1-\bar{x}_2}$$
$$= (1845 - 0)/\sqrt{(2960)^2/40 + (3065)^2/35}$$
$$= 1845/698.17 = 2.643$$

conclusion:
   Reject H_o; there is sufficient evidence to conclude that $\mu_1-\mu_2 \neq 0$ (in fact, $\mu_1-\mu_2 > 0$).
The conclusion that the mean pay is more at American than TWA is not necessarily relevant
for an individual considering employment. Both companies might have exactly the same pay
scale, for example, but the American attendants might have more accumulated experience and
hence be paid more.

10. Refer to the notation and data summary for exercise #9.
   $(\bar{x}_1-\bar{x}_2) \pm z_{.05}\sqrt{\sigma_1^2/n_1 + \sigma_2^2/n_2}$
   $1845 \pm 1.645 \cdot \sqrt{(2960)^2/40 + (3065)^2/35}$
   1845 ± 1148
   $697 < \mu_1-\mu_2 < 2993$

11. Let Tom's class be group 1.
   original claim: $\mu_1-\mu_2 = 0$ [$n_1 > 30$ and $n_2 > 30$, use z (with s's for σ's)]
   $\bar{x}_1-\bar{x}_2 = 76.0 - 73.4 = 2.6$
   H_o: $\mu_1-\mu_2 = 0$
   H_1: $\mu_1-\mu_2 \neq 0$
   $\alpha = .05$
   C.R. $z < -z_{.025} = -1.960$
   $z > z_{.025} = 1.960$
   calculations:

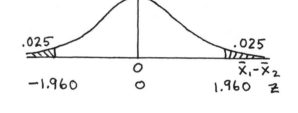

$$z_{\bar{x}_1-\bar{x}_2} = (\bar{x}_1-\bar{x}_2 - \mu_{\bar{x}_1-\bar{x}_2})/\sigma_{\bar{x}_1-\bar{x}_2}$$
$$= (2.6 - 0)/\sqrt{(14.1)^2/35 + (13.5)^2/40}$$
$$= 2.6/3.199 = .812$$

conclusion:
   Do not reject H_o; there is not sufficient evidence to conclude that $\mu_1-\mu_2 \neq 0$.

12. Refer to the notation and data summary for exercise #11.
   $(\bar{x}_1-\bar{x}_2) \pm z_{.025}\sqrt{\sigma_1^2/n_1 + \sigma_2^2/n_2}$
   $2.6 \pm 1.960 \cdot \sqrt{(14.1)^2/35 + (13.5)^2/40}$
   2.6 ± 6.3
   $-3.7 < \mu_1-\mu_2 < 8.9$

13. Let the East be group 1.
   $(\bar{x}_1-\bar{x}_2) \pm z_{.025}\sqrt{\sigma_1^2/n_1 + \sigma_2^2/n_2}$
   $(421-347) \pm 1.960 \cdot \sqrt{(122)^2/35 + (85)^2/50}$
   74 ± 47
   $27 < \mu_1-\mu_2 < 121$
   No, the confidence interval does <u>not</u> include zero. This suggests that there <u>is</u> a significant
   difference between the two means (in fact, that μ_{East} is larger than μ_{West}).

14. Refer to the notation and data summary for exercise #13. original claim: $\mu_1 - \mu_2 = 0$

H_o: $\mu_1 - \mu_2 = 0$

H_1: $\mu_1 - \mu_2 \neq 0$

$\alpha = .05$

C.R. $z < -z_{.025} = -1.960$

$\quad\ z > z_{.025} = 1.960$

calculations:

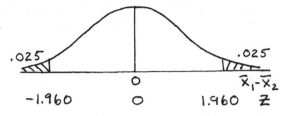

$$z_{\bar{x}_1 - \bar{x}_2} = (\bar{x}_1 - \bar{x}_2 - \mu_{\bar{x}_1 - \bar{x}_2})/\sigma_{\bar{x}_1 - \bar{x}_2}$$

$$= (74 - 0)/\sqrt{(122)^2/35 + (85)^2/50}$$

$$= 74/23.870 = 3.100$$

conclusion:

Reject H_o; there is sufficient evidence to conclude that $\mu_1 - \mu_2 \neq 0$ (in fact, $\mu_1 - \mu_2 > 0$).

15. Let Weston be group 1.

original claim: $\mu_1 - \mu_2 = 0$ [$n_1 > 30$ and $n_2 > 30$, use z (with s's for σ's)]

$\bar{x}_1 - \bar{x}_2 = 12.2 - 14.0 = -1.8$

H_o: $\mu_1 - \mu_2 = 0$

H_1: $\mu_1 - \mu_2 \neq 0$

$\alpha = .05$

C.R. $z < -z_{.025} = -1.960$

$\quad\ z > z_{.025} = 1.960$

calculations:

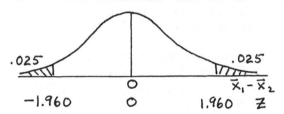

$$z_{\bar{x}_1 - \bar{x}_2} = (\bar{x}_1 - \bar{x}_2 - \mu_{\bar{x}_1 - \bar{x}_2})/\sigma_{\bar{x}_1 - \bar{x}_2}$$

$$= (-1.8 - 0)/\sqrt{(1.5)^2/50 + (2.1)^2/50}$$

$$= -1.8/.365 = -4.932$$

conclusion:

Reject H_o; there is sufficient evidence to conclude that $\mu_1 - \mu_2 \neq 0$ (in fact, $\mu_1 - \mu_2 < 0$).

The conclusion that the mean response time is less for Weston than for Mid-Valley is not necessarily relevant for an individual considering calling an ambulance. Both companies might have exactly the same response time over the same distance, for example, but the Weston sample might have involved smaller distances. Similarly, the most important criterion for any one particular person in need of an ambulance to consider would probably be the distance from the ambulance company to the scene of the emergency.

16. Let those not wearing seat belts be group 1.

original claim: $\mu_1 - \mu_2 > 0$ [$n_1 > 30$ and $n_2 > 30$, use z (with s's for σ's)]

$\bar{x}_1 - \bar{x}_2 = 1.39 - .83 = .56$

H_o: $\mu_1 - \mu_2 \leq 0$

H_1: $\mu_1 - \mu_2 > 0$

$\alpha = .01$

C.R. $z > z_{.01} = 2.327$

calculations:

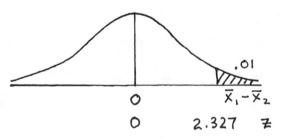

$$z_{\bar{x}_1 - \bar{x}_2} = (\bar{x}_1 - \bar{x}_2 - \mu_{\bar{x}_1 - \bar{x}_2})/\sigma_{\bar{x}_1 - \bar{x}_2}$$

$$= (.56 - 0)/\sqrt{(3.06)^2/290 + (1.77)^2/123}$$

$$= .56/.240 = 2.330$$

conclusion:

Reject H_o; there is sufficient evidence to conclude that $\mu_1 - \mu_2 > 0$.

Yes; based on this result, there is significant evidence in favor of seat belt use among children.

17. Let the 18-24 year olds be group 1.
original claim: $\mu_1-\mu_2 = 0$ [$n_1 > 30$ and $n_2 > 30$, use z (with s's for σ's)]

<u>group 1: 18-24 inclusive</u> <u>group 2: 25 and older</u>

n = 37	n = 56
$\Sigma x = 3634.3$	$\Sigma x = 5491.2$
$\Sigma x^2 = 356{,}992.32$	$\Sigma x^2 = 538{,}473.56$
$\bar{x} = 98.2243$	$\bar{x} = 98.0571$
$s^2 = .4349$	$s^2 = .4032$

$\bar{x}_1-\bar{x}_2 = 98.2243 - 98.0571 = .1672$

a. H_o: $\mu_1-\mu_2 = 0$
H_1: $\mu_1-\mu_2 \neq 0$
$\alpha = .05$
C.R. $z < -z_{.025} = -1.960$
$\quad z > z_{.025} = 1.960$
calculations:

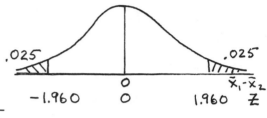

$z_{\bar{x}_1-\bar{x}_2} = (\bar{x}_1-\bar{x}_2 - \mu_{\bar{x}_1-\bar{x}_2})/\sigma_{\bar{x}_1-\bar{x}_2}$
$\quad = (.1672 - 0)/\sqrt{.4349/37 + .4032/56}$
$\quad = .1672/.1377 = 1.214$
conclusion:
Do not reject H_o; there is not sufficient evidence to conclude that $\mu_1-\mu_2 \neq 0$.

b. H_o: $\mu_1-\mu_2 = 0$
H_1: $\mu_1-\mu_2 \neq 0$
$\alpha = .05$
C.R. P-value $< .05$
calculations:

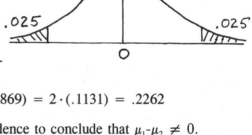

$z_{\bar{x}_1-\bar{x}_2} = (\bar{x}_1-\bar{x}_2 - \mu_{\bar{x}_1-\bar{x}_2})/\sigma_{\bar{x}_1-\bar{x}_2}$
$\quad = (.1672 - 0)/\sqrt{.4349/37 + .4032/56}$
$\quad = .1672/.1377 = 1.214$
P-value $= 2 \cdot P(z > 1.21) = 2 \cdot (.5000 - .3869) = 2 \cdot (.1131) = .2262$
conclusion:
Do not reject H_o; there is not sufficient evidence to conclude that $\mu_1-\mu_2 \neq 0$.

c. $(\bar{x}_1-\bar{x}_2) \pm z_{.025}\sqrt{\sigma_1^2/n_1 + \sigma_2^2/n_2}$
$(98.2243 - 98.0571) \pm 1.960 \cdot \sqrt{.4349/37 + .4032/56}$
$.17 \pm .27$
$-.10 < \mu_1-\mu_2 < .44$
Since the confidence interval includes the value zero, there is not sufficient evidence to conclude that $\mu_1-\mu_2 \neq 0$.

18. Refer to the notation and data summary for exercise #7. original claim: $\mu_1-\mu_2 > 0$
H_o: $\mu_1-\mu_2 \leq 0$
H_1: $\mu_1-\mu_2 > 0$
$\alpha = .01$
C.R. $z > z_{.01} = 2.327$
calculations:

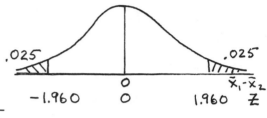

$z_{\bar{x}_1-\bar{x}_2} = (\bar{x}_1-\bar{x}_2 - \mu_{\bar{x}_1-\bar{x}_2})/\sigma_{\bar{x}_1-\bar{x}_2}$
$2.327 = (\bar{x}_1-\bar{x}_2 - 0)/1.401$
$\bar{x}_1-\bar{x}_2 = 3.26$

$\beta = P(\bar{x}_1-\bar{x}_2 < 3.26 \mid \mu_1-\mu_2 = 6.0)$
$\quad = P(z < -1.95)$
$\quad = .5000 - .4744$
$\quad = .0256$

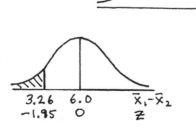

19. a. $x = 5,10,15$
$\mu = \Sigma x/n = 30/3 = 10$
$\sigma^2 = \Sigma(x-\mu)^2/n = [(-5)^2 + (0)^2 + (5)^2]/3 = 50/3$
   b. $y = 1,2,3$
$\mu = \Sigma y/n = 6/3 = 2$
$\sigma^2 = \Sigma(y-\mu)^2/n = [(-1)^2 + (0)^2 + (1)^2]/3 = 2/3$
   c. $z = x-y = 4,3,2,9,8,7,14,13,12$
$\mu = \Sigma z/n = 72/9 = 8$
$\sigma^2 = \Sigma(z-\mu)^2/n$
$= [(-4)^2 + (-5)^2 + (-6)^2 + (1)^2 + (0)^2 + (-1)^2 + (6)^2 + (5)^2 + (4)^2]/9 = 156/9 = 52/3$
   d. $\sigma^2_{x-y} = \sigma^2_x + \sigma^2_y$
$52/3 = 50/3 + 2/3$
$52/3 = 52/3$

20. Let R stand for range. **NOTE:** The problem refers to all possible x-y differences (where n_x and n_y might even be different) and not to x-y differences for paired data.
$R_{x-y} = \text{highest}_{x-y} - \text{lowest}_{x-y}$
$= (\text{highest x - lowest y}) - (\text{lowest x - highest y})$
$= \text{highest x - lowest y - lowest x + highest y}$
$= (\text{highest x - lowest x}) + (\text{highest y - lowest y})$
$= R_x + R_y$
That is, the range of all possible x-y values is the sum of the individual ranges of x and y.

8-4 Comparing Two Variances

NOTE: The following conventions are used in this manual regarding the F test.
* The set of scores with the larger sample variance is designated group 1.
* Even though always designating the scores with the larger sample variance as group 1 makes lower critical values are unnecessary in two-tailed tests, the lower critical value will be calculated (using the method given in exercise #19) and included (in brackets) for completeness and for consistency with the other tests.
* The degrees of freedom for group 1 (numerator) and group 2 (denominator) will be given with the F as a superscript and subscript respectively.
* If the desired degrees of freedom does not appear in Table A-5, the closest entry will be used. If the desired degrees of freedom is exactly halfway between two tabled values, the conservative approach of using the smaller degrees of freedom is employed. Since any finite number is closer to 120 than ∞, 120 is used for all degrees of freedom larger than 120.
* Since all hypotheses in the text question the equality of σ^2_1 and σ^2_2, the calculation of F [which is statistically defined to be $F = (s^2_1/\sigma^2_1)/(s^2_2/\sigma^2_2)$] is shortened to $F = s^2_1/s^2_2$.

1. Let sample A be group 1. original claim: $\sigma^2_1 = \sigma^2_2$
$H_o: \sigma^2_1 = \sigma^2_2$
$H_1: \sigma^2_1 \neq \sigma^2_2$
$\alpha = .05$
C.R. $F < F^9_{9,.975} = [.2484]$
$F > F^9_{9,.025} = 4.0260$
calculations:
$F = s^2_1/s^2_2$
$= 50/25 = 2.0000$
conclusion:
Do not reject H_o; there is not sufficient evidence to conclude that $\sigma^2_1 \neq \sigma^2_2$.

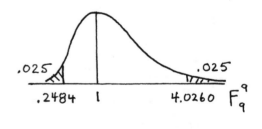

2. Let sample A be group 1. original claim: $\sigma_1^2 = \sigma_2^2$
H_o: $\sigma_1^2 = \sigma_2^2$
H_1: $\sigma_1^2 \neq \sigma_2^2$
$\alpha = .05$
C.R. $F < F_{14,.975}^9 = [.2653]$
$\quad\;\; F > F_{14,.025}^9 = 3.2093$
calculations:
$\quad F = s_1^2/s_2^2$
$\quad\quad = 50/12 = 4.1667$
conclusion:
$\quad$ Reject H_o; there is sufficient evidence to conclude that $\sigma_1^2 \neq \sigma_2^2$ (in fact, $\sigma_1^2 > \sigma_2^2$).

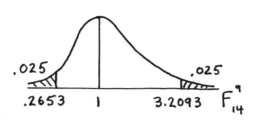

3. Let sample A be group 1. original claim: $\sigma_1^2 > \sigma_2^2$
H_o: $\sigma_1^2 \leq \sigma_2^2$
H_1: $\sigma_1^2 > \sigma_2^2$
$\alpha = .05$
C.R. $F > F_{199,.05}^{15} = 1.7505$
calculations:
$\quad F = s_1^2/s_2^2$
$\quad\quad = (15.00)^2/(12.00)^2 = 1.4038$
conclusion:
$\quad$ Do not reject H_o; there is not sufficient evidence to conclude that $\sigma_1^2 > \sigma_2^2$.

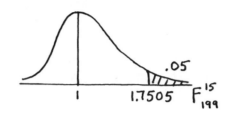

4. Let sample A be group 1. original claim: $\sigma_1^2 > \sigma_2^2$
H_o: $\sigma_1^2 \leq \sigma_2^2$
H_1: $\sigma_1^2 > \sigma_2^2$
$\alpha = .05$
C.R. $F > F_{24,.05}^{34} = 1.9390$
calculations:
$\quad F = s_1^2/s_2^2$
$\quad\quad = 42.3/16.2 = 2.6111$
conclusion:
$\quad$ Reject H_o; there is sufficient evidence to conclude that $\sigma_1^2 > \sigma_2^2$.

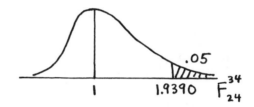

5. Let the traditional method be group 1. original claim: $\sigma_1^2 = \sigma_2^2$
H_o: $\sigma_1^2 = \sigma_2^2$
H_1: $\sigma_1^2 \neq \sigma_2^2$
$\alpha = .05$
C.R. $F < F_{29,.975}^{24} = [.4527]$
$\quad\;\; F > F_{29,.025}^{24} = 2.1540$
calculations:
$\quad F = s_1^2/s_2^2$
$\quad\quad = (.37)^2/(.31)^2 = 1.4246$
conclusion:
$\quad$ Do not reject H_o; there is not sufficient evidence to conclude that $\sigma_1^2 \neq \sigma_2^2$.
No; one cannot be 95% certain that there is really any difference in the variations.

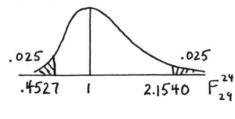

6. Let the first college be group 1. original claim: $\sigma_1^2 = \sigma_2^2$
  H_o: $\sigma_1^2 = \sigma_2^2$
  H_1: $\sigma_1^2 \neq \sigma_2^2$
  $\alpha = .02$
  C.R. $F < F_{29,.99}^{59} = [.4930]$
      $F > F_{29,.01}^{59} = 2.2344$
  calculations:
      $F = s_1^2/s_2^2$
      $= (14.2)^2/(9.8)^2 = 2.0995$
  conclusion:
      Do not reject H_o; there is not sufficient evidence to conclude that $\sigma_1^2 \neq \sigma_2^2$.

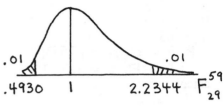

7. Let the first time period be group 1. original claim: $\sigma_1^2 = \sigma_2^2$
  H_o: $\sigma_1^2 = \sigma_2^2$
  H_1: $\sigma_1^2 \neq \sigma_2^2$
  $\alpha = .05$
  C.R. $F < F_{11,.975}^{11} = [.2836]$
      $F > F_{11,.025}^{11} = 3.5257$
  calculations:
      $F = s_1^2/s_2^2$
      $= (11.07)^2/(10.39)^2 = 1.1352$
  conclusion:
      Do not reject H_o; there is not sufficient evidence to conclude that $\sigma_1^2 \neq \sigma_2^2$.

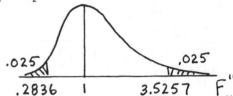

8. Let the brown ones be group 1. original claim: $\sigma_1^2 = \sigma_2^2$
  H_o: $\sigma_1^2 = \sigma_2^2$
  H_1: $\sigma_1^2 \neq \sigma_2^2$
  $\alpha = .05$
  C.R. $F < F_{8,.975}^{29} = [.3747]$
      $F > F_{8,.025}^{29} = 3.8940$
  calculations:
      $F = s_1^2/s_2^2$
      $= (.0521)^2/(.0465)^2 = 1.2554$
  conclusion:
      Do not reject H_o; there is not sufficient evidence to conclude that $\sigma_1^2 \neq \sigma_2^2$.
  No; the evidence does not indicate a significant difference in the standard deviations of the green and brown candies.

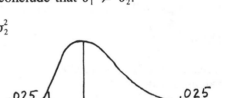

9. Let TWA be group 1. original claim: $\sigma_1^2 = \sigma_2^2$
  H_o: $\sigma_1^2 = \sigma_2^2$
  H_1: $\sigma_1^2 \neq \sigma_2^2$
  $\alpha = .10$
  C.R. $F < F_{39,.95}^{34} = [.5581]$
      $F > F_{39,.05}^{34} = 1.7444$
  calculations:
      $F = s_1^2/s_2^2$
      $= (3065)^2/(2960)^2 = 1.0722$
  conclusion:
      Do not reject H_o; there is not sufficient evidence to conclude that $\sigma_1^2 \neq \sigma_2^2$.

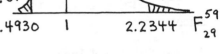

10. Let the 25 to 34 year olds be group 1. original claim: $\sigma_1^2 > \sigma_2^2$

H_o: $\sigma_1^2 \leq \sigma_2^2$
H_1: $\sigma_1^2 > \sigma_2^2$
$\alpha = .01$
C.R. $F > F_{1656,.01}^{803} = 1.5330$
calculations:
$F = s_1^2/s_2^2$
$\quad = (35.0)^2/(27.0)^2 = 1.6804$

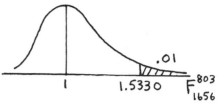

conclusion:
  Reject H_o; there is sufficient evidence to conclude that $\sigma_1^2 > \sigma_2^2$.

11. Let the manual keying be group 1. original claim: $\sigma_1^2 = \sigma_2^2$

H_o: $\sigma_1^2 = \sigma_2^2$
H_1: $\sigma_1^2 \neq \sigma_2^2$
$\alpha = .02$
C.R. $F < F_{9,.99}^{15} = [.2568]$
$\quad\;\, F > F_{9,.01}^{15} = 4.9621$
calculations:
$F = s_1^2/s_2^2$
$\quad = 225.0/56.0 = 4.0179$

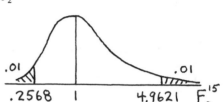

conclusion:
  Do not reject H_o; there is not sufficient evidence to conclude that $\sigma_1^2 \neq \sigma_2^2$.

12. Let sample A be group 1. original claim: $\sigma_1^2 = \sigma_2^2$

H_o: $\sigma_1^2 = \sigma_2^2$
H_1: $\sigma_1^2 \neq \sigma_2^2$
$\alpha = .05$
C.R. $F < F_{24,.975}^{29} = [.4643]$
$\quad\;\, F > F_{24,.025}^{29} = 2.2090$
calculations:
$F = s_1^2/s_2^2$
$\quad = (.22)^2/(.14)^2 = 2.4694$

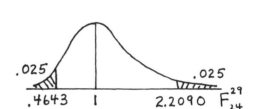

conclusion:
  Reject H_o; there is sufficient evidence to conclude that $\sigma_1^2 \neq \sigma_2^2$ (in fact, $\sigma_1^2 > \sigma_2^2$).
  Yes; corrective action should be taken to achieve a lower standard deviation for the night shift.

13. experimental   control
    $n = 20$ $n = 16$
    $\Sigma x = 1327.61$ $\Sigma x = 1660.28$
    $\Sigma x^2 = 96477.3313$ $\Sigma x^2 = 181167.5442$
    $\bar{x} = 66.3805$ $\bar{x} = 103.7675$
    $s^2 = 439.469$ $s^2 = 592.296$

Let the controls be group 1. original claim: $\sigma_1^2 = \sigma_2^2$

H_o: $\sigma_1^2 = \sigma_2^2$
H_1: $\sigma_1^2 \neq \sigma_2^2$
$\alpha = .05$
C.R. $F < F_{19,.975}^{15} = [.3629]$
$\quad\;\, F > F_{19,.025}^{15} = 2.6171$
calculations:
$F = s_1^2/s_2^2$
$\quad = 592.296/439.469 = 1.3478$

conclusion:
  Do not reject H_o; there is not sufficient evidence to conclude that $\sigma_1^2 \neq \sigma_2^2$.

14. <u>easy to difficult</u>   <u>difficult to easy</u>

$n = 25$ $n = 16$
$\Sigma x = 677.88$ $\Sigma x = 507.65$
$\Sigma x^2 = 19,509.3278$ $\Sigma x^2 = 16,379.0161$
$\bar{x} = 27.115$ $\bar{x} = 31.728$
$s^2 = 47.0198$ $s^2 = 18.1489$

Let the easy to difficult sample be group 1. original claim: $\sigma_1^2 = \sigma_2^2$
H_o: $\sigma_1^2 = \sigma_2^2$
H_1: $\sigma_1^2 \neq \sigma_2^2$
$\alpha = .05$
C.R. F $< F_{15,.975}^{24} = [.4103]$
       F $> F_{15,.025}^{24} = 2.7006$
calculations:
       F $= s_1^2/s_2^2$

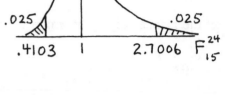

       $= 47.0198/18.1489 = 2.5908$
conclusion:
       Do not reject H_o; there is not sufficient evidence to conclude that $\sigma_1^2 \neq \sigma_2^2$.

15. <u>placebo</u>   <u>calcium</u>

$n = 13$ $n = 15$
$\Sigma x = 1490.5$ $\Sigma x = 1740.4$
$\Sigma x^2 = 171965.47$ $\Sigma x^2 = 202936.92$
$\bar{x} = 114.65$ $\bar{x} = 116.03$
$s^2 = 89.493$ $s^2 = 71.722$

Let the placebos be group 1. original claim: $\sigma_1^2 = \sigma_2^2$
H_o: $\sigma_1^2 = \sigma_2^2$
H_1: $\sigma_1^2 \neq \sigma_2^2$
$\alpha = .05$
C.R. F $< F_{14,.975}^{12} = [.3147]$
       F $> F_{14,.025}^{12} = 3.0502$
calculations:
       F $= s_1^2/s_2^2$

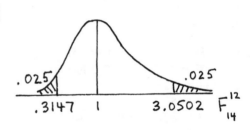

       $= 89.493/71.722 = 1.2478$
conclusion:
       Do not reject H_o; there is not sufficient evidence to conclude that $\sigma_1^2 \neq \sigma_2^2$.

16. Let the zone 7 values be group 1. original claim: $\sigma_1^2 = \sigma_2^2$
H_o: $\sigma_1^2 = \sigma_2^2$
H_1: $\sigma_1^2 \neq \sigma_2^2$
$\alpha = .05$
C.R. F $< F_{13,.975}^{10} = [.2762]$
       F $> F_{13,.025}^{10} = 3.2497$
calculations:
       F $= s_1^2/s_2^2$

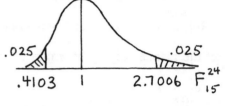

       $= 2122/455 = 4.6637$
conclusion:
       Reject H_o; there is sufficient evidence to conclude that $\sigma_1^2 \neq \sigma_2^2$ (in fact, $\sigma_1^2 > \sigma_2^2$).

17. Let the first professor's scores be group 1.
original claim: $\sigma_1^2 > \sigma_2^2$
H_o: $\sigma_1^2 \leq \sigma_2^2$
H_1: $\sigma_1^2 > \sigma_2^2$
$\alpha = .05$
C.R. $F > F_{24,.05}^{24} = 1.9838$
calculations:
$F = s_1^2/s_2^2$
$= 103.4/39.7$
$= 2.6045$

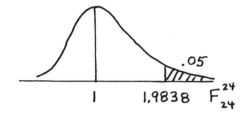

conclusion:
Reject H_o; there is sufficient evidence to conclude that $\sigma_1^2 > \sigma_2^2$.
A student very weak in organic chemistry should choose the second professor -- i.e., the one whose grades have the smaller variance. The potentially extremely low scores (and -- to the disadvantage of very good students -- the potentially extremely high scores) will tend to turn out closer to the mean of the scores.

18. Since the calculated F_{19}^{24} in the two-tailed test is $3.8643 > F_{19,.01}^{24} = 2.9249$, it follows that
$$\text{P-value} < 2(.01)$$
$$\text{P-value} < .02$$

19. a. $F_L = F_{9,.975}^9 = 1/F_{9,.025}^9 = 1/4.0260 = .2484$
$F_R = F_{9,.025}^9 = 4.0260$
b. $F_L = F_{6,.975}^9 = 1/F_{9,.025}^6 = 1/4.3197 = .2315$
$F_R = F_{6,.025}^9 = 5.5234$
c. $F_L = F_{9,.975}^6 = 1/F_{6,.025}^9 = 1/5.5234 = .1810$
$F_R = F_{9,.025}^6 = 4.3197$
d. $F_L = F_{9,.99}^{24} = 1/F_{24,.025}^9 = 1/3.2560 = .3071$
$F_R = F_{9,.01}^{24} = 4.7290$
e. $F_L = F_{24,.99}^9 = 1/F_{9,.025}^{24} = 1/4.7290 = .2115$
$F_R = F_{24,.01}^9 = 3.2560$

20. Let the experimentals be group 1.
$F_L = F_{15,.975}^{19} = 1/F_{19,.025}^{15} = 1/2.6171 = .3821$
$F_R = F_{15,.025}^{19} = 2.7559$
$$(s_1^2/s_2^2) \cdot (1/F_R) < \sigma_1^2/\sigma_2^2 < (s_1^2/s_2^2) \cdot (1/F_L)$$
$$(439.469/592.296) \cdot (1/2.7559) < \sigma_1^2/\sigma_2^2 < (439.469/592.296) \cdot (1/.3821)$$
$$.269 < \sigma_1^2/\sigma_2^2 < 1.942$$

21. a. No. Adding a constant to each score does not affect the spread of the scores. All the standard deviations and variances remain the same, and so the F statistic (i.e., the ratio of the variances) is unchanged.
b. No. Multiplying each score by a constant multiplies the standard deviation of those scores by that constant and the variance of those scores by the square of that constant. If this is done to both groups, so that each variance (i.e., the numerator and the denominator of the F statistic) is multiplied by the square of that constant, then the F statistic is unchanged.
c. No. The change from Fahrenheit to Celsius is done by multiplication and addition. As noted in part (a), the additive constant does not affect the variances and, therefore, does not affect the F statistic. As noted in part (b), the multiplicative constant affects both the numerator and denominator of the F statistic in the same manner and, therefore, does not affect the value of the F statistic.

22. a. Let the second sample be group 1. original claim: $\sigma_1^2 > \sigma_2^2$

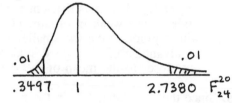

H_o: $\sigma_1^2 \leq \sigma_2^2$
H_1: $\sigma_1^2 > \sigma_2^2$
$\alpha = .05$
C.R. F > ?
calculations:
  $F = s_1^2/s_2^2$
  $= 57/37 = 1.5404$
If $n_1 = n_2 = 61$, we have C.R. $F > F_{60,.05}^{60} = 1.5343$ and barely reject H_o.
The approximate minimum size of each sample, therefore, is 61.

b. Let the first sample be group 1. original claim: $\sigma_1^2 = \sigma_2^2$

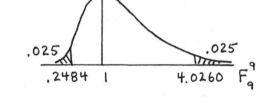

H_o: $\sigma_1^2 = \sigma_2^2$
H_1: $\sigma_1^2 \neq \sigma_2^2$
$\alpha = .02$
C.R. $F < F_{24,.99}^{20} = [.3497]$
    $F > F_{24,.01}^{20} = 2.7380$
calculations:
  $F = s_1^2/s_2^2 = 67.2/s_2^2 > 2.7380$
        $s_2^2 < 67.2/2.7380$
        $s_2^2 < 24.54$ rounded down to 24.5
The maximum variance of the second sample that causes H_o to be rejected as stated is 24.5.

8-5 Inferences about Two Means: Independent and Small Samples

1. Let Sample A be group 1.
original claim: $\mu_1-\mu_2 = 0$ [small samples and σ unknown, first test H_o: $\sigma_1^2 = \sigma_2^2$]

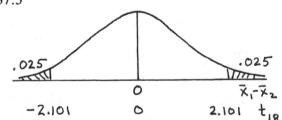

H_o: $\sigma_1^2 = \sigma_2^2$
H_1: $\sigma_1^2 \neq \sigma_2^2$
$\alpha = .05$
C.R. $F < F_{9,.975}^9 = [.2484]$
    $F > F_{9,.025}^9 = 4.0260$
calculations:
  $F = s_1^2/s_2^2$
     $= 50/25 = 2.0000$
conclusion:
    Do not reject H_o; there is not sufficient evidence to conclude that $\sigma_1^2 \neq \sigma_2^2$.
Now proceed using s_p^2 for both s_1^2 and s_2^2
$\bar{x}_1-\bar{x}_2 = 200 - 185 = 15$
$s_p^2 = (df_1 \cdot s_1^2 + df_2 \cdot s_2^2)/(df_1 + df_2)$
  $= (9 \cdot 50 + 9 \cdot 25)/(9 + 9) = 675/18 = 37.5$
H_o: $\mu_1-\mu_2 = 0$
H_1: $\mu_1-\mu_2 \neq 0$
$\alpha = .05$
C.R. $t < -t_{18,.025} = -2.101$
    $t > t_{18,.025} = 2.101$
calculations:
$t_{\bar{x}_1-\bar{x}_2} = (\bar{x}_1-\bar{x}_2 - \mu_{\bar{x}_1-\bar{x}_2})/s_{\bar{x}_1-\bar{x}_2}$
    $= (15 - 0)/\sqrt{37.5/10 + 37.5/10}$
    $= 15/2.739 = 5.477$
conclusion:
    Reject H_o; there is sufficient evidence to conclude that $\mu_1-\mu_2 \neq 0$ (in fact, $\mu_1-\mu_2 > 0$).

2. Let Sample A be group 1.
    original claim: $\mu_1 - \mu_2 = 0$ [small samples and σ unknown, first test H_o: $\sigma_1^2 = \sigma_2^2$]
        H_o: $\sigma_1^2 = \sigma_2^2$
        H_1: $\sigma_1^2 \neq \sigma_2^2$
        $\alpha = .05$
        C.R. $F < F_{14,.975}^9 = [.2653]$
             $F > F_{14,.025}^9 = 3.2093$

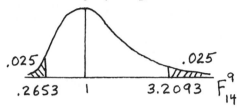

        calculations:
           $F = s_1^2/s_2^2$
              $= 50/12 = 4.1667$
        conclusion:
           Reject H_o; there is sufficient evidence to conclude that $\sigma_1^2 \neq \sigma_2^2$ (in fact, $\sigma_1^2 > \sigma_2^2$).
    Now proceed using s_1^2 and s_2^2 and the smaller degrees of freedom (viz., $df_1 = 9$).
        $\bar{x}_1 - \bar{x}_2 = 255 - 212 = 43$
    H_o: $\mu_1 - \mu_2 = 0$
    H_1: $\mu_1 - \mu_2 \neq 0$
    $\alpha = .05$
    C.R. $t < -t_{9,.025} = -2.262$
        $t > t_{9,.025} = 2.262$

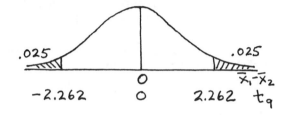

    calculations:
        $t_{\bar{x}_1-\bar{x}_2} = (\bar{x}_1-\bar{x}_2 - \mu_{\bar{x}_1-\bar{x}_2})/s_{\bar{x}_1-\bar{x}_2}$
           $= (43 - 0)/\sqrt{50/10 + 12/15}$
           $= 43/4.408 = 17.855$
    conclusion:
        Reject H_o; there is sufficient evidence to conclude that $\mu_1 - \mu_2 \neq 0$ (in fact, $\mu_1 - \mu_2 > 0$).

3. Refer to exercise #1.
    $(\bar{x}_1-\bar{x}_2) \pm t_{18,.025}\sqrt{s_p^2/n_1 + s_p^2/n_2}$
    $15 \pm 2.101 \cdot \sqrt{37.5/10 + 37.5/10}$
    15 ± 6
    $9 < \mu_1 - \mu_2 < 21$

4. Refer to exercise #2.
    $(\bar{x}_1-\bar{x}_2) \pm t_{9,.025}\sqrt{s_1^2/n_1 + s_2^2/n_2}$
    $43 \pm 2.262 \cdot \sqrt{50/10 + 12/15}$
    43 ± 5
    $38 < \mu_1 - \mu_2 < 48$

5. Let the traditional method be group 1.
    original claim: $\mu_1 - \mu_2 = 0$ [small samples and σ unknown, first test H_o: $\sigma_1^2 = \sigma_2^2$]
        H_o: $\sigma_1^2 = \sigma_2^2$
        H_1: $\sigma_1^2 \neq \sigma_2^2$
        $\alpha = .05$
        C.R. $F < F_{29,.975}^{24} = [.4527]$
             $F > F_{29,.025}^{24} = 2.1540$

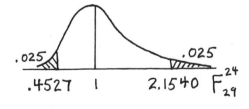

        calculations:
           $F = s_1^2/s_2^2$
              $= (.37)^2/(.31)^2 = 1.4246$
        conclusion:
           Do not reject H_o; there is not sufficient evidence to conclude that $\sigma_1^2 \neq \sigma_2^2$.

Now proceed using s_p^2 for both s_1^2 and s_2^2

$\bar{x}_1 - \bar{x}_2 = 4.31 - 4.07 = .24$

$s_p^2 = (df_1 \cdot s_1^2 + df_2 \cdot s_2^2)/(df_1 + df_2)$
$= (24 \cdot (.37)^2 + 29 \cdot (.31)^2)/(24 + 29) = 6.0725/53 = .1146$

H_o: $\mu_1 - \mu_2 = 0$
H_1: $\mu_1 - \mu_2 \neq 0$
$\alpha = .05$
C.R. $t < -t_{53,.025} = -1.960$
$\quad t > t_{53,.025} = 1.960$
calculations:

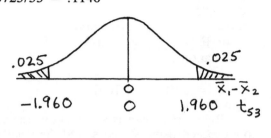

$t_{\bar{x}_1 - \bar{x}_2} = (\bar{x}_1 - \bar{x}_2 - \mu_{\bar{x}_1 - \bar{x}_2})/s_{\bar{x}_1 - \bar{x}_2}$
$= (.24 - 0)/\sqrt{.1146/25 + .1146/30}$
$= .24/.0917 = 2.618$

conclusion:
Reject H_o; there is sufficient evidence to conclude that $\mu_1 - \mu_2 \neq 0$ (in fact, $\mu_1 - \mu_2 > 0$).
A person buying a battery should prefer one manufactured by the traditional method.

6. Refer to the notation, F test and data summary for exercise #5.
$(\bar{x}_1 - \bar{x}_2) \pm t_{53,.025}\sqrt{s_p^2/n_1 + s_p^2/n_2}$
$.24 \pm 1.960\sqrt{.1146/25 + .1146/30}$
$.24 \pm .18$
$.06 < \mu_1 - \mu_2 < .42$

7. Let the first time period be group 1.
original claim: $\mu_1 - \mu_2 = 0$ [small samples and σ unknown, first test H_o: $\sigma_1^2 = \sigma_2^2$]
H_o: $\sigma_1^2 = \sigma_2^2$
H_1: $\sigma_1^2 \neq \sigma_2^2$
$\alpha = .05$
C.R. $F < F_{11,.975}^{11} = [.2836]$
$\quad F > F_{11,.025}^{11} = 3.5257$
calculations:

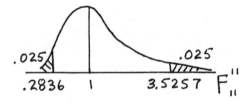

$F = s_1^2/s_2^2$
$= (11.07)^2/(10.39)^2 = 1.1352$
conclusion:
Do not reject H_o; there is not sufficient evidence to conclude that $\sigma_1^2 \neq \sigma_2^2$.
Now proceed using s_p^2 for both s_1^2 and s_2^2
$\bar{x}_1 - \bar{x}_2 = 46.42 - 51.00 = -4.58$
$s_p^2 = (df_1 \cdot s_1^2 + df_2 \cdot s_2^2)/(df_1 + df_2)$
$= (11 \cdot (11.07)^2 + 11 \cdot (10.39)^2)/(11 + 11) = 2535.467/22 = 115.25$
H_o: $\mu_1 - \mu_2 = 0$
H_1: $\mu_1 - \mu_2 \neq 0$
$\alpha = .05$
C.R. $t < -t_{22,.025} = -2.074$
$\quad t > t_{22,.025} = 2.074$
calculations:

$t_{\bar{x}_1 - \bar{x}_2} = (\bar{x}_1 - \bar{x}_2 - \mu_{\bar{x}_1 - \bar{x}_2})/s_{\bar{x}_1 - \bar{x}_2}$
$= (-4.58 - 0)/\sqrt{115.25/12 + 115.25/12}$
$= -4.58/4.383 = -1.045$
conclusion:
Do not reject H_o; there is not sufficient evidence to conclude that $\mu_1 - \mu_2 \neq 0$.

8. Let the brown ones be group 1.
   original claim: $\mu_1-\mu_2 = 0$ [small samples and σ unknown, first test H_o: $\sigma_1^2 = \sigma_2^2$]
      H_o: $\sigma_1^2 = \sigma_2^2$
      H_1: $\sigma_1^2 \neq \sigma_2^2$
      $\alpha = .05$
      C.R. $F < F_{8,.975}^{29} = [.3747]$
          $F > F_{8,.025}^{29} = 3.8940$

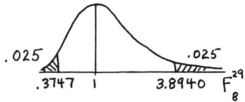

      calculations:
         $F = s_1^2/s_2^2$
            $= (.0521)^2/(.0465)^2 = 1.2554$
      conclusion:
         Do not reject H_o; there is not sufficient evidence to conclude that $\sigma_1^2 \neq \sigma_2^2$.
   Now proceed using s_p^2 for both s_1^2 and s_2^2
      $\bar{x}_1-\bar{x}_2 = .9256 - .8901 = .0355$
      $s_p^2 = (df_1 \cdot s_1^2 + df_2 \cdot s_2^2)/(df_1 + df_2)$
          $= (29 \cdot (.0521)^2 + 8 \cdot (.0465)^2)/(29 + 8) = .0960/37 = .00260$
     H_o: $\mu_1-\mu_2 = 0$
     H_1: $\mu_1-\mu_2 \neq 0$
     $\alpha = .05$
     C.R. $t < -t_{37,.025} = -1.960$
        $t > t_{37,.025} = 1.960$

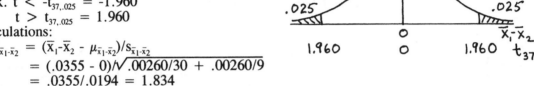

     calculations:
      $t_{\bar{x}_1-\bar{x}_2} = (\bar{x}_1-\bar{x}_2 - \mu_{\bar{x}_1-\bar{x}_2})/s_{\bar{x}_1-\bar{x}_2}$
        $= (.0355 - 0)/\sqrt{.00260/30 + .00260/9}$
        $= .0355/.0194 = 1.834$
   conclusion:
     Do not reject H_o; there is not sufficient evidence to conclude that $\mu_1-\mu_2 \neq 0$.
     No; no corrective action is necessary.

9. Let the manual keying be group 1.
   original claim: $\mu_1-\mu_2 = 0$ [small samples and σ unknown, first test H_o: $\sigma_1^2 = \sigma_2^2$]
      H_o: $\sigma_1^2 = \sigma_2^2$
      H_1: $\sigma_1^2 \neq \sigma_2^2$
      $\alpha = .02$
      C.R. $F < F_{9,.99}^{15} = [.2568]$
          $F > F_{9,.01}^{15} = 4.9621$

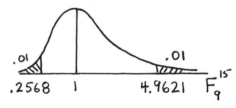

      calculations:
         $F = s_1^2/s_2^2$
            $= 225.0/56.0 = 4.0179$
      conclusion:
         Do not reject H_o; there is not sufficient evidence to conclude that $\sigma_1^2 \neq \sigma_2^2$.
   Now proceed using s_p^2 for both s_1^2 and s_2^2
      $\bar{x}_1-\bar{x}_2 = 157.6 - 112.4 = 45.2$
      $s_p^2 = (df_1 \cdot s_1^2 + df_2 \cdot s_2^2)/(df_1 + df_2)$
         $= (15 \cdot 225.0 + 9 \cdot 56.0)/(15 + 9) = 3879/24 = 161.625$

H_o: $\mu_1-\mu_2 = 0$
H_1: $\mu_1-\mu_2 \neq 0$
$\alpha = .02$
C.R. t < $-t_{24,.01} = -2.492$
    t > $t_{24,.01} = 2.492$
calculations:

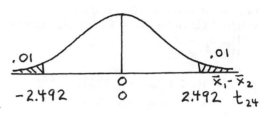

$$t_{\bar{x}_1-\bar{x}_2} = (\bar{x}_1-\bar{x}_2 - \mu_{\bar{x}_1-\bar{x}_2})/s_{\bar{x}_1-\bar{x}_2}$$
$$= (45.2 - 0)/\sqrt{161.625/16 + 161.625/10}$$
$$= 45.2/5.125 = 8.820$$

conclusion:
    Reject H_o; there is sufficient evidence to conclude that $\mu_1-\mu_2 \neq 0$ (in fact, $\mu_1-\mu_2>0$).

10. Refer to the notation, F test and data summary for exercise #9.
$(\bar{x}_1-\bar{x}_2) \pm t_{24,.01}\sqrt{s_p^2/n_1 + s_p^2/n_2}$
$45.2 \pm 2.492\sqrt{161.625/16 + 161.625/10}$
45.2 ± 12.8
$32.4 < \mu_1-\mu_2 < 58.0$

11. Let the night shift be group 1.
    confidence interval: $\mu_1-\mu_2$ [small samples and σ unknown, first test H_o: $\sigma_1^2 = \sigma_2^2$]
        H_o: $\sigma_1^2 = \sigma_2^2$
        H_1: $\sigma_1^2 \neq \sigma_2^2$
        $\alpha = .05$
        C.R. F < $F_{24,.975}^{29} = [.4643]$
            F > $F_{24,.025}^{29} = 2.2090$

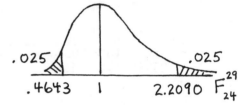

        calculations:
            $F = s_1^2/s_2^2$
                $= (.22)^2/(.14)^2 = 2.4694$
        conclusion:
            Reject H_o; there is sufficient evidence to conclude that $\sigma_1^2 \neq \sigma_2^2$ (in fact, $\sigma_1^2 > \sigma_2^2$).
    Now proceed using s_1^2 and s_2^2 and the smaller degrees of freedom (viz., $df_2 = 24$).
        $\bar{x}_1-\bar{x}_2 = 11.85 - 12.02 = -.17$
        $(\bar{x}_1-\bar{x}_2) \pm t_{24,.025}\sqrt{s_1^2/n_1 + s_2^2/n_2}$
        $-.17 \pm 2.064\cdot\sqrt{(.22)^2/30 + (.14)^2/25}$
        $-.17 \pm .10$
        $-.27 < \mu_1-\mu_2 < -.07$

12. Refer to the notation, F test and data summary for exercise #11.
    H_o: $\mu_1-\mu_2 = 0$
    H_1: $\mu_1-\mu_2 \neq 0$
    $\alpha = .05$
    C.R. t < $-t_{24,.025} = -2.064$
        t > $t_{24,.025} = 2.064$
    calculations:

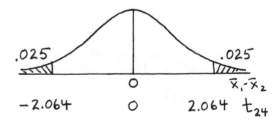

        $t_{\bar{x}_1-\bar{x}_2} = (\bar{x}_1-\bar{x}_2 - \mu_{\bar{x}_1-\bar{x}_2})/s_{\bar{x}_1-\bar{x}_2}$
            $= (-.17 - 0)/\sqrt{(.22)^2/30 + (.14)^2/25}$
            $= -.17/.0490 = -3.472$
    conclusion:
        Reject H_o; there is sufficient evidence to conclude that $\mu_1-\mu_2 \neq 0$ (in fact, $\mu_1-\mu_2 < 0$).

13. Let the O-C patients be group 1.
  original claim: $\mu_1 - \mu_2 = 0$ [small samples and σ unknown, first test H_o: $\sigma_1^2 = \sigma_2^2$]
    H_o: $\sigma_1^2 = \sigma_2^2$
    H_1: $\sigma_1^2 \neq \sigma_2^2$
    $\alpha = .05$ [or any other appropriate value]
    C.R. $F < F_{9,.975}^{9} = [.2484]$
         $F > F_{9,.025}^{9} = 4.0260$
    calculations:
      $F = s_1^2/s_2^2$
        $= (.08)^2/(.08)^2 = 1.0000$

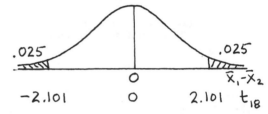

  conclusion:
    Do not reject H_o; there is not sufficient evidence to conclude that $\sigma_1^2 \neq \sigma_2^2$.
  Now proceed using s_p^2 for both s_1^2 and s_2^2
  $\overline{x}_1 - \overline{x}_2 = .34 - .45 = -.11$
  $s_p^2 = (df_1 \cdot s_1^2 + df_2 \cdot s_2^2)/(df_1 + df_2)$
    $= (9 \cdot (.08)^2 + 9 \cdot (.08)^2)/(9 + 9) = .1152/18 = .0064$
  H_o: $\mu_1 - \mu_2 = 0$
  H_1: $\mu_1 - \mu_2 \neq 0$
  $\alpha = .05$ [or any other appropriate value]
  C.R. $t < -t_{18,.025} = -2.101$
      $t > t_{18,.025} = 2.101$
  calculations:

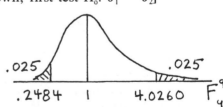

    $t_{\overline{x}_1-\overline{x}_2} = (\overline{x}_1 - \overline{x}_2 - \mu_{\overline{x}_1-\overline{x}_2})/s_{\overline{x}_1-\overline{x}_2}$
        $= (-.11 - 0)/\sqrt{.0064/10 + .0064/10}$
        $= -.11/.0358 = -3.075$
  conclusion:
    Reject H_o; there is sufficient evidence to conclude that $\mu_1 - \mu_2 \neq 0$ (in fact, $\mu_1 - \mu_2 < 0$).
  Based on this result, it does appear that obsessive-compulsive disorders have a biological
  indicator.

14. Refer to the notation, F test and data summary for exercise #13.
  $(\overline{x}_1-\overline{x}_2) \pm t_{18,.005}\sqrt{s_p^2/n_1 + s_p^2/n_2}$
  $-.11 \pm 2.878\sqrt{.0064/10 + .0064/10}$
  $-.11 \pm .10$
  $-.21 < \mu_1 - \mu_2 < -.01$

15. Let the O-C patients be group 1.
  confidence interval: $\mu_1 - \mu_2$ [small samples and σ unknown, first test H_o: $\sigma_1^2 = \sigma_2^2$]
    H_o: $\sigma_1^2 = \sigma_2^2$
    H_1: $\sigma_1^2 \neq \sigma_2^2$
    $\alpha = .05$
    C.R. $F < F_{9,.975}^{9} = [.2484]$
         $F > F_{9,.025}^{9} = 4.0260$
    calculations:
      $F = s_1^2/s_2^2$
        $= (156.84)^2/(137.97)^2 = 1.2922$

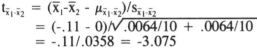

  conclusion:
    Do not reject H_o; there is not sufficient evidence to conclude that $\sigma_1^2 \neq \sigma_2^2$.

Now proceed using s_p^2 for both s_1^2 and s_2^2

$\bar{x}_1 - \bar{x}_2 = 1390.03 - 1268.41 = 121.62$

$s_p^2 = (df_1 \cdot s_1^2 + df_2 \cdot s_2^2)/(df_1 + df_2)$

$= (9 \cdot (156.84)^2 + 9 \cdot (137.97)^2)/(9 + 9) = 43634.5065/18 = 21817.3$

$(\bar{x}_1 - \bar{x}_2) \pm t_{18,.025}\sqrt{s_p^2/n_1 + s_p^2/n_2}$

$121.62 \pm 2.101 \cdot \sqrt{21817.3/10 + 21817.3/10}$

121.62 ± 138.78

$-17.16 < \mu_1 - \mu_2 < 260.40$

16. Refer to the notation, F test and data summary for exercise #15.

H_o: $\mu_1 - \mu_2 = 0$

H_1: $\mu_1 - \mu_2 \neq 0$

$\alpha = .05$

C.R. $t < -t_{18,.025} = -2.101$

$\quad\quad t > t_{18,.025} = 2.101$

calculations:

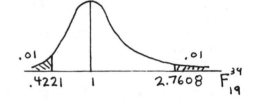

$t_{\bar{x}_1 - \bar{x}_2} = (\bar{x}_1 - \bar{x}_2 - \mu_{\bar{x}_1 - \bar{x}_2})/s_{\bar{x}_1 - \bar{x}_2}$

$= (121.62 - 0)/\sqrt{21817.3/10 + 21817.3/10}$

$= 121.62/104.44 = 1.164$

conclusion:

Do not reject H_o; there is not sufficient evidence to conclude that $\mu_1 - \mu_2 \neq 0$.

17. Let the sales division be group 1.

original claim: $\mu_1 - \mu_2 = 0$ [small samples and σ unknown, first test H_o: $\sigma_1^2 = \sigma_2^2$]

H_o: $\sigma_1^2 = \sigma_2^2$

H_1: $\sigma_1^2 \neq \sigma_2^2$

$\alpha = .02$

C.R. $F < F_{19,.99}^{39} = [.4221]$

$\quad\quad F > F_{19,.01}^{39} = 2.7608$

calculations:

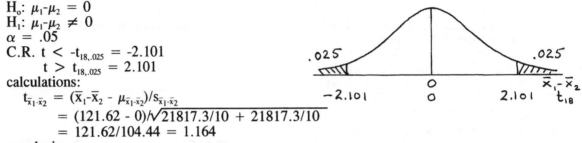

$F = s_1^2/s_2^2$

$= (8.65)^2/(4.93)^2 = 3.0785$

conclusion:

Reject H_o; there is sufficient evidence to conclude that $\sigma_1^2 \neq \sigma_2^2$ (in fact, $\sigma_1^2 > \sigma_2^2$).

Now proceed using s_1^2 and s_2^2 and the smaller degrees of freedom (viz., $df_2 = 19$).

$\bar{x}_1 - \bar{x}_2 = 10.26 - 6.93 = 3.33$

H_o: $\mu_1 - \mu_2 = 0$

H_1: $\mu_1 - \mu_2 \neq 0$

$\alpha = .02$

C.R. $t < -t_{19,.01} = -2.540$

$\quad\quad t > t_{19,.01} = 2.540$

calculations:

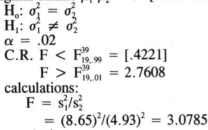

$t_{\bar{x}_1 - \bar{x}_2} = (\bar{x}_1 - \bar{x}_2 - \mu_{\bar{x}_1 - \bar{x}_2})/s_{\bar{x}_1 - \bar{x}_2}$

$= (3.33 - 0)/\sqrt{(8.65)^2/40 + (4.93)^2/20}$

$= 3.33/1.757 = 1.896$

conclusion:

Do not reject H_o; there is not sufficient evidence to conclude that $\mu_1 - \mu_2 \neq 0$.

18. Refer to the notation, F test and data summary for exercise #17.

$(\bar{x}_1 - \bar{x}_2) \pm t_{19,.01}\sqrt{s_p^2/n_1 + s_p^2/n_2}$

$3.33 \pm 2.540\sqrt{(8.65)^2/40 + (4.93)^2/20}$

3.33 ± 4.46

$-1.13 < \mu_1 - \mu_2 < 7.79$

19. <u>red</u>   <u>brown</u>
     $n = 17$ $n = 30$
     $\Sigma x = 15.373$ $\Sigma x = 27.769$
     $\Sigma x^2 = 13.929193$ $\Sigma x^2 = 25.782620$
     $\bar{x} = .9043$ $\bar{x} = .9256$
     $s^2 = .001717$ $s^2 = .002714$ Let the brown be group 1.

original claim: $\mu_1 - \mu_2 = 0$ [small samples and σ unknown, first test H_o: $\sigma_1^2 = \sigma_2^2$]

H_o: $\sigma_1^2 = \sigma_2^2$
H_1: $\sigma_1^2 \neq \sigma_2^2$
$\alpha = .05$ [or any other appropriate value]
C.R. $F < F_{16,.975}^{29} = [.4301]$
     $F > F_{16,.025}^{29} = 2.5678$
calculations:
     $F = s_1^2/s_2^2$
       $= .002714/.001717 = 1.5802$

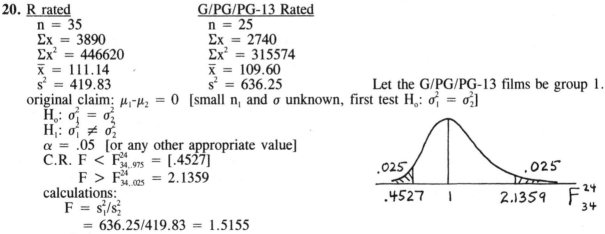

conclusion:
     Do not reject H_o; there is not sufficient evidence to conclude that $\sigma_1^2 \neq \sigma_2^2$.
Now proceed using s_p^2 for both s_1^2 and s_2^2
     $\bar{x}_1 - \bar{x}_2 = .9256 - .9043 = .0213$
     $s_p^2 = (df_1 \cdot s_1^2 + df_2 \cdot s_2^2)/(df_1 + df_2)$
         $= (29 \cdot (.002714) + 16 \cdot (.001717))/(29 + 16) = .106187/45 = .002360$

H_o: $\mu_1 - \mu_2 = 0$
H_1: $\mu_1 - \mu_2 \neq 0$
$\alpha = .05$ [or any other appropriate value]
C.R. $t < -t_{45,.025} = -1.960$
     $t > t_{45,.025} = 1.960$
calculations:
     $t_{\bar{x}1-\bar{x}2} = (\bar{x}_1 - \bar{x}_2 - \mu_{\bar{x}_1 - \bar{x}_2})/s_{\bar{x}_1 - \bar{x}_2}$
       $= (.0213 - 0)/\sqrt{.002360/30 + .002360/17}$
       $= .0213/.0147 = 1.447$

conclusion:
     Do not reject H_o; there is not sufficient evidence to conclude that $\mu_1 - \mu_2 \neq 0$.
No. One cannot be 95% certain that the differences observed reflect anything other than the normal variability expected to occur when the process is in control as specified.

20. <u>R rated</u>   <u>G/PG/PG-13 Rated</u>
     $n = 35$ $n = 25$
     $\Sigma x = 3890$ $\Sigma x = 2740$
     $\Sigma x^2 = 446620$ $\Sigma x^2 = 315574$
     $\bar{x} = 111.14$ $\bar{x} = 109.60$
     $s^2 = 419.83$ $s^2 = 636.25$ Let the G/PG/PG-13 films be group 1.

original claim: $\mu_1 - \mu_2 = 0$ [small n_1 and σ unknown, first test H_o: $\sigma_1^2 = \sigma_2^2$]

H_o: $\sigma_1^2 = \sigma_2^2$
H_1: $\sigma_1^2 \neq \sigma_2^2$
$\alpha = .05$ [or any other appropriate value]
C.R. $F < F_{34,.975}^{24} = [.4527]$
     $F > F_{34,.025}^{24} = 2.1359$
calculations:
     $F = s_1^2/s_2^2$
       $= 636.25/419.83 = 1.5155$
conclusion:
     Do not reject H_o; there is not sufficient evidence to conclude that $\sigma_1^2 \neq \sigma_2^2$.

Now proceed using s_p^2 for both s_1^2 and s_2^2

$\bar{x}_1 - \bar{x}_2 = 109.60 - 111.14 = -1.543$

$s_p^2 = (df_1 \cdot s_1^2 + df_2 \cdot s_2^2)/(df_1 + df_2)$
$= (24 \cdot (636.25) + 34 \cdot (419.83))/(24 + 34) = 29544.29/58 = 509.38$

H_o: $\mu_1 - \mu_2 = 0$
H_1: $\mu_1 - \mu_2 \neq 0$
$\alpha = .05$ [or any other appropriate value]
C.R. $t < -t_{58,.025} = -1.960$
      $t > t_{58,.025} = 1.960$

calculations:

$t_{\bar{x}_1 - \bar{x}_2} = (\bar{x}_1 - \bar{x}_2 - \mu_{\bar{x}_1 - \bar{x}_2})/s_{\bar{x}_1 - \bar{x}_2}$
$= (-1.543 - 0)/\sqrt{509.38/25 + 509.38/35}$
$= -1.543/5.910 = .261$

conclusion:
Do not reject H_o; there is not sufficient evidence to conclude that $\mu_1 - \mu_2 \neq 0$.

21. $A = s_1^2/n_1 = (.22)^2/30 = .001613$
$B = s_2^2/n_2 = (.14)^2/25 = .000784$
$df = (A + B)^2/(A^2/df_1 + B^2/df_2)$
$= (.002397)^2/(.000002602/29 + .000000614/24)$
$= .000005745/.000000115$
$= 49.8$

$(\bar{x}_1 - \bar{x}_2) \pm t_{49.8,.025}\sqrt{s_1^2/n_1 + s_2^2/n_2}$
$-.17 \pm 1.960 \cdot \sqrt{(.22)^2/30 + (.14)^2/25}$
$-.17 \pm .10$
$-.27 < \mu_1 - \mu_2 < -.07$
This is the same result obtained in exercise #11.

22. a. The calculated t_{18} is 5.477 a two-tailed test.
   Since $5.477 > t_{18,.005} = 2.878$, P-value $< 2(.005) = .01$.
  b. The calculated t_{53} is 2.618 a two-tailed test.
   Using the z Table since df = 53, P-value $= 2 \cdot P(z > 2.62) = 2 \cdot (.5000 - .4956) = .0088$.
  c. The calculated t_{24} is 8.820 a two-tailed test.
   Since $8.820 > t_{24,.005} = 2.797$, P-value $< 2(.005) = .01$.
  d. The calculated t_{24} is -3.472 a two-tailed test.
   Since $-3.472 < -t_{24,.005} = -2.797$, P-value $< 2(.005) = .01$.

23. a. No really. If $s_1 = s_2$, then $F = s_1^2/s_2^2 = 1.0000$. Since all upper tail F critical values are greater than 1.0000 and all lower tail F critical values are less than 1.0000, the value $F = 1.0000$ will not be in the critical region for any df_1 and df_2.
  b. Since s_p^2 is a weighted average of s_1^2 and s_2^2, $s_1 = s_2 = s$ means that $s_p^2 = s^2$.

8-6 Inferences about Two Proportions

NOTE: To be consistent with the notation of the previous sections, thereby reinforcing the patterns and concepts presented there, the manual uses the "usual" z formula written to apply to $\hat{p}_1 - \hat{p}_2$'s

$z_{\hat{p}_1 - \hat{p}_2} = (\hat{p}_1 - \hat{p}_2 - \mu_{\hat{p}_1 - \hat{p}_2})/\sigma_{\hat{p}_1 - \hat{p}_2}$

with $\mu_{\hat{p}_1 - \hat{p}_2} = p_1 - p_2$
and $\sigma_{\hat{p}_1 - \hat{p}_2} = \sqrt{\bar{p}\bar{q}/n_1 + \bar{p}\bar{q}/n_2}$ [when H_o includes $p_1 = p_2$]
where $\bar{p} = (x_1 + x_2)/(n_1 + n_2)$

And so the formula for the z statistic may also be written as

$z_{\hat{p}_1 - \hat{p}_2} = ((\hat{p}_1 - \hat{p}_2) - (p_1 - p_2))/\sqrt{\bar{p}\bar{q}/n_1 + \bar{p}\bar{q}/n_2}$

1. $\hat{p}_1 = x_1/n_1 = 45/100 = .450$
   $\hat{p}_2 = x_2/n_2 = 115/200 = .575$
   $\hat{p}_1-\hat{p}_2 = .450 - .575 = -.125$
   a. $\bar{p} = (x_1 + x_2)/(n_1 + n_2) = (45 + 115)/(100 + 200) = 160/300 = .533$
   b. $z_{\hat{p}_1-\hat{p}_2} = (\hat{p}_1-\hat{p}_2 - \mu_{\hat{p}_1-\hat{p}_2})/\sigma_{\hat{p}_1-\hat{p}_2}$
   $\quad\quad = (-.125 - 0)/\sqrt{(.533)(.467)/100 + (.533)(.467)/200}$
   $\quad\quad = -.125/.0611 = -2.046$
   c. $\pm z_{.025} = \pm 1.960$
   d. P-value $= 2 \cdot P(z < -2.05) = 2 \cdot (.5000 - .4798) = 2 \cdot (.0202) = .0404$

NOTE: Since $\bar{p}$ is the weighted average of $\hat{p}_1$ and $\hat{p}_2$, it must always fall between those two values. If it does not, then an error has been made that must be corrected before proceeding. Calculation of $\sigma_{\hat{p}_1-\hat{p}_2} = \sqrt{\bar{p}\bar{q}/n_1 + \bar{p}\bar{q}/n_2}$ can be accomplished with no round-off loss on most calculators by calculating $\bar{p}$ and proceeding as follows: STORE 1-RECALL = * RECALL = STORE RECALL ÷ n_1 + RECALL ÷ n_2 = $\sqrt{}$. The quantity $\sigma_{\hat{p}_1-\hat{p}_2}$ may then be STORED for future use. Each calculator is different -- learn how your calculator works, and do the homework on the same calculator you will use for the exam. If you have any questions about performing/storing calculations on your calculator, check with your instructor or class assistant.

2. $\hat{p}_1 = x_1/n_1 = 30/250 = .120$
   $\hat{p}_2 = x_2/n_2 = 44/800 = .055$
   $\hat{p}_1-\hat{p}_2 = .120 - .055 = .065$
   a. $\bar{p} = (x_1 + x_2)/(n_1 + n_2) = (30 + 44)/(250 + 800) = 74/1050 = .0705$
   b. $z_{\hat{p}_1-\hat{p}_2} = (\hat{p}_1-\hat{p}_2 - \mu_{\hat{p}_1-\hat{p}_2})/\sigma_{\hat{p}_1-\hat{p}_2}$
   $\quad\quad = (.065 - 0)/\sqrt{(.0705)(.9295)/250 + (.0705)(.9295)/800}$
   $\quad\quad = .065/.0185 = 3.505$
   c. $\pm z_{.025} = \pm 1.960$
   d. P-value $= 2 \cdot P(z > 3.50) = 2 \cdot (.5000 - .4999) = 2 \cdot (.0001) = .0002$

3. Let the Democrats be group 1. original claim: $p_1-p_2 = 0$
   $\hat{p}_1 = x_1/n_1 = .35$ [see NOTE below]
   $\hat{p}_2 = x_2/n_2 = .41$ [see NOTE below]
   $\hat{p}_1-\hat{p}_2 = .35 - .41 = -.06$
   $\bar{p} = (x_1 + x_2)/(n_1 + n_2) = (.35 \cdot 552 + .41 \cdot 417)/(552 + 417) = 364.17/969 = .376$
   $H_0: p_1-p_2 = 0$
   $H_1: p_1-p_2 \neq 0$
   $\alpha = .05$
   C.R. $z < -z_{.025} = -1.960$
   $\quad\quad z > z_{.025} = 1.960$
   calculations:
   $z_{\hat{p}_1-\hat{p}_2} = (\hat{p}_1-\hat{p}_2 - \mu_{\hat{p}_1-\hat{p}_2})/\sigma_{\hat{p}_1-\hat{p}_2}$
   $\quad\quad = (-.06 - 0)/\sqrt{(.376)(.624)/552 + (.376)(.624)/417}$
   $\quad\quad = -.06/.0314 = -1.909$

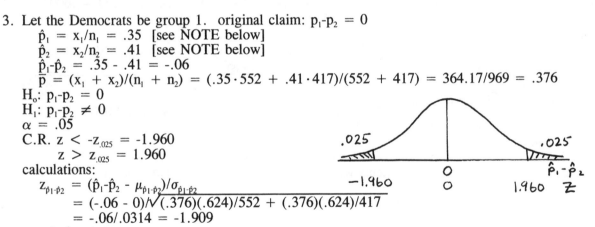

   conclusion:
   Do not reject H_0; there is not sufficient evidence to conclude that $p_1-p_2 \neq 0$.

NOTE: In the preceding problem x_1 and x_2 were not given and must be deduced from the values of $\hat{p}_1 = 35\%$ and $\hat{p}_2 = 41\%$. Unfortunately any x_1 between 191 and 195 inclusive and any x_2 between 169 and 173 inclusive produces the given percents. Whenever this occurs (i.e., x_1 and x_2 cannot be determined exactly), the estimate $x = \hat{p} \cdot n$ will be employed without further manipulation -- even if that estimate is not a whole number.

4. Use the notation preliminary calculations of exercise #3.

$(\hat{p}_1-\hat{p}_2) \pm z_{.025}\sqrt{\hat{p}_1\hat{q}_1/n_1 + \hat{p}_2\hat{q}_2/n_2}$

$.06 \pm 1.960\cdot\sqrt{(.35)(.65)/552 + (.41)(.59)/417}$

$.06 \pm .06$

$-.12 < p_1-p_2 < .00$

NOTE: Since the third decimal place of $\hat{p}_1-\hat{p}_2 = .06$ cannot be determined, the interval is limited to two decimal accuracy.

5. Let those not wearing seat belts be group 1.   original claim: $p_1-p_2 > 0$

$\hat{p}_1 = x_1/n_1 = 50/290 = .1724$

$\hat{p}_2 = x_2/n_2 = 16/123 = .1301$

$\hat{p}_1-\hat{p}_2 = .1724 - .1301 = .0423$

$\bar{p} = (x_1 + x_2)/(n_1 + n_2) = (50 + 16)/(290 + 123) = 66/413 = .160$

$H_0: p_1-p_2 \leq 0$

$H_1: p_1-p_2 > 0$

$\alpha = .05$

C.R. $z > z_{.05} = 1.645$

calculations:

$z_{\hat{p}_1-\hat{p}_2} = (\hat{p}_1-\hat{p}_2 - \mu_{\hat{p}_1-\hat{p}_2})/\sigma_{\hat{p}_1-\hat{p}_2}$

$= (.0423 - 0)/\sqrt{(.160)(.840)/290 + (.160)(.840)/123}$

$= .0423/.0394 = 1.074$

conclusion:

Do not reject H_0; there is not sufficient evidence to conclude that $p_1-p_2 > 0$.

6. Let those receiving the Salk vaccine be group 1.   original claim: $p_1-p_2 < 0$

$\hat{p}_1 = x_1/n_1 = 33/200,000 = .000165$

$\hat{p}_2 = x_2/n_2 = 115/200,000 = .000575$

$\hat{p}_1-\hat{p}_2 = .000165 - .000575 = .000410$

$\bar{p} = (x_1 + x_2)/(n_1 + n_2) = (33 + 115)/(200,000 + 200,000) = 148/400,000 = .00037$

$H_0: p_1-p_2 \geq 0$

$H_1: p_1-p_2 < 0$

$\alpha = .01$

C.R. $z < -z_{.01} = -2.327$

calculations:

$z_{\hat{p}_1-\hat{p}_2} = (\hat{p}_1-\hat{p}_2 - \mu_{\hat{p}_1-\hat{p}_2})/\sigma_{\hat{p}_1-\hat{p}_2}$

$= (.000410 - 0)/\sqrt{(.00037)(.99963)/200,000 + (.00037)(.99963)/200,000}$

$= .000410/.0000608 = -6.742$

conclusion:

Reject H_0; there is sufficient evidence to conclude that $p_1-p_2 < 0$.

Yes; assuming the group assignments were made at random, the vaccine appears to be effective.

7. Let the 18-24 year olds be group 1.

$\hat{p}_1 = x_1/n_1 = .360$

$\hat{p}_2 = x_2/n_2 = .540$

$\hat{p}_1-\hat{p}_2 = .360 - .540 = -.180$

$(\hat{p}_1-\hat{p}_2) \pm z_{.025}\sqrt{\hat{p}_1\hat{q}_1/n_1 + \hat{p}_2\hat{q}_2/n_2}$

$-.180 \pm 1.960\cdot\sqrt{(.360)(.640)/200 + (.540)(.460)/250}$

$-.180 \pm .091$

$-.271 < p_1-p_2 < -.089$

8. Use the notation preliminary calculations of exercise #7. original claim: $p_1-p_2 = 0$
$\bar{p} = (x_1 + x_2)/(n_1 + n_2) = (.360 \cdot 200 + .540 \cdot 250)/(200 + 250) = 207/450 = .460$
H_o: $p_1-p_2 = 0$
H_1: $p_1-p_2 \neq 0$
$\alpha = .05$
C.R. $z < -z_{.025} = -1.960$
$z > z_{.025} = 1.960$
calculations:

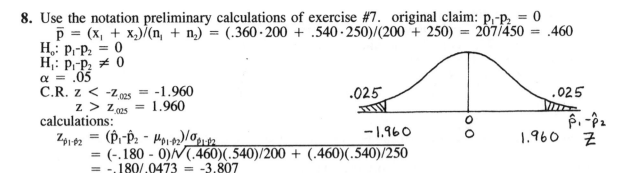

$z_{\hat{p}_1-\hat{p}_2} = (\hat{p}_1-\hat{p}_2 - \mu_{\hat{p}_1-\hat{p}_2})/\sigma_{\hat{p}_1-\hat{p}_2}$
$= (-.180 - 0)/\sqrt{(.460)(.540)/200 + (.460)(.540)/250}$
$= -.180/.0473 = -3.807$
conclusion:
Reject H_o; there is sufficient evidence to conclude that $p_1-p_2 \neq 0$ (in fact, $p_1-p_2 < 0$).

9. Let the vinyl gloves be group 1. original claim: $p_1-p_2 > 0$
$\hat{p}_1 = x_1/n_1 = .63$
$\hat{p}_2 = x_2/n_2 = .07$
$\hat{p}_1-\hat{p}_2 = .63 - .07 = .56$
$\bar{p} = (x_1 + x_2)/(n_1 + n_2) = (.63 \cdot 240 + .07 \cdot 240)/(240 + 240) = 168/480 = .350$
H_o: $p_1-p_2 \leq 0$
H_1: $p_1-p_2 > 0$
$\alpha = .005$
C.R. $z > z_{.005} = 2.575$
calculations:
$z_{\hat{p}_1-\hat{p}_2} = (\hat{p}_1-\hat{p}_2 - \mu_{\hat{p}_1-\hat{p}_2})/\sigma_{\hat{p}_1-\hat{p}_2}$
$= (.56 - 0)/\sqrt{(.350)(.650)/240 + (.350)(.650)/240}$
$= .56/.0435 = 12.861$
conclusion:
Reject H_o; there is sufficient evidence to conclude that $p_1-p_2 > 0$.

10. Use the notation preliminary calculations of exercise #9.
$(\hat{p}_1-\hat{p}_2) \pm z_{.005}\sqrt{\hat{p}_1\hat{q}_1/n_1 + \hat{p}_2\hat{q}_2/n_2}$
$.56 \pm 2.575 \cdot \sqrt{(.63)(.37)/240 + (.07)(.93)/240}$
$.56 \pm .09$
$.47 < p_1-p_2 < .65$
Yes; the difference warrants a decision to use latex gloves. **NOTE:** Since the third decimal place of $\hat{p}_1-\hat{p}_2 = .56$ cannot be determined, the interval is limited to two decimal accuracy.

11. Let the central city be group 1. original claim: $p_1-p_2 = 0$
$\hat{p}_1 = x_1/n_1 = .289$
$\hat{p}_2 = x_2/n_2 = .171$
$\hat{p}_1-\hat{p}_2 = .289 - .171 = .118$
$\bar{p} = (x_1 + x_2)/(n_1 + n_2) = (.289 \cdot 294 + .171 \cdot 1015)/(294 + 1015) = 258.5/1309 = .198$
H_o: $p_1-p_2 = 0$
H_1: $p_1-p_2 \neq 0$
$\alpha = .01$
C.R. $z < -z_{.005} = -2.575$
$z > z_{.005} = 2.575$
calculations:
$z_{\hat{p}_1-\hat{p}_2} = (\hat{p}_1-\hat{p}_2 - \mu_{\hat{p}_1-\hat{p}_2})/\sigma_{\hat{p}_1-\hat{p}_2}$
$= (.118 - 0)/\sqrt{(.198)(.802)/294 + (.198)(.802)/1015}$
$= .118/.0264 = 4.475$
conclusion:
Reject H_o; there is sufficient evidence to conclude that $p_1-p_2 \neq 0$ (in fact, $p_1-p_2 > 0$).

12. Let the 18-24 year olds be group 1.

$\hat{p}_1 = x_1/n_1 = .0425$ [with this accuracy, x_1 must be 117]

$\hat{p}_2 = x_2/n_2 = .0455$ [with this accuracy, x_2 must be 100]

$\hat{p}_1 - \hat{p}_2 = .0425 - .0455 = -.0029$

$(\hat{p}_1 - \hat{p}_2) \pm z_{.025}\sqrt{\hat{p}_1\hat{q}_1/n_1 + \hat{p}_2\hat{q}_2/n_2}$

$-.0029 \pm 1.960 \cdot \sqrt{(.0425)(.9575)/2750 + (.0455)(.9545)/2200}$

$-.0029 \pm .0115$

$-.0144 < p_1 - p_2 < .0086$

Yes; the interval contains zero, indicating no significant difference between the two rates of crime.

13. Let the public colleges be group 1. original claim: $p_1 - p_2 = 0$

$\hat{p}_1 = x_1/n_1 = .30$

$\hat{p}_2 = x_2/n_2 = .26$

$\hat{p}_1 - \hat{p}_2 = .30 - .26 = .04$

$\bar{p} = (x_1 + x_2)/(n_1 + n_2) = (.30 \cdot 1000 + .26 \cdot 500)/(1000 + 500) = 430/1500 = .287$

$H_o: p_1 - p_2 = 0$

$H_1: p_1 - p_2 \neq 0$

$\alpha = .05$

C.R. $z < -z_{.025} = -1.960$

     $z > z_{.025} = 1.960$

calculations:

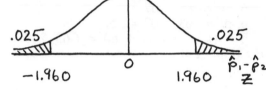

$z_{\hat{p}_1 - \hat{p}_2} = (\hat{p}_1 - \hat{p}_2 - \mu_{\hat{p}_1 - \hat{p}_2})/\sigma_{\hat{p}_1 - \hat{p}_2}$

     $= (.04 - 0)/\sqrt{(.287)(.713)/1000 + (.287)(.713)/500}$

     $= .04/.0248 = 1.615$

conclusion:

Do not reject H_o; there is not sufficient evidence to conclude that $p_1 - p_2 \neq 0$.

14. Use the notation preliminary calculations of exercise #13.

$(\hat{p}_1 - \hat{p}_2) \pm z_{.025}\sqrt{\hat{p}_1\hat{q}_1/n_1 + \hat{p}_2\hat{q}_2/n_2}$

$.04 \pm 1.960 \cdot \sqrt{(.30)(.70)/500 + (.26)(.74)/1000}$

$.04 \pm .05$

$-.01 < p_1 - p_2 < .09$

No; the difference does not appear to be significant. **NOTE:** Since the third decimal place of $\hat{p}_1 - \hat{p}_2 = .04$ cannot be determined, the interval is limited to two decimal accuracy.

15. Let the males be group 1. original claim: $p_1 - p_2 = 0$

$\hat{p}_1 = x_1/n_1 = 25/86 = .29070$

$\hat{p}_2 = x_2/n_2 = 11/39 = .28205$

$\hat{p}_1 - \hat{p}_2 = .29070 - .28205 = .00865$

$\bar{p} = (x_1 + x_2)/(n_1 + n_2) = (25 + 11)/(86 + 39) = 36/125 = .288$

$H_o: p_1 - p_2 = 0$

$H_1: p_1 - p_2 \neq 0$

$\alpha = .05$

C.R. $z < -z_{.025} = -1.960$

     $z > z_{.025} = 1.960$

calculations:

$z_{\hat{p}_1 - \hat{p}_2} = (\hat{p}_1 - \hat{p}_2 - \mu_{\hat{p}_1 - \hat{p}_2})/\sigma_{\hat{p}_1 - \hat{p}_2}$

     $= (.00865 - 0)/\sqrt{(.288)(.712)/86 + (.288)(.712)/39}$

     $= .0865/.0874 = .099$

conclusion:

Do not reject H_o; there is not sufficient evidence to conclude that $p_1 - p_2 \neq 0$.

16. Let the R movies be group 1.
$\hat{p}_1 = x_1/n_1 = 29/35$
$\hat{p}_2 = x_2/n_2 = 20/25$
$\bar{p} = (x_1 + x_2)/(n_1 + n_2) = (29 + 20)/(35 + 25) = 49/60$
In general, to use the normal approximation it must be true that $n_i p_i \geq 5$ and $n_i q_i \geq 5$ for $i=1,2$.
Testing $H_o: p_1-p_2 = 0$ requires that $n_i p \geq 5$ and $n_i q \geq 5$ -- where p is the common proportion.
Using $\bar{p}$ to estimate the common proportion,
$n_1 p = 35 \cdot (49/60) = 28.6 \geq 5$
$n_1 q = 35 \cdot (11/60) = 6.4 \geq 5$
$n_2 p = 25 \cdot (49/60) = 20.4 \geq 5$
$n_2 q = 25 \cdot (11/60) = 4.6 < 5$ -- the normal approximation is not valid.

17. Let the Californians be group 1. original claim: $p_1-p_2 = .25$
$\hat{p}_1 = x_1/n_1 = 210/500 = .42$
$\hat{p}_2 = x_2/n_2 = 120/500 = .24$
$\hat{p}_1-\hat{p}_2 = .42 - .24 = .18$
$H_o: p_1-p_2 = .25$
$H_1: p_1-p_2 \neq .25$
$\alpha = .05$
C.R. $z < -z_{.025} = -1.960$
$\quad z > z_{.025} = 1.960$
calculations:

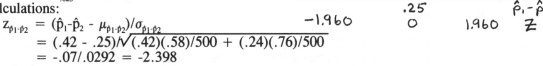

$z_{\hat{p}_1-\hat{p}_2} = (\hat{p}_1-\hat{p}_2 - \mu_{\hat{p}_1-\hat{p}_2})/\sigma_{\hat{p}_1-\hat{p}_2}$
$\quad = (.42 - .25)/\sqrt{(.42)(.58)/500 + (.24)(.76)/500}$
$\quad = -.07/.0292 = -2.398$
conclusion:
Reject H_o; there is sufficient evidence to conclude that $p_1-p_2 \neq .25$ (in fact, $p_1-p_2 < .25$).

18. $\hat{p}_1 = x_1/n_1 = 40/100 = .40$ for groups 1 and 2, $\bar{p} = 70/200 = .35$
$\hat{p}_2 = x_2/n_2 = 30/100 = .30$ for groups 2 and 3, $\bar{p} = 50/200 = .25$
$\hat{p}_3 = x_3/n_3 = 20/100 = .20$ for groups 1 and 3, $\bar{p} = 50/200 = .30$
a. $H_o: p_1-p_2 = 0$
$H_1: p_1-p_2 \neq 0$
$\alpha = .05$
C.R. $z < -z_{.025} = -1.960$
$\quad z > z_{.025} = 1.960$
calculations:
$z_{\hat{p}_1-\hat{p}_2} = (\hat{p}_1-\hat{p}_2 - \mu_{\hat{p}_1-\hat{p}_2})/\sigma_{\hat{p}_1-\hat{p}_2}$
$\quad = (.10 - 0)/\sqrt{(.35)(.65)/100 + (.35)(.65)/100}$
$\quad = .10/.0675 = 1.482$
conclusion:
Do not reject H_o; there is not sufficient evidence to conclude that $p_1-p_2 \neq 0$.
b. $H_o: p_2-p_3 = 0$
$H_1: p_2-p_3 \neq 0$
$\alpha = .05$
C.R. $z < -z_{.025} = -1.960$
$\quad z > z_{.025} = 1.960$
calculations:

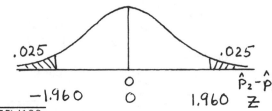

$z_{\hat{p}_2-\hat{p}_3} = (\hat{p}_2-\hat{p}_3 - \mu_{\hat{p}_2-\hat{p}_3})/\sigma_{\hat{p}_2-\hat{p}_3}$
$\quad = (.10 - 0)/\sqrt{(.25)(.75)/100 + (.25)(.75)/100}$
$\quad = .10/.0612 = 1.633$
conclusion:
Do not reject H_o; there is not sufficient evidence to conclude that $p_2-p_3 \neq 0$.

c. H_o: $p_1 - p_3 = 0$
  H_1: $p_1 - p_3 \neq 0$
  $\alpha = .05$
  C.R. $z < -z_{.025} = -1.960$
       $z > z_{.025} = 1.960$
  calculations:
    $z_{p_1-p_3} = (\hat{p}_1 - \hat{p}_3 - \mu_{p_1-p_3})/\sigma_{p_1-p_3}$
        $= (.20 - 0)/\sqrt{(.30)(.70)/100 + (.30)(.70)/100}$
        $= .20/.0648 = 3.086$
  conclusion:
    Reject H_o; there is sufficient evidence to conclude that $p_1 - p_3 \neq 0$ (in fact, $p_1 - p_3 > 0$).

d. No; failing to find a difference between population 1 and population 2, and between population 2 and population 3, does not necessarily mean one will fail to find a difference between population 1 and population 3.

19. $E^2 = (z_{\alpha/2})^2 (p_1 q_1/n_1 + p_2 q_2/n_2)$ squaring the original equation
   $E^2 = (z_{\alpha/2})^2 (.25/n_1 + .25/n_2)$ setting the unknown proportions to .5
   $E^2 = (z_{\alpha/2})^2 (.25/n + .25/n)$ requiring $n_1 = n_2 = n$
   $E^2 = (z_{\alpha/2})^2 (.50/n)$ addition
    $n = (z_{\alpha/2})^2 (.50)/E^2$ solving for n
     $= (1.960)^2 (.50)/(.03)^2$ for $\alpha = .05$ and $E = .03$
     $= 2134.2$ rounded up to 2135

20. The test in exercise #9 was a right-tailed one at the .005 level of significance. To test the claim using a confidence interval, construct a 99% confidence interval and reject the null hypothesis if the interval falls entirely above the hypothesized difference of zero.
Such an interval is constructed in exercise #10: $.47 < p_1 - p_2 < .65$. The appropriate conclusion is to reject H_o and conclude that $p_1 - p_2 > 0$.

Review Exercises

1. Let Orange County be group 1.
  concerns $\mu_1 - \mu_2$ [$n_1 > 30$ and $n_2 > 30$, use z (with s's for σ's)]
  $\bar{x}_1 - \bar{x}_2 = 183.0 - 253.1 = -70.1$

  a. original claim: $\mu_1 - \mu_2 = 0$

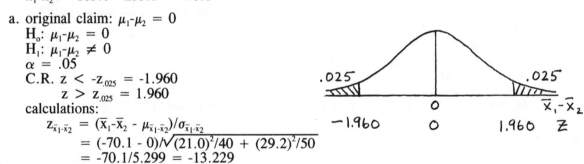

   H_o: $\mu_1 - \mu_2 = 0$
   H_1: $\mu_1 - \mu_2 \neq 0$
   $\alpha = .05$
   C.R. $z < -z_{.025} = -1.960$
       $z > z_{.025} = 1.960$
   calculations:
    $z_{\bar{x}_1-\bar{x}_2} = (\bar{x}_1 - \bar{x}_2 - \mu_{\bar{x}_1-\bar{x}_2})/\sigma_{\bar{x}_1-\bar{x}_2}$
        $= (-70.1 - 0)/\sqrt{(21.0)^2/40 + (29.2)^2/50}$
        $= -70.1/5.299 = -13.229$
   conclusion:
    Reject H_o; there is sufficient evidence to conclude that $\mu_1 - \mu_2 \neq 0$ (in fact, $\mu_1 - \mu_2 < 0$).

  b. $(\bar{x}_1 - \bar{x}_2) \pm z_{.025} \sqrt{\sigma_1^2/n_1 + \sigma_2^2/n_2}$
    $-70.1 \pm 1.960 \cdot \sqrt{(21.0)^2/40 + (29.2)^2/50}$
    -70.1 ± 10.4
    $-80.5 < \mu_1 - \mu_2 < -59.7$

2. Let the market region A be group 1.

$\hat{p}_1 = x_1/n_1 = .38$

$\hat{p}_2 = x_2/n_2 = .14$

$\hat{p}_1-\hat{p}_2 = .38 - .14 = .24$

a. $(\hat{p}_1-\hat{p}_2) \pm z_{.025}\sqrt{\hat{p}_1\hat{q}_1/n_1 + \hat{p}_2\hat{q}_2/n_2}$

$.24 \pm 2.327 \cdot \sqrt{(.38)(.62)/250 + (.14)(.86)/400}$

$.24 \pm .08$

$.16 < p_1-p_2 < .32$

NOTE: Since the third decimal place of $\hat{p}_1-\hat{p}_2 = .06$ cannot be determined, the interval is limited to two decimal accuracy.

b. original claim: $p_1-p_2 > 0$

$\bar{p} = (x_1 + x_2)/(n_1 + n_2) = (.38 \cdot 250 + .14 \cdot 400)/(250 + 400) = 151/650 = .232$

$H_o: p_1-p_2 \le 0$

$H_1: p_1-p_2 > 0$

$\alpha = .02$

C.R. $z > z_{.02} = 2.05$

calculations:

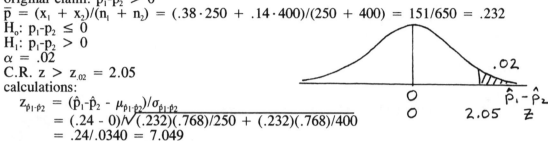

$z_{\hat{p}_1-\hat{p}_2} = (\hat{p}_1-\hat{p}_2 - \mu_{\hat{p}_1-\hat{p}_2})/\sigma_{\hat{p}_1-\hat{p}_2}$

$= (.24 - 0)/\sqrt{(.232)(.768)/250 + (.232)(.768)/400}$

$= .24/.0340 = 7.049$

conclusion:

Reject H_o; there is sufficient evidence to conclude that $p_1-p_2 > 0$.

3. original claim $\mu_d > 0$ [$n \le 30$ and σ_d unknown, use t]

$d = x_{new} - x_{ord}$: 2.1 2.1 .5 -.1 -.6 3.2 .4 1.5 1.8

$n = 9$

$\Sigma d = 10.9$ $\bar{d} = 1.21$

$\Sigma d^2 = 25.33$ $s_d = 1.231$

$H_o: \mu_d \le 0$

$H_1: \mu_d > 0$

$\alpha = .05$

C.R. $t > t_{8,.05} = 1.860$

calculations:

$t_{\bar{d}} = (\bar{d} - \mu_{\bar{d}})/s_{\bar{d}}$

$= (1.21 - 0)/(1.231/\sqrt{9})$

$= 1.21/.410 = 2.951$

conclusion:

Reject H_o; there is sufficient evidence to conclude that $\mu_d > 0$.

No; the results do not give the postal service evidence of fraud, but they do give evidence that Minton's claim is valid.

4.

Dozenol	Niteze
$n = 12$	$n = 12$
$\Sigma x = 5936$	$\Sigma x = 6284$
$\Sigma x^2 = 2,938,906$	$\Sigma x^2 = 3,294,058$
$\bar{x} = 494.667$	$\bar{x} = 523.667$
$s^2 = 233.152$	$s^2 = 303.333$

Let Niteze be group 1.

a. original claim: $\mu_1-\mu_2 = 0$ [small samples and σ unknown, first test H_o: $\sigma_1^2 = \sigma_2^2$]

H_o: $\sigma_1^2 = \sigma_2^2$
H_1: $\sigma_1^2 \neq \sigma_2^2$
$\alpha = .05$
C.R. $F < F_{11,.975}^{11} = [.2836]$ [OR .2916]
     $F > F_{11,.025}^{11} = 3.5257$ [OR 3.4296]
calculations:
     $F = s_1^2/s_2^2$
       $= 303.333/233.152 = 1.3010$

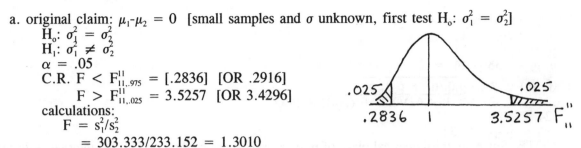

conclusion:
     Do not reject H_o; there is not sufficient evidence to conclude that $\sigma_1^2 \neq \sigma_2^2$.
Now proceed using s_p^2 for both s_1^2 and s_2^2
     $\bar{x}_1-\bar{x}_2 = 523.667 - 494.667 = 29.00$
     $s_p^2 = (df_1 \cdot s_1^2 + df_2 \cdot s_2^2)/(df_1 + df_2)$
         $= (11 \cdot (303.333) + 11 \cdot (233.152))/(11 + 11) = 5901.33/22 = 268.24$
H_o: $\mu_1-\mu_2 = 0$
H_1: $\mu_1-\mu_2 \neq 0$
$\alpha = .05$
C.R. $t < -t_{22,.025} = -2.074$
     $t > t_{22,.025} = 2.074$
calculations:
     $t_{\bar{x}_1-\bar{x}_2} = (\bar{x}_1-\bar{x}_2 - \mu_{\bar{x}_1-\bar{x}_2})/s_{\bar{x}_1-\bar{x}_2}$
         $= (29.00 - 0)/\sqrt{268.24/12 + 268.24/12}$
         $= 29.00/6.686 = 4.337$
conclusion:
     Reject H_o; there is sufficient evidence to conclude that $\mu_1-\mu_2 \neq 0$ (in fact, $\mu_1-\mu_2 > 0$).

b. $(\bar{x}_1-\bar{x}_2) \pm t_{22,.025}\sqrt{s_p^2/n_1 + s_p^2/n_2}$
   $29.00 \pm 2.074\sqrt{268.24/12 + 268.24/12}$
   29.00 ± 13.87
   $15.1 < \mu_1-\mu_2 < 42.9$

5. Let Albany County be group 1. original claim: $p_1-p_2 > 0$ [concerns p's, use z]
   $\hat{p}_1 = x_1/n_1 = 558/24,384 = .02288$
   $\hat{p}_2 = x_2/n_2 = 1214/166,197 = .00730$
   $\hat{p}_1-\hat{p}_2 = .02288 - .00730 = .01558$
   $\bar{p} = (x_1 + x_2)/(n_1 + n_2) = (558 + 1214)/(24,384 + 166,197) = 1772/190,581 = .00930$
H_o: $p_1-p_2 \leq 0$
H_1: $p_1-p_2 > 0$
$\alpha = .01$
C.R. $z > z_{.01} = 2.327$
calculations:
   $z_{\hat{p}_1-\hat{p}_2} = (\hat{p}_1-\hat{p}_2 - \mu_{\hat{p}_1-\hat{p}_2})/\sigma_{\hat{p}_1-\hat{p}_2}$
       $= (.01558 - 0)/\sqrt{(.00930)(.99070)/24,384 + (.00930)(.99070)/166,197}$
       $= .01558/.0006582 = 23.671$
conclusion:
   Reject H_o; there is sufficient evidence to conclude that $p_1-p_2 > 0$.

6. Let the women be group 1.
H_o: $\sigma_1^2 = \sigma_2^2$
H_1: $\sigma_1^2 \neq \sigma_2^2$
$\alpha = .02$
C.R. $F < F_{85,.99}^{67} = [.5446]$
    $F > F_{85,.01}^{67} = 1.8363$
calculations:
    $F = s_1^2/s_2^2$
       $= (114.16)^2/(97.23)^2 = 1.3786$
conclusion:
    Do not reject H_o; there is not sufficient evidence to conclude that $\sigma_1^2 \neq \sigma_2^2$.

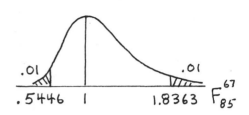

7. Let the women be group 1.
original claim: $\mu_1-\mu_2 = 0$ [$n_1 > 30$ and $n_2 > 30$, use z (with s's for σ's)]
    $\bar{x}_1-\bar{x}_2 = 538.82 - 525.23 = 13.59$
H_o: $\mu_1-\mu_2 = 0$
H_1: $\mu_1-\mu_2 \neq 0$
$\alpha = .02$
C.R. $z < -z_{.01} = -2.327$
    $z > z_{.01} = 2.327$
calculations:
    $z_{\bar{x}_1-\bar{x}_2} = (\bar{x}_1-\bar{x}_2 - \mu_{\bar{x}_1-\bar{x}_2})/\sigma_{\bar{x}_1-\bar{x}_2}$
       $= (13.59 - 0)/\sqrt{(114.16)^2/68 + (97.23)^2/86}$
       $= 13.59/17.366$
       $= .783$
conclusion:
    Do not reject H_o; there is not sufficient evidence to conclude that $\mu_1-\mu_2 \neq 0$.

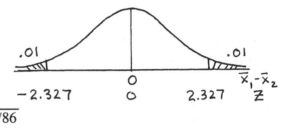

8.
System 1	System 2
n = 15	n = 15
$\Sigma x = 919$	$\Sigma x = 966$
$\Sigma x^2 = 56799$	$\Sigma x^2 = 62282$
$\bar{x} = 61.267$	$\bar{x} = 64.400$
$s^2 = 35.3524$	$s^2 = 5.1143$

Let System 1 be group 1.

a. original claim: $\mu_1-\mu_2 < 0$ [small samples and σ unknown, first test H_o: $\sigma_1^2 = \sigma_2^2$]
    H_o: $\sigma_1^2 = \sigma_2^2$
    H_1: $\sigma_1^2 \neq \sigma_2^2$
    $\alpha = .05$
    C.R. $F < F_{14,.975}^{14} = [.3391]$
       $F > F_{14,.025}^{14} = 2.9493$
    calculations:
       $F = s_1^2/s_2^2$
          $= 35.3524/5.1143 = 6.9125$
    conclusion:
       Reject H_o; there is sufficient evidence to conclude that $\sigma_1^2 \neq \sigma_2^2$ (in fact, $\sigma_1^2 > \sigma_2^2$).
Now proceed using s_1^2 and s_2^2 and the smaller degrees of freedom (viz., $df_1 = df_2 = 14$).
    $\bar{x}_1-\bar{x}_2 = 61.267 - 64.400 = -3.133$

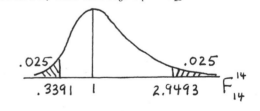

H_o: $\mu_1 - \mu_2 \geq 0$
H_1: $\mu_1 - \mu_2 < 0$
$\alpha = .05$
C.R. $t < -t_{14,.05} = -1.761$
calculations:

$t_{\bar{x}_1-\bar{x}_2} = (\bar{x}_1 - \bar{x}_2 - \mu_{\bar{x}_1-\bar{x}_2})/s_{\bar{x}_1-\bar{x}_2}$

$= (-3.133 - 0)/\sqrt{35.3524/15 + 5.1143/15}$

$= -3.133/1.642 = 1.908$

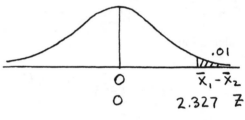

conclusion:
Reject H_o; there is sufficient evidence to conclude that $\mu_1 - \mu_2 < 0$.

b. $(\bar{x}_1 - \bar{x}_2) \pm t_{14,.025}\sqrt{s_1^2/n_1 + s_2^2/n_2}$

$-3.133 \pm 2.145\sqrt{35.3524/15 + 5.1143/15}$

-3.133 ± 3.523

$-6.7 < \mu_1 - \mu_2 < .4$

9. Let the six-week program be group 1.
original claim: $\mu_1 - \mu_2 > 0$ [$n_1 > 30$ and $n_2 > 30$, use z (with s's for σ's)]
$\bar{x}_1 - \bar{x}_2 = 83.5 - 79.8 = 3.7$
H_o: $\mu_1 - \mu_2 \leq 0$
H_1: $\mu_1 - \mu_2 > 0$
$\alpha = .01$
C.R. $z > z_{.01} = 2.327$
calculations:

$z_{\bar{x}_1-\bar{x}_2} = (\bar{x}_1 - \bar{x}_2 - \mu_{\bar{x}_1-\bar{x}_2})/\sigma_{\bar{x}_1-\bar{x}_2}$

$= (3.7 - 0)/\sqrt{(16.3)^2/60 + (19.2)^2/35}$

$= 3.7/3.868 = .957$

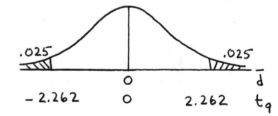

conclusion:
Do not reject H_o; there is not sufficient evidence to conclude that $\mu_1 - \mu_2 > 0$.

10. original claim $\mu_d = 0$ [$n \leq 30$ and σ_d unknown, use t]
d = $x_{sitting} - x_{supine}$: .03 -.64 -.35 .11 .17 .47 .49 .40 .29 .44
n = 10
$\Sigma d = 1.41$ $\bar{d} = .141$
$\Sigma d^2 = 1.4727$ $s_d = .376$
H_o: $\mu_d = 0$
H_1: $\mu_d \neq 0$
$\alpha = .05$
C.R. $t < -t_{9,.025} = -2.262$
    $t > t_{9,.025} = 2.262$
calculations:

$t_{\bar{d}} = (\bar{d} - \mu_{\bar{d}})/s_{\bar{d}}$

$= (.141 - 0)/(.376/\sqrt{10})$

$= .141/.119 = 1.185$

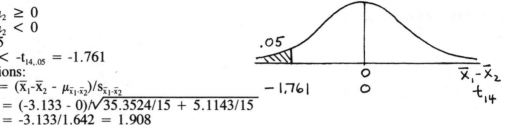

conclusion:
Do not reject H_o; there is not sufficient evidence to conclude that $\mu_d \neq 0$.

11. Let the prepared students be group 1.
   original claim: $p_1-p_2 > 0$ [concerns p's, use z]
   $\hat{p}_1 = x_1/n_1 = 62/80 = .775$
   $\hat{p}_2 = x_2/n_2 = 23/50 = .460$
   $\hat{p}_1-\hat{p}_2 = .775 - .460 = .315$
   $\bar{p} = (x_1 + x_2)/(n_1 + n_2) = (62 + 23)/(80 + 50) = 85/130 = .654$
   H_o: $p_1-p_2 \le 0$
   H_1: $p_1-p_2 > 0$
   $\alpha = .05$
   C.R. $z > z_{.05} = 1.645$
   calculations:

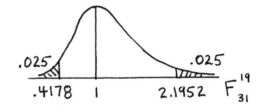

   $z_{\hat{p}_1-\hat{p}_2} = (\hat{p}_1-\hat{p}_2 - \mu_{\hat{p}_1-\hat{p}_2})/\sigma_{\hat{p}_1-\hat{p}_2}$
   $= (.315 - 0)/\sqrt{(.654)(.346)/80 + (.654)(.346)/50}$
   $= .315/.0858 = 3.673$
   conclusion:
   Reject H_o; there is sufficient evidence to conclude that $p_1-p_2 > 0$.

12. Let the Dayton Machine sample be group 1.
   a. original claim: $\sigma_1 = \sigma_2$ [concerns σ's, use F]
      H_o: $\sigma_1^2 = \sigma_2^2$
      H_1: $\sigma_1^2 \ne \sigma_2^2$
      $\alpha = .05$
      C.R. $F < F^{19}_{31,.975} = [.4178]$
          $F > F^{19}_{31,.025} = 2.1952$
      calculations:

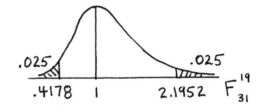

      $F = s_1^2/s_2^2$
          $= (2.1)^2/(.4)^2 = 27.5625$
      conclusion:
      Reject H_o; there is sufficient evidence to conclude that $\sigma_1^2 \ne \sigma_2^2$ (in fact, $\sigma_1^2 > \sigma_2^2$).
   b. Now proceed using s_1^2 and s_2^2 and the smaller degrees of freedom (viz., $df_1 = 19$).
      original claim: $\mu_1-\mu_2 = 0$
      $\bar{x}_1-\bar{x}_2 = 66.0 - 68.3 = -2.3$
      H_o: $\mu_1-\mu_2 = 0$
      H_1: $\mu_1-\mu_2 \ne 0$
      $\alpha = .05$
      C.R. $t < -t_{19,.025} = -2.093$
          $t > t_{19,.025} = 2.093$
      calculations:

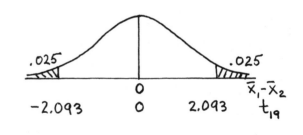

      $t_{\bar{x}_1-\bar{x}_2} = (\bar{x}_1-\bar{x}_2 - \mu_{\bar{x}_1-\bar{x}_2})/s_{\bar{x}_1-\bar{x}_2}$
          $= (-2.3 - 0)/\sqrt{(2.1)^2/20 + (.4)^2/32}$
          $= -2.3/.464 = -4.955$
      conclusion:
      Reject H_o; there is sufficient evidence to conclude that $\mu_1-\mu_2 \ne 0$ (in fact, $\mu_1-\mu_2 < 0$).
   c. $(\bar{x}_1-\bar{x}_2) \pm t_{19,.025}\sqrt{s_1^2/n_1 + s_2^2/n_2}$
      $-2.3 \pm 2.093\sqrt{(2.1)^2/20 + (.4)^2/32}$
      -2.3 ± 1.0
      $-3.3 < \mu_1-\mu_2 < -1.3$

13. a. original claim $\mu_d < 0$ [n ≤ 30 and σ_d unknown, use t]
   $d = x_{bef} - x_{aft}$: -3 -2 -3 -3 -3 -3 -3 -3
   $n = 8$
   $\Sigma d = -23$ $\bar{d} = -2.875$
   $\Sigma d^2 = 67$ $s_d = .354$
   H_o: $\mu_d \geq 0$
   H_1: $\mu_d < 0$
   $\alpha = .025$
   C.R. $t < -t_{7,.025} = -2.365$
   calculations:

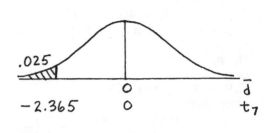

   $t_{\bar{d}} = (\bar{d} - \mu_d)/s_{\bar{d}}$
   $\quad = (-2.875 - 0)/(.354/\sqrt{8})$
   $\quad = -2.875/.125 = -23.000$
   conclusion:
       Reject H_o; there is sufficient evidence to conclude that $\mu_d < 0$.
   b. $\bar{d} \pm t_{7,.025} \cdot s_d/\sqrt{n}$
   $-2.9 \pm 2.365 \cdot (.125)/\sqrt{8}$
   $-2.9 \pm .3$
   $-3.2 < \mu_d < -2.6$

14. Let Dover be group 1. [$n_1 > 30$ and $n_2 > 30$, use z (with s's for σ's]
   a. original claim: $\mu_1-\mu_2 = 0$
   $\bar{x}_1-\bar{x}_2 = 43.7 - 48.2 = -4.5$
   H_o: $\mu_1-\mu_2 = 0$
   H_1: $\mu_1-\mu_2 \neq 0$
   $\alpha = .05$
   C.R. $z < -z_{.025} = -1.960$
   $\quad\quad z > z_{.025} = 1.960$
   calculations:

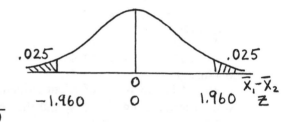

   $z_{\bar{x}_1-\bar{x}_2} = (\bar{x}_1-\bar{x}_2 - \mu_{\bar{x}_1-\bar{x}_2})/\sigma_{\bar{x}_1-\bar{x}_2}$
   $\quad = (-4.5 - 0)/\sqrt{(16.2)^2/40 + (16.5)^2/60}$
   $\quad = -4.5/3.331 = -1.351$
   conclusion:
       Do not reject H_o; there is not sufficient evidence to conclude that $\mu_1-\mu_2 \neq 0$.
   b. $(\bar{x}_1-\bar{x}_2) \pm z_{.025}\sqrt{\sigma_1^2/n_1 + \sigma_2^2/n_2}$
   $-4.5 \pm 1.960\sqrt{(16.2)^2/40 + (16.5)^2/60}$
   -4.5 ± 6.5
   $-11.0 < \mu_1-\mu_2 < 2.0$

15. Let Barrington be group 1.
   original claim: $\mu_1-\mu_2 = 0$ [small samples and σ unknown, first test H_o: $\sigma_1^2 = \sigma_2^2$]
   H_o: $\sigma_1^2 = \sigma_2^2$
   H_1: $\sigma_1^2 \neq \sigma_2^2$
   $\alpha = .05$
   C.R. $F < F_{17,.975}^{23} = [.4054]$
   $\quad\quad F > F_{17,.025}^{23} = 2.5598$
   calculations:

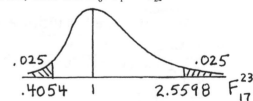

   $F = s_1^2/s_2^2$
   $\quad = (6.1)^2/(5.8)^2 = 1.1061$
   conclusion:
       Do not reject H_o; there is not sufficient evidence to conclude that $\sigma_1^2 \neq \sigma_2^2$.

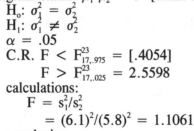

Now proceed using s_p^2 for both s_1^2 and s_2^2

$\bar{x}_1-\bar{x}_2 = 80.6 - 85.7 = -5.1$

$s_p^2 = (df_1 \cdot s_1^2 + df_2 \cdot s_2^2)/(df_1 + df_2)$
$= (23 \cdot (6.1)^2 + 17 \cdot (5.8)^2)/(23 + 17)$
$= 1427.71/40 = 35.69$

$H_o: \mu_1-\mu_2 = 0$
$H_1: \mu_1-\mu_2 \neq 0$
$\alpha = .05$
C.R. $t < -t_{40,.025} = -1.960$
$t > t_{40,.025} = 1.960$
calculations:

$t_{\bar{x}_1-\bar{x}_2} = (\bar{x}_1-\bar{x}_2 - \mu_{\bar{x}_1-\bar{x}_2})/s_{\bar{x}_1-\bar{x}_2}$
$= (-5.1 - 0)/\sqrt{35.69/24 + 35.69/18}$
$= -5.1/1.863 = -2.738$

conclusion:
Reject H_o; there is sufficient evidence to conclude that $\mu_1-\mu_2 \neq 0$ (in fact, $\mu_1-\mu_2 < 0$).

16. Let the men be group 1. [concerns p's, use z]
a. original claim: $p_1-p_2 > 0$

$\hat{p}_1 = x_1/n_1 = 52/500 = .104$
$\hat{p}_2 = x_2/n_2 = 27/500 = .054$
$\hat{p}_1-\hat{p}_2 = .104 - .054 = .050$
$\bar{p} = (x_1 + x_2)/(n_1 + n_2) = (52 + 27)/(500 + 500) = 79/1000 = .079$

$H_o: p_1-p_2 \leq 0$
$H_1: p_1-p_2 > 0$
$\alpha = .01$
C.R. $z > z_{.01} = 2.327$
calculations:

$z_{\hat{p}_1-\hat{p}_2} = (\hat{p}_1-\hat{p}_2 - \mu_{\hat{p}_1-\hat{p}_2})/\sigma_{\hat{p}_1-\hat{p}_2}$
$= (.050 - 0)/\sqrt{(.079)(.921)/500 + (.079)(.921)/500}$
$= .050/.01706 = 2.931$

conclusion:
Reject H_o; there is sufficient evidence to conclude that $p_1-p_2 > 0$.

b. $(\hat{p}_1-\hat{p}_2) \pm z_{.005}\sqrt{\hat{p}_1\hat{q}_1/n_1 + \hat{p}_2\hat{q}_2/n_2}$
$.050 \pm 2.575 \cdot \sqrt{(.104)(.896)/500 + (.054)(.946)/500}$
$.050 \pm .044$
$.006 < p_1-p_2 < .094$

Chapter 9

Correlation and Regression

9-2 Correlation

1. The critical values below are taken from Table A-6.
   a. CV = ±.444; r = .502 indicates a significant (positive) linear correlation
   b. CV = ±.444; r = .203 indicates no significant linear correlation
   c. CV = ±.279; r = -.281 indicates a significant (negative) linear correlation

2. The critical values below are taken from Table A-6.
   a. CV = ±.220; r = -.351 indicates a significant (negative) linear correlation
   b. CV = ±.444; r = -.370 indicates no significant linear correlation
   c. CV = ±.294; r = .312 indicates a significant (positive) linear correlation

3. a. y

x	y	xy	x^2	y^2
1	1	1	1	1
1	5	5	1	25
2	4	8	4	16
3	2	6	9	4
7	12	20	15	46

b. n = 4
   $\Sigma x = 7$
   $\Sigma x^2 = 15$
   $(\Sigma x)^2 = (7)^2 = 49$
   $\Sigma xy = 20$

c. $n(\Sigma xy) - (\Sigma x)(\Sigma y) = 4(20) - (7)(12) = -4$
   $n(\Sigma x^2) - (\Sigma x)^2 = 4(15) - (7)^2 = 11$
   $n(\Sigma y^2) - (\Sigma y)^2 = 4(46) - (12)^2 = 40$
   $r = [n(\Sigma xy) - (\Sigma x)(\Sigma y)]/[\sqrt{n(\Sigma x^2) - (\Sigma x)^2} \cdot \sqrt{n(\Sigma y^2) - (\Sigma y)^2}]$
   $= [-4]/[\sqrt{11} \cdot \sqrt{40}] = -.191$

4. a. y

x	y	xy	x^2	y^2
0	3	0	0	9
1	3	3	1	9
1	4	4	1	16
2	5	10	4	25
5	6	30	25	36
9	21	47	31	95

b. n = 5
   $\Sigma x = 9$
   $\Sigma x^2 = 31$
   $(\Sigma x)^2 = (9)^2 = 81$
   $\Sigma xy = 47$

c. $n(\Sigma xy) - (\Sigma x)(\Sigma y) = 5(47) - (9)(21) = 46$
   $n(\Sigma x^2) - (\Sigma x)^2 = 5(31) - (9)^2 = 74$
   $n(\Sigma y^2) - (\Sigma y)^2 = 5(95) - (21)^2 = 34$
   $r = [n(\Sigma xy) - (\Sigma x)(\Sigma y)]/[\sqrt{n(\Sigma x^2) - (\Sigma x)^2} \cdot \sqrt{n(\Sigma y^2) - (\Sigma y)^2}]$
   $= [46]/[\sqrt{74} \cdot \sqrt{34}] = .917$

NOTE: In each of problems 5-20, the first variable listed is given the designation x, and the second variable listed is given the designation y. In correlation problems, the designation of x and y is arbitrary -- so long as a person remains consistent after making the designation. For part (d) of each problem, the following summary statistics should be saved: n, Σx, Σy, Σx^2, Σy^2, Σxy.

5. **a.**

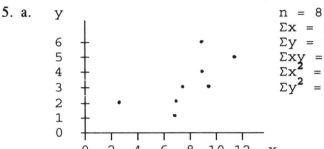

$$n = 8$$
$$\Sigma x = 62.28$$
$$\Sigma y = 26$$
$$\Sigma xy = 221.83$$
$$\Sigma x^2 = 533.6532$$
$$\Sigma y^2 = 104$$

b. $n(\Sigma xy) - (\Sigma x)(\Sigma y) = 8(221.83) - (62.28)(26) = 155.36$
$n(\Sigma x^2) - (\Sigma x)^2 = 8(533.6532) - (62.28)^2 = 390.4272$
$n(\Sigma y^2) - (\Sigma y)^2 = 8(104) - (26)^2 = 156$
$r = [n(\Sigma xy) - (\Sigma x)(\Sigma y)]/[\sqrt{n(\Sigma x^2) - (\Sigma x)^2} \cdot \sqrt{n(\Sigma y^2) - (\Sigma y)^2}]$
$= 155.36/[\sqrt{390.4272} \cdot \sqrt{156}] = .630$

c. $H_o: \rho = 0$
$H_1: \rho \neq 0$
$\alpha = .05$

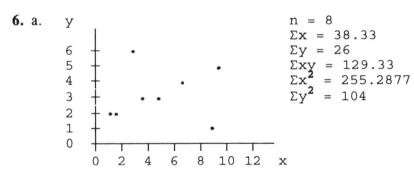

C.R. $r < -.707$ OR C.R. $t < -t_{6,.025} = -2.447$
     $r > .707$ $t > t_{6,.025} = 2.447$
calculations: calculations:
    $r = .630$ $t_r = (r - \mu_r)/s_r$
                      $= (.630 - 0)/\sqrt{(1-(.630)^2)/6}$
                      $= .630/.317 = 1.985$

conclusion:
    Do not reject H_o; there is not sufficient evidence to conclude that $\rho \neq 0$.

6. **a.**

y

```
6  |      •
5  |              •
4  |          •
3  |      • •
2  | ••
1  |              •
0  +--+--+--+--+--+--+--
   0  2  4  6  8 10 12  x
```

$$n = 8$$
$$\Sigma x = 38.33$$
$$\Sigma y = 26$$
$$\Sigma xy = 129.33$$
$$\Sigma x^2 = 255.2877$$
$$\Sigma y^2 = 104$$

b. $n(\Sigma xy) - (\Sigma x)(\Sigma y) = 8(129.33) - (38.33)(26) = 38.06$
$n(\Sigma x^2) - (\Sigma x)^2 = 8(255.2877) - (38.33)^2 = 573.1127$
$n(\Sigma y^2) - (\Sigma y)^2 = 8(104) - (26)^2 = 156$
$r = [n(\Sigma xy) - (\Sigma x)(\Sigma y)]/[\sqrt{n(\Sigma x^2) - (\Sigma x)^2} \cdot \sqrt{n(\Sigma y^2) - (\Sigma y)^2}]$
$= 38.06/[\sqrt{573.1127} \cdot \sqrt{156}] = .127$

c. H_o: $\rho = 0$
H_1: $\rho \neq 0$
$\alpha = .05$
C.R. r < -.707   OR C.R. t < -$t_{6,.025}$ = -2.447
r > .707   t > $t_{6,.025}$ = 2.447
calculations: calculations:
r = .127 $t_r = (r - \mu_r)/s_r$
                          $= (.127 - 0)/\sqrt{(1-(.127)^2)/6}$
                          $= .127/.405 = .314$

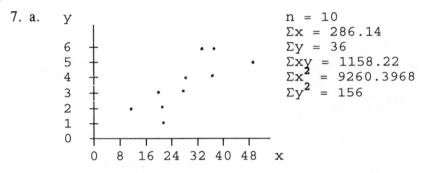

conclusion:
   Do not reject H_o; there is not sufficient evidence to conclude that $\rho \neq 0$.

7. a.

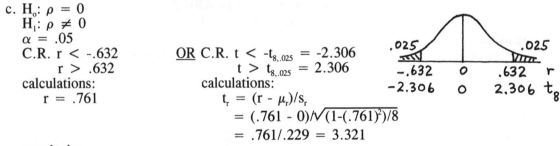

n = 10
Σx = 286.14
Σy = 36
Σxy = 1158.22
Σx^2 = 9260.3968
Σy^2 = 156

b. $n(\Sigma xy) - (\Sigma x)(\Sigma y) = 10(1158.22) - (286.14)(36) = 1281.16$
$n(\Sigma x^2) - (\Sigma x)^2 = 10(9260.3968) - (286.14)^2 = 10727.8684$
$n(\Sigma y^2) - (\Sigma y)^2 = 10(156) - (36)^2 = 264$
$r = [n(\Sigma xy) - (\Sigma x)(\Sigma y)]/[\sqrt{n(\Sigma x^2) - (\Sigma x)^2} \cdot \sqrt{n(\Sigma y^2) - (\Sigma y)^2}]$
$= 1281.16/[\sqrt{10727.8684} \cdot \sqrt{264}] = .761$

c. H_o: $\rho = 0$
H_1: $\rho \neq 0$
$\alpha = .05$
C.R. r < -.632   OR C.R. t < -$t_{8,.025}$ = -2.306
r > .632   t > $t_{8,.025}$ = 2.306
calculations: calculations:
r = .761 $t_r = (r - \mu_r)/s_r$
                          $= (.761 - 0)/\sqrt{(1-(.761)^2)/8}$
                          $= .761/.229 = 3.321$

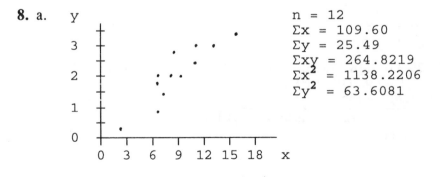

conclusion:
   Reject H_o; there is sufficient evidence to conclude that $\rho \neq 0$ (in fact, $\rho > 0$).

8. a.

n = 12
Σx = 109.60
Σy = 25.49
Σxy = 264.8219
Σx^2 = 1138.2206
Σy^2 = 63.6081

b. $n(\Sigma xy) - (\Sigma x)(\Sigma y) = 12(264.8219) - (109.60)(25.49) = 384.1588$
   $n(\Sigma x^2) - (\Sigma x)^2 = 12(1138.2206) - (109.60)^2 = 1646.4872$
   $n(\Sigma y^2) - (\Sigma y)^2 = 12(63.6081) - (25.49)^2 = 113.5571$
   $r = [n(\Sigma xy) - (\Sigma x)(\Sigma y)]/[\sqrt{n(\Sigma x^2) - (\Sigma x)^2} \cdot \sqrt{n(\Sigma y^2) - (\Sigma y)^2}]$
   $= 384.1588/[\sqrt{1646.4872} \cdot \sqrt{113.5571}] = .888$

c. $H_o: \rho = 0$
   $H_1: \rho \neq 0$
   $\alpha = .05$
   C.R. $r < -.576$ <u>OR</u> C.R. $t < -t_{10,.025} = -2.228$
   $r > .576$ $t > t_{10,.025} = 2.228$
   calculations: calculations:
   $r = .630$ $t_r = (r - \mu_r)/s_r$
   $$ $= (.888 - 0)/\sqrt{(1-(.888)^2)/10}$
   $$ $= .888/.145 = 6.121$

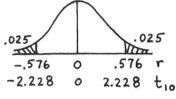

.025   .025
−.576   0   .576 r
−2.228 0   2.228 t_{10}

conclusion:
   Reject H_o; there is sufficient evidence to conclude that $\rho \neq 0$ (in fact $\rho > 0$).

9. a.   y

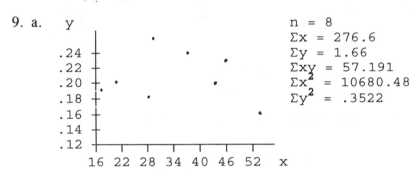

.24
.22
.20
.18
.16
.14
.12
   16 22 28 34 40 46 52   x

$n = 8$
$\Sigma x = 276.6$
$\Sigma y = 1.66$
$\Sigma xy = 57.191$
$\Sigma x^2 = 10680.48$
$\Sigma y^2 = .3522$

b. $n(\Sigma xy) - (\Sigma x)(\Sigma y) = 8(57.191) - (276.6)(1.66) = -1.628$
   $n(\Sigma x^2) - (\Sigma x)^2 = 8(10680.48) - (276.6)^2 = 8936.28$
   $n(\Sigma y^2) - (\Sigma y)^2 = 8(.3522) - (1.66)^2 = .062$
   $r = [n(\Sigma xy) - (\Sigma x)(\Sigma y)]/[\sqrt{n(\Sigma x^2) - (\Sigma x)^2} \cdot \sqrt{n(\Sigma y^2) - (\Sigma y)^2}]$
   $= -1.628/[\sqrt{8936.28} \cdot \sqrt{.062}] = -.069$

c. $H_o: \rho = 0$
   $H_1: \rho \neq 0$
   $\alpha = .05$
   C.R. $r < -.707$ <u>OR</u> C.R. $t < -t_{6,.025} = -2.447$
   $r > .707$ $t > t_{6,.025} = 2.447$
   calculations: calculations:
   $r = -.069$ $t_r = (r - \mu_r)/s_r$
   $$ $= (-.069 - 0)/\sqrt{(1-(-.069)^2)/6}$
   $$ $= -.069/.407 = -.170$

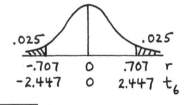

.025   .025
−.707   0   .707 r
−2.447 0   2.447 t_6

conclusion:
   Do not reject H_o; there is not sufficient evidence to conclude that $\rho \neq 0$.

10. a.

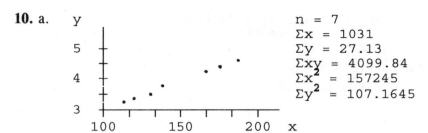

$$n = 7$$
$$\Sigma x = 1031$$
$$\Sigma y = 27.13$$
$$\Sigma xy = 4099.84$$
$$\Sigma x^2 = 157245$$
$$\Sigma y^2 = 107.1645$$

b. $n(\Sigma xy) - (\Sigma x)(\Sigma y) = 7(4099.84) - (1031)(27.13) = 727.85$
$n(\Sigma x^2) - (\Sigma x)^2 = 7(157245) - (1031)^2 = 37754$
$n(\Sigma y^2) - (\Sigma y)^2 = 7(107.1645) - (27.13)^2 = 14.1146$
$r = [n(\Sigma xy) - (\Sigma x)(\Sigma y)]/[\sqrt{n(\Sigma x^2) - (\Sigma x)^2} \cdot \sqrt{n(\Sigma y^2) - (\Sigma y)^2}]$
$\quad = 727.85/[\sqrt{37754} \cdot \sqrt{14.1146}\,] = .997$

c. $H_o: \rho = 0$
$H_1: \rho \neq 0$
$\alpha = .05$

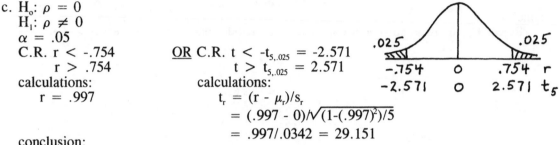

C.R. $r < -.754$ OR C.R. $t < -t_{5,.025} = -2.571$
$\quad\;\; r > .754$ $t > t_{5,.025} = 2.571$
calculations: calculations:
$\quad r = .997$ $t_r = (r - \mu_r)/s_r$
$\qquad\qquad\qquad\qquad = (.997 - 0)/\sqrt{(1-(.997)^2)/5}$
$\qquad\qquad\qquad\qquad = .997/.0342 = 29.151$
conclusion:
   Reject H_o; there is sufficient evidence to conclude that $\rho \neq 0$ (in fact, $\rho > 0$).

11. a.

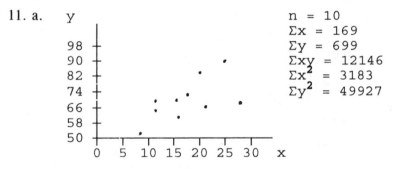

$$n = 10$$
$$\Sigma x = 169$$
$$\Sigma y = 699$$
$$\Sigma xy = 12146$$
$$\Sigma x^2 = 3183$$
$$\Sigma y^2 = 49927$$

b. $n(\Sigma xy) - (\Sigma x)(\Sigma y) = 10(12146) - (169)(699) = 3329$
$n(\Sigma x^2) - (\Sigma x)^2 = 10(3183) - (169)^2 = 3269$
$n(\Sigma y^2) - (\Sigma y)^2 = 10(49927) - (699)^2 = 10669$
$r = [n(\Sigma xy) - (\Sigma x)(\Sigma y)]/[\sqrt{n(\Sigma x^2) - (\Sigma x)^2} \cdot \sqrt{n(\Sigma y^2) - (\Sigma y)^2}]$
$\quad = 3329/[\sqrt{3269} \cdot \sqrt{10669}\,] = .564$

c. $H_o: \rho = 0$
$H_1: \rho \neq 0$
$\alpha = .05$

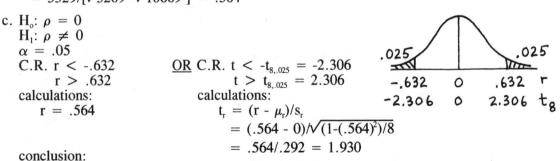

C.R. $r < -.632$ OR C.R. $t < -t_{8,.025} = -2.306$
$\quad\;\; r > .632$ $t > t_{8,.025} = 2.306$
calculations: calculations:
$\quad r = .564$ $t_r = (r - \mu_r)/s_r$
$\qquad\qquad\qquad\qquad = (.564 - 0)/\sqrt{(1-(.564)^2)/8}$
$\qquad\qquad\qquad\qquad = .564/.292 = 1.930$
conclusion:
   Do not reject H_o; there is not sufficient evidence to conclude that $\rho \neq 0$.

12. a.

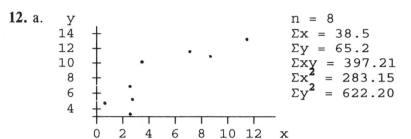

$$n = 8$$
$$\Sigma x = 38.5$$
$$\Sigma y = 65.2$$
$$\Sigma xy = 397.21$$
$$\Sigma x^2 = 283.15$$
$$\Sigma y^2 = 622.20$$

b. $n(\Sigma xy) - (\Sigma x)(\Sigma y) = 8(397.21) - (38.5)(65.2) = 667.48$
$n(\Sigma x^2) - (\Sigma x)^2 = 8(283.15) - (38.5)^2 = 782.95$
$n(\Sigma y^2) - (\Sigma y)^2 = 8(622.20) - (65.2)^2 = 726.56$
$r = [n(\Sigma xy) - (\Sigma x)(\Sigma y)]/[\sqrt{n(\Sigma x^2) - (\Sigma x)^2} \cdot \sqrt{n(\Sigma y^2) - (\Sigma y)^2}]$
$= 667.48/[\sqrt{782.95} \cdot \sqrt{726.56}] = .885$

c. H_o: $\rho = 0$
H_1: $\rho \neq 0$
$\alpha = .05$

C.R. $r < -.707$	OR C.R. $t < -t_{6,.025} = -2.447$
$r > .707$	$t > t_{6,.025} = 2.447$
calculations:	calculations:
$r = .885$	$t_r = (r - \mu_r)/s_r$
	$= (.885 - 0)/\sqrt{(1-(.885)^2)/6}$
	$= .885/.190 = 4.656$

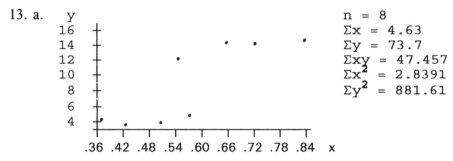

conclusion:
Reject H_o; there is sufficient evidence to conclude that $\rho \neq 0$ (in fact, $\rho > 0$).

13. a.

$$n = 8$$
$$\Sigma x = 4.63$$
$$\Sigma y = 73.7$$
$$\Sigma xy = 47.457$$
$$\Sigma x^2 = 2.8391$$
$$\Sigma y^2 = 881.61$$

b. $n(\Sigma xy) - (\Sigma x)(\Sigma y) = 8(47.457) - (4.63)(73.7) = 38.425$
$n(\Sigma x^2) - (\Sigma x)^2 = 8(2.8391) - (4.63)^2 = 1.2759$
$n(\Sigma y^2) - (\Sigma y)^2 = 8(881.61) - (73.7)^2 = 1621.19$
$r = [n(\Sigma xy) - (\Sigma x)(\Sigma y)]/[\sqrt{n(\Sigma x^2) - (\Sigma x)^2} \cdot \sqrt{n(\Sigma y^2) - (\Sigma y)^2}]$
$= 38.425/[\sqrt{1.2759} \cdot \sqrt{1621.19}] = .845$

c. H_o: $\rho = 0$
H_1: $\rho \neq 0$
$\alpha = .05$

C.R. $r < -.707$	OR C.R. $t < -t_{6,.025} = -2.447$
$r > .707$	$t > t_{6,.025} = 2.447$
calculations:	calculations:
$r = .845$	$t_r = (r - \mu_r)/s_r$
	$= (.845 - 0)/\sqrt{(1-(.845)^2)/6}$
	$= .845/.218 = 3.868$

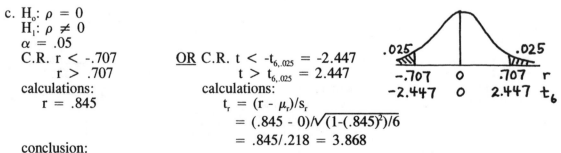

conclusion:
Reject H_o; there is sufficient evidence to conclude that $\rho \neq 0$ (in fact, $\rho > 0$).

14. a.

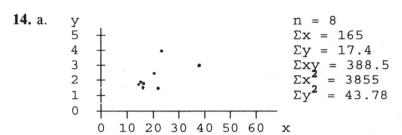

n = 8
$\Sigma x = 165$
$\Sigma y = 17.4$
$\Sigma xy = 388.5$
$\Sigma x^2 = 3855$
$\Sigma y^2 = 43.78$

b. $n(\Sigma xy) - (\Sigma x)(\Sigma y) = 8(388.5) - (165)(17.4) = 237.0$
$n(\Sigma x^2) - (\Sigma x)^2 = 8(3855) - (165)^2 = 3615$
$n(\Sigma y^2) - (\Sigma y)^2 = 8(43.78) - (17.4)^2 = 47.48$
$r = [n(\Sigma xy) - (\Sigma x)(\Sigma y)]/[\sqrt{n(\Sigma x^2) - (\Sigma x)^2} \cdot \sqrt{n(\Sigma y^2) - (\Sigma y)^2}]$
$= 237.0/[\sqrt{3615} \cdot \sqrt{47.48}] = .572$

c. $H_o: \rho = 0$
$H_1: \rho \neq 0$
$\alpha = .05$

C.R. r < -.707   <u>OR</u> C.R. t < -$t_{6,.025}$ = -2.447
    r > .707   t > $t_{6,.025}$ = 2.447
calculations: calculations:
    r = .572 $t_r = (r - \mu_r)/s_r$
                                    $= (.572 - 0)/\sqrt{(1-(.572)^2)/6}$
                                    $= .572/.335 = 1.708$

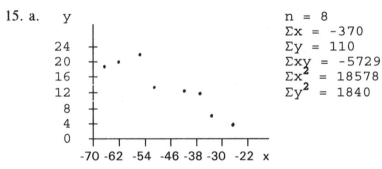

conclusion:
    Do not reject H_o; there is not sufficient evidence to conclude that $\rho \neq 0$.

15. a.

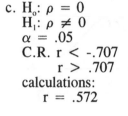

n = 8
$\Sigma x = -370$
$\Sigma y = 110$
$\Sigma xy = -5729$
$\Sigma x^2 = 18578$
$\Sigma y^2 = 1840$

b. $n(\Sigma xy) - (\Sigma x)(\Sigma y) = 8(-5729) - (-370)(110) = -5132$
$n(\Sigma x^2) - (\Sigma x)^2 = 8(18578) - (-370)^2 = 11724$
$n(\Sigma y^2) - (\Sigma y)^2 = 8(1840) - (110)^2 = 2620$
$r = [n(\Sigma xy) - (\Sigma x)(\Sigma y)]/[\sqrt{n(\Sigma x^2) - (\Sigma x)^2} \cdot \sqrt{n(\Sigma y^2) - (\Sigma y)^2}]$
$= -5132/[\sqrt{11724} \cdot \sqrt{2620}] = -.926$

c. $H_o: \rho = 0$
$H_1: \rho \neq 0$
$\alpha = .05$

C.R. r < -.707   <u>OR</u> C.R. t < -$t_{6,.025}$ = -2.447
    r > .707   t > $t_{6,.025}$ = 2.447
calculations: calculations:
    r = -.926 $t_r = (r - \mu_r)/s_r$
                                    $= (-.926 - 0)/\sqrt{(1-(-.926)^2)/6}$
                                    $= -.926/.154 = -6.007$

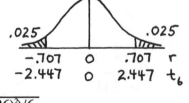

conclusion:
    Reject H_o; there is sufficient evidence to conclude that $\rho \neq 0$ (in fact, $\rho < 0$).

16. a.

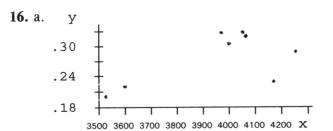

$n = 8$
$\Sigma x = 31607$
$\Sigma y = 2.23$
$\Sigma xy = 8868.91$
$\Sigma x^2 = 125,348,161$
$\Sigma y^2 = .6417$

b. $n(\Sigma xy) - (\Sigma x)(\Sigma y) = 8(8868.91) - (31607)(2.23) = 467.67$
$n(\Sigma x^2) - (\Sigma x)^2 = 8(125,348,161) - (31607)^2 = 3,782,839$
$n(\Sigma y^2) - (\Sigma y)^2 = 8(.6147) - (2.23)^2 = .1607$
$r = [n(\Sigma xy) - (\Sigma x)(\Sigma y)]/[\sqrt{n(\Sigma x^2) - (\Sigma x)^2} \cdot \sqrt{n(\Sigma y^2) - (\Sigma y)^2}]$
$= 467.67/[\sqrt{3,782,839} \cdot \sqrt{.1607}\ ] = .600$

c. $H_o: \rho = 0$
$H_1: \rho \neq 0$
$\alpha = .05$
C.R. $r < -.707$ $\underline{OR}$ C.R. $t < -t_{6,.025} = -2.447$
    $r > .707$ $t > t_{6,.025} = 2.447$
calculations: calculations:
    $r = .600$ $t_r = (r - \mu_r)/s_r$
                              $= (.600 - 0)/\sqrt{(1-(.600)^2)/6}$
                              $= .600/.327 = 1.836$

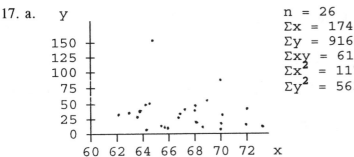

conclusion:
Do not reject H_o; there is not sufficient evidence to conclude that $\rho \neq 0$.

17. a.

$n = 26$
$\Sigma x = 1747.8$
$\Sigma y = 916.2$
$\Sigma xy = 61354.14$
$\Sigma x^2 = 117715.86$
$\Sigma y^2 = 56357.08$

b. $n(\Sigma xy) - (\Sigma x)(\Sigma y) = 26(61354.14) - (1747.8)(916.2) = -6126.72$
$n(\Sigma x^2) - (\Sigma x)^2 = 26(117715.86) - (1747.8)^2 = 5807.52$
$n(\Sigma y^2) - (\Sigma y)^2 = 26(56357.08) - (916.2)^2 = 625861.64$
$r = [n(\Sigma xy) - (\Sigma x)(\Sigma y)]/[\sqrt{n(\Sigma x^2) - (\Sigma x)^2} \cdot \sqrt{n(\Sigma y^2) - (\Sigma y)^2}]$
$= -6126.72/[\sqrt{5807.52} \cdot \sqrt{625861.64}] = -.102$

c. $H_o: \rho = 0$
$H_1: \rho \neq 0$
$\alpha = .05$
C.R. $r < -.396$ $\underline{OR}$ C.R. $t < -t_{24,.025} = -2.064$
    $r > .396$ $t > t_{24,.025} = 2.064$
calculations: calculations:
    $r = -.102$ $t_r = (r - \mu_r)/s_r$
                              $= (-.102 - 0)/\sqrt{(1-(-.102)^2)/24}$
                              $= -.102/.203 = -.500$

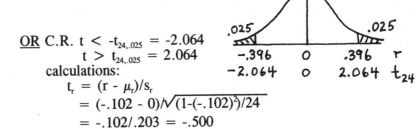

conclusion:
Do not reject H_o; there is not sufficient evidence to conclude that $\rho \neq 0$.

18. a.

y
150 ┼ •
130 ┼
110 ┼
90 ┼   •
70 ┼
50 ┼   •
30 ┼   •   • •
10 ┼   •• •• • •
   └─┼────┼────┼────┼────┼──
    100 125 150 175 200   x

$n = 17$
$\Sigma x = 2644$
$\Sigma y = 673.6$
$\Sigma xy = 97338.2$
$\Sigma x^2 = 416292$
$\Sigma y^2 = 47856.88$

b. $n(\Sigma xy) - (\Sigma x)(\Sigma y) = 17(97338.2) - (2644)(673.6) = -126249.0$
$n(\Sigma x^2) - (\Sigma x)^2 = 17(416292) - (2644)^2 = 86228$
$n(\Sigma y^2) - (\Sigma y)^2 = 17(47856.88) - (673.6)^2 = 359830.00$
$r = [n(\Sigma xy) - (\Sigma x)(\Sigma y)]/[\sqrt{n(\Sigma x^2) - (\Sigma x)^2} \cdot \sqrt{n(\Sigma y^2) - (\Sigma y)^2}]$
$= -126249/[\sqrt{86228} \cdot \sqrt{359830.00}] = -.717$

c. $H_o: \rho = 0$
$H_1: \rho \neq 0$
$\alpha = .05$

C.R. $r < -.482$ OR C.R. $t < -t_{15,.025} = -2.132$
    $r > .482$ $t > t_{15,.025} = 2.132$
calculations: calculations:
    $r = -.717$ $t_r = (r - \mu_r)/s_r$
                                        $= (-.717 - 0)/\sqrt{(1-(.-.717)^2)/15}$
                                        $= -.717/.180 = -3.981$

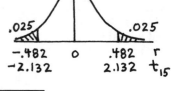

.025   .025
−.482   0   .482   r
−2.132   2.132 t_{15}

conclusion:
    Reject H_o; there is sufficient evidence to conclude that $\rho \neq 0$ (in fact, $\rho < 0$).

19. a.

y
230 ┼
215 ┼   •
200 ┼   •••
185 ┼   •••
170 ┼   •   •
155 ┼
140 ┼ •
   └─┼────┼────┼────┼────┼────┼────┼──
   1500 2000 2500 3000 3500 4000 4500   X

$n = 32$
$\Sigma x = 96654$
$\Sigma y = 6034.5$
$\Sigma xy = 18,431,272.0$
$\Sigma x^2 = 300,403,872$
$\Sigma y^2 = 1,144,240.56$

b. $n(\Sigma xy) - (\Sigma x)(\Sigma y) = 32(18,431,272.0) - (96654)(6034.5) = 6542141.0$
$n(\Sigma x^2) - (\Sigma x)^2 = 32(300,403,872) - (96654)^2 = 270928188$
$n(\Sigma y^2) - (\Sigma y)^2 = 32(1,144,240.56) - (6034.5)^2 = 200507.67$
$r = [n(\Sigma xy) - (\Sigma x)(\Sigma y)]/[\sqrt{n(\Sigma x^2) - (\Sigma x)^2} \cdot \sqrt{n(\Sigma y^2) - (\Sigma y)^2}]$
$= 6542141.0/[\sqrt{270928188} \cdot \sqrt{200507.67}] = .888$

c. $H_o: \rho = 0$
$H_1: \rho \neq 0$
$\alpha = .05$
C.R. $r < -.361$ OR C.R. $t < -t_{30,.025} = -1.960$
$r > .361$ $t > t_{30,.025} = 1.960$
calculations: calculations:
$r = .888$ $t_r = (r - \mu_r)/s_r$
                        $= (.888 - 0)/\sqrt{(1-(.888)^2)/30}$
                        $= .888/.0841 = 10.556$

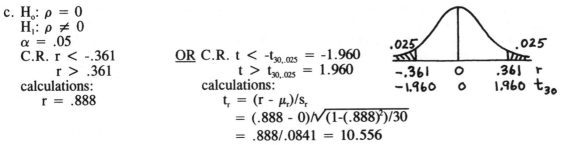

conclusion:
Reject H_o; there is sufficient evidence to conclude that $\rho \neq 0$ (in fact, $\rho > 0$).

20. a.

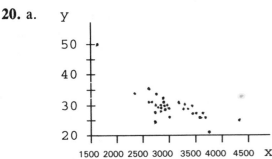

$n = 32$
$\Sigma x = 96654$
$\Sigma y = 932$
$\Sigma xy = 2,761,896$
$\Sigma x^2 = 300,403,872$
$\Sigma y^2 = 27842$

b. $n(\Sigma xy) - (\Sigma x)(\Sigma y) = 32(2,761,896) - (96654)(932) = -1,700,856$
$n(\Sigma x^2) - (\Sigma x)^2 = 32(300,403,872) - (96654)^2 = 270,928,188$
$n(\Sigma y^2) - (\Sigma y)^2 = 32(27842) - (932)^2 = 22320$
$r = [n(\Sigma xy) - (\Sigma x)(\Sigma y)]/[\sqrt{n(\Sigma x^2) - (\Sigma x)^2} \cdot \sqrt{n(\Sigma y^2) - (\Sigma y)^2}]$
$= -1,700,856[\sqrt{270928188} \cdot \sqrt{22320}] = -.692$

c. $H_o: \rho = 0$
$H_1: \rho \neq 0$
$\alpha = .05$
C.R. $r < -.361$ OR C.R. $t < -t_{30,.025} = -1.960$
$r > .361$ $t > t_{30,.025} = 1.960$
calculations: calculations:
$r = -.692$ $t_r = (r - \mu_r)/s_r$
                        $= (-.692 - 0)/\sqrt{(1-(-.692)^2)/30}$
                        $= -.692/.132 = -5.245$

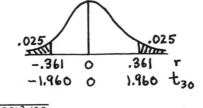

conclusion:
Reject H_o; there is sufficient evidence to conclude that $\rho \neq 0$ (in fact, $\rho < 0$).

21. A linear correlation coefficient very close to zero indicates <u>no</u> significant linear correlation and no tendencies can be inferred.

22. A significant linear correlation indicates that the factors are associated, not that there is a cause and effect relationship. Even if there is a cause and effect relationship, correlation analysis cannot identify which factor is the cause and which factor is the effect.

23. A linear correlation coefficient very close to zero indicates no significant <u>linear</u> correlation, but there may some other type of relationship between the variables.

24. A significant linear correlation between group averages indicates nothing about the relationship between the individual scores -- which may be uncorrelated, correlated in the opposite direction, or have different correlations within each of the groups.

25. The following table gives the values for y, x, x^2, log x, $\sqrt{x}$ and $1/x$. The rows at the bottom of the table give the sum of the values (i.e., Σv), the sum of squares of the values (i.e., Σv^2), the sum of each value times the corresponding y value (i.e., Σvy), and the quantity $n\Sigma v^2 - (\Sigma v)^2$ needed in subsequent calculations.

y	x	x^2	log x	$\sqrt{x}$	$1/x$
.11	1.3	1.69	.1139	1.1402	.7692
.38	2.4	5.76	.3802	1.5492	.4167
.41	2.6	6.76	.4150	1.6125	.3846
.45	2.8	7.84	.4472	1.6733	.3571
.39	2.4	5.76	.3802	1.5492	.4167
.48	3.0	9.00	.4771	1.7321	.3333
.61	4.1	16.81	.6128	2.0248	.2439
Σv 2.83	18.6	53.62	2.8264	11.2814	2.9216
Σv^2 1.2817	53.62	539.95	1.2774	18.6000	1.3850
Σvy	8.258	25.495	1.2795	4.7989	1.0326
$n\Sigma v^2 - (\Sigma v)^2$ .9630	29.38	904.55	.9533	2.9300	1.1593

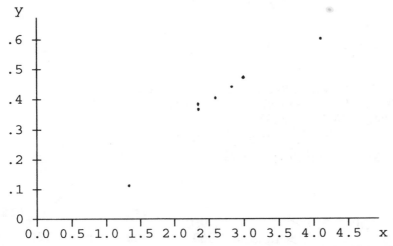

In general, $r = [n(\Sigma vy) - (\Sigma v)(\Sigma y)]/[\sqrt{n(\Sigma v^2) - (\Sigma v)^2} \cdot \sqrt{n(\Sigma y^2) - (\Sigma y)^2}]$

a. For v = x, $r = [7(8.258) - (18.6)(2.83)]/[\sqrt{29.38} \cdot \sqrt{.9630}\,] = .9716$

b. For v = x^2, $r = [7(25.495) - (53.62)(2.83)]/[\sqrt{904.55} \cdot \sqrt{.9630}\,] = .9053$

c. For v = log x, $r = [7(1.2795) - (2.8264)(2.83)]/[\sqrt{.9533} \cdot \sqrt{.9630}\,] = .9996$

d. For v = $\sqrt{x}$, $r = [7(4.7989) - (11.2814)(2.83)]/[\sqrt{2.9300} \cdot \sqrt{.9630}\,] = .9918$

e. For v = $1/x$, $r = [7(1.0326) - (2.9216)(2.83)]/[\sqrt{29.38} \cdot \sqrt{1.1593}] = -.9842$

In each case above, the critical values from Table A-6 for testing significance at the .05 level are $\pm.754$. While all the correlations are significant, the largest value for r was obtained in part (c).

26. a. For $\pm t_{48,.025} = \pm1.960$, the critical values are $r = \pm1.960/\sqrt{(\pm1.960)^2 + 48} = \pm.272$.

b. For $\pm t_{73,.05} = \pm1.645$, the critical values are $r = \pm1.645/\sqrt{(\pm1.645)^2 + 73} = \pm.189$.

c. For $-t_{18,.05} = -1.734$, the critical value is $r = -1.734/\sqrt{(-1.734)^2 + 18} = -.378$.

d. For $t_{8,.05} = 1.860$, the critical value is $r = 1.860/\sqrt{(1.860)^2 + 8} = .549$.

e. For $t_{10,.01} = 2.764$, the critical value is $r = 2.764/\sqrt{(2.764)^2 + 10} = .658$.

27.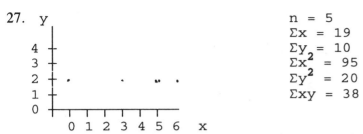

$n = 5$
$\Sigma x = 19$
$\Sigma y = 10$
$\Sigma x^2 = 95$
$\Sigma y^2 = 20$
$\Sigma xy = 38$

$r = [n(\Sigma xy) - (\Sigma x)(\Sigma y)]/[\sqrt{n(\Sigma x^2) - (\Sigma x)^2} \cdot \sqrt{n(\Sigma y^2) - (\Sigma y)^2}]$

$\quad = [5(38) - (19)(10)]/[\sqrt{5(95) - (19)^2} \cdot \sqrt{5(20) - (10)^2}]$

$\quad = [0]/[\sqrt{114} \cdot \sqrt{0}]$

$\quad = 0/0$ which is not defined (officially called an "indeterminate form")

This is an interesting exercise that illustrates what typically happens in mathematics when two opposing lines of reasoning apply.
  * *Since all the points are on the same straight line, the linear correlation is perfect and r must equal 1.00 (or -1.00?).*
  * *Since y is always equal to 2 regardless of x, there is no relationship of any kind between the variables and r must equal 0.*
The result is an indeterminate form, in this case 0/0, and a problem that requires advanced mathematical considerations. While the proper answer is "the linear correlation coefficient does not exist for these data," the generally accepted practical answer is "the linear correlation coefficient for these data may be defined to be zero."
NOTE: The form 0/0 is called "indeterminate" and cannot be permanently assigned a mathematical value because it combines the two opposing truths "0/a = 0" and "a/0 = ∞." Another indeterminate form that cannot be permanently assigned a mathematical value is 0^0, which combines the two opposing truths "$0^a = 0$" and "$a^0 = 1$."

28. None of the changes discussed affect the numerical value of the linear correlation coefficient for the following reasons.
  (1) The formula $r = [n(\Sigma xy) - (\Sigma x)(\Sigma y)]/[\sqrt{n(\Sigma x^2) - (\Sigma x)^2} \cdot \sqrt{n(\Sigma y^2) - (\Sigma y)^2}]$ is symmetric in x and y -- i.e., reversing x and y in the formula does not algebraically change the formula.
  (2) The right side of the formula is a fraction. Multiplying each x [or each y] by a constant will multiply both the numerator and the denominator by that constant, but will not change the value of the fraction.
  (3) Adding a constant c to each x will not change the value of r because
  $n \cdot \Sigma(x+c)y - \Sigma(x+c) \cdot \Sigma y = n \cdot \Sigma(xy + cy) - (\Sigma x + \Sigma c) \cdot \Sigma y$
  $\qquad\qquad\qquad\qquad\qquad\quad = n \cdot \Sigma xy + n \cdot \Sigma cy - \Sigma x \cdot \Sigma y - \Sigma c \cdot \Sigma y$
  $\qquad\qquad\qquad\qquad\qquad\quad = n \cdot \Sigma xy + nc \cdot \Sigma y - \Sigma x \cdot \Sigma y - nc \cdot \Sigma y$
  $\qquad\qquad\qquad\qquad\qquad\quad = n \cdot \Sigma xy - \Sigma x \cdot \Sigma y$
  -- i.e., the numerator of the fraction for r is not changed -- and
  $n \cdot \Sigma(x+c)^2 - [\Sigma(x+c)]^2 = n \cdot \Sigma(x^2 + 2xc + c^2) - [\Sigma x + \Sigma c]^2$
  $\qquad\qquad\qquad\qquad\qquad\quad = n \cdot \Sigma x^2 + 2nc \cdot \Sigma x + n \cdot \Sigma c^2 - [(\Sigma x)^2 + 2 \cdot \Sigma c\Sigma x + (\Sigma c)^2]$
  $\qquad\qquad\qquad\qquad\qquad\quad = n \cdot \Sigma x^2 + 2nc \cdot \Sigma x + n \cdot nc^2 - (\Sigma x)^2 - 2nc \cdot \Sigma x + (nc)^2$
  $\qquad\qquad\qquad\qquad\qquad\quad = n \cdot \Sigma x^2 - (\Sigma x)^2$
  -- i.e., the denominator of the fraction for r is not changed.

29. a.

```
  y
     |
190 -+ •
150 -+
120 -+
 90 -+
 60 -+
 30 -+     .   .    . .
  0 -+--+--+--+--+--+--+--+-
     -70 -62 -54 -46 -38 -30 -22  x
```

$n = 8$
$\Sigma x = -370$
$\Sigma y = 281$
$\Sigma xy = -17015$
$\Sigma x^2 = 18578$
$\Sigma y^2 = 37579$

b. $n(\Sigma xy) - (\Sigma x)(\Sigma y) = 8(-17015) - (-370)(281) = -32150$
$n(\Sigma x^2) - (\Sigma x)^2 = 8(18578) - (-370)^2 = 11724$
$n(\Sigma y^2) - (\Sigma y)^2 = 8(37579) - (281)^2 = 221671$

$r = [n(\Sigma xy) - (\Sigma x)(\Sigma y)]/[\sqrt{n(\Sigma x^2) - (\Sigma x)^2} \cdot \sqrt{n(\Sigma y^2) - (\Sigma y)^2}]$
$= -32150/[\sqrt{11724} \cdot \sqrt{221671}] = -.631$

c. $H_o: \rho = 0$
$H_1: \rho \neq 0$
$\alpha = .05$
C.R. $r < -.707$ OR C.R. $t < -t_{6,.025} = -2.447$
$\quad r > .707$ $\qquad t > t_{6,.025} = 2.447$
calculations: calculations:
$\quad r = -.631$ $\quad t_r = (r - \mu_r)/s_r$
$\qquad = (-.631 - 0)/\sqrt{(1-(-.631)^2)/6}$
$\qquad = -.631/.317 = -1.991$

[graph: normal curve with .025 shaded in each tail, marked -.707, 0, .707 r and -2.447, 0, 2.447 t_6]

conclusion:
Do not reject H_o; there is not sufficient evidence to conclude that $\rho \neq 0$.

In this instance, the extreme score lowered to calculated value of r [from -.926] so that it was no longer statistically significant. In general, an extreme score can either lower or raise the calculated value of r -- and the effect can be anywhere from dramatic to minimal.

30. a. Multiplying the numerator of the left side of the equation by n yields
$n \cdot \Sigma(x-\bar{x})(y-\bar{y}) = n \cdot \Sigma(xy - x\bar{y} - \bar{x}y + \bar{x}\bar{y})$
$= n\Sigma xy - n\bar{y}\Sigma x - n\bar{x}\Sigma y + n \cdot n\bar{x}\bar{y}$
$= n\Sigma xy - (\Sigma y)(\Sigma x) - (\Sigma x)(\Sigma y) + (\Sigma x)(\Sigma y)$
$= n\Sigma xy - (\Sigma x)(\Sigma y)$
-- which is the desired numerator.
Multiplying the denominator of the left side of the equation by n yields
$n(n-1)s_x s_y = \sqrt{n(n-1)s_x^2} \sqrt{n(n-1)s_y^2}$
$= \sqrt{n(\Sigma x^2) - (\Sigma x)^2} \cdot \sqrt{n(\Sigma y^2) - (\Sigma y)^2}$
-- which is the desired denominator.
Since multiplying the numerator and the denominator by the same factor does not change the value of the fraction, the two formulas are algebraically equivalent.

b. Multiplying the numerator of the right side of Formula 9-1 by $1/n^2$ yields

$(1/n^2) \cdot [n(\Sigma xy) - (\Sigma x)(\Sigma y)] = [(\Sigma xy)/n] - [(\Sigma x)/n] \cdot [(\Sigma y)/n]$

-- which is the desired numerator.

Multiplying the denominator of the right side of Formula 9-1 by $1/n^2$ yields

$(1/n^2) \cdot [\sqrt{n(\Sigma x^2) - (\Sigma x)^2} \cdot \sqrt{n(\Sigma y^2) - (\Sigma y)^2}]$

$= (1/n)\sqrt{n(\Sigma x^2) - (\Sigma x)^2} \cdot (1/n)\sqrt{n(\Sigma y^2) - (\Sigma y)^2}$

$= \sqrt{(1/n)(\Sigma x^2) - (1/n^2)(\Sigma x)^2} \cdot \sqrt{(1/n)(\Sigma y^2) - (1/n^2)(\Sigma y)^2}$

$= \sqrt{(\Sigma x^2/n) - (\Sigma x/n)^2} \cdot \sqrt{(\Sigma y^2)/n - (\Sigma y/n)^2}$

-- which is the desired denominator.

Since multiplying the numerator and the denominator by the same factor does not change the value of the fraction, the two formulas are algebraically equivalent.

31.

x	y
1	1
2	4
3	9
4	16
5	25
6	36
7	49
8	64
9	81
10	100

$n = 11$
$\Sigma x = 55$
$\Sigma y = 385$
$\Sigma x^2 = 385$
$\Sigma y^2 = 25333$
$\Sigma xy = 3025$

$r = [n(\Sigma xy) - (\Sigma x)(\Sigma y)]/[\sqrt{n(\Sigma x^2) - (\Sigma x)^2} \cdot \sqrt{n(\Sigma y^2) - (\Sigma y)^2}]$

$= [11(3025) - (55)(385)]/[\sqrt{11(385) - (55)^2} \cdot \sqrt{11(25333) - (385)^2}]$

$= [12100]/[\sqrt{1210} \cdot \sqrt{130438}] = .963$

The $r = .963$ indicates a significant linear relationship -- at the .05 level, the critical values from Table A-6 are $\pm.602$. This section of the parabola, which does not include points on both sides of the minimum at (0,0), is very close to being a straight line.

32. For $r = .600$, $(1+r)/(1-r) = (1.6)/(.4) = 4$. Following the procedure outlined in the text,

Step a. $z_{.025} = 1.960$

Step b. $w_L = \frac{1}{2} \cdot \ln(4) - 1.960/\sqrt{47} = .407$

$w_R = \frac{1}{2} \cdot \ln(4) + 1.960/\sqrt{47} = .979$

Step c. $(e^{.814} - 1)/(e^{.814} + 1) < \rho < (e^{1.958} - 1)/(e^{1.958} + 1)$

$(1.258)/(3.258) < \rho < (6.086)/(8.086)$

$.386 < \rho < .753$

NOTE: While the distribution of r is not normal, $\tanh^{-1}(r)$ [i.e., the inverse hyperbolic tangent of r] follows a normal distribution with $\mu_r = \tanh^{-1}(\rho)$ and $\sigma_r = 1/\sqrt{n-3}$. The steps above are equivalent to finding $\tanh^{-1}(r)$, constructing a confidence interval for $\tanh^{-1}(\rho)$, and then applying the hyperbolic tangent function to the endpoints to produce a confidence interval for ρ.

9-3 Regression

NOTE: For exercises 1-20, the exact summary statistics (i.e., without any rounding) are given on the right. While the intermediate calculations on the left are presented rounded to various degrees of accuracy, the entire unrounded values were preserved in the calculator until the very end.

1. $\bar{x} = 3.50$
    $\bar{y} = 2.17$
    $\hat{\beta}_1 = [n(\Sigma xy) - (\Sigma x)(\Sigma y)]/[n(\Sigma x^2) - (\Sigma x)^2]$
    $= [6(33)-(21)(13)]/[6(85)-(21)^2]$
    $= -75/69$
    $= -1.09$
    $\hat{\beta}_0 = \bar{y} - \hat{\beta}_1 \bar{x}$
    $= 2.17 - (-1.09)(3.50) = 5.97$
    $\hat{y} = \hat{\beta}_0 + \hat{\beta}_1 x = 5.97 - 1.09x$

    $n = 6$
    $\Sigma x = 21$
    $\Sigma y = 13$
    $\Sigma x^2 = 85$
    $\Sigma y^2 = 43$
    $\Sigma xy = 33$

2. $\bar{x} = 10.0$
    $\bar{y} = 20.9$
    $\hat{\beta}_1 = [n(\Sigma xy) - (\Sigma x)(\Sigma y)]/[n(\Sigma x^2) - (\Sigma x)^2]$
    $= 8260/4140 = 2.00$
    $\hat{\beta}_0 = \bar{y} - \hat{\beta}_1 \bar{x}$
    $= 20.9 - (2.00)(10.0) = .948$
    $\hat{y} = \hat{\beta}_0 + \hat{\beta}_1 x = .948 + 2.00x$

    $n = 10$
    $\Sigma x = 100$
    $\Sigma y = 209$
    $\Sigma x^2 = 1414$
    $\Sigma y^2 = 6017$
    $\Sigma xy = 2916$

3. $\bar{x} = 1.75$
    $\bar{y} = 3.00$
    $\hat{\beta}_1 = [n(\Sigma xy) - (\Sigma x)(\Sigma y)]/[n(\Sigma x^2) - (\Sigma x)^2]$
    $= [4(20)-(7)(12)]/[4(15)-(7)^2]$
    $= -4/11$
    $= -.364$
    $\hat{\beta}_0 = \bar{y} - \hat{\beta}_1 \bar{x}$
    $= 3.00 - (-.364)(1.75) = 3.64$
    $\hat{y} = \hat{\beta}_0 + \hat{\beta}_1 x = 3.64 - .364x$

    $n = 4$
    $\Sigma x = 7$
    $\Sigma y = 12$
    $\Sigma x^2 = 15$
    $\Sigma y^2 = 46$
    $\Sigma xy = 20$

4. $\bar{x} = 1.8$
    $\bar{y} = 4.2$
    $\hat{\beta}_1 = [n(\Sigma xy) - (\Sigma x)(\Sigma y)]/[n(\Sigma x^2) - (\Sigma x)^2]$
    $= 46/74 = .622$
    $\hat{\beta}_0 = \bar{y} - \hat{\beta}_1 \bar{x}$
    $= 4.2 - (.622)(1.8) = 3.08$
    $\hat{y} = \hat{\beta}_0 + \hat{\beta}_1 x = 3.08 + .622x$

    $n = 5$
    $\Sigma x = 9$
    $\Sigma y = 21$
    $\Sigma x^2 = 31$
    $\Sigma y^2 = 95$
    $\Sigma xy = 47$

5. $\bar{x} = 7.785$
    $\bar{y} = 3.250$
    $\hat{\beta}_1 = [n(\Sigma xy) - (\Sigma x)(\Sigma y)]/[n(\Sigma x^2) - (\Sigma x)^2]$
    $= [8(221.83)-(62.28)(26)]/[8(533.65)-(62.28)^2]$
    $= 155.36/390.43$
    $= .398$
    $\hat{\beta}_0 = \bar{y} - \hat{\beta}_1 \bar{x}$
    $= 3.250 - (.398)(7.785) = .152$
    $\hat{y} = \hat{\beta}_0 + \hat{\beta}_1 x = .152 + .398x$

    $n = 8$
    $\Sigma x = 62.28$
    $\Sigma y = 26$
    $\Sigma x^2 = 533.6532$
    $\Sigma y^2 = 104$
    $\Sigma xy = 221.83$

6. $\bar{x} = 4.79$

$\bar{y} = 3.25$

$\hat{\beta}_1 = [n(\Sigma xy) - (\Sigma x)(\Sigma y)]/[n(\Sigma x^2) - (\Sigma x)^2]$

$\quad = 38.06/573.1127 = .0664$

$\hat{\beta}_0 = \bar{y} - \hat{\beta}_1\bar{x}$

$\quad = 3.25 - (.0664)(4.79) = 2.93$

$\hat{y} = \hat{\beta}_0 + \hat{\beta}_1 x = 2.93 + .0664x$

$n = 8$

$\Sigma x = 38.33$

$\Sigma y = 26$

$\Sigma x^2 = 255.2877$

$\Sigma y^2 = 104$

$\Sigma xy = 129.33$

7. $\bar{x} = 28.614$

$\bar{y} = 3.600$

$\hat{\beta}_1 = [n(\Sigma xy) - (\Sigma x)(\Sigma y)]/[n(\Sigma x^2) - (\Sigma x)^2]$

$\quad = [10(1158.22)-(286.14)(36)]/[10(9260.39)-(286.14)^2]$

$\quad = 1281.16/10727.87$

$\quad = .119$

$\hat{\beta}_0 = \bar{y} - \hat{\beta}_1\bar{x}$

$\quad = 3.600 - (.119)(28.614) = .183$

$\hat{y} = \hat{\beta}_0 + \hat{\beta}_1 x = .183 + .119x$

$n = 10$

$\Sigma x = 286.14$

$\Sigma y = 36$

$\Sigma x^2 = 9260.3968$

$\Sigma y^2 = 156$

$\Sigma xy = 1158.22$

8. $\bar{x} = 9.13$

$\bar{y} = 2.12$

$\hat{\beta}_1 = [n(\Sigma xy) - (\Sigma x)(\Sigma y)]/[n(\Sigma x^2) - (\Sigma x)^2]$

$\quad = 384.1588/1646.4872 = .233$

$\hat{\beta}_0 = \bar{y} - \hat{\beta}_1\bar{x}$

$\quad = 2.12 - (.233)(9.13) = -.00682$

$\hat{y} = \hat{\beta}_0 + \hat{\beta}_1 x = -.00682 + .233x$

$n = 12$

$\Sigma x = 109.60$

$\Sigma y = 25.49$

$\Sigma x^2 = 1138.2206$

$\Sigma y^2 = 63.6081$

$\Sigma xy = 264.8219$

9. $\bar{x} = 34.575$

$\bar{y} = .2075$

$\hat{\beta}_1 = [n(\Sigma xy) - (\Sigma x)(\Sigma y)]/[n(\Sigma x^2) - (\Sigma x)^2]$

$\quad = [8(57.19)-(276.6)(1.66)]/[8(10680.48)-(276.6)^2]$

$\quad = -1.63/8936.28$

$\quad = -.000182$

$\hat{\beta}_0 = \bar{y} - \hat{\beta}_1\bar{x}$

$\quad = .2075 - (-.000182)(34.575) = .214$

$\hat{y} = \hat{\beta}_0 + \hat{\beta}_1 x = .214 - .000182x$

$n = 8$

$\Sigma x = 276.6$

$\Sigma y = 1.66$

$\Sigma x^2 = 10680.48$

$\Sigma y^2 = .3522$

$\Sigma xy = 57.191$

10. $\bar{x} = 147$

$\bar{y} = 3.88$

$\hat{\beta}_1 = [n(\Sigma xy) - (\Sigma x)(\Sigma y)]/[n(\Sigma x^2) - (\Sigma x)^2]$

$\quad = 727.85/37754 = .0193$

$\hat{\beta}_0 = \bar{y} - \hat{\beta}_1\bar{x}$

$\quad = 3.88 - (.0193)(147) = 1.04$

$\hat{y} = \hat{\beta}_0 + \hat{\beta}_1 x = 1.04 + .0193x$

$n = 7$

$\Sigma x = 1031$

$\Sigma y = 27.13$

$\Sigma x^2 = 157245$

$\Sigma y^2 = 107.1645$

$\Sigma xy = 4099.84$

11. $\bar{x} = 16.9$

$\bar{y} = 69.9$

$\hat{\beta}_1 = [n(\Sigma xy) - (\Sigma x)(\Sigma y)]/[n(\Sigma x^2) - (\Sigma x)^2]$

$\quad = [10(12146)-(169)(699)]/[10(3183)-(169)^2]$

$\quad = 3329/3269$

$\quad = 1.02$

$\hat{\beta}_0 = \bar{y} - \hat{\beta}_1\bar{x}$

$\quad = 69.9 - (1.02)(16.9) = 52.7$

$\hat{y} = \hat{\beta}_0 + \hat{\beta}_1 x = 52.7 - 1.02x$

$n = 10$

$\Sigma x = 169$

$\Sigma y = 699$

$\Sigma x^2 = 3183$

$\Sigma y^2 = 49927$

$\Sigma xy = 12146$

12. $\bar{x} = 4.81$ $n = 8$

$\bar{y} = 8.15$ $\Sigma x = 38.5$

$\hat{\beta}_1 = [n(\Sigma xy) - (\Sigma x)(\Sigma y)]/[n(\Sigma x^2) - (\Sigma x)^2]$ $\Sigma y = 65.2$

    $= 667.48/782.95 = .853$ $\Sigma x^2 = 283.15$

$\hat{\beta}_0 = \bar{y} - \hat{\beta}_1\bar{x}$ $\Sigma y^2 = 622.20$

    $= 8.15 - (.853)(4.81) = 4.05$ $\Sigma xy = 397.21$

$\hat{y} = \hat{\beta}_0 + \hat{\beta}_1 x = 4.05 + .853x$

13. $\bar{x} = .579$ $n = 8$

$\bar{y} = 9.2125$ $\Sigma x = 4.63$

$\hat{\beta}_1 = [n(\Sigma xy) - (\Sigma x)(\Sigma y)]/[n(\Sigma x^2) - (\Sigma x)^2]$ $\Sigma y = 73.7$

    $= [8(47.46)-(4.63)(73.7)]/[8(2.84)-(4.63)^2]$ $\Sigma x^2 = 2.8391$

    $= 38.425/1.276$ $\Sigma y^2 = 881.61$

    $= 30.12$ $\Sigma xy = 47.457$

$\hat{\beta}_0 = \bar{y} - \hat{\beta}_1\bar{x}$

    $= 9.2125 - (30.12)(.579) = -8.22$

$\hat{y} = \hat{\beta}_0 + \hat{\beta}_1 x = -8.22 + 30.12x$

$\hat{y}_{.75} = -8.22 + 30.12(.75) = 14.4$

14. $\bar{x} = 20.6$ $n = 8$

$\bar{y} = 2.175$ $\Sigma x = 165$

$\hat{\beta}_1 = [n(\Sigma xy) - (\Sigma x)(\Sigma y)]/[n(\Sigma x^2) - (\Sigma x)^2]$ $\Sigma y = 17.4$

    $= 237.0/3615 = .0656$ $\Sigma x^2 = 3855$

$\hat{\beta}_0 = \bar{y} - \hat{\beta}_1\bar{x}$ $\Sigma y^2 = 43.78$

    $= 2.175 - (.0656)(20.6) = .823$ $\Sigma xy = 388.5$

$\hat{y} = \hat{\beta}_0 + \hat{\beta}_1 x = .823 + .0656x$

$\hat{y}_{18} = .823 + .0656(18) = 2.0$ [i.e., \$2000]

15. $\bar{x} = -46.25$ $n = 8$

$\bar{y} = 13.75$ $\Sigma x = -370$

$\hat{\beta}_1 = [n(\Sigma xy) - (\Sigma x)(\Sigma y)]/[n(\Sigma x^2) - (\Sigma x)^2]$ $\Sigma y = 110$

    $= [8(-5729)-(-370)(110)]/[8(18578)-(-370)^2]$ $\Sigma x^2 = 18578$

    $= -5132/11724$ $\Sigma y^2 = 1840$

    $= -.438$ $\Sigma xy = -5729$

$\hat{\beta}_0 = \bar{y} - \hat{\beta}_1\bar{x}$

    $= 13.75 - (-.438)(-46.25) = -6.50$

$\hat{y} = \hat{\beta}_0 + \hat{\beta}_1 x = -6.50 - .438x$

$\hat{y}_{-60} = -6.50 - .438(-60) = 19.8$

16. $\bar{x} = 3950$ $n = 8$

$\bar{y} = .279$ $\Sigma x = 31607$

$\hat{\beta}_1 = [n(\Sigma xy) - (\Sigma x)(\Sigma y)]/[n(\Sigma x^2) - (\Sigma x)^2]$ $\Sigma y = 2.23$

    $= 467.67/3782839 = .000124$ $\Sigma x^2 = 125,348,161$

$\hat{\beta}_0 = \bar{y} - \hat{\beta}_1\bar{x}$ $\Sigma y^2 = .6417$

    $= .279 - (.000124)(3950) = -.210$ $\Sigma xy = 8868.91$

$\hat{y} = \hat{\beta}_0 + \hat{\beta}_1 x = -.210 + .000124x$

$\hat{y}_{3650} = -.210 + .000124(3650) = .24$

17. $\bar{x} = 67.22$
$\bar{y} = 35.24$
$\hat{\beta}_1 = [n(\Sigma xy) - (\Sigma x)(\Sigma y)]/[n(\Sigma x^2) - (\Sigma x)^2]$
   $= [26(61354)-(1747)(916)]/[26(56357)-(1747)^2]$
   $= -6126.72/5807.52$
   $= -1.05$
$\hat{\beta}_0 = \bar{y} - \hat{\beta}_1 \bar{x}$
   $= 35.24 - (-1.05)(67.22) = 106.2$
$\hat{y} = \hat{\beta}_0 + \hat{\beta}_1 x = 106.2 - 1.05x$

       $n = 26$
       $\Sigma x = 1747.8$
       $\Sigma y = 916.2$
       $\Sigma x^2 = 117715.86$
       $\Sigma y^2 = 56357.08$
       $\Sigma xy = 61354.14$

18. $\bar{x} = 156$
$\bar{y} = 39.6$
$\hat{\beta}_1 = [n(\Sigma xy) - (\Sigma x)(\Sigma y)]/[n(\Sigma x^2) - (\Sigma x)^2]$
   $= -126249.9/86228 = -1.46$
$\hat{\beta}_0 = \bar{y} - \hat{\beta}_1 \bar{x}$
   $= 39.6 - (-1.46)(156) = 267$
$\hat{y} = \hat{\beta}_0 + \hat{\beta}_1 x = 267 - 1.46x$

       $n = 17$
       $\Sigma x = 2644$
       $\Sigma y = 673.6$
       $\Sigma x^2 = 416292$
       $\Sigma y^2 = 47856.88$
       $\Sigma xy = 3397338.2$

19. $\bar{x} = 188.6$
$\bar{y} = 3020.4$
$\hat{\beta}_1 = [n(\Sigma xy) - (\Sigma x)(\Sigma y)]/[n(\Sigma x^2) - (\Sigma x)^2]$
   $= [32(18,431,272)-(6034)(96654)]/[32(1,144,240)...$
   $= 6542141/200508$ $...-(6034)^2]$
   $= 32.6$
$\hat{\beta}_0 = \bar{y} - \hat{\beta}_1 \bar{x}$
   $= 3020.4 - (32.6)(188.6) = 3132.5$
$\hat{y} = \hat{\beta}_0 + \hat{\beta}_1 x = 3132.5 + 32.6x$

       $n = 32$
       $\Sigma x = 6034.5$
       $\Sigma y = 96654$
       $\Sigma x^2 = 1,144,240.56$
       $\Sigma y^2 = 300,403,872$
       $\Sigma xy = 18,431,272.0$

20. $\bar{x} = 3020$
$\bar{y} = 29.1$
$\hat{\beta}_1 = [n(\Sigma xy) - (\Sigma x)(\Sigma y)]/[n(\Sigma x^2) - (\Sigma x)^2]$
   $= -1,700,856/270,928,188 = -.00628$
$\hat{\beta}_0 = \bar{y} - \hat{\beta}_1 \bar{x}$
   $= 29.1 - (-.00628)(3020) = 48.1$
$\hat{y} = \hat{\beta}_0 + \hat{\beta}_1 x = 48.1 - .00628x$

       $n = 32$
       $\Sigma x = 96654$
       $\Sigma y = 932$
       $\Sigma x^2 = 300,403,872$
       $\Sigma y^2 = 27842$
       $\Sigma xy = 2,761,896$

21. The .05 critical values for r are taken from Table A-6.
   a. $CV = \pm.196$; $r=.999$ is significant
      use $\hat{y} = 10.0 + 50x$
         $\hat{y}_{2.0} = 10.0 + 50(2.0) = 110.0$
   b. $CV = \pm.632$; $r=.005$ is not significant
      use $\hat{y} = \bar{y}$
         $\hat{y}_{2.0} = \bar{y} = 25.0$
   c. $CV = \pm.514$; $r=.519$ is significant
      use $\hat{y} = 10.0 + 50x$
         $\hat{y}_{2.0} = 10.0 + 50(2.0) = 110.0$
   d. $CV = \pm.396$; $r=.393$ is not significant
      use $\hat{y} = \bar{y}$
         $\hat{y}_{2.0} = \bar{y} = 25.0$
   e. $CV = \pm.444$; $r=.567$ is significant
      use $\hat{y} = 10.0 + 50x$
         $\hat{y}_{2.0} = 10.0 + 50(2.0) = 110.0$

22. The .05 critical values for r are taken from Table A-6.
  a. CV $= \pm.878$; r$=-.102$ is not significant
    use $\hat{y} = \bar{y}$
       $\hat{y}_{1.0} = \bar{y} = 40.0$
  b. CV $= \pm.279$; r$=-.997$ is significant
    use $\hat{y} = 50.0 - 20.0x$
       $\hat{y}_{1.0} = 50.0 - 20.0(1.0) = 30.0$
  c. CV $= \pm.444$; r$=-.403$ is not significant
    use $\hat{y} = \bar{y}$
       $\hat{y}_{1.0} = \bar{y} = 40.0$
  d. CV $= \pm.444$; r$=-.449$ is significant
    use $\hat{y} = 50.0 - 20.0x$
       $\hat{y}_{1.0} = 50.0 - 20.0(1.0) = 30.0$
  e. CV $= \pm.254$; r$=-.229$ is not significant
    use $\hat{y} = \bar{y}$
       $\hat{y}_{1.0} = \bar{y} = 40.0$

23. The .01 critical values for r are taken from Table A-6.
  a. CV $= \pm.402$; r$=.01$ is not significant
    use $\hat{y} = \bar{y}$
       $\hat{y}_5 = \bar{y} = 6$
  b. CV $= \pm.402$; r$=.93$ is significant
    use $\hat{y} = 2 + 3x$
       $\hat{y}_5 = 2 + 3(5) = 17$
  c. CV $= \pm.561$; r$=-.654$ is significant
    use $\hat{y} = 2 - 3x$
       $\hat{y}_6 = 2 - 3(6) = -16$
  d. CV $= \pm.561$; r$=.432$ is not significant
    use $\hat{y} = \bar{y}$
       $\hat{y}_5 = \bar{y} = 6$
  e. CV $= \pm.256$; r$=-.175$ is not significant
    use $\hat{y} = \bar{y}$
       $\hat{y}_5 = \bar{y} = 6$

24. The .05 critical values for r are taken from Table A-6.
  a. CV $= \pm.632$; r$=-.236$ is not significant
    use $\hat{y} = \bar{y}$
       $\hat{y}_{8.0} = \bar{y} = 8.40$
  b. CV $= \pm.632$; r$=-.654$ is significant
    use $\hat{y} = 3.5 - 2.0x$
       $\hat{y}_{8.0} = 3.5 - 2.0(8.0) = -12.5$
  c. CV $= \pm.632$; r$=.602$ is not significant
    use $\hat{y} = \bar{y}$
       $\hat{y}_{8.0} = \bar{y} = 8.40$
  d. CV $= \pm.361$; r$=-.304$ is not significant
    use $\hat{y} = \bar{y}$
       $\hat{y}_{8.0} = \bar{y} = 8.40$
  e. CV $= \pm.236$; r$=.257$ is significant
    use $\hat{y} = 3.5 + 2.0x$
       $\hat{y}_{8.0} = 3.5 + 2.0(8.0) = 19.5$

25. <u>original data</u>

$n = 5$
$\Sigma x = 4{,}234{,}178$
$\Sigma y = 576$
$\Sigma x^2 = 3{,}595{,}324{,}583{,}102$
$\Sigma y^2 = 67552$
$\Sigma xy = 491{,}173{,}342$

$\bar{x} = 846835.6$
$\bar{y} = 115.2$
$n\Sigma xy - (\Sigma x)(\Sigma y) = 16{,}980{,}182$
$n\Sigma x^2 - (\Sigma x)^2 = 48{,}459{,}579{,}826$
$\hat{\beta}_1 = 16{,}980{,}182/48{,}459{,}579{,}826$
    $= .0003504$
$\hat{\beta}_0 = \bar{y} - \hat{\beta}_1\bar{x}$
    $= 115.2 - .0003504(846835.6)$
    $= -181.53$
$\hat{y} = \hat{\beta}_0 + \hat{\beta}_1 x$
    $= -181.53 + .0003504x$

<u>original data divided by 1000</u>

$n = 5$
$\Sigma x = 4{,}234.178$
$\Sigma y = 576$
$\Sigma x^2 = 3{,}595{,}324.583102$
$\Sigma y^2 = 67552$
$\Sigma xy = 491{,}173.342$

$\bar{x} = 846.8356$
$\bar{y} = 115.2$
$n\Sigma xy - (\Sigma x)(\Sigma y) = 16{,}980.182$
$n\Sigma x^2 - (\Sigma x)^2 = 48{,}459.579826$
$\hat{\beta}_1 = 16{,}980.182/48{,}459.579826$
    $= .3504$
$\hat{\beta}_0 = \bar{y} - \hat{\beta}_1\bar{x}$
    $= 115.2 - .3504(846.8356)$
    $= -181.53$
$\hat{y} = \hat{\beta}_0 + \hat{\beta}_1 x$
    $= -181.53 + .3504x$

Dividing each x by 1000 multiplies $\hat{\beta}_1$, the coefficient of x in the regression equation, by 1000; multiplying the x coefficient by 1000 and dividing x by 1000 will "cancel out" and all predictions remain the same.

Dividing each y by 1000 divides both $\hat{\beta}_1$ and $\hat{\beta}_0$ by 1000; consistent with the new "units" for y, all predictions will also turn out divided by 1000.

26.

x	y	using $\hat{y} = 7.25 - 2.00x$ $\hat{y}$	$y-\hat{y}$	$(y-\hat{y})^2$	using $\hat{y} = 6 - \frac{x}{2}$ $\hat{y}$	$y-\hat{y}$	$(y-\hat{y})^2$
1	5	5.25	-.25	.0625	5	0	0
2	4	3.25	.75	.5625	4	0	0
2	3	3.25	-.25	.0625	4	-1	1
3	1	1.25	-.25	.0625	3	-2	4
8	13	13.00	0	.7500			5

$\bar{x} = 2.00$
$\bar{y} = 3.25$
$\hat{\beta}_1 = [n(\Sigma xy) - (\Sigma x)(\Sigma y)]/[n(\Sigma x^2) - (\Sigma x)^2]$
    $= -16/8 = -2.00$
$\hat{\beta}_0 = \bar{y} - \hat{\beta}_1\bar{x}$
    $= 3.25 - (-2.00)(2.00) = 7.25$
$\hat{y} = \hat{\beta}_0 + \hat{\beta}_1 x = 7.25 - 2.00x$

$n = 4$
$\Sigma x = 8$
$\Sigma y = 13$
$\Sigma x^2 = 18$
$\Sigma y^2 = 51$
$\Sigma xy = 22$

As expected $\Sigma(y-\hat{y})^2$ is less [.75 < 5.00] using the regression equation $\hat{y} = 7.25 - 2.00x$.

27. Replace the values x: 4 5 6 7
              with (x^2): 16 25 36 49

Use the same format employed for this section's previous exercises substituting (x^2) for x.

$\overline{(x^2)} = 31.5$
$\bar{y} = 10.25$
$\hat{\beta}_1 = [n(\Sigma(x^2)y) - (\Sigma(x^2))(\Sigma y)]/[n(\Sigma(x^2)^2 - (\Sigma(x^2))^2]$
    $= [4(1599)-(126)(41)]/[4(4578)-(126)^2]$
    $= 1230/2436 = .505$
$\hat{\beta}_0 = \bar{y} - \hat{\beta}_1\overline{(x^2)} = 10.25 - (.505)(31.5) = -5.66$
$\hat{y} = \hat{\beta}_0 + \hat{\beta}_1 x^2 = -5.66 + .505x^2$

$n = 4$
$\Sigma(x^2) = 126$
$\Sigma y = 41$
$\Sigma(x^2)^2 = 4578$
$\Sigma y^2 = 579$
$\Sigma(x^2)y = 1599$

28. Since $r = [n(\Sigma xy) - (\Sigma x)(\Sigma y)]/[n(n-1)s_x s_y]$ and $\beta_1 = [n(\Sigma xy) - (\Sigma x)(\Sigma y)]/[n(n-1)s_x^2$, and since standard deviations are positive, both expressions will have the same sign as the quantity $[n(\Sigma xy) - (\Sigma x)(\Sigma y)]$.

29. NOTE: Because the formula $\hat{\beta}_0 = \bar{y} - \hat{\beta}_1\bar{x}$ was derived assuming that $\bar{y}$ was the correct predicted value for $\bar{x}$, it cannot be used to prove that fact. One must refer back to the original definition of $\hat{\beta}_0$.

$$\hat{y} = \hat{\beta}_0 + \hat{\beta}_1 x$$
$$\hat{y}_{\bar{x}} = \hat{\beta}_0 + \hat{\beta}_1 \bar{x}$$

$$= \frac{(\Sigma y)(\Sigma x^2) - (\Sigma x)(\Sigma xy)}{n(\Sigma x^2) - (\Sigma x)^2} + \frac{n(\Sigma xy) - (\Sigma x)(\Sigma y)}{n(\Sigma x^2) - (\Sigma x)^2} \cdot (\Sigma x)/n$$

$$= \frac{(\Sigma y)(\Sigma x^2) - (\Sigma x)(\Sigma xy) + [n(\Sigma xy) - (\Sigma x)(\Sigma y)] \cdot (\Sigma x)/n}{n(\Sigma x^2) - (\Sigma x)^2}$$

$$= \frac{(\Sigma y)(\Sigma x^2) - (\Sigma x)(\Sigma xy) + n(\Sigma xy)(\Sigma x)/n - (\Sigma x)(\Sigma y)(\Sigma x)/n}{n(\Sigma x^2) - (\Sigma x)^2}$$

$$= \frac{(\Sigma y)(\Sigma x^2) - (\Sigma x)(\Sigma xy) + (\Sigma xy)(\Sigma x) - (\Sigma y)(\Sigma x)^2/n}{n(\Sigma x^2) - (\Sigma x)^2}$$

$$= \frac{(\Sigma y)(\Sigma x^2) - (\Sigma y)(\Sigma x)^2/n}{n(\Sigma x^2) - (\Sigma x)^2}$$

$$= \frac{(\Sigma y)[(\Sigma x^2) - (\Sigma x)^2/n]}{n[(\Sigma x^2) - (\Sigma x)^2/n]}$$

$$= (\Sigma y)/n$$

$$= \bar{y}$$

30. If $H_o: \rho = 0$ is true, there is no linear correlation between x and y and $\hat{y} = \bar{y}$ is the appropriate prediction for y for any x. If $H_o: \beta_1 = 0$ is true then $\hat{y} = \beta_0 = \bar{y}$. Since both null hypotheses imply precisely the same result, they are equivalent.

9-4 Variation and Prediction Intervals

1. The coefficient of determination is $r^2 = (.333)^2 = .111$.
   The portion of the total variation explained by the regression line is $r^2 = .111 = 11.1\%$.

2. The coefficient of determination is $r^2 = (-.444)^2 = .197$.
   The portion of the total variation explained by the regression line is $r^2 = .197 = 19.7\%$.

3. The coefficient of determination is $r^2 = (.800)^2 = .640$.
   The portion of the total variation explained by the regression line is $r^2 = .640 = 64.0\%$.

4. The coefficient of determination is $r^2 = (-.700)^2 = .490$.
   The portion of the total variation explained by the regression line is $r^2 = .490 = 49.0\%$.

5. The predicted values were calculated using the regression line $\hat{y} = 2 + 3x$.

x	y	$\hat{y}$	$\bar{y}$	$\hat{y}-\bar{y}$	$(\hat{y}-\bar{y})^2$	$y-\hat{y}$	$(y-\hat{y})^2$	$y-\bar{y}$	$(y-\bar{y})^2$
1	5	5	12.2	-7.2	51.84	0	0	-7.2	51.84
2	8	8	12.2	-4.2	17.64	0	0	-4.2	17.64
3	11	11	12.2	-1.2	1.44	0	0	-1.2	1.44
5	17	17	12.2	4.8	23.04	0	0	4.8	23.04
6	20	20	12.2	7.8	60.84	0	0	7.8	60.84
17	61	61	61.0	0	154.80	0	0	0	154.80

NOTE: A table such as the one above organizes the work and provides all the values needed to discuss variation. In such a table, the following must always be true and can be used as a check before proceeding.

* $\Sigma y = \Sigma \hat{y} = \Sigma \bar{y}$
* $\Sigma(\hat{y}-\bar{y}) = \Sigma(y-\hat{y}) = \Sigma(y-\bar{y}) = 0$
* $\Sigma(y-\bar{y})^2 + \Sigma(y-\hat{y})^2 = \Sigma(y-\bar{y})^2$

a. The explained variation is $\Sigma(\hat{y}-\bar{y})^2 = 154.80$

b. The unexplained variation is $\Sigma(y-\hat{y})^2 = 0$

c. The total variation is $\Sigma(y-\bar{y})^2 = 154.80$

d. $r^2 = \Sigma(\hat{y}-\bar{y})^2/\Sigma(y-\bar{y})^2 = 154.80/154.80 = 1.00$

e. $s_e^2 = \Sigma(y-\hat{y})^2/(n-2) = 0/3 = 0$
   $s_e = 0$

6. The predicted values were calculated using the regression line $\hat{y} = .182817 + .119423x$.

x	y	$\hat{y}$	$\bar{y}$	$\hat{y}-\bar{y}$	$(\hat{y}-\bar{y})^2$	$y-\hat{y}$	$(y-\hat{y})^2$	$y-\bar{y}$	$(y-\bar{y})^2$
10.67	2	1.468	3.6	-2.132	4.546	.532	.283	-1.6	2.56
19.96	3	2.567	3.6	-1.033	1.068	.433	.188	-.6	.36
27.60	3	3.479	3.6	-.121	.015	-.479	.229	-.6	.36
38.11	6	4.734	3.6	1.134	1.286	1.266	1.603	2.4	5.76
27.90	4	3.515	3.6	-.085	.007	.485	.235	.4	.16
21.90	2	2.789	3.6	-.802	.643	-.798	.637	-1.6	2.56
21.83	1	2.790	3.6	-.810	.656	-1.790	3.203	-2.6	6.76
49.27	5	6.067	3.6	2.467	6.085	-1.067	1.138	1.4	1.96
33.27	6	4.156	3.6	.556	.309	1.844	3.400	2.4	5.76
35.54	4	4.427	3.6	.827	.684	-.427	.182	.4	.16
286.14	36	36.000	36.0	0	15.299	0	11.098	0	26.40

a. The explained variation is $\Sigma(\hat{y}-\bar{y})^2 = 11.10$

b. The unexplained variation is $\Sigma(y-\hat{y})^2 = 15.30$

c. The total variation is $\Sigma(y-\bar{y})^2 = 26.40$

d. $r^2 = \Sigma(\hat{y}-\bar{y})^2/\Sigma(y-\bar{y})^2 = 15.30/26.40 = .5795$

e. $s_e^2 = \Sigma(y-\hat{y})^2/(n-2) = 11.10/8 = 1.3875$
   $s_e = 1.1779$

7. The predicted values were calculated using the regression line $\hat{y}$ = -8.21713 + 30.1160x.

x	y	$\hat{y}$	$\overline{y}$	$\hat{y}-\overline{y}$	$(\hat{y}-\overline{y})^2$	$y-\hat{y}$	$(y-\hat{y})^2$	$y-\overline{y}$	$(y-\overline{y})^2$
.65	14.7	11.358	9.2125	2.146	4.604	3.342	11.167	5.4875	30.113
.55	12.3	8.347	9.2125	-.866	.750	3.953	15.629	3.0875	9.533
.72	14.6	13.466	9.2125	4.254	18.096	1.134	1.285	5.3875	29.025
.83	15.1	16.779	9.2125	7.567	57.254	-1.679	2.820	5.8875	34.663
.57	5.0	8.949	9.2125	-.264	.069	-3.949	15.595	-4.2125	17.745
.51	4.1	7.142	9.2125	-2.070	4.287	-3.042	9.254	-5.1125	26.138
.43	3.8	4.733	9.2125	-4.480	20.068	-.933	.870	-5.4125	29.295
.37	4.1	2.926	9.2125	-6.287	39.523	1.174	1.379	-5.1125	26.138
4.63	73.7	73.700	73.0000	0	144.651	0	57.998	0	202.649

a. The explained variation is $\Sigma(\hat{y}-\overline{y})^2$ = 144.651

b. The unexplained variation is $\Sigma(y-\hat{y})^2$ = 57.998

c. The total variation is $\Sigma(y-\overline{y})^2$ = 202.649

d. $r^2 = \Sigma(\hat{y}-\overline{y})^2/\Sigma(y-\overline{y})^2$ = 144.651/202.649 = .714

e. $s_e^2 = \Sigma(y-\hat{y})^2/(n-2)$ = 57.998/6 = 9.666
   s_e = 3.109

8. The predicted values were calculated using the regression line $\hat{y}$ = .182817 + .119423x.

x	y	$\hat{y}$	$\overline{y}$	$\hat{y}-\overline{y}$	$(\hat{y}-\overline{y})^2$	$y-\hat{y}$	$(y-\hat{y})^2$	$y-\overline{y}$	$(y-\overline{y})^2$
-62	21	20.644	13.75	6.894	47.532	.356	.126	7.25	52.5625
-41	13	11.452	13.75	-2.298	5.281	1.548	2.397	-.75	.5625
-36	12	9.263	13.75	-4.487	20.131	2.737	7.490	-1.75	3.0625
-26	3	4.886	13.75	-8.864	78.572	-1.886	3.557	-10.75	115.5625
-33	6	7.950	13.75	-5.800	33.640	-1.950	3.803	-7.75	60.0625
-56	22	18.018	13.75	4.268	18.215	3.982	15.857	8.25	68.0625
-50	14	15.392	13.75	1.642	2.695	-1.392	1.936	.25	.0625
-66	19	22.395	13.75	8.645	74.741	-3.395	11.528	5.25	27.5625
-370	110	110.000	110.00	0	280.807	0	46.694	0	327.5000

a. The explained variation is $\Sigma(\hat{y}-\overline{y})^2$ = 280.81

b. The unexplained variation is $\Sigma(y-\hat{y})^2$ = 46.69

c. The total variation is $\Sigma(y-\overline{y})^2$ = 327.50

d. $r^2 = \Sigma(\hat{y}-\overline{y})^2/\Sigma(y-\overline{y})^2$ = 280.81/327.50 = .8574

e. $s_e^2 = \Sigma(y-\hat{y})^2/(n-2)$ = 46.69/6 = 7.7817
   s_e = 2.7896

9. a. $\hat{y}$ = 2 + 3x
   $\hat{y}_4$ = 2 + 3(4) = 14

   b. $\hat{y} \pm t_{n-2,\alpha/2}s_e\sqrt{1 + 1/n + n(x_o-\overline{x})^2/[n\Sigma x^2-(\Sigma x)^2]}$
   $\hat{y} \pm 0$, since s_e = 0
   The prediction "interval" in this case shrinks to a single point. Since r^2 = 1.00 (i.e., 100% of the variability in the y's can be explained by the regression), a perfect prediction can be made. For a practical example of such a situation, consider the regression line $\hat{y}$ = 1.099x for predicting the amount of money y due for purchasing x gallons of gasoline at $1.099 per gallon -- the "prediction" will be exactly correct every time because of the perfect correlation between the number of gallons purchased and the amount of money due.

10. a. $\hat{y} = .182817 + .119423x$

$\hat{y}_{20.0} = .182817 + .119423(20.0) = 2.57$

b. preliminary calculations

n = 10

$\Sigma x = 286.14 \qquad \bar{x} = (\Sigma x)/n = 28.614$

$\Sigma x^2 = 9260.3976 \qquad n\Sigma x^2-(\Sigma x)^2 = 10727.8764$

$\hat{y} \pm t_{n-2,\alpha/2}s_e\sqrt{1 + 1/n + n(x_o-\bar{x})^2/[n\Sigma x^2-(\Sigma x)^2]}$

$\hat{y}_{20.0} \pm t_{8,.005}(1.1779)\sqrt{1 + 1/10 + 10(20.0-28.614)^2/[10727.8764]}$

$2.57 \pm (3.355)(1.1779)\sqrt{1.1692}$

2.57 ± 4.27

$-1.70 < y_{20.0} < 6.84$

11. a. $\hat{y} = -89.21713 + 30.1160x$

$\hat{y}_{.75} = -8.21713 + 30.1160(.75) = 14.37$

b. preliminary calculations

n = 8

$\Sigma x = 4.63 \qquad \bar{x} = (\Sigma x)/n = 4.63/8 = .57875$

$\Sigma x^2 = 2.8391 \qquad n\Sigma x^2-(\Sigma x)^2 = 8(2.8391)-(4.63)^2 = 1.2759$

$\hat{y} \pm t_{n-2,\alpha/2}s_e\sqrt{1 + 1/n + n(x_o-\bar{x})^2/[n\Sigma x^2-(\Sigma x)^2]}$

$\hat{y}_{.75} \pm t_{6,.025}(3.109)\sqrt{1 + 1/8 + 8(.75-.57875)^2/[1.2759]}$

$14.37 \pm (2.447)(3.109)\sqrt{1.30888}$

14.37 ± 8.70

$5.67 < y_{.75} < 23.07$

12. a. $\hat{y} = -6.49522 - .437735x$

$\hat{y}_{-50} = -6.49522 - .437735(-50) = 15.39$

b. preliminary calculations

n = 8

$\Sigma x = -370 \qquad \bar{x} = -46.25$

$\Sigma x^2 = 18578 \qquad n\Sigma x^2-(\Sigma x)^2 = 11724$

$\hat{y} \pm t_{n-2,\alpha/2}s_e\sqrt{1 + 1/n + n(x_o-\bar{x})^2/[n\Sigma x^2-(\Sigma x)^2]}$

$\hat{y}_{-50} \pm t_{6,.005}(2.7896)\sqrt{1 + 1/8 + 8(-50+46.25)^2/[11724]}$

$15.39 \pm (3.707)(2.7896)\sqrt{1.1346}$

15.39 ± 11.02

$4.38 < y_{-50} < 26.41$

Exercises 13-16 refer to the chapter problem of Table 9-1. They use the following, which are calculated and/or discussed at various places in the text,

n = 8

$\Sigma x = 14.60 \qquad\qquad \hat{y} = .549 + 1.480x$

$\Sigma x^2 = 32.9632 \qquad\qquad s_e = .971554$

and the values obtained below.

$\bar{x} = (\Sigma x)/n = 14.60/8 = 1.825$

$n\Sigma x^2-(\Sigma x)^2 = 8(32.9632)-(14.60)^2 = 50.5456$

13. $\hat{y} \pm t_{n-2,\alpha/2}s_e\sqrt{1 + 1/n + n(x_o-\bar{x})^2/[n\Sigma x^2-(\Sigma x)^2]}$

$\hat{y}_{1.00} \pm t_{6,.025}(.971554)\sqrt{1 + 1/8 + 8(1.00-1.825)^2/[50.5456]}$

$2.029 \pm (2.447)(.971554)\sqrt{1.23272}$

2.029 ± 2.640

$-.61 < y_{1.00} < 4.67$

14. $\hat{y} \pm t_{n-2,\alpha/2}s_e\sqrt{1 + 1/n + n(x_o-\bar{x})^2/[n\Sigma x^2-(\Sigma x)^2]}$

$\hat{y}_{3.00} \pm t_{6,.05}(.971554)\sqrt{1 + 1/8 + 8(3.00-1.825)^2/[50.5456]}$

$4.989 \pm (1.943)(.971554)\sqrt{1.3435}$

4.989 ± 2.188

$2.80 < y_{3.00} < 7.18$

15. $\hat{y} \pm t_{n-2,\alpha/2}s_e\sqrt{1 + 1/n + n(x_o-\bar{x})^2/[n\Sigma x^2-(\Sigma x)^2]}$

$\hat{y}_{1.50} \pm t_{6,.05}(.971554)\sqrt{1 + 1/8 + 8(1.50-1.825)^2/[50.5456]}$

$2.769 \pm (1.943)(.971554)\sqrt{1.14172}$

2.769 ± 2.017

$.75 < y_{1.50} < 4.79$

16. $\hat{y} \pm t_{n-2,\alpha/2}s_e\sqrt{1 + 1/n + n(x_o-\bar{x})^2/[n\Sigma x^2-(\Sigma x)^2]}$

$\hat{y}_{.75} \pm t_{6,.005}(.971554)\sqrt{1 + 1/8 + 8(.75-1.825)^2/[50.5456]}$

$1.659 \pm (3.707)(.971554)\sqrt{1.3079}$

1.659 ± 4.119

$-2.46 < y_{.75} < 5.78$

17. This exercise uses the following values from the chapter problem of Table 9-1, which are calculated and/or discussed at various places in the text,

$n = 8$ $\hat{\beta}_o = .549$

$\Sigma x = 14.60$ $\hat{\beta}_1 = 1.480$

$\Sigma x^2 = 32.9632$ $s_e = .971554$

and the values obtained below.

$\bar{x} = (\Sigma x)/n = 14.60/8 = 1.825$

$\Sigma x^2-(\Sigma x)^2/n = (32.9632)-(14.60)^2/8 = 6.3182$

a. $\hat{\beta}_o \pm t_{n-2,\alpha/2}s_e\sqrt{1/n + \bar{x}^2/[\Sigma x^2-(\Sigma x)^2/n]}$

$.549 \pm t_{6,.025}(.971554)\sqrt{1/8 + (1.825)^2/[6.3182]}$

$.549 \pm (2.447)(.971554)\sqrt{.65215}$

$.549 \pm 1.920$

$-1.37 < \beta_o < 2.47$

b. $\hat{\beta}_1 \pm t_{n-2,\alpha/2}s_e/\sqrt{\Sigma x^2-(\Sigma x)^2/n}$

$1.480 \pm t_{6,.025}(.971554)/\sqrt{6.3182}$

$1.480 \pm (2.447)(.971554)/\sqrt{6.3182}$

$1.480 \pm .946$

$.53 < \beta_1 < 2.43$

18. a. If $s_e = 0$, then $\Sigma(y-\hat{y})^2 = 0$ -- which means that all observed data lie on the regression line and $r = 1.00$ or $r = -1.00$.

b. If $\Sigma(\hat{y}-\bar{y})^2 = 0$, then $\hat{y} = -\bar{y}$ for all x -- which means the regression line is parallel to the x axis and has slope 0.

19. a. Since $s_e^2 = \Sigma(y-\hat{y})^2/(n-2)$, then $(n-2)s_e^2 = \Sigma(y-\hat{y})^2$.

The unexplained variation is equal to $(n-2)s_e^2$.

b. $r^2 = $ (explained variation)/(total variation)

(total variation) $\cdot r^2 = $ (explained variation)

[(explained variation) + (unexplained variation)] $\cdot r^2 = $ (explained variation)

(explained variation) $\cdot r^2 + $ (unexplained variation) $\cdot r^2 = $ (explained variation)

(unexplained variation) $\cdot r^2 = $ (explained variation) - (explained variation) $\cdot r^2$

(unexplained variation) $\cdot r^2 = $ (explained variation)$(1 - r^2)$

(unexplained variation) $\cdot r^2/(1 - r^2) = $ (explained variation)

The explained variation is equal to (unexplained variation) $\cdot r^2/(1 - r^2)$.

c. If $r^2 = .900$, then $r = \pm.949$.
Since the regression line has a negative slope (i.e., $\hat{\beta}_1 = -2$), we choose the negative root. The linear correlation coefficient, therefore, is $r = -.949$.

20. The following values are given prior to the solutions for exercises #13-16.
$n = 8$
$\bar{x} = 1.825$ $\hat{y} = .549 + 1.480x$
$n\Sigma x^2-(\Sigma x)^2 = 50.5456$ $s_e = .971554$
$\hat{y} \pm t_{n-2,\alpha/2}s_e\sqrt{1/n + n(x_o-\bar{x})^2/[n\Sigma x^2-(\Sigma x)^2]}$
$\hat{y}_{2.50} \pm t_{6,.025}(.971554)\sqrt{1/8 + 8(2.50-1.825)^2/[50.5456]}$
$4.249 \pm (2.447)(.971554)\sqrt{.1971}$
4.249 ± 1.056
$3.19 < \mu_{y \mid x=3.00} < 5.30$

9-5 Multiple Regression

1. $\hat{y} = 34.8 + 1.21x_1 + .23x_2$
$\hat{y}_{24,92} = 34.8 + 1.21(24) + .23(92) = 85.0$

2. $\hat{y} = 34.8 + 1.21x_1 + .23x_2$
$\hat{y}_{12,71} = 34.8 + 1.21(12) + .23(71) = 65.65$

3. $\hat{y} = 34.8 + 1.21x_1 + .23x_2$
$\hat{y}_{18,81} = 34.8 + 1.21(18) + .23(81) = 75.2$

4. $\hat{y} = 34.8 + 1.21x_1 + .23x_2$
$\hat{y}_{31,99} = 34.8 + 1.21(31) + .23(99) = 95.1$

5. Let y be the household size.
Let x_1, x_2 and x_3 be the weights of discarded metal, plastic and food.
$\hat{y} = .92 - .244x_1 + 1.75x_2 - .073x_3$

6. a. P-value $= .116$
b. $R^2 = 73.9\% = .739$
c, adjusted $R^2 = 54.4\% = .544$

7. No; since the P-value of .116 is greater than .05, there is not statistical significance.
The equation $\hat{y} = \bar{y}$ should be used for making predictions.

8. a. $\hat{y} = .924 - .2440x_1 + 1.7476x_2 - .0732x_3$
$\hat{y}_{2.25,2.95,4.27} = .924 - .2440(2.25) + 1.7476(2.95) - .0732(4.27) = 5.22$
b. $\hat{y} = .549 + 1.48x$
$\hat{y}_{2.95} = .549 + 1.48(2.95) = 4.915$
c. R^2 cannot decrease as variables, no matter how uninformative, are added to the regression. The adjusted R^2 gives the best picture of the reliability of the prediction.
For the regression in part (a), the adjusted R^2 is given as .544.
For the regression in part (b), the adjusted R^2 must be calculated.
   $R^2 = (r)^2 = (.842)^2 = .709$
   $n = 8$
   $k = 1$
   adjusted $R^2 = 1 - [(n-1)/(n-k-1)] \cdot (1-R^2) = 1 - [7/6] \cdot (.291) = 1 - .340 = .660$
Since the regression in part (b) has the higher adjusted R^2, the better predicted household size is 4.915.

9. Following is the complete Minitab input and output required for the problem.
NOTE: It is often easier to input the data as whole numbers and then to use a LET statement to move the decimal point accordingly. In addition, it is a good idea to PRINT the data before analyzing it -- just to make certain that Minitab is talking about the same data that you are.

a. $\hat{y} = .804 - .363x_2 + 1.76x_4$

b. P-value $= .040$

c. $R^2 = 72.3\% = .723$

d. adjusted $R^2 = 61.3\% = .613$

e. yes; since P-value $= .040 < .05$

```
MTB > SET C1
DATA> 2 3 3 6 4 2 1 5
DATA> END
MTB > SET C2
DATA> 109 104 257 302 150 210 193 357
DATA> END
MTB > LET C2=C2/100
MTB > SET C4
DATA> 27 141 219 283 219 181 85 305
DATA> END
MTB > LET C4=C4/100
MTB > NAME C1'Y' C2'X2' C4'X4'
MTB > PRINT C1 C2 C4
```

```
 ROW    Y      X2      X4

  1     2     1.09    0.27
  2     3     1.04    1.41
  3     3     2.57    2.19
  4     6     3.02    2.83
  5     4     1.50    2.19
  6     2     2.10    1.81
  7     1     1.93    0.85
  8     5     3.57    3.05
```

```
MTB > REGRESS C1 2 C2 C4
```

```
The regression equation is
Y = 0.804 - 0.363 X2 + 1.76 X4
```

Predictor	Coef	Stdev	t-ratio	p
Constant	0.8038	0.9820	0.82	0.450
X2	-0.3635	0.7288	-0.50	0.639
X4	1.7591	0.6959	2.53	0.053

```
s = 1.039      R-sq = 72.3%      R-sq(adj) = 61.3%
```

Analysis of Variance

SOURCE	DF	SS	MS	F	p
Regression	2	14.105	7.052	6.54	0.040
Error	5	5.395	1.079		
Total	7	19.500			

SOURCE	DF	SEQ SS
X2	1	7.211
X4	1	6.894

10. Following is the complete Minitab input and output required for the problem.

a. $\hat{y} = 2.00 - .454x_3 + 2.62x_4$

b. P-value $= .017$

c. $R^2 = 80.4\% = .804$

d. adjusted $R^2 = 72.6\% = .726$

e. yes; since P-value $= .017 < .05$

```
MTB > SET C1
DATA> 2 3 3 6 4 2 1 5
DATA> SET C3
DATA> 241 757 955 882 872 696 683 1142
DATA> END
MTB > LET C3=C3/100
MTB > SET C4
DATA> 27 141 219 283 219 181 85 305
DATA> END
MTB > LET C4=C4/100
MTB > NAME C1'Y' C3'X3' C4'X4'
MTB > PRINT C1 C3 C4
```

ROW	Y	X3	X4
1	2	2.41	0.27
2	3	7.57	1.41
3	3	9.55	2.19
4	6	8.82	2.83
5	4	8.72	2.19
6	2	6.96	1.81
7	1	6.83	0.85
8	5	11.42	3.05

```
MTB > REGRESS C1 2 C3 C4
```

The regression equation is
$Y = 2.00 - 0.454 X3 + 2.62 X4$

Predictor	Coef	Stdev	t-ratio	p
Constant	2.004	1.171	1.71	0.148
X3	-0.4544	0.2921	-1.56	0.181
X4	2.6212	0.8119	3.23	0.023

$s = 0.8737$ R-sq = 80.4%   R-sq(adj) = 72.6%

Analysis of Variance

SOURCE	DF	SS	MS	F	p
Regression	2	15.6835	7.8417	10.27	0.017
Error	5	3.8165	0.7633		
Total	7	19.5000			

SOURCE	DF	SEQ SS
X3	1	7.7277
X4	1	7.9558

11. Following is the complete Minitab input and output required for the problem.

    NOTE: When a command is given after DATA>, Minitab assumes the ENDOFDATA and executes the command. It is a good idea to PRINT the data before analyzing it -- just to make certain that Minitab is talking about the same data that you are.

The regression equation is $\hat{y} = 2.17 + 2.44x + .464x^2$.

The multiple coefficient of determination $R^2 = 100.0\%$ means that all the points lie exactly (i.e., within the accuracy of the problem) on the parabola.

```
MTB > SET C1
DATA> 5 14 19 42 26
DATA> SET C2
DATA> 1 3 4 7 5
DATA> LET C3=C2*C2
MTB > NAME C1'Y'
MTB > NAME C2'X'
MTB > NAME C3'X2'
MTB > PRINT C1-C3

 ROW     Y      X     X2

  1       5      1      1
  2      14      3      9
  3      19      4     16
  4      42      7     49
  5      26      5     25

MTB > REGRESS C1 2 C2 C3

The regression equation is
Y = 2.17 + 2.44 X + 0.464 X2


Predictor        Coef       Stdev      t-ratio        p
Constant       2.1714      0.5741        3.78     0.063
X              2.4357      0.3140        7.76     0.016
X2            0.46429     0.03802       12.21     0.007

s = 0.3485      R-sq = 100.0%     R-sq(adj) = 99.9%

Analysis of Variance

SOURCE          DF          SS         MS        F          p
Regression       2      774.56     387.28   3189.35    0.000
Error            2        0.24       0.12
Total            4      774.80

SOURCE          DF      SEQ SS
X                1      756.45
X2               1       18.11
```

12. Use the following commands to enter the problem into Minitab.
```
MTB > SET C1
DATA> -2.0 -1.0 0.0 1.0 2.0 3.0
DATA> END
MTB > SET C2
DATA> 13.0 4.0 5.0 4.0 13.0 68.0
DATA> END
MTB > LET C3=C1*C1
MTB > LET C4=C1*C1*C1
MTB > LET C5=C1*C1*C1*C1
MTB > NAME C1'X' C2'Y' C3'X2' C4'X3' C5'X4'
```

The following commands produce the desired output for parts (a) -(d). The answers for parts (a)-(d) are given in the table at the end of the exercise.

a. `MTB > REGRESS C2 1 C1`

b. `MTB > REGRESS C2 2 C1 C3`

c. `MTB > REGRESS C2 3 C1 C3 C4`

d. `MTB > REGRESS C2 4 C1 C3 C4 C5`

e. Based on the results of the regressions, the equation in part (d) with and adjusted R^2 of $1.000 = 100.0\%$ seems to give the best results.

part	regression equation	P-value	R^2	adjusted R^2
a.	$\hat{y} = 13.53 + 8.60x$	.167	.416	.270
b.	$\hat{y} = -3.23 + 2.31x + 6.29x^2$	.036	.890	.817
c.	$\hat{y} = 1.57 - 6.29x + 3.29x^2 + 2.00x^3$	.039	.974	.934
d.	$\hat{y} = 5.00 + 0.00x - 2.00x^2 + 0.00x^3 + 1.00x^4$	.000	1.000	1.000

Review Exercises

1. Let x be the cost and y be the miles.

$n = 8$ $n(\Sigma xy) - (\Sigma x)(\Sigma y) = 8(7073.09) - (190.3)(281.9)$
$\Sigma x = 190.3$ $= 2939.15$
$\Sigma y = 281.9$ $n(\Sigma x^2) - (\Sigma x)^2 = 8(5332.07) - (190.3)^2$
$\Sigma x^2 = 5332.07$ $= 6642.47$
$\Sigma y^2 = 11616.25$ $n(\Sigma y^2) - (\Sigma y)^2 = 8(11616.25) - (281.9)^2$
$\Sigma xy = 7073.09$ $= 13462.39$

a. $r = [n(\Sigma xy) - (\Sigma x)(\Sigma y)]/[\sqrt{n(\Sigma x^2) - (\Sigma x)^2} \cdot \sqrt{n(\Sigma y^2) - (\Sigma y)^2}]$
$= [2939.15]/[\sqrt{6442.47} \cdot \sqrt{13462.39}] = .316$

b. $CV = \pm.707$

c. no significant linear correlation

d. $\hat{\beta}_1 = [n(\Sigma xy) - (\Sigma x)(\Sigma y)]/[n(\Sigma x^2) - (\Sigma x)^2]$
$= 2939.15/6442.47 = .456$
$\hat{\beta}_0 = \bar{y} - \hat{\beta}_1\bar{x}$
$= (281.9/8) - (.456)(190.3/8) = 24.4$
$\hat{y} = \hat{\beta}_0 + \hat{\beta}_1 x$
$= 24.4 + .456x$

e.   y

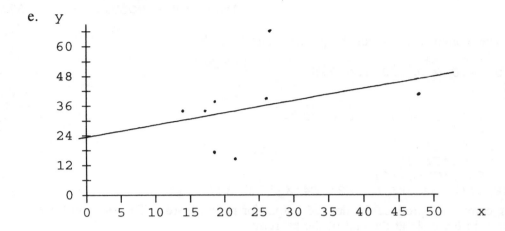

2. Let x be the age and y be the score.

n = 6

$\Sigma x = 47.0$ $n(\Sigma x^2) - (\Sigma x)^2 = 159.80$

$\Sigma x^2 = 394.80$

$\Sigma y = 273.3$ $n(\Sigma y^2) - (\Sigma y)^2 = 24962.61$

$\Sigma y^2 = 16609.25$

$\Sigma xy = 2459.87$ $n(\Sigma xy) - (\Sigma x)(\Sigma y) = 1914.12$

a. $r = [n(\Sigma xy) - (\Sigma x)(\Sigma y)]/[\sqrt{n(\Sigma x^2) - (\Sigma x)^2} \cdot \sqrt{n(\Sigma y^2) - (\Sigma y)^2}]$

   $= [1914.12]/[\sqrt{159.80}\sqrt{24962.61}\ ] = .958$

b. $CV = \pm.811$

c. significant (positive) linear correlation

d. $\hat{\beta}_1 = [n(\Sigma xy) - (\Sigma x)(\Sigma y)]/[n(\Sigma x^2) - (\Sigma x)^2]$

   $= 1914.12/159.80 = 11.98$

   $\hat{\beta}_0 = \bar{y} - \hat{\beta}_1\bar{x}$

   $= (273.3/6) - (11.98)(47.0/6) = -48.28$

   $\hat{y} = \hat{\beta}_0 + \hat{\beta}_1 x$

   $= -48.28 + 11.98x$

e.   y

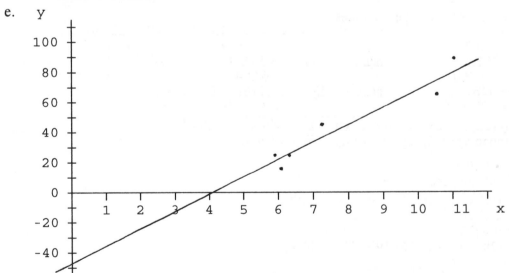

3. Let x be the minutes and y be the points.

$n = 9$ $n(\Sigma xy) - (\Sigma x)(\Sigma y) = 9(4{,}581{,}424) - (7238)(3741)$
$\Sigma x = 7238$ $= 14{,}155{,}458$
$\Sigma y = 3741$ $n(\Sigma x^2) - (\Sigma x)^2 = 9(8{,}267{,}024) - (7238)^2$
$\Sigma x^2 = 8{,}267{,}024$ $= 22{,}014{,}572$
$\Sigma y^2 = 2{,}637{,}573$ $n(\Sigma y^2) - (\Sigma y)^2 = 9(2{,}637{,}573) - (3741)^2$
$\Sigma xy = 4{,}581{,}424$ $= 9{,}743{,}076$

a. $r = [n(\Sigma xy) - (\Sigma x)(\Sigma y)]/[\sqrt{n(\Sigma x^2) - (\Sigma x)^2} \cdot \sqrt{n(\Sigma y^2) - (\Sigma y)^2}]$
 $= [14{,}155{,}458]/[\sqrt{22{,}014{,}572}\ \sqrt{9{,}743{,}076}] = .967$

b. $CV = \pm.666$

c. significant (positive) linear correlation

d. $\hat{\beta}_1 = [n(\Sigma xy) - (\Sigma x)(\Sigma y)]/[n(\Sigma x^2) - (\Sigma x)^2]$
 $= 14{,}155{,}458/22{,}014{,}572 = .643$
$\hat{\beta}_0 = \bar{y} - \hat{\beta}_1\bar{x}$
 $= (3741/9) - (.643)(7238/9) = -101$
$\hat{y} = \hat{\beta}_0 + \hat{\beta}_1 x$
 $= -101 + .643x$

e.

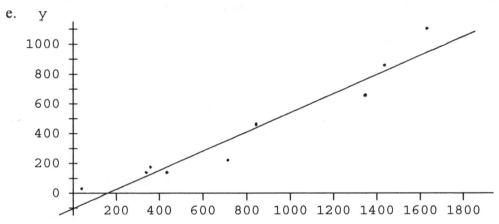

4. Let x be the productivity and y be the dexterity.

$n = 15$
$\Sigma x = 1168$ $n(\Sigma x^2) - (\Sigma x)^2 = 96416$
$\Sigma x^2 = 97376$
$\Sigma y = 98$ $n(\Sigma y^2) - (\Sigma y)^2 = 1166$
$\Sigma y^2 = 718$
$\Sigma xy = 7850$ $n(\Sigma xy) - (\Sigma x)(\Sigma y) = 3286$

a. $r = [n(\Sigma xy) - (\Sigma x)(\Sigma y)]/[\sqrt{n(\Sigma x^2) - (\Sigma x)^2} \cdot \sqrt{n(\Sigma y^2) - (\Sigma y)^2}]$
 $= [3286]/[\sqrt{96416}\ \sqrt{1166}] = .310$

b. $CV = \pm.514$

c. no significant linear correlation

d. $\hat{\beta}_1 = [n(\Sigma xy) - (\Sigma x)(\Sigma y)]/[n(\Sigma x^2) - (\Sigma x)^2]$
 $= 3286/96416 = .0341$
$\hat{\beta}_0 = \bar{y} - \hat{\beta}_1\bar{x}$
 $= (98/15) - .0341(1168/15) = 3.88$
$\hat{y} = \hat{\beta}_0 + \hat{\beta}_1 x$
 $= 3.88 + .0341x$

e.

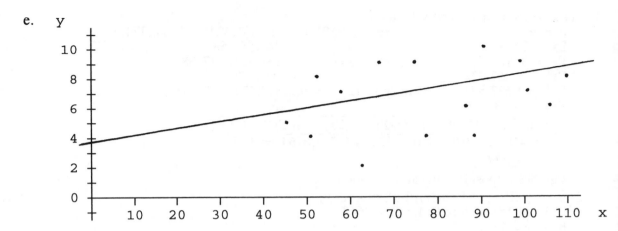

5. Let x be the minutes and y be the savings.

$n = 9$ $n(\Sigma xy) - (\Sigma x)(\Sigma y) = 9(909.927) - (75.42)(101.2)$
$\Sigma x = 75.42$ $= 556.839$
$\Sigma y = 101.2$ $n(\Sigma x^2) - (\Sigma x)^2 = 9(729.3314) - (75.42)^2$
$\Sigma x^2 = 729.3314$ $= 875.8062$
$\Sigma y^2 = 1177.36$ $n(\Sigma y^2) - (\Sigma y)^2 = 9(1177.36) - (101.2)^2$
$\Sigma xy = 909.927$ $= 354.8$

a. $r = [n(\Sigma xy) - (\Sigma x)(\Sigma y)]/[\sqrt{n(\Sigma x^2) - (\Sigma x)^2} \cdot \sqrt{n(\Sigma y^2) - (\Sigma y)^2}]$
$= [556.839]/[\sqrt{875.8062} \cdot \sqrt{354.8}\,] = .999$

H_o: $\rho = 0$
H_1: $\rho \neq 0$
$\alpha = .05$
C.R. $r < -.666$ OR C.R. $t < -t_{7,.025} = -2.365$
$r > .666$ $t > t_{7,.025} = 2.365$
calculations: calculations:
$r = .999$ $t_r = (r - \mu_r)/s_r$
                                    $= (.999 - 0)/\sqrt{(1-(.999)^2)/7}$
                                    $= .999/.0175$
                                    $= 57.072$

conclusion:
   Reject H_o; there is sufficient evidence to conclude that $\rho \neq 0$ (in fact, $\rho > 0$).

b. $\hat{\beta}_1 = [n(\Sigma xy) - (\Sigma x)(\Sigma y)]/[n(\Sigma x^2) - (\Sigma x)^2]$
$= 556.839/875.8062 = .636$
$\hat{\beta}_o = \bar{y} - \hat{\beta}_1\bar{x}$
$= (101.2/9) - (.636)(75.42/9) = 5.92$
$\hat{y} = \hat{\beta}_o + \hat{\beta}_1 x$
$= 5.92 + .636x$

c. $\hat{y} = \hat{\beta}_o + \hat{\beta}_1 x$
$\hat{y}_{5.0} = 5.92 + (.636)(5) = 9.10$

6. Let x be the periodical values and y be the savings.

n = 9	n(Σxy) - (Σx)(Σy) = 9(11.230) - (1.04)(101.2)
Σx = 1.04	= -4.178
Σy = 101.2	n(Σx^2) - (Σx)^2 = 9(.1436) - (1.04)^2
Σx^2 = .1436	= .2108
Σy^2 = 1177.36	n(Σy^2) - (Σy)^2 = 9(1177.36) - (101.2)^2
Σxy = 11.230	= 354.8

a. r = [n(Σxy) - (Σx)(Σy)]/[$\sqrt{n(\Sigma x^2) - (\Sigma x)^2}$ ·$\sqrt{n(\Sigma y^2) - (\Sigma y)^2}$]
  = [-4.178]/[$\sqrt{.2108}$ $\sqrt{354.8}$] = -.483
H_0: ρ = 0
H_1: $\rho \neq 0$
α = .05

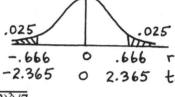

C.R. r < -.666   <u>OR</u> C.R. t < -$t_{7,.025}$ = -2.365
    r > .666   t > $t_{7,.025}$ = 2.365
calculations: calculations:
    r = -.483 t_r = (r - μ_r)/s_r
                 = (-.483 - 0)/$\sqrt{(1-(-.483)^2)/7}$
                 = -.483/.3309
                 = -1.460

conclusion:
    Do not reject H_0; there is not sufficient evidence to conclude that $\rho \neq 0$.

b. $\hat{\beta}_1$ = [n(Σxy) - (Σx)(Σy)]/[n(Σx^2) - (Σx)^2]
   = -4.178/.2108 = -19.82
$\hat{\beta}_0$ = $\bar{y}$ - $\hat{\beta}_1\bar{x}$
   = (101.2/9) - (-19.82)(1.04/9) = 13.53
$\hat{y}$ = $\hat{\beta}_0$ + $\hat{\beta}_1 x$
   = 13.53 - 19.82x

c. $\hat{y}$ = $\hat{\beta}_0$ + $\hat{\beta}_1 x$
$\hat{y}_{.20}$ = 13.53 + (-19.82)(.20) = 9.6

7. Refer to the notation and summary calculations of exercise #5. This exercise requires the calculation of s_e. Since r = .999 is very close to 1.000, there is very little variability unexplained by the regression line and s_e will be very close to zero. This necessitates extra care in the calculations for s_e -- STORE and RECALL with complete accuracy any intermediate values other than n or the primary summations given exactly above.

s_e^2 = [Σy^2 - $\hat{\beta}_0(\Sigma y)$ - $\hat{\beta}_1(\Sigma xy)$]/(n-2)
   = [1177.36 - (5.9164)(101.2) - (.6358)(909.927)]/7 = .084540/7 = .012077
s_e = .1099

$\hat{y}$ ± $t_{n-2,\alpha/2}s_e\sqrt{1 + 1/n + n(x_o-\bar{x})^2/[n\Sigma x^2-(\Sigma x)^2]}$
$\hat{y}_{5.0}$ ± $t_{7,.025}$(.1099)$\sqrt{1 + 1/9 + 9(5.0-8.38)^2/[875.8062]}$
9.095 ± (2.365)(.1099)$\sqrt{1.2285}$
9.095 ± .288
8.81 < $y_{5.0}$ < 9.38

8. Place the minutes, periodicals and savings values in Minitab's C1, C2 and C3.
The line *MTB > REGRESS C3 2 C1 C2* produces output indicating the following values.
   regression equation: $\hat{y}$ = 6.060 + .630x_1 - .815x_2
   R^2: .998
   adjusted R^2: .998
   P-value: .000

Chapter 10

Multinomial Experiments and Contingency Tables

10-2 Multinomial Experiments

NOTE: In multinomial problems, always verify that $\Sigma E = \Sigma O$ before proceeding. If these sums are not equal, then an error has been made and further calculations have no meaning.

1. H_o: $p_0 = p_1 = p_2 = ... = p_9 = .10$
    H_1: at least one of the proportions is different from .10
    $\alpha = .01$
    C.R. $\chi^2 > \chi^2_{9,.01} = 21.666$
    calculations:

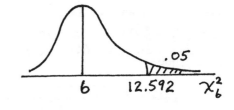

digit	O	E	$(O-E)^2/E$
0	35	8	91.125
1	0	8	8.000
2	2	8	4.500
3	1	8	6.125
4	4	8	2.000
5	24	8	32.000
6	1	8	6.125
7	4	8	2.000
8	7	8	.125
9	2	8	5.400
	80	80	153.500

$\chi^2 = \Sigma[(O-E)^2/E]$
$= 153.500$

conclusion:
    Reject H_o; there is sufficient evidence to conclude that at least one of the proportions is different from .10.

2. H_o: $p_{Sun} = p_{Mon} = p_{Tue} = ... = p_{Sat} = 1/7$
    H_1: at least one of the proportions is different from 1/7
    $\alpha = .05$
    C.R. $\chi^2 > \chi^2_{6,.05} = 12.592$
    calculations:

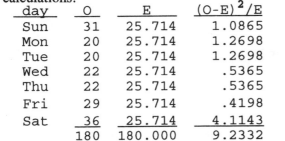

day	O	E	$(O-E)^2/E$
Sun	31	25.714	1.0865
Mon	20	25.714	1.2698
Tue	20	25.714	1.2698
Wed	22	25.714	.5365
Thu	22	25.714	.5365
Fri	29	25.714	.4198
Sat	36	25.714	4.1143
	180	180.000	9.2332

$\chi^2 = \Sigma[(O-E)^2/E]$
$= 9.233$

conclusion:
    Do not reject H_o; there is not sufficient evidence to conclude that at least one of the proportions is different from 1/7.

3. H_o: $p_{Sun} = p_{Mon} = p_{Tue} = \ldots = p_{Sat} = 1/7$
   H_1: at least one of the proportions is different from 1/7
   $\alpha = .05$
   C.R. $\chi^2 > \chi^2_{6,.05} = 12.592$
   calculations:

day	O	E	$(O-E)^2/E$
Sun	40	30.857	2.709
Mon	24	30.857	1.524
Tue	25	30.857	1.112
Wed	28	30.857	.265
Thu	29	30.857	.112
Fri	32	30.857	.042
Sat	38	30.857	1.653
	216	216.000	7.417

$\chi^2 = \Sigma[(O-E)^2/E]$
$= 7.417$

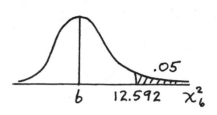

conclusion:
   Do not reject H_o; there is not sufficient evidence to conclude that at least one of the proportions is different from 1/7.

4. H_o: $p_1 = p_2 = p_3 = \ldots = p_6 = 1/6$
   H_1: at least one of the proportions is different from 1/6
   $\alpha = .05$ [or any other appropriate value]
   C.R. $\chi^2 > \chi^2_{5,.05} = 11.071$
   calculations:

die	O	E	$(O-E)^2/E$
1	6	10	1.600
2	16	10	3.600
3	11	10	.100
4	12	10	.400
5	8	10	.400
6	7	10	.900
	60	60	7.000

$\chi^2 = \Sigma[(O-E)^2/E]$
$= 7.000$

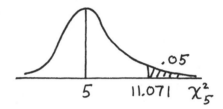

conclusion:
   Do not reject H_o; there is not sufficient evidence to conclude that at least one of the proportions is different from 1/6.

5. H_o: $p_{Mon} = .30$, $p_{Tue} = .15$, $p_{Wed} = .15$, $p_{Thu} = .20$, $p_{Fri} = .20$
   H_1: at least one of the proportions is different from what is claimed
   $\alpha = .05$
   C.R. $\chi^2 > \chi^2_{4,.05} = 9.488$
   calculations:

day	O	E	$(O-E)^2/E$
Mon	31	44.10	3.891
Tue	42	22.05	18.050
Wed	18	22.05	.744
Thu	25	29.40	.659
Fri	31	29.40	.087
	147	147.00	23.431

$\chi^2 = \Sigma[(O-E)^2/E]$
$= 23.431$

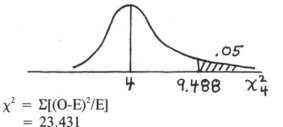

conclusion:
   Reject H_o; there is sufficient evidence to conclude that at least one of the proportions is different from what is claimed.
   Rejection of this claim may provide indirect help in correcting the industrial accident problem. It may indicate that the safety expert does not know what he talking about and that a different such person should be employed. If a single day (in this case, Tuesday) makes an

unusually large contribution to $\Sigma[(O-E)^2/E]$, it may mean that the safety expert knows what he is talking about and that the test has identified correctable circumstances unique to the plant being studied. At the very least, rejection of the hypothesis indicates that medical staffing should not be determined based on the proportions claimed.

6. H_o: $p_{van} = .62$, $p_{cho} = .18$, $p_{neo} = .12$, $p_{fud} = .08$
H_1: at least one of the proportions is different from what is claimed
$\alpha = .05$
C.R. $\chi^2 > \chi^2_{3,.05} = 7.815$
calculations:

type	O	E	$(O-E)^2/E$
van	120	124	.129
cho	40	36	.444
neo	18	24	1.500
fud	22	16	2.250
	200	200	4.323

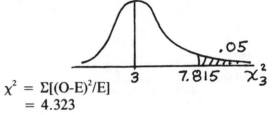

$\chi^2 = \Sigma[(O-E)^2/E]$
$= 4.323$

conclusion:
Do not reject H_o; there is not sufficient evidence to conclude that at least one of the proportions is different from what is claimed.

7. H_o: $p_{Sun} = .07$, $p_{Mon} = .05$ $p_{Tue} = .09$, $p_{Wed} = .11$, $p_{Thu} = .19$, $p_{Fri} = .24$, $p_{Sat} = .25$
H_1: at least one of the proportions is different from what is claimed
$\alpha = .05$
C.R. $\chi^2 > \chi^2_{6,.05} = 12.592$
calculations:

day	O	E	$(O-E)^2/E$
Sun	9	7.21	.444
Mon	6	5.15	.140
Tue	10	9.27	.057
Wed	8	11.33	.979
Thu	19	9.57	.017
Fri	23	24.72	.120
Sat	28	25.75	.197
	103	103.00	1.954

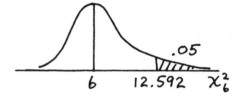

$\chi^2 = \Sigma[(O-E)^2/E]$
$= 1.954$

conclusion:
Do not reject H_o; there is not sufficient evidence to conclude that at least one of the proportions is different from what is claimed.

8. H_o: $p_A = .10$, $p_B = .25$, $p_C = .35$, $p_D = .20$, $p_F = .10$
H_1: at least one of the proportions is different from the past
$\alpha = .05$
C.R. $\chi^2 > \chi^2_{4,.05} = 9.488$
calculations:

grade	O	E	$(O-E)^2/E$
A	15	7.50	7.500
B	22	18.75	.563
C	25	26.25	.060
D	8	15.00	3.267
F	5	7.50	.833
	75	75.00	12.223

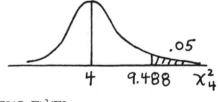

$\chi^2 = \Sigma[(O-E)^2/E]$
$= 12.223$

conclusion:
Reject H_o; there is sufficient evidence to conclude that at least one of the proportions is different from the past.

9. H_o: $p_0 = p_1 = p_2 = ... = p_9 = .10$
   H_1: at least one of the proportions is different from .10
   $\alpha = .05$
   C.R. $\chi^2 > \chi^2_{9,.05} = 16.919$
   calculations:

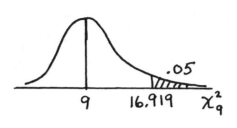

digit	O	E	$(O-E)^2/E$
0	8	10	.400
1	8	10	.400
2	12	10	.400
3	11	10	.100
4	10	10	.000
5	8	10	.400
6	9	10	.100
7	8	10	.400
8	12	10	.400
9	14	10	1.600
	100	100	4.200

$\chi^2 = \Sigma[(O-E)^2/E]$
$= 4.200$

conclusion:
Do not reject H_o; there is not sufficient evidence to conclude that at least one of the proportions is different from .10.

10. H_o: $p_0 = p_1 = p_2 = ... = p_9 = .10$
    H_1: at least one of the proportions is different from .10
    $\alpha = .05$
    C.R. $\chi^2 > \chi^2_{9,.05} = 16.919$
    calculations:

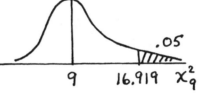

digit	O	E	$(O-E)^2/E$
0	0	10	10.000
1	17	10	4.900
2	17	10	4.900
3	1	10	8.100
4	17	10	4.900
5	16	10	3.600
6	0	10	10.000
7	16	10	3.600
8	16	10	3.600
9	0	10	10.000
	100	100	63.600

$\chi^2 = \Sigma[(O-E)^2/E]$
$= 63.600$

conclusion:
Reject H_o; there is sufficient evidence to conclude that at least one of the proportions is different from .10.
Yes; the result differs dramatically from that found in exercise #9. Since 22/7 is a repeating decimal, once the repetitions begin only those digits will appear and other digits will not appear at all.

11. NOTE: Usually the hypothesized p_i's are given and the formula $E_i = np_i$ is used to find the individual expected values. Here, the individual expected values are given (and they sum to $n = 320$) and the formula must be used "in reverse" to solve for the hypothesized p_i's.
H_o: $p_A = .0625$, $p_B = .0625$, $p_C = .1250$, $p_D = .3750$, $p_E = .3750$
H_1: at least one of the proportions is different from what is claimed
$\alpha = .01$
C.R. $\chi^2 > \chi^2_{4,.01} = 13.277$
calculations:

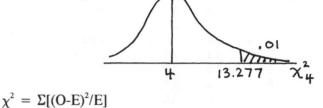

type	O	E	$(O-E)^2/E$
A	30	20	5.000
B	15	20	1.250
C	58	40	8.100
D	83	120	11.408
E	134	120	1.633
	320	320	27.392

$$\chi^2 = \Sigma[(O-E)^2/E]$$
$$= 27.392$$

conclusion:
   Reject H_o; there is sufficient evidence to conclude that at least one of the proportions is different from what is claimed.

12. H_o: $p_1 = .16$, $p_2 = .44$, $p_3 = .27$, $p_4 = .13$
H_1: at least one of the proportions is different from the license proportions
$\alpha = .05$
C.R. $\chi^2 > \chi^2_{3,.05} = 7.815$
calculations:

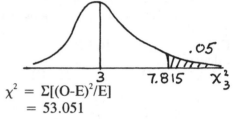

group	O	E	$(O-E)^2/E$
1: < 25	36	14.08	34.125
2: 25-44	21	38.72	8.109
3: 45-64	12	23.76	5.821
4: > 64	19	11.44	4.996
	88	88.00	53.051

$$\chi^2 = \Sigma[(O-E)^2/E]$$
$$= 53.051$$

conclusion:
   Reject H_o; there is sufficient evidence to conclude that at least one of the proportions is different from the license proportions.
Yes; the "under 25" group appears to have a disproportionate number of crashes. **NOTE:** A fairer, but much more difficult to set up, test would be to base the E values on the proportion of miles driven and not on the proportion of licenses possessed.

13. H_o: $p_{bro} = .30$, $p_{yel} = .20$, $p_{red} = .20$, $p_{ora} = .10$, $p_{gre} = .10$, $p_{tan} = .10$
H_1: at least one of the proportions is different from what is claimed
$\alpha = .05$
C.R. $\chi^2 > \chi^2_{5,.05} = 11.071$
calculations:

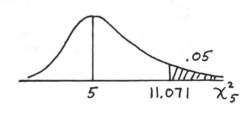

color	O	E	$(O-E)^2/E$
bro	30	30	.000
yel	24	20	.800
red	17	20	.450
ora	9	10	.100
gre	9	10	.100
tan	11	10	.100
	100	100	1.550

$$\chi^2 = \Sigma[(O-E)^2/E]$$
$$= 1.550$$

conclusion:
   Do not reject H_o; there is not sufficient evidence to conclude that at least one of the proportions is different from what is claimed.

14. H_o: $p_1 = .28$, $p_2 = .51$, $p_3 = .21$
H_1: at least one of the proportions is different from those of the general population
$\alpha = .05$
C.R. $\chi^2 > \chi^2_{2,.05} = 5.991$
calculations:

educ	O	E	$(O-E)^2/E$
1: elem	11	35.00	16.457
2: H.S.	78	63.75	3.185
3: coll	36	26.25	3.621
	125	125.00	23.264

$\chi^2 = \Sigma[(O-E)^2/E]$
$= 23.264$

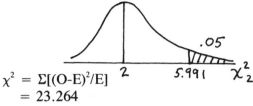

conclusion:
Reject H_o; there is sufficient evidence to conclude that at least one of the proportions is different from those of the general population.
No; based on this result, it does not appear that the sample is educationally representative of the U.S. population.

15. H_o: $p_{poor} = p_{fair} = p_{good} = p_{exce} = .25$
H_1: at least one of the proportions is different from .25
$\alpha = .05$
C.R. $\chi^2 > \chi^2_{3,.05} = 7.815$
calculations:

type	O	E	$(O-E)^2/E$
poor	4	15	8.067
fair	13	15	.267
good	29	15	13.067
exce	14	15	.067
	60	60	21.467

$\chi^2 = \Sigma[(O-E)^2/E]$
$= 21.467$

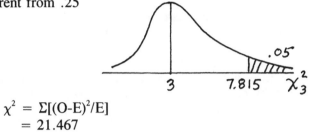

conclusion:
Reject H_o; there is sufficient evidence to conclude that at least one of the proportions is different from .25.
NOTE: Before giving movie critic James Harrington two thumbs down on the basis of the above test, one should determine exactly how the list of movies in the appendix was obtained and whether it is a fair test of the critic's claim. Is it possible, for instance, that the given list was obtained from a list of movies being shown at some point in time? Since poor movies have shorter runs, they would be under represented in the sample. Or is it possible that the critic as talking of "all films produced" and not merely those that make it onto some list. As a recognized critic with inside connections, he might be aware of films so poor that they never make it to the box office or onto lists that are compiled.

16. H_o: $p_0 = p_1 = p_2 = ... = p_9 = .10$
H_1: at least one of the proportions is different from .10
$\alpha = .05$
C.R. $\chi^2 > \chi^2_{9,.05} = 16.919$

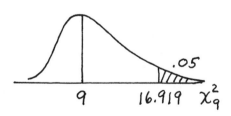

calculations:

digit	O	E	$(O-E)^2/E$
0	13	15	.267
1	16	15	.067
2	11	15	1.067
3	17	15	.267
4	14	15	.067
5	18	15	.600
6	18	15	.600
7	18	15	.600
8	12	15	.600
9	13	15	.267
	150	150	4.400

$$\chi^2 = \Sigma[(O-E)^2/E]$$
$$= 4.400$$

conclusion:
Do not reject H_o; there is not sufficient evidence to conclude that at least one of the proportions is different from .10.

17. Because $4.168 < 4.400 < 14.684$
    [i.e., $\chi^2_{9,.90} < 4.400 < \chi^2_{9,.10}$],
    it must be $.10 < $ P-value $ < .90$ [and the P-value is very close to .90].

18. Multiplying each O value by a constant c will not change the number of categories and, therefore, will not change the critical value. Multiplying each O value by c multiplies each E value by c, since $\Sigma O = \Sigma E$ must always be true. Furthermore, the calculated test statistic is multiplied by c since $\chi^2 = \Sigma[(cO-cE)^2/cE] = \Sigma[c^2(O-E)^2/cE] = \Sigma[c(O-E)^2/E] = c\Sigma[(O-E)^2/E]$.

19. NOTE: Both outcomes having the same expected frequency is equivalent to $p_1 = p_2 = .5$.

a. H_o: $p_1 = p_2 = .5$
   H_1: at least one of the proportions is different from .5
   $\alpha = .05$
   C.R. $\chi^2 > \chi^2_{1,.05} = 3.841$
   calculations

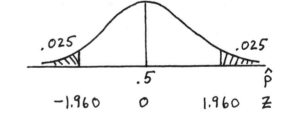

type	O	E	O-E	$(O-E)^2$	$(O-E)^2/E$
A	f_1	$(f_1+f_2)/2$	$(f_1-f_2)/2$	$(f_1-f_2)^2/4$	$[(f_1-f_2)^2/4]/[(f_1+f_2)/2]$
B	f_2	$(f_1+f_2)/2$	$(f_2-f_1)/2$	*$(f_1-f_2)^2/4$	$[(f_1-f_2)^2/4]/[(f_1+f_2)/2]$
	f_1+f_2	f_1+f_2			$[(f_1-f_2)^2/2]/[(f_1+f_2)/2]$

$$\chi^2 = \Sigma[(O-E)^2/E]$$ *NOTE: $(f_2-f_1)^2 = (f_1-f_2)^2$
$$= [(f_1-f_2)^2/2]/[(f_1+f_2)/2]$$
$$= (f_1-f_2)^2/(f_1+f_2)$$

b. H_o: $p = .5$
   H_1: $p \neq .5$
   $\alpha = .05$
   C.R. $z < -z_{.025} = -1.960$
   $z > z_{.025} = 1.960$
   calculations:
   $$z_{\hat{p}} = (\hat{p} - \mu_{\hat{p}})/\sigma_{\hat{p}}$$
   $$= [f_1/(f_1+f_2) - .5]/\sqrt{(.5)(.5)/(f_1+f_2)}$$
   $$= [.5(f_1-f_2)/(f_1+f_2)]/\sqrt{(.5)(.5)/(f_1+f_2)}$$
   $$= [(f_1-f_2)/(f_1+f_2)]/\sqrt{1/(f_1+f_2)}$$
   $$= (f_1-f_2)/\sqrt{f_1+f_2}$$

Note that $z^2 = \chi^2$ since $[(f_1-f_2)/\sqrt{f_1+f_2}]^2 = (f_1-f_2)^2/(f_1+f_2)$ and $(\pm1.960)^2 = 3.841$.

20. a. Use the binomial formula $P(x) = [n!/x!(n-x)!] \cdot p^x \cdot (1-p)^{n-x}$ with n=3 and p=1/3.

$P(x=0) = 1(1/3)^0(2/3)^3 = 8/27 = .296$
$P(x=1) = 3(1/3)^1(2/3)^2 = 12/27 = .444$
$P(x=2) = 3(1/3)^2(2/3)^1 = 6/27 = .222$
$P(x=3) = 1(1/3)^3(2/3)^0 = 1/27 = .037$

b.

x	O	E	$(O-E)^2/E$
0	89	88.89	.000
1	133	133.33	.001
2	52	66.67	3.227
3	26	11.11	19.951
	300	300.00	23.179

c. H_o: there is goodness of fit to the binomial distribution with n=3 and p=1/3
H_1: there is not goodness of fit
$\alpha = .05$
C.R. $\chi^2 > \chi^2_{3,.05} = 7.815$
calculations:
$\chi^2 = \Sigma[(O-E)^2/E] = 23.179$
conclusion:
Reject H_o; there is sufficient evidence to conclude that the observed frequencies do not fit a binomial distribution with n=3 and p=1/3.
NOTE: The phrase "goodness of fit" is the accepted statistical terminology in this context. The test could also be expressed as H_o: $p_0 = 8/27$, $p_1 = 12/27$, $p_2 = 6/27$, $p_3 = 1/27$
H_1: at least proportion is not as claimed

21. NOTE: Usually the hypothesized p_i's are given and the formula $E_i = np_i$ is used to find the individual expected values. Here, the individual expected values are given (and they sum to n = 53) and the formula must be used "in reverse" to solve for the hypothesized p_i's.
H_o: $p_{1,2} = .17$, $p_3 = .15$, $p_4 = .13$, $p_{5,6} = .19$, $p_{7,8} = .13$, $p_{9,10} = .23$
H_1: at least one of the proportions is different from what is claimed
$\alpha = .05$
C.R. $\chi^2 > \chi^2_{5,.05} = 11.071$
calculations:

slot	O	E	$(O-E)^2/E$
1,2	10	9	.111
3	8	8	.000
4	9	7	.571
5,6	8	10	.400
7,8	3	7	2.286
9,10	15	12	.750
	53	53	4.118

$\chi^2 = \Sigma[(O-E)^2/E]$
$= 4.118$

conclusion:
Do not reject H_o; there is not sufficient evidence to conclude that at least one of the proportions is different from what is claimed.

22. a. Refer to the illustration at the right.
$P(x < 79.5) = .5000 - .4147 = .0853$
$P(79.5 < x < 95.5) - .4147 - .1179 = .2968$
$P(95.5 < x < 110.5) = .1179 + .2580 = .3759$
$P(110.5 < x < 120.5) = .4147 - .2580 = .1567$
$P(x > 120.5) = .5000 - .4147 = .0853$

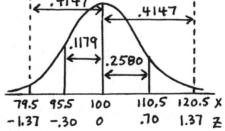

b.

score	O	E	$(O-E)^2/E$
< 80	20	17.06	.507
80- 95	20	59.36	26.099
96-110	80	75.18	.309
111-120	40	31.34	2.393
>120	40	17.06	30.847
	200	200.00	60.154

c. H_o: there is goodness of fit to the normal distribution with $\mu=100$ and $\sigma=15$
   H_1: there is not goodness of fit
   $\alpha = .01$
   C.R. $\chi^2 > \chi^2_{4,.01} = 13.277$
   calculations:
   $\chi^2 = \Sigma[(O-E)^2/E] = 60.154$
   conclusion:
      Reject H_o; there is sufficient evidence to conclude that the observed frequencies do not fit a normal distribution with $\mu=100$ and $\sigma=15$.
   NOTE: The phrase "goodness of fit" is the accepted statistical terminology in this context.

10-3 Contingency Tables

NOTE: For each row and each column it must be true that $\Sigma O = \Sigma E$. After the marginal row and column totals are calculated, both the row totals and the column totals must sum to produce the same grand total. If either of the preceding is not true, then an error has been made and further calculations have no meaning. In addition, the following are true for all χ^2 contingency table analyses in this manual.
* The E values for each cell are given in parentheses below the O values.
* The addends used to calculate the χ^2 test statistic follow the physical arrangement of the cells in the original contingency table. This practice makes it easier to monitor the large number of intermediate steps involved and helps to prevent errors caused by missing or double-counting cells.
* The accompanying chi-square illustration follows the "usual" shape as pictured with Table A-6, even though that shape is not correct for df=1 or df=2.

1. H_o: type of crime and criminal/victim connection are independent
   H_1: type of crime and criminal/victim connection are related
   $\alpha = .05$
   C.R. $\chi^2 > \chi^2_{2,.05} = 5.991$
   calculations:

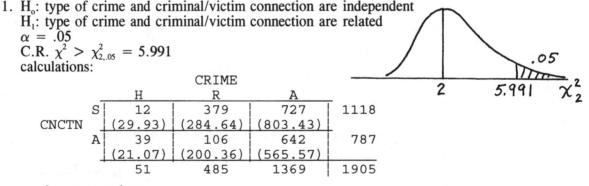

		CRIME			
		H	R	A	
CNCTN	S	12	379	727	1118
		(29.93)	(284.64)	(803.43)	
	A	39	106	642	787
		(21.07)	(200.36)	(565.57)	
		51	485	1369	1905

   $\chi^2 = \Sigma[(O-E)^2/E]$
      $= 10.7418 + 31.2847 + 7.2715$
      $15.2600 + 44.4425 + 10.3298$
      $= 119.30$
   conclusion:
      Reject H_o; there is sufficient evidence to conclude that the type of crime and the criminal/victim connection are related.

2. H_o: sex of shopper and purchase category are independent
H_1: sex of shopper and purchase category are related
$\alpha = .05$
C.R. $\chi^2 > \chi^2_{2,.05} = 5.991$
calculations:

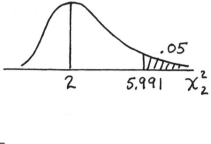

		CATEGORY			
		F	D	S	
	F	203	73	142	418
SEX		(209)	(69.67)	(139.33)	
	M	97	27	57	182
		(91)	(30.33)	(60.67)	
		300	100	200	600

$\chi^2 = \Sigma[(O-E)^2/E]$
$\quad = .1722 + .1595 + .0510$
$\quad\quad .3956 + .3663 + .1172$
$\quad = 1.262$

conclusion:
Do not reject H_o; there is not sufficient evidence to conclude that the sex of the shopper and the purchase category are related.

3. H_o: drug treatment and oral reaction are independent
H_1: drug treatment and oral reaction are related
$\alpha = .05$
C.R. $\chi^2 > \chi^2_{1,.05} = 3.841$
calculations:

		TREATMENT		
		D	P	
	S	43	35	78
RCTN		(38.87)	(39.13)	
	N	109	118	227
		(113.13)	(113.87)	
		152	153	305

$\chi^2 = \Sigma[(O-E)^2/E]$
$\quad = .4383 + .4355$
$\quad\quad .1506 + .1496$
$\quad = 1.174$

conclusion:
Do not reject H_o; there is not sufficient evidence to conclude that the drug treatment and the oral reaction are related.
A person thinking about using Nicorette might still want to be concerned about mouth soreness. While he cannot be 95% sure that there is soreness associated with Nicorette, neither can he be sure that there is not such soreness.

4. H_o: sentence and plea are independent
H_1: sentence and plea are related
$\alpha = .05$
C.R. $\chi^2 > \chi^2_{1,.05} = 3.841$
calculations:

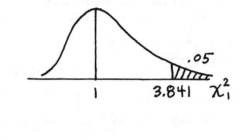

.05

3.841 χ^2_1

PLEA

		G	NG	
	P	392	58	450
SNTCE		(418.48)	(31.52)	
	NP	564	14	578
		(537.52)	(40.48)	
		956	72	1028

$\chi^2 = \Sigma[(O-E)^2/E]$
$= 1.6759 + 22.2518$
$\quad 1.3047 + 17.3241$
$= 42.557$

conclusion:
Reject H_o; there is sufficient evidence to conclude that a person's sentence and his original plea are related.
Yes; these results suggest that a guilty plea should be encouraged.

5. H_o: success on the test and group membership are independent
H_1: success on the test and group membership are related
$\alpha = .05$
C.R. $\chi^2 > \chi^2_{1,.05} = 3.841$
calculations:

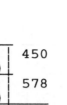

.05

3.841 χ^2_1

TEST

		P	F	
	A	10	14	24
GROUP		(17.49)	(6.51)	
	B	417	145	562
		(409.51)	(152.49)	
		427	159	586

$\chi^2 = \Sigma[(O-E)^2/E]$
$= 3.2062 + 8.6105$
$\quad .1369 + .3677$
$= 12.321$

conclusion:
Reject H_o; there is sufficient evidence to conclude that success on the test and group membership are related.
NOTE: This does not necessarily mean that anything unfair or illegal is occurring (e.g., that the minority group's tests are graded improperly or that the test is deliberately or unintentionally culturally biased).

6. H_o: conviction and prior sentences are independent
H_1: conviction and prior sentences are related
$\alpha = .05$
C.R. $\chi^2 > \chi^2_{3,.05} = 7.815$
calculations:

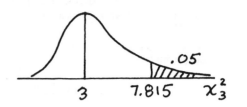

		CONVICTION		
		N	Y	
PRIOR	0	100	672	772
		(99.56)	(677.44)	
	1	56	394	450
		(55.12)	(394.88)	
SNTCE	2	16	173	189
		(23.15)	(165.85)	
	more	11	72	83
		(10.17)	(72.83)	
		183	1311	1494

$$\chi^2 = \Sigma[(O-E)^2/E]$$
$$= .3127 + .0436$$
$$.0140 + .0020$$
$$2.2086 + .3083$$
$$.0683 + .0095$$
$$= 2.967$$

conclusion:
Do not reject H_o; there is not sufficient evidence to conclude that the likelihood of a conviction and the number of prior sentences are related.
No; convictions do not seem to be affected by prior sentences.

7. H_o: wearing a helmet and receiving facial injuries are independent
H_1: wearing a helmet and receiving facial injuries are related
$\alpha = .05$
C.R. $\chi^2 > \chi^2_{1,.05} = 3.841$
calculations:

		HELMET		
		Y	N	
INJR	Y	30	182	212
		(45.11)	(166.89)	
	N	83	236	319
		(67.89)	(251.11)	
		113	418	531

$$\chi^2 = \Sigma[(O-E)^2/E]$$
$$= 5.0640 + 1.3690$$
$$3.3654 + .9098$$
$$= 10.708$$

conclusion:
Reject H_o; there is sufficient evidence to conclude that wearing a helmet and receiving facial injuries are related.
NOTE: There is a statistical relationship, but not necessarily a cause-and-effect relationship. This test does not really address the question "Does a helmet seem to be effective in helping to prevent facial injuries in a crash?" The fact that people who wear helmets receive statistically fewer facial injuries could be due to other factors than the helmet preventing the injury (e.g., people who wear helmets might be safer people who go slower and whose accidents are less serious).

8. H_o: college type and dropout rate are independent
H_1: college type and dropout rate are related
$\alpha = .05$
C.R. $\chi^2 > \chi^2_{3,.05} = 7.815$
calculations:

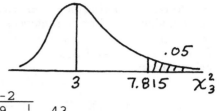

		pub-4	pri-4	pub-2	pri-2	
	Y	10	9	15	9	43
DROPOUT		(10.90)	(11.20)	(9.99)	(10.90)	
	N	26	28	18	27	99
		(25.10)	(25.80)	(23.01)	(25.10)	
		36	37	33	36	142

COLLEGE TYPE (spanning header above pub-4, pri-4, pub-2, pri-2)

$\chi^2 = \Sigma[(O-E)^2/E]$
$= .0745 + .4336 + 2.5098 + .3316$
$.0324 + .1883 + 1.0897 + .1440$
$= 4.803$

conclusion:
Do not reject H_o; there is not sufficient evidence to conclude that a college's type and its dropout rate are related.
No; in this context, the type of college does not appear to be a factor for concern.

9. H_o: having an accident and using a cellular phone are independent
H_1: having an accident and using a cellular phone are related
$\alpha = .05$
C.R. $\chi^2 > \chi^2_{1,.05} = 3.841$
calculations:

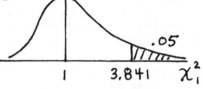

		ACCIDENT		
		Y	N	
	Y	23	282	305
PHONE		(27.76)	(277.24)	
	N	46	407	453
		(41.24)	(411.76)	
		69	689	758

$\chi^2 = \Sigma[(O-E)^2/E]$
$= .8174 + .0819$
$.5503 + .0551$
$= 1.505$

conclusion:
Do not reject H_o; there is not sufficient evidence to conclude that having an accident and using a cellular phone are related.
No; based on these results it does not appear that the use of cellular phones affects driving safety.
NOTE: The exercise's statement that the data are from "a study of randomly selected car accidents and drivers who use cellular phones" does not explain how the 407 persons who did neither appear in the survey. Even rejecting the above hypothesis, however, only is evidence that having accidents and having cellular phones are related, not that the phones cause accidents. Any cause-and-effect relationship could even go the other way: drivers tending to have accidents might tend to get cellular phones to be able to call for help.

10. H_o: amount of smoking and seat belt use are independent
H_1: amount of smoking and seat belt use are related
$\alpha = .05$
C.R. $\chi^2 > \chi^2_{3,.05} = 7.815$
calculations:

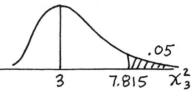

		NUMBER OF CIGARETTES				
		0	1-14	15-34	35+	
	Y	175	20	42	6	243
BELT USE		(171.5)	(19.59)	(43.94)	(7.941)	
	N	149	17	41	9	216
		(152.5)	(17.41)	(39.06)	(7.059)	
		324	37	83	15	459

$$\chi^2 = \Sigma[(O-E)^2/E]$$
$$= .0702 + .0087 + .0858 + .4745$$
$$.0790 + .0097 + .0965 + .5338$$
$$= 1.358$$

conclusion:
Do not reject H_o; there is not sufficient evidence to conclude that amount of smoking and seat belt use are related.
No; the theory receives no support from the sample data.

11. H_o: cause of death and determining agent are independent
H_1: cause of death and determining agent are related
$\alpha = .05$
C.R. $\chi^2 > \chi^2_{3,.05} = 7.815$
calculations:

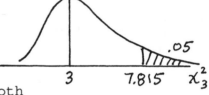

		CAUSE				
		natural	vehicle	suicide	hom/oth	
	DC	102	130	54	140	426
AGENCY		(101.5)	(131)	(57)	(136.5)	
	MP	101	132	60	133	426
		(101.5)	(131)	(57)	(136.5)	
		203	262	114	273	852

$$\chi^2 = \Sigma[(O-E)^2/E]$$
$$= .0025 + .0076 + .1579 + .0897$$
$$.0025 + .0076 + .1579 + .0897$$
$$= .515$$

conclusion:
Do not reject H_o; there is not sufficient evidence to conclude that the cause of death and the determining agency are related.

12. H_o: department and grade distribution are independent
H_1: department and grade distribution are related
$\alpha = .10$
C.R. $\chi^2 > \chi^2_{4,.10} = 7.779$
calculations:

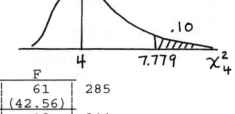

		GRADE					
		A	B	C	D	F	
DEPT	M	49	69	64	42	61	285
		(54.41)	(76.50)	(76.50)	(35.02)	(42.56)	
	P	52	73	78	23	18	244
		(46.59)	(65.50)	(65.50)	(29.98)	(36.44)	
		101	142	142	65	79	529

$$\chi^2 = \Sigma[(O-E)^2/E]$$
$$= .5387 + .7358 + 2.0433 + 1.3917 + 7.9880$$
$$.6292 + .8595 + 2.3867 + 1.6255 + 9.3302$$
$$= 27.529$$

conclusion:
Reject H_o; there is sufficient evidence to conclude that department and grade distribution are related.

13. H_o: smoking and age are independent
H_1: smoking and age are related
$\alpha = .05$
C.R. $\chi^2 > \chi^2_{3,.05} = 7.815$
calculations:

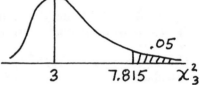

		AGE				
		20-24	25-34	35-44	45-64	
SMOKE	S	18	15	17	15	65
		(16.25)	(16.25)	(16.25)	(16.25)	
	N	32	35	33	35	135
		(33.75)	(33.75)	(33.75)	(33.75)	
		50	50	50	50	200

$$\chi^2 = \Sigma[(O-E)^2/E]$$
$$= .1885 + .0963 + .0346 + .0962$$
$$.0907 + .0463 + .0167 + .0463$$
$$= .615$$

conclusion:
Do not reject H_o; there is not sufficient evidence to conclude that smoking and age are related.
At present, there are not a higher percentage of smokers in one age group than in another.
Targeting cigarette advertising to younger smokers, however, will reach persons who will be around longer and (if consistently successful) will ultimately raise the percentages in all age groups.

14. H_o: age and cooperation are independent
H_1: age and cooperation are related
$\alpha = .01$
C.R. $\chi^2 > \chi^2_{5,.01} = 15.086$
calculations:

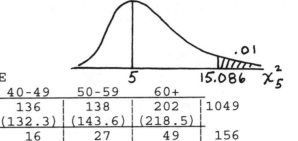

| | | \multicolumn{6}{c}{AGE} | |
		18-21	22-29	30-39	40-49	50-59	60+	
CO-OP	Y	73	255	245	136	138	202	1049
		(73.13)	(239.4)	(242.0)	(132.3)	(143.6)	(218.5)	
	N	11	20	33	16	27	49	156
		(10.87)	(35.6)	(36.0)	(19.7)	(21.4)	(33.5)	
		84	275	278	152	165	251	1205

$$\chi^2 = \Sigma[(O-E)^2/E]$$
$$= .0002 + 1.0168 + .0369 + .1022 + .2214 + 1.2468$$
$$.0014 + 6.8371 + .2484 + .6875 + 1.4886 + 8.3838$$
$$= 20.271$$

conclusion:
Reject H_o; there is sufficient evidence to conclude that a person's age and level of cooperation are related.

15. H_o: time and location of fatal accidents are independent
H_1: time and location of fatal accidents are related
$\alpha = .10$
C.R. $\chi^2 > \chi^2_{7,.10} = 12.017$
calculations:

| LOC | \multicolumn{8}{c}{TIME} | |
	1-4	4-7	7-10	10-1	1-4	4-7	7-10	10-1	
nyc	73	60	53	68	80	67	87	81	569
	(70.68)	(47.58)	(51.38)	(62.26)	(75.03)	(88.08)	(88.08)	(85.91)	
oth	187	115	136	161	196	257	237	235	1524
	(189.32)	(127.42)	(137.62)	(166.74)	(200.97)	(235.92)	(235.92)	(230.09)	
	260	175	189	229	276	324	324	326	2093

$$\chi^2 = \Sigma[(O-E)^2/E]$$
$$= .0759 + 3.2448 + .0510 + .5300 + .3288 + 5.0459 + .0133 + .2803$$
$$.0284 + 1.2115 + .0190 + .1979 + .1228 + 1.8840 + .0050 + .1047$$
$$= 13.143$$

conclusion:
Reject H_o; there is sufficient evidence to conclude that the time and the location of fatal accidents are related.

16. H_o: product opinion and geographic region are independent
H_1: product opinion and geographic region are related
$\alpha = .01$
C.R. $\chi^2 > \chi^2_{4,.01} = 13.277$
calculations:

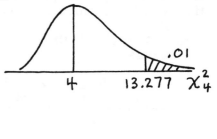

		OPINION			
		L	D	U	
	N	30	15	15	60
		(20.43)	(26.81)	(12.77)	
REGION	S	10	30	20	60
		(20.43)	(26.81)	(12.77)	
	W	40	60	15	115
		(39.15)	(51.38)	(24.47)	
		80	100	50	235

$$\chi^2 = \Sigma[(O\text{-}E)^2/E]$$
$$= 4.4889 + 5.2014 + .3910$$
$$5.3214 + .3799 + 4.0933$$
$$.0185 + 1.4461 + 3.6637$$
$$= 25.009$$

conclusion:
   Reject H_o; there is sufficient evidence to conclude that product opinion and geographic region are related.
   No; based on the above results it would probably not be wise to use the same marketing strategy in each region.

17. Since $25.009 > 14.869$
   [i.e., $25.009 > \chi^2_{4,.005}$ -- the largest χ^2 in the row]
   then P-value $< .005$.

18. H_o: having an accident and using a cellular phone are independent
   H_1: having an accident and using a cellular phone are related
   $\alpha = .05$
   C.R. $\chi^2 > \chi^2_{1,.05} = 3.841$
   calculations:

		ACCIDENT		
		Y	N	
	Y	23	282	305
PHONE		(27.76)	(277.24)	
	N	46	407	453
		(41.24)	(411.76)	
		69	689	758

$$\chi^2 = \Sigma[(\,|\,O\text{-}E\,|\,-.5)^2/E]$$
$$= .6548 + .0656$$
$$.4409 + .0442$$
$$= 1.205$$

conclusion:
   Do not reject H_o; there is not sufficient evidence to conclude that having an accident and using a cellular phone are related.
   Without the correction for continuity [see the solution for exercise #9 for the details] the calculated test statistic is 1.505. Since $(\,|\,O\text{-}E\,|\,-.5)^2 < (O\text{-}E)^2$ whenever $|\,O\text{-}E\,| > .25$, Yates' correction generally lowers the calculated test statistic.

19. Multiplying each O value by positive constant k multiplies all the totals (and hence each E value) by that same positive constant k. The calculated statistic becomes

$$
\begin{aligned}
(\chi^2 \text{ new}) &= \Sigma[(kO\text{-}kE)^2/kE] \\
&= \Sigma[k^2(O\text{-}E)^2/kE] \\
&= \Sigma[k(O\text{-}E)^2/E] \\
&= k \cdot \Sigma[(O\text{-}E)^2/E] \\
&= k \cdot (\chi^2 \text{ old})
\end{aligned}
$$

Review Exercises

1. H_o: $p_{Mon} = p_{Tue} = p_{Wed} = \ldots = p_{Sun} = 1/7$
   H_1: at least one of the proportions is different from 1/7
   $\alpha = .05$
   C.R. $\chi^2 > \chi^2_{6,.05} = 12.592$
   calculations:

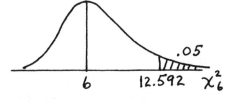

day	O	E	$(O\text{-}E)^2/E$
Mon	74	66.286	.8978
Tue	60	66.286	.5961
Wed	66	66.286	.0012
Thu	71	66.286	.3353
Fri	51	66.286	3.5249
Sat	66	66.286	.0012
Sun	76	66.286	1.4236
	464	464.000	6.7801

$\chi^2 = \Sigma[(O\text{-}E)^2/E]$
$= 6.780$

conclusion:
Do not reject H_o; there is not sufficient evidence to conclude that at least one of the proportions is different from 1/7.

2. H_o: $p_{bro} = .22$, $p_{whi} = .22$, $p_{ltb} = .20$, $p_{dkb} = .18$, $p_{red} = .18$
   H_1: at least one of the proportions is not as claimed
   $\alpha = .05$
   C.R. $\chi^2 > \chi^2_{4,.05} = 9.488$
   calculations:

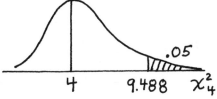

color	O	E	$(O\text{-}E)^2/E$
bro	60	59.4	.0061
whi	61	59.4	.0431
ltb	43	54.0	2.2407
dkb	41	48.6	1.1885
red	65	48.6	5.5342
	270	270.0	9.0126

$\chi^2 = \Sigma[(O\text{-}E)^2/E]$
$= 9.013$

conclusion:
Do not reject H_o; there is not sufficient evidence to conclude that at least one of the proportions is not as claimed.

3. H_o: $p_{Jan} = p_{Feb} = p_{Mar} = \ldots = p_{Dec} = 1/12$
   H_1: at least one of the proportions is different from 1/12
   $\alpha = .05$
   C.R. $\chi^2 > \chi^2_{11,.05} = 19.675$
   calculations:

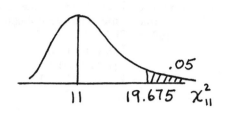

month	O	E	$(O-E)^2/E$
Jan	8	10	.400
Feb	12	10	.400
Mar	9	10	.100
Apr	15	10	2.500
May	6	10	1.600
Jun	12	10	.400
Jul	4	10	3.600
Aug	7	10	.900
Sep	11	10	.100
Oct	11	10	.100
Nov	5	10	2.500
Dec	20	10	10.000
	120	120	22.600

$\chi^2 = \Sigma[(O-E)^2/E]$
$= 22.600$

conclusion:
   Reject H_o; there is sufficient evidence to conclude that at least one of the proportions is different from 1/12.
   No; the fact that some months were offered as answers more than other months says nothing about how often the offered answers were correct and whether or not the subjects have ESP.

4. H_o: headache occurrence and drug usage are independent
   H_1: headache occurrence and drug usage are related
   $\alpha = .05$
   C.R. $\chi^2 > \chi^2_{2,.05} = 5.991$
   calculations:

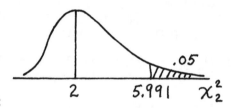

		DRUG			
		Seldane	Placebo	Control	
	Y	49	49	24	122
HDCHE		(45.99)	(39.16)	(36.86)	
	N	732	616	602	1950
		(735.01)	(625.84)	(589.14)	
		781	665	626	2072

$\chi^2 = \Sigma[(O-E)^2/E]$
$= .1976 + 2.4752 + 4.4862$
$ .0124 + .1549 + .2807$
$= 7.607$

conclusion:
   Reject H_o; there is sufficient evidence to conclude that headache occurrence and drug usage are related.
   Careful examination of the table indicates that while Seldane users have more headaches than those taking no medication, they have fewer headaches than those taking a placebo. This suggests that Seldane usage is associated with but not necessarily responsible for the headaches.

5. H_o: county of violation and type of violation are independent
H_1: county of violation and type of violation are related
$\alpha = .05$
C.R. $\chi^2 > \chi^2_{6,.05} = 12.592$
calculations:

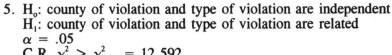

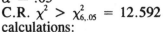

		COUNTY			
		A	M	O	W	
	dwi	6	19	6	10	41
		(5.76)	(14.74)	(6.51)	(13.99)	
TYPE	spd	103	261	160	226	750
		(105.45)	(269.55)	(119.18)	(255.82)	
	dtd	60	152	25	174	411
		(57.79)	(147.71)	(65.31)	(140.19)	
		169	432	191	410	1202

$\chi^2 = \Sigma[(O-E)^2/E]$
$= .0096 + 1.2342 + .0407 + 1.1355$
$\quad .0569 + .2712 + 13.9841 + 3.4768$
$\quad .0848 + .1244 + 24.8786 + 8.1533$
$= 53.450$

conclusion:
Reject H_o; there is sufficient evidence to conclude that the county of the violation and the type of the violation are related.

6. H_o: for correct responses, sensory state and ball type are independent
H_1: for correct responses, sensory state and ball type are related
$\alpha = .05$
C.R. $\chi^2 > \chi^2_{6,.05} = 12.592$
calculations:

		BALL TYPE			
		B-3	S-3	S-2	
	N	18	19	16	53
SNSRY		(20.63)	(17.19)	(15.18)	
	D	19	18	9	46
		(17.90)	(14.92)	(13.18)	
	B	19	12	19	50
STATE		(19.46)	(16.22)	(14.32)	
	DB	16	11	9	36
		(14.01)	(11.68)	(10.31)	
		72	60	53	185

$\chi^2 = \Sigma[(O-E)^2/E]$
$= .3346 + .1908 + .0439$
$\quad .0673 + .6363 + 1.3248$
$\quad .0108 + 1.0962 + 1.5262$
$\quad .2824 + .0391 + .1673$
$= 5.720$

conclusion:
Do not reject H_o; there is not sufficient evidence to conclude that, for correct responses, sensory state and ball type are related.
Advertising claims of a different "feel" cannot be evaluated with the given information. Assuming having a different "feel" means balls will be correctly identified, the relevant information is the proportion correct responses. Whether the "feel" of some balls is more often detected in different sensory states does not address the question of whether some balls have a different "feel."

7. H_o: $p_0 = .80$, $p_1 = .16$, $p_{more} = .04$
H_1: at least one of the proportions is different from what is claimed
$\alpha = .05$
C.R. $\chi^2 > \chi^2_{2,.05} = 5.991$
calculations:

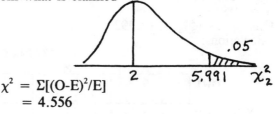

count	O	E	$(O-E)^2/E$
0	172	160	.9000
1	23	32	2.5312
more	5	8	1.1250
	200	200	4.5562

$\chi^2 = \Sigma[(O-E)^2/E]$
$= 4.556$

conclusion:
Do not reject H_o; there is not sufficient evidence to conclude that at least one of the proportions is different from what is claimed.

8. H_o: $p_0 = p_1 = p_2 = \ldots = p_9 = .10$
H_1: at least one of the proportions is different from .10
$\alpha = .05$ [or any other appropriate value]
C.R. $\chi^2 > \chi^2_{9,.05} = 16.919$
calculations:

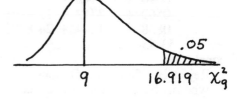

digit	O	E	$(O-E)^2/E$
0	10	18	3.556
1	29	18	6.722
2	22	18	.888
3	28	18	5.556
4	19	18	.056
5	12	18	2.000
6	15	18	.500
7	16	18	.222
8	13	15	1.389
9	16	18	.222
	180	180	21.111

$\chi^2 = \Sigma[(O-E)^2/E]$
$= 21.111$

conclusion:
Reject H_o; there is sufficient evidence to conclude that at least one of the proportions is different from .10.
No; the students were not successful in choosing their own random numbers. The above result is consistent with the fact that people asked to give random numbers tend to supply fewer zero digits than predicted by chance.

Chapter 11

Analysis of Variance

11-2 One-Way ANOVA with Equal Sample Sizes

NOTE: This section is calculation-oriented. Do not get so involved with the formulas that you miss concepts. This manual arranges the calculations to promote both computational efficiency and understanding of the underlying principles. The following notation is used in this section.

k = the number of groups
n = the number of scores in each group
$\bar{x}_i$ = the mean of group i (where i = 1,2,...,k)
s_i^2 = the variance of group i (where i = 1,2,...,k)
$\bar{\bar{x}} = \Sigma\bar{x}_i/k$ = the mean of the group means
          = the overall mean of all the scores in all the groups
$s_{\bar{x}}^2 = \Sigma(\bar{x}_i-\bar{\bar{x}})^2/(k-1)$ = the variance of the group means
$s_p^2 = \Sigma s_i^2/k$ = the mean of the group variances

* Exercise 1 is worked in complete detail, showing even work done on the calculator without having to be written down. Subsequent exercises are worked showing all intermediate steps, but without writing down detail for routine work done on the calculator.

* While the manual typically shows only three decimal places for the intermediate steps, all decimal places were carried in the calculator. DO NOT ROUND OFF INTERMEDIATE ANSWERS. SAVE calculated values that will be used again. See your instructor or class assistant if you need help using your calculator accurately and efficiently.

1. The following preliminary values are identified and/or calculated.
   $k = 3$ numerator df = k-1 = 2
   $n = 5$ denominator df = k(n-1) = 3(4) = 12
   $\bar{\bar{x}} = \Sigma\bar{x}_i/k = (98.940 + 98.580 + 97.800)/3 = 98.440$
   $s_{\bar{x}}^2 = \Sigma(\bar{x}_i-\bar{\bar{x}})^2/(k-1) = [(98.940-98.440)^2 + (98.580-98.440)^2 + (97.800-98.400)^2]/2$
   $\qquad = [(.500)^2 + (.140)^2 + (-.640)^2]/2 = .6792/2 = .3396$
   $s_p^2 = \Sigma s_i^2/k = [(.568)^2 + (.701)^2 + (.752)^2]/3 = 1.379529/3 = .459843$

   $H_o: \mu_1 = \mu_2 = \mu_3$
   $H_1:$ at least one mean is different
   $\alpha = .05$
   C.R. $F > F_{12,.05}^2 = 3.8853$
   calculations:
   $\quad F = ns_{\bar{x}}^2/s_p^2$
   $\qquad = 5(.3396)/.459843 = 3.6926$
   conclusion:
   Do not reject H_o; there is not sufficient evidence to conclude that at least one mean is different.

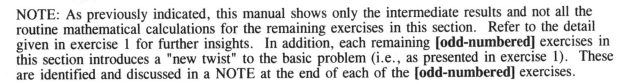

NOTE: As previously indicated, this manual shows only the intermediate results and not all the routine mathematical calculations for the remaining exercises in this section. Refer to the detail given in exercise 1 for further insights. In addition, each remaining [odd-numbered] exercises in this section introduces a "new twist" to the basic problem (i.e., as presented in exercise 1). These are identified and discussed in a NOTE at the end of each of the [odd-numbered] exercises.

2. The following preliminary values are identified and/or calculated.

$k = 3$

$n = 25$

$s_{\bar{x}}^2 = \Sigma(\bar{x}_i-\bar{\bar{x}})^2/(k-1)$

$\quad = 7.98/2 = 3.99$

$\bar{\bar{x}} = \Sigma\bar{x}_i/k$

$\quad = 78.9$

$s_p^2 = \Sigma s_i^2/k$

$\quad = 732.24/3 = 244.04$

$H_o: \mu_1 = \mu_2 = \mu_3$

$H_1:$ at least one mean is different

$\alpha = .05$

C.R. $F > F_{72,.05}^2 = 3.1504$

calculations:

$\quad F = ns_{\bar{x}}^2/s_p^2$

$\quad\quad = 25(3.99)/244.08 = .4087$

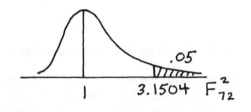

conclusion:

Do not reject H_o; there is not sufficient evidence to conclude that at least one mean is different.

3. The following preliminary values are identified and/or calculated.

$k = 4$

$n = 62$

$s_{\bar{x}}^2 = \Sigma(\bar{x}_i-\bar{\bar{x}})^2/(k-1)$

$\quad = 36.636/3 = 12.212$

$\bar{\bar{x}} = \Sigma\bar{x}_i/k$

$\quad = 4.327$

$s_p^2 = \Sigma s_i^2/k$

$\quad = 29.356/4 = 7.339$

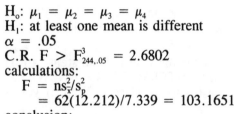

$H_o: \mu_1 = \mu_2 = \mu_3 = \mu_4$

$H_1:$ at least one mean is different

$\alpha = .05$

C.R. $F > F_{244,.05}^3 = 2.6802$

calculations:

$\quad F = ns_{\bar{x}}^2/s_p^2$

$\quad\quad = 62(12.212)/7.339 = 103.1651$

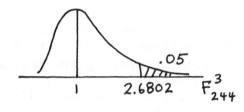

conclusion:

Reject H_o; there is sufficient evidence to conclude that at least one mean is different.

NOTE: The "new twist" is the fact that the accumulation of relatively large sample sizes for several groups pushes the denominator degrees of freedom [df = k(n-1)] past 120, the highest finite entry in the F table. When this occurs, we follow the pattern of using the closest entry and [since 120 is closer to any finite number than ∞ (i.e., infinity)] choose the entry for df = 120.

4. The following preliminary values are identified and/or calculated.

$k = 3$

$n = 5$

$s_{\bar{x}}^2 = \Sigma(\bar{x}_i-\bar{\bar{x}})^2/(k-1)$

$\quad = .8936/2 = .4468$

$\bar{\bar{x}} = \Sigma\bar{x}_i/k$

$\quad = 27.92$

$s_p^2 = \Sigma s_i^2/k$

$\quad = 4.96/3 = 1.653$

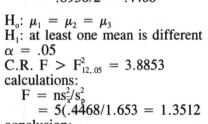

$H_o: \mu_1 = \mu_2 = \mu_3$

$H_1:$ at least one mean is different

$\alpha = .05$

C.R. $F > F_{12,.05}^2 = 3.8853$

calculations:

$\quad F = ns_{\bar{x}}^2/s_p^2$

$\quad\quad = 5(.4468/1.653 = 1.3512$

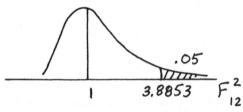

conclusion:

Do not reject H_o; there is not sufficient evidence to conclude that at least one mean is different.

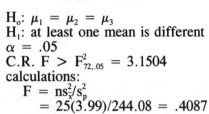

5. The following preliminary values are identified and/or calculated.

$k = 5$ $\bar{\bar{x}} = \Sigma\bar{x}_i/k$

$n = 10$ $= 102.4$

$s_{\bar{x}}^2 = \Sigma(\bar{x}_i - \bar{\bar{x}})^2/(k-1)$ $s_p^2 = \Sigma s_i^2/k$

$= 93.20/4 = 23.30$ $= 755/5 = 151$

$H_o: \mu_1 = \mu_2 = \mu_3 = \mu_4 = \mu_5$
$H_1:$ at least one mean is different
$\alpha = .05$
C.R. $F > F_{45,.05}^4 = 2.6060$
calculations:

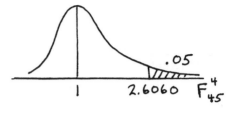

$F = ns_{\bar{x}}^2/s_p^2$

$= 10(23.30)/151 = 1.5430$

conclusion:

Do not reject H_o; there is not sufficient evidence to conclude that at least one mean is different.

NOTE: The "new twist" is the fact that the variability for each group was given in terms of variance (i.e., s^2) and not standard deviation (i.e., s). This means that the given values do not have to be squared again when finding $s_p^2 = \Sigma s_i^2/k$. In problems for which the summary statistics are not given (and for which the reader must make all calculations from the raw data), calculating s^2 instead of s (since it is s^2 that is utilized in subsequent calculations) avoids extra work and the possible introduction of round-off errors due to taking the square root.

6. The following preliminary values are identified and/or calculated.

$k = 5$ $\bar{\bar{x}} = \Sigma\bar{x}_i/k$

$n = 20$ $= 73.8$

$s_{\bar{x}}^2 = \Sigma(\bar{x}_i - \bar{\bar{x}})^2/(k-1)$ $s_p^2 = \Sigma s_i^2/k$

$= 20.8/4 = 5.2$ $= 286/5 = 57.2$

$H_o: \mu_1 = \mu_2 = \mu_3 = \mu_4 = \mu_5$
$H_1:$ at least one mean is different
$\alpha = .05$
C.R. $F > F_{95,.05}^4 = 2.4472$
calculations:

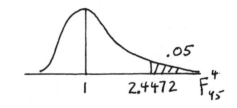

$F = ns_{\bar{x}}^2/s_p^2$

$= 20(5.2)/57.2 = 1.8182$

conclusion:

Do not reject H_o; there is not sufficient evidence to conclude that at least one mean is different.

7. The following preliminary values are identified and/or calculated.

	A	B	C	D	E
n	4	4	4	4	4
Σx	72	76	76	84	68
Σx^2	1318	1478	1448	1770	1170
$\bar{x}$	18	19	19	21	17
s^2	7.333	11.333	1.333	2.000	4.667

$k = 5$ $\bar{\bar{x}} = \Sigma\bar{x}_i/k$

$n = 4$ $= 18.8$

$s_{\bar{x}}^2 = \Sigma(\bar{x}_i - \bar{\bar{x}})^2/(k-1)$ $s_p^2 = \Sigma s_i^2/k$

$= 8.80/4 = 2.20$ $= 26.667/5 = 5.333$

H_o: $\mu_A = \mu_B = \mu_C = \mu_D = \mu_E$
H_1: at least one mean is different
$\alpha = .05$
C.R. $F > F^4_{15,.05} = 3.0556$
calculations:
$\quad F = ns^2_{\bar{x}}/s^2_p$
$\qquad = 4(2.20)/5.333 = 1.6500$
conclusion:

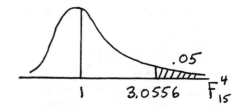

Do not reject H_o; there is not sufficient evidence to conclude that at least one mean is different.

NOTE: The "new twist" is the fact that the group means and variances are not given and need to be calculated from the raw data.

8. The following preliminary values are identified and/or calculated.

	1: 2-par	2: 1-par	3: trans
n	5	5	5
Σx	680	565	540
Σx^2	92850	64275	59150
$\bar{x}$	136	113	108
s^2	92.5	107.5	207.5

$k = 3$
$n = 5$
$s^2_{\bar{x}} = \Sigma(\bar{x}_i - \bar{\bar{x}})^2/(k-1)$
$\quad = 446/2 = 223$

$\bar{\bar{x}} = \Sigma\bar{x}_i/k$
$\quad = 119$
$s^2_p = \Sigma s^2_i/k$
$\quad = 407.5/3 = 135.83$

H_o: $\mu_1 = \mu_2 = \mu_3$
H_1: at least one mean is different
$\alpha = .05$
C.R. $F > F^2_{12,.05} = 3.8853$
calculations:
$\quad F = ns^2_{\bar{x}}/s^2_p$
$\qquad = 5(223)/135.83 = 8.2086$
conclusion:

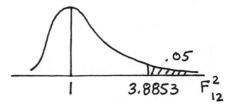

Reject H_o; there is sufficient evidence to conclude that at least one mean is different.

9. The following preliminary values are identified and/or calculated.

	1	4	7
n	10	10	10
Σx	1493	1445	1387
Σx^2	225,541	236,287	210,189
$\bar{x}$	149.3	144.5	138.7
s^2	292.900	3053.833	1979.122

$k = 3$
$n = 10$
$s^2_{\bar{x}} = \Sigma(\bar{x}_i - \bar{\bar{x}})^2/(k-1)$
$\quad = 56.347/2 = 28.173$

$\bar{\bar{x}} = \Sigma\bar{x}_i/k$
$\quad = 144.167$
$s^2_p = \Sigma s^2_i/k$
$\quad = 5325.856/3 = 1775.285$

H_o: $\mu_1 = \mu_4 = \mu_7$
H_1: at least one mean is different
$\alpha = .05$
C.R. $F > F^2_{27,.05} = 3.3541$
calculations:
$\quad F = ns^2_{\bar{x}}/s^2_p$
$\quad\quad = 10(28.173)/1775.185 = .1587$

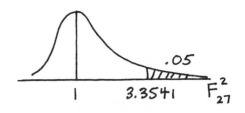

conclusion:
Do not reject H_o; there is not sufficient evidence to conclude that at least one mean is different.
No, no zone seems to have homes with higher (or lower) selling prices.

NOTE: The "new twist" is the fact that the selling prices (whose overall average is \$144,167) can be analyzed in terms of 1000's of dollars instead of dollars. This is so because dividing all the selling prices by 1000 also divides all the group means by 1000. The variances (of the sample means and the pooled within group estimate) in the numerator and the denominator of the F statistic are each divided by 1000^2, and the F statistic is precisely what it would have been had we worked with dollars instead of \$1000's.

10. The following preliminary values are identified and/or calculated.

	1	4	7
n	10	10	10
Σx	185	161	172
Σx^2	3581	2887	3064
$\bar{x}$	18.5	16.1	17.2
s^2	17.611	32.766	11.733

$k = 3$ $\quad\quad\quad\quad\quad\quad \bar{\bar{x}} = \Sigma\bar{x}_i/k$
$n = 10$ $\quad\quad\quad\quad\quad\quad\quad = 17.267$
$s^2_{\bar{x}} = \Sigma(\bar{x}_i - \bar{\bar{x}})^2/(k-1) \quad s^2_p = \Sigma s^2_i/k$
$\quad = 2.8867/2 = 1.4433 \quad\quad = 62.1111/3 = 20.7037$

H_o: $\mu_1 = \mu_4 = \mu_7$
H_1: at least one mean is different
$\alpha = .05$
C.R. $F > F^2_{27,.05} = 3.3541$
calculations:
$\quad F = ns^2_{\bar{x}}/s^2_p$
$\quad\quad = 10(1.4433)/20.7037 = .6971$

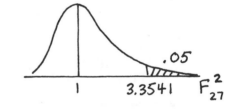

conclusion:
Do not reject H_o; there is not sufficient evidence to conclude that at least one mean is different.
No; no one zone seems to have larger homes.

11. The following preliminary values are identified and/or calculated.

	1	4	7
n	10	10	10
Σx	8.65	5.71	21.39
Σx^2	9.4507	5.0457	78.9271
$\bar{x}$	.865	.571	2.139
s^2	.219	.198	3.686

$k = 3$ $\quad\quad\quad\quad\quad\quad \bar{\bar{x}} = \Sigma\bar{x}_i/k$
$n = 10$ $\quad\quad\quad\quad\quad\quad\quad = 1.192$
$s^2_{\bar{x}} = \Sigma(\bar{x}_i - \bar{\bar{x}})^2/(k-1) \quad s^2_p = \Sigma s^2_i/k$
$\quad = 1.389/2 = .695 \quad\quad\quad = 4.103/3 = 1.368$

H_o: $\mu_1 = \mu_4 = \mu_7$
H_1: at least one mean is different
$\alpha = .05$
C.R. $F > F^2_{27,.05} = 3.3541$
calculations:
   $F = ns^2_{\bar{x}}/s^2_p$
      $= 10(.695)/1.368$
      $= 5.0793$
conclusion:

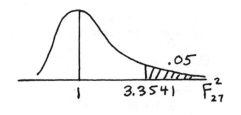

   Reject H_o; there is sufficient evidence to conclude that at least one mean is different.
Yes, one zone (viz., zone 7) does appear to have larger lots.

NOTE: The "new twist" is the fact that the author expects a specific deduction on the basis of rejecting the null hypothesis. In truth, the only conclusion possible from the ANOVA test is the one stated, that "at least one mean is different." Considering the ordering of the sample means, there are three distinct possibilities: $\mu_7 > \mu_1 > \mu_4$, $\mu_7 > \mu_1 = \mu_4$, $\mu_7 = \mu_1 > \mu_4$. There are further statistical procedures that can be employed to determine which of the preceding cases should be concluded, but they are beyond the scope of the text.

NOTE ALSO: One of the assumptions of the ANOVA test is that all the groups have the same variance (and that common variance is estimated by s^2_p -- the average of the group variances). In this exercise, one of the sample variances is almost 20 times larger than the others, and it is questionable whether the population variances are truly equal. This means that perhaps the ANOVA test is not appropriate and that a professional statistician should be consulted about a more advanced procedure.

12. The following preliminary values are identified and/or calculated.

	1	4	7
n	10	10	10
Σx	17.4	22.7	19.9
Σx^2	31.24	58.43	41.21
$\bar{x}$	1.74	2.27	1.99
s^2	.10711	.76677	.17877

$k = 3$ $\bar{\bar{x}} = \Sigma\bar{x}_i/k$
$n = 10$ $\quad\; = 2.00$
$s^2_{\bar{x}} = \Sigma(\bar{x}_i - \bar{\bar{x}})^2/(k-1)$ $s^2_p = \Sigma s^2_i/k$
$\quad\; = .1406/2 = .0703$ $\quad\; = 1.05267/3 = .3509$

H_o: $\mu_1 = \mu_4 = \mu_7$
H_1: at least one mean is different
$\alpha = .05$
C.R. $F > F^2_{27,.05} = 3.3541$
calculations:
   $F = ns^2_{\bar{x}}/s^2_p$
      $= 10(.0703)/.3509 = 2.0035$
conclusion:

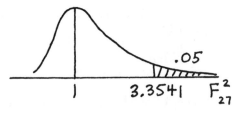

   Do not reject H_o; there is not sufficient evidence to conclude that at least one mean is different.
No; no one zone seems to have higher taxes.

13. a. $n_1 > 30$ and $n_2 > 30$, use z (with s's for σ's): $\bar{x}_1 - \bar{x}_2 = 197 - 202 = -5$

    $H_0: \mu_1 - \mu_2 = 0$
    $H_1: \mu_1 - \mu_2 \neq 0$
    $\alpha = .05$
    C.R. $z < -z_{.025} = -1.960$
         $z > z_{.025} = 1.960$
    calculations:

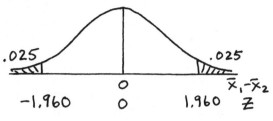

$$z_{\bar{x}_1 - \bar{x}_2} = (\bar{x}_1 - \bar{x}_2 - \mu_{\bar{x}_1 - \bar{x}_2})/\sigma_{\bar{x}_1 - \bar{x}_2}$$
$$= (-5 - 0)/\sqrt{(18)^2/50 + (20)^2/50}$$
$$= -5/3.805 = -1.314$$

    conclusion:
        Do not reject H_0; there is not sufficient evidence to conclude that $\mu_1 - \mu_2 \neq 0$.

b. $n_2 > 30$ and $n_3 > 30$, use z (with s's for σ's): $\bar{x}_2 - \bar{x}_3 = 202 - 208 = -6$

    $H_0: \mu_2 - \mu_3 = 0$
    $H_1: \mu_2 - \mu_3 \neq 0$
    $\alpha = .05$
    C.R. $z < -z_{.025} = -1.960$
         $z > z_{.025} = 1.960$
    calculations:

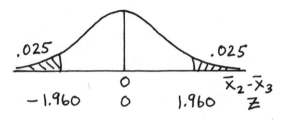

$$z_{\bar{x}_2 - \bar{x}_3} = (\bar{x}_2 - \bar{x}_3 - \mu_{\bar{x}_2 - \bar{x}_3})/\sigma_{\bar{x}_2 - \bar{x}_3}$$
$$= (-6 - 0)/\sqrt{(20)^2/50 + (23)^2/50}$$
$$= -6/4.310 = -1.392$$

    conclusion:
        Do not reject H_0; there is not sufficient evidence to conclude that $\mu_1 - \mu_2 \neq 0$.

c. $n_1 > 30$ and $n_3 > 30$, use z (with s's for σ's): $\bar{x}_1 - \bar{x}_3 = 197 - 208 = -11$

    $H_0: \mu_1 - \mu_3 = 0$
    $H_1: \mu_1 - \mu_3 \neq 0$
    $\alpha = .05$
    C.R. $z < -z_{.025} = -1.960$
         $z > z_{.025} = 1.960$
    calculations:

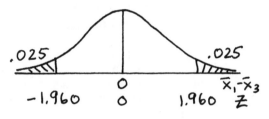

$$z_{\bar{x}_1 - \bar{x}_3} = (\bar{x}_1 - \bar{x}_3 - \mu_{\bar{x}_1 - \bar{x}_3})/\sigma_{\bar{x}_1 - \bar{x}_3}$$
$$= (-11 - 0)/\sqrt{(18)^2/50 + (23)^2/50}$$
$$= -11/4.130 = -2.663$$

    conclusion:
        Reject H_0; there is sufficient evidence to conclude that $\mu_1 - \mu_2 \neq 0$ (in fact, $\mu_1 - \mu_3 < 0$).

d. The following preliminary values are identified and/or calculated.

    $k = 3$ $\bar{\bar{x}} = \Sigma\bar{x}_i/k$
    $n = 50$ $= 202.333$
    $s_{\bar{x}}^2 = \Sigma(\bar{x}_i - \bar{\bar{x}})^2/(k-1)$ $s_p^2 = \Sigma s_i^2/k$
        $= 60.667/2 = 30.333$ $= 1253/3 = 417.667$

    $H_0: \mu_1 = \mu_2 = \mu_3$
    $H_1:$ at least one mean is different
    $\alpha = .05$
    C.R. $F > F_{147,.05}^2 = 3.0718$
    calculations:

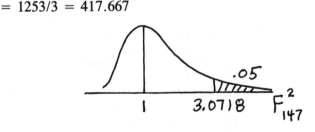

       $F = ns_{\bar{x}}^2/s_p^2$
           $= 50(30.333)/417.667 = 3.6313$

    conclusion:
        Reject H_0; there is sufficient evidence to conclude that at least one mean is different.

e. This is a complicated question whose complete answer requires more than can be ascertained from this exercise and the scope of this text. Each approach has its advantages.

In support of the approach of part (d), it may be said that this approach
* seems more efficient, since it requires only a single test.
* provides 95% confidence. The approach of parts (a),(b),(c) involves three tests, each of which gives 95% confidence separately -- but 95% confidence in three different tests leads to an overall confidence less than 95%. [For independent tests (which these are <u>not</u>), for example, the overall confidence rate would be $(.95)^3 = .857 = 85.7\%$.

In support of the approach of parts (a),(b),(c), it may be said that this approach
* can use either one-tailed or two-tailed tests.
* can identify precisely which means are significantly different and which are not, while the approach of part (d) can only say that "at least one mean is different."
* seems more certain. The -2.663 of part (c) was farther into the critical region determined by -1.960 than the 3.6313 of part (d) was into the critical region determined by 3.0718. For part (c), P-value = 2(.0039) = .0078; for part (d), .01 < P-value < .025. The approach of part (c) seems to give stronger evidence of a difference.

Because the advantages given above for the approach of part (d) are significant, and because the limited general conclusion of the approach of part (d) can be improved by additional procedures (beyond the scope of the text), that approach is generally considered the better one.

NOTE: The "new twist" is the fact that two different methods (those of chapter 8 and those of this chapter) can be used to analyze the same problem and that these different methods each have their own advantages and disadvantages.

14. a. $_5C_2 = 5!/2!3! = 10$
b. P(no type I error in one t test) = .95
assuming independence, P(no Type I error in ten t tests) = $(.95)^{10} = .599$
c. P(no type I error in one F test) = .95
d. the analysis of variance test

15. a. Adding a constant to each of the sample means (because, it is assumed, that constant was added to each of the individual scores) does not change the calculated $F = ns_{\bar{x}}^2/s_p^2$. Since all the means increase by the same amount, the variance among them (i.e., $s_{\bar{x}}^2$) is not affected; since all the individual scores increase by the same amount, the group variances and hence the average group variance (i.e., s_p^2) is not affected.

b. This question has two possible interpretations:
* Multiplying each sample mean by a constant (because, it is assumed, each individual scores was multiplied by that constant) does not change the calculated $F = ns_{\bar{x}}^2/s_p^2$. Since all the means are multiplied by a constant, the variance among them (i.e., $s_{\bar{x}}^2$) is multiplied by the square of that constant; since all the individual scores are multiplied by a constant, the group variances and hence the average group variance (i.e., s_p^2) is multiplied by the square of that constant. Since both the numerator and the denominator are multiplied by the same amount, the calculated F is not affected.
* Multiplying each sample mean by a constant (because, it is assumed, we were drawing from populations whose means were a constant multiple of the original means but whose variances were the same as the original variances) multiples the calculated $F = ns_{\bar{x}}^2/s_p^2$ by the square of that constant. Since all the means are multiplied by a constant, the variance among them (i.e., $s_{\bar{x}}^2$) is multiplied by the square of that constant; since variance of the individual scores is not changed, neither is s_p^2. Since the numerator is multiplied by the square of that constant and the denominator is not affected, the calculated F is multiplied by the square of that constant.

c. If the 5 samples have the same mean, then there is no variability among the means and $s_{\bar{x}}^2 = 0$. Assuming there was variability among the individual scores, $s_p^2 \neq 0$. Since the numerator is 0 and the denominator is not, the calculated F statistic equals 0.

NOTE: The "new twist" is the fact that questions are posed without specific numerical data. This tests both the reader's ability to reason abstractly and his understanding of the underlying statistical principles.

16. The following preliminary values are identified and/or calculated.

	1	4
n	10	10
Σx	1493	1445
Σx^2	225,541	236,287
$\bar{x}$	149.3	144.5
s^2	292.900	3053.833

a. $n_1 \leq 30$ and $n_4 \leq 30$, use t (assume $\sigma_1 = \sigma_4$ and use s_p^2)
   $\bar{x}_1 - \bar{x}_4 = 149.3 - 144.5 = 4.8$
   $s_p^2 = [9(292.000) + 9(2053.833)]/(9 + 9) = 1673$

   $H_o: \mu_1 - \mu_4 = 0$
   $H_1: \mu_1 - \mu_4 \neq 0$
   $\alpha = .05$
   C.R. $t < -t_{18,.025} = -2.101$
       $t > t_{18,.025} = 2.101$
   calculations:

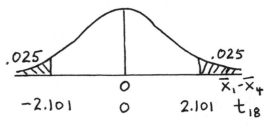

   $t_{\bar{x}_1 - \bar{x}_4} = (\bar{x}_1 - \bar{x}_4 - \mu_{\bar{x}_1 - \bar{x}_4})/s_{\bar{x}_1 - \bar{x}_4}$
       $= (4.8 - 0)/\sqrt{1673/10 + 1673/10}$
       $= 4.8/18.294 = .262$
   conclusion:
       Do not reject H_o; there is not sufficient evidence to conclude that $\mu_1 - \mu_2 \neq 0$.

b. The following preliminary values are identified and/or calculated.
   k = 2 $\bar{\bar{x}} = \Sigma \bar{x}_i/k$
   n = 10 $= 146.9$
   $s_{\bar{x}}^2 = \Sigma(\bar{x}_i - \bar{\bar{x}})^2/(k-1)$ $s_p^2 = \Sigma s_i^2/k$
       $= 11.52/1 = 11.52$ $= 3346.733/2 = 1673.3365$

   $H_o: \mu_1 = \mu_2$
   $H_1:$ at least one mean is different
   $\alpha = .05$
   C.R. $F > F_{18,.05}' = 4.4139$
   calculations:

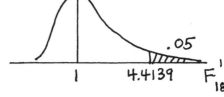

       $F = n s_{\bar{x}}^2/s_p^2$
         $= 10(11.52)/1673.3665 = .0688$
   conclusion:
       Do not reject H_o; there is not sufficient evidence to conclude that at least one mean is different.

c. critical value: $t^2 = (\pm 2.101)^2 = 4.414 = F$
   calculated statistic: $t^2 = (.262)^2 = .069 = F$

11-3 One-Way ANOVA with Unequal Sample Sizes

NOTE: This section is calculation-oriented. Do not get so involved with the formulas that you miss concepts. This manual arranges the calculations to promote both computational efficiency and understanding of the underlying principles. The following notation is used in this section.

k = the number of groups

n_i = the number of scores in group i (where i = 1,2,...k)

$\bar{x}_i$ = the mean of group i (where i = 1,2,...,k)

s_i^2 = the variance of group i (where i = 1,2,...,k)

The following items are calculated as in the previous section except that the unequal sample sizes require a weighting using either n_i or $(n_i-1) = df_i$.

$\bar{\bar{x}} = \Sigma\bar{x}_i/k$ becomes $\Sigma n_i\bar{x}_i/\Sigma n_i$

   = the overall weighted mean of all the scores in all the groups

$ns_{\bar{x}}^2 = n\Sigma(\bar{x}_i-\bar{\bar{x}})^2/(k-1)$ becomes $\Sigma n_i(\bar{x}_i-\bar{\bar{x}})^2/(k-1)$

   = the weighted variance estimate from the group means

$s_p^2 = \Sigma s_i^2/k$ becomes $\Sigma(n_i-1)s_i^2/\Sigma(n_i-1) = \Sigma df_i s_i^2/\Sigma df_i$

   = the weighted mean of the group variances

* All exercises are worked in complete detail, with the work arranged in an ANOVA table. In addition, NOTES are given commenting on the order in which the values in the table were filled in and on how new values were obtained from previous ones.

* While the manual typically shows only a certain number of decimal places (depending on the exercise) for the intermediate steps, all decimal places were carried in the calculator. DO NOT ROUND OFF INTERMEDIATE ANSWERS. SAVE calculated values that will be used again. See your instructor or class assistant if you need help using your calculator accurately and efficiently.

* Use of the ANOVA table to organize the calculations is strongly recommended. The calculated values appear in the ANOVA table as follows.

source	SS	df	MS	F
treatment	$\Sigma n_i(\bar{x}_i-\bar{\bar{x}})^2$	$k-1$	SS_{Trt}/df_{Trt}	MS_{Trt}/MS_E
error	$\Sigma df_i s_i^2$	$\Sigma df_i = \Sigma n_i-k$	SS_E/df_E	
total	$\Sigma(x-\bar{\bar{x}})^2$	Σn_i-1		

The SS and df columns are additive as indicated. If $MS_{Tot} = \Sigma(x-\bar{\bar{x}})^2/(\Sigma n_i-1)$ is calculated, it gives s^2 (i.e., the variance for all the scores considered as one big group) and not the sum of the MS column.

1. The following preliminary values are identified and/or calculated.

	G	F	P	total
n	5	7	6	18
Σx	437	484	357	1278
Σx^2	38289	34168	21469	93926
$\bar{x}$	87.400	69.143	59.500	71.000
s^2	23.800	117.143	45.500	

$\bar{\bar{x}} = \Sigma n_i\bar{x}_i/\Sigma n_i$

 = [5(87.400) + 7(69.143) + 6(59.500)]/(5 + 7 + 6) = 1278/18

 = 71.000 [NOTE: This must always agree with the $\bar{x}$ in the "total" column.]

$\Sigma n_i(\bar{x}_i-\bar{\bar{x}})^2 = 5(87.400-71.000)^2 + 7(69.143-71.000)^2 + 6(59.500-71.000)^2$

  = 2162.443

$\Sigma df_i s_i^2 = 4(23.800) + 6(117.142) + 5(45.500)$

  = 1025.558

$H_o: \mu_G = \mu_F = \mu_P$
$H_1:$ at least one mean is different
$\alpha = .05$
C.R. $F > F^2_{15,.05} = 3.6823$
calculations:

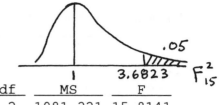

source	SS	df	MS	F
Trt	2162.443	2	1081.221	15.8141
Error	1025.558	15	68.371	
Total	3188.000	17		

$F = MS_{Trt}/MS_E$
$= 1081.221/68.371$
$= 15.8141$

conclusion:
Reject H_o; there is sufficient evidence to conclude that at least one mean is different.

NOTE: Complete the ANOVA table as follows:
(1) Enter $SS_{Trt} = \Sigma n_i(\bar{x}_i - \bar{\bar{x}})^2$ and $SS_E = \Sigma df_i s_i^2$ values from the preliminary calculations.
(2) Enter $df_{Trt} = k-1$ and $df_E = \Sigma df_i = \Sigma(n_i-1) = \Sigma n_i - k$.
(3) Add the SS and df columns to find SS_{Tot} and df_{Tot}. [The df_{Tot} must equal $\Sigma n_i - 1$.]
(4) Calculate $MS_{Trt} = SS_{Trt}/df_{Trt}$ and $MS_E = SS_E/df_E$.
(5) Calculate $F = MS_{Trt}/MS_E$.
As a final check, calculate s^2 (i.e., the variance of all the scores in one large group) two different ways as indicated below. If these answers agree, the problem is probably correct.
  * from the "total" column in the table for the preliminary calculations:
    $s^2 = [n\Sigma x^2 - (\Sigma x)^2]/[n(n-1)]$
       $= [18(93926)-(1278)^2]/[18(17)] = 57384/306 = 187.53$
  * from the "total" row of the ANOVA table
    $s^2 = SS_{Tot}/df_{Tot}$
       $= 3188.000/17 = 187.53$

2. The following preliminary values are identified and/or calculated.

	A	B	C	total
n	4	6	8	18
Σx	72	96	162	330
Σx^2	1304	1546	3310	6160
$\bar{x}$	18.00	16.00	20.25	18.33
s^2	2.667	2.000	4.212	

$\bar{\bar{x}} = \Sigma n_i \bar{x}_i / \Sigma n_i = 18.33$
$\Sigma n_i(\bar{x}_i - \bar{\bar{x}})^2 = 4(18.00-18.33)^2 + 6(16.00-18.33)^2 + 8(20.25-18.33)^2 = 62.5$
$\Sigma df_i s_i^2 = 3(2.667) + 5(2.000) + 7(4.214) = 47.5$

$H_o: \mu_A = \mu_B = \mu_C$
$H_1:$ at least one mean is different
$\alpha = .05$
C.R. $F > F^2_{15,.05} = 3.6823$
calculations:

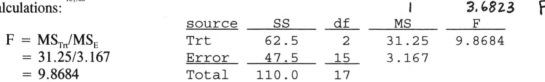

source	SS	df	MS	F
Trt	62.5	2	31.25	9.8684
Error	47.5	15	3.167	
Total	110.0	17		

$F = MS_{Trt}/MS_E$
$= 31.25/3.167$
$= 9.8684$

conclusion:
Reject H_o; there is sufficient evidence to conclude that at least one mean is different.

3. The following preliminary values are identified and/or calculated.

	A	B	C	D	total
n	6	10	8	5	29
Σx	93	74	147	92	406
Σx^2	1537	684	2951	1710	6882
$\bar{x}$	15.500	7.400	18.375	18.400	14.000
s^2	19.100	15.156	35.696	4.300	

$\bar{\bar{x}} = \Sigma n_i \bar{x}_i / \Sigma n_i$
$= [6(15.500) + 10(7.400) + 8(18.375) + 5(18.400]/(6 + 10 + 8 + 5) = 406/29$
$= 14.000$ [NOTE: This must always agree with the $\bar{x}$ in the "total" column.]
$\Sigma n_i(\bar{x}_i - \bar{\bar{x}})^2 = 6(15.500-14.000)^2 + 10(7.400-14.000)^2 + 8(18.375-14.000)^2$
$+ 5(18.400-14.000)^2$
$= 699.025$
$\Sigma df_i s_i^2 = 5(19.100) + 9(15.156) + 7(35.696) + 4(4.300) = 498.975$

$H_0: \mu_A = \mu_B = \mu_C = \mu_D$
H_1: at least one mean is different
$\alpha = .01$
C.R. $F > F^3_{25,.01} = 4.6755$
calculations:

source	SS	df	MS	F
Trt	699.025	3	233.008	11.6743
Error	498.975	25	19.959	
Total	1198.000	28		

$F = MS_{Trt}/MS_E$
$= 233.008/19.959$
$= 11.6743$

conclusion:
Reject H_0; there is sufficient evidence to conclude that at least one mean is different.

NOTE: Complete the ANOVA table as follows:
(1) Enter $SS_{Trt} = \Sigma n_i(\bar{x}_i - \bar{\bar{x}})^2$ and $SS_E = \Sigma df_i s_i^2$ values from the preliminary calculations.
(2) Enter $df_{Trt} = k-1$ and $df_E = \Sigma df_i = \Sigma(n_i-1) = \Sigma n_i - k$.
(3) Add the SS and df columns to find SS_{Tot} and df_{Tot}. [The df_{Tot} must equal $\Sigma n_i - 1$.]
(4) Calculate $MS_{Trt} = SS_{Trt}/df_{Trt}$ and $MS_E = SS_E/df_E$.
(5) Calculate $F = MS_{Trt}/MS_E$.
As a final check, calculate s^2 (i.e., the variance of all the scores in one large group) two different ways as indicated below. If these answers agree, the problem is probably correct.
   * from the "total" column in the table for the preliminary calculations:
      $s^2 = [n\Sigma x^2 - (\Sigma x)^2]/[n(n-1)]$
      $= [29(6882)-(406)^2]/[29(28)] = 34742/812 = 42.79$
   * from the "total" row of the ANOVA table
      $s^2 = SS_{Tot}/df_{Tot}$
      $= 1198.000/28 = 42.79$

4. The following preliminary values are identified and/or calculated.

	1: D&T	2: M	3: KG&L	total
n	10	10	12	32
Σx	763	792	902	2457
Σx^2	58669	63472	68600	190741
$\bar{x}$	76.30	79.20	75.17	76.78
s^2	50.233	82.844	72.697	

$\bar{\bar{x}} = \Sigma n_i \bar{x}_i / \Sigma n_i = 76.78$
$\Sigma n_i(\bar{x}_i - \bar{\bar{x}})^2 = 10(76.30-76.78)^2 + 10(79.20-76.78)^2 + 12(75.17-76.78)^2 = 92.10$
$\Sigma df_i s_i^2 = 9(50.233) + 9(82.444) + 11(72.697) = 1997.37$

H_o: $\mu_1 = \mu_2 = \mu_3$
H_1: at least one mean is different
$\alpha = .05$
C.R. $F > F^2_{29,.05} = 3.3277$
calculations:

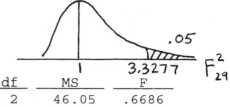

	source	SS	df	MS	F
$F = MS_{Trt}/MS_E$	Trt	92.10	2	46.05	.6686
$= 46.05/68.87$	Error	1197.37	29	68.87	
$= .6686$	Total	1289.47	31		

conclusion:
  Do not reject H_o; there is not sufficient evidence to conclude that at least one mean is different.

5. The following preliminary values are identified and/or calculated.

	1	2	3	4	5	total
n	11	11	10	9	7	48
Σx	36.7	39.6	32.5	27.6	25.5	161.9
Σx^2	124.49	144.68	106.73	85.22	94.65	555.77
$\bar{x}$	3.3364	3.6000	3.2500	3.0667	3.6429	3.3729
s^2	.2045	.2120	.1228	.0725	.2929	

$\bar{\bar{x}} = \Sigma n_i \bar{x}_i / \Sigma n_i$
  $= [11(3.3364) + 11(3.6000) + 10(3.2500) + 9(3.0667) + 7(3.6429)]/48 = 161.9/48$
  $= 3.3729$ [NOTE: This must always agree with the $\bar{x}$ in the "total" column.]
$\Sigma n_i(\bar{x}_i - \bar{\bar{x}})^2 = 11(3.3364-3.3729)^2 + 11(3.6000-3.3729)^2 + 10(3.2500-3.3729)^2$
                    $+ 9(3.0667-3.3729)^2 + 7(3.6429-3.3729)^2$
  $= 2.087$
$\Sigma df_i s_i^2 = 10(.2045) + 10(.2120) + 9(.1228) + 8(.0725) + 6(.2929) = 7.608$

H_o: $\mu_1 = \mu_2 = \mu_3 = \mu_4 = \mu_5$
H_1: at least one mean is different
$\alpha = .05$
C.R. $F > F^4_{43,.05} = 2.6060$
calculations:

	source	SS	df	MS	F
$F = MS_{Trt}/MS_E$	Trt	2.087	4	.522	2.9491
$= .522/.177$	Error	7.608	43	.177	
$= 2.9491$	Total	9.694	47		

conclusion:
  Reject H_o; there is sufficient evidence to conclude that at least one mean is different.
NOTE: Complete the ANOVA table as follows:
(1) Enter $SS_{Trt} = \Sigma n_i(\bar{x}_i - \bar{\bar{x}})^2$ and $SS_E = \Sigma df_i s_i^2$ values from the preliminary calculations.
(2) Enter $df_{Trt} = k-1$ and $df_E = \Sigma df_i = \Sigma(n_i-1) = \Sigma n_i - k$.
(3) Add the SS and df columns to find SS_{Tot} and df_{Tot}. [The df_{Tot} must equal $\Sigma n_i - 1$.]
(4) Calculate $MS_{Trt} = SS_{Trt}/df_{Trt}$ and $MS_E = SS_E/df_E$.
(5) Calculate $F = MS_{Trt}/MS_E$.
As a final check, calculate s^2 (i.e., the variance of all the scores in one large group) two different ways as indicated below. If these answers agree, the problem is probably correct.
  * from the "total" column in the table for the preliminary calculations:
    $s^2 = [n\Sigma x^2 - (\Sigma x)^2]/[n(n-1)]$
      $= [48(555.77)-(161.9)^2]/[48(47)] = 465.35/2256 = .206$
  * from the "total" row of the ANOVA table
    $s^2 = SS_{Tot}/df_{Tot}$
      $= 9.694/47 = .206$

6. The following preliminary values are identified and/or calculated.

	1: GTP	2: BST	3: TGS	total
n	12	12	8	32
Σx	470	480	480	1430
Σx^2	28700	43600	36200	108500
$\overline{x}$	39.17	40.00	60.00	44.69
s^2	935.606	2218.182	1057.143	

$\overline{\overline{x}} = \Sigma n_i \overline{x}_i / \Sigma n_i = 44.69$

$\Sigma n_i (\overline{x}_i - \overline{\overline{x}})^2 = 12(39.17-44.69)^2 + 12(40.00-44.69)^2 + 8(60.00-44.69)^2 = 2505.21$

$\Sigma df_i s_i^2 = 11(935.606) + 11(2218.182) + 7(1057.143) = 42091.67$

H_o: $\mu_1 = \mu_2 = \mu_3$
H_1: at least one mean is different
$\alpha = .05$
C.R. $F > F^2_{29,.05} = 3.3277$
calculations:

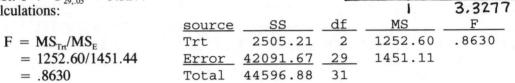

source	SS	df	MS	F
Trt	2505.21	2	1252.60	.8630
Error	42091.67	29	1451.11	
Total	44596.88	31		

$F = MS_{Trt}/MS_E$
$= 1252.60/1451.44$
$= .8630$

conclusion:

Do not reject H_o; there is not sufficient evidence to conclude that at least one mean is different.

It appears that the course results are basically the same. To test the effectiveness of the courses, one needs a control group to take the SAT a second time with no additional studies -- an analysis of variance test could then be used on the four groups.

7. The following preliminary values are identified and/or calculated.

	R	O	Y	B	T	G	total
n	17	9	24	30	11	9	100
Σx	15.373	8.267	21.939	27.769	10.242	8.011	91.601
Σx^2	13.929193	7.596853	20.085206	25.782618	9.551278	7.147987	84.093135
$\overline{x}$	.90429	.91856	.91413	.92563	.93109	.89011	.91601
s^2	.001717	.000394	.001314	.002714	.001504	.002163	

$\overline{\overline{x}} = \Sigma n_i \overline{x}_i / \Sigma n_i$
$= [17(.90429) + 9(.91856) + 24(.91413) + 30(.92562) + 11(.93109) + 9(.89011)]/100$
$= 91.601/100$
$= .91601$ [NOTE: This must always agree with the $\overline{x}$ in the "total" column.]

$\Sigma n_i(\overline{x}_i - \overline{\overline{x}})^2 = 17(.90429-.91601)^2 + 9(.91856-.91601)^2 + 24(.91413-.91601)^2$
$\qquad + 30(.92563-.91601)^2 + 11(.93109-.91601)^2 + 9(.89011-.91601)^2$
$\qquad = .01379$

$\Sigma df_i s_i^2 = 16(.001717) + 8(.000394) + 23(.001314)$
$\qquad + 29(.002714) + 10(.001504) + 8(.002163)$
$\qquad = .17190$

H_o: $\mu_R = \mu_O = \mu_Y = \mu_B = \mu_T = \mu_G$
H_1: at least one mean is different
$\alpha = .05$
C.R. F > $F^5_{94,.05} = 2.2899$
calculations:

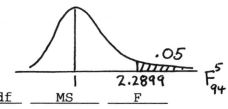

source	SS	df	MS	F
Trt	.01379	5	.00276	1.5082
Error	.17190	94	.00183	
Total	.18569	99		

F = MS_{Trt}/MS_E
  = .00276/.00183
  = 1.5082

conclusion:
 Do not reject H_o; there is not sufficient evidence to conclude that at least one mean is different.
No; we cannot be 95% certain that the company has a problem that requires corrective action.

NOTE: Complete the ANOVA table as follows:
(1) Enter $SS_{Trt} = \Sigma n_i(\bar{x}_i - \bar{\bar{x}})^2$ and $SS_E = \Sigma df_i s_i^2$ values from the preliminary calculations.
(2) Enter $df_{Trt} = k-1$ and $df_E = \Sigma df_i = \Sigma(n_i - 1) = \Sigma n_i - k$.
(3) Add the SS and df columns to find SS_{Tot} and df_{Tot}. [The df_{Tot} must equal $\Sigma n_i - 1$.]
(4) Calculate $MS_{Trt} = SS_{Trt}/df_{Trt}$ and $MS_E = SS_E/df_E$.
(5) Calculate F = MS_{Trt}/MS_E.
As a final check, calculate s^2 (i.e., the variance of all the scores in one large group) two different ways as indicated below. If these answers agree, the problem is probably correct.
 * from the "total" column in the table for the preliminary calculations:
  $s^2 = [n\Sigma x^2 - (\Sigma x)^2]/[n(n-1)]$
     = $[100(84.093135) - (91.601)^2]/[100(99)] = 18.571799/9900 = .001876$
 * from the "total" row of the ANOVA table
  $s^2 = SS_{Tot}/df_{Tot}$
     = $.18569/99 = .001876$

8. The following preliminary values are identified and/or calculated.

	1: <2750	2: middle	3: >3000	total
n	9	10	13	32
Σx	299	281	352	932
Σx^2	10301	7943	9598	27842
$\bar{x}$	32.22	28.10	27.08	29.125
s^2	45.944	5.211	5.577	

$\bar{\bar{x}} = \Sigma n_i \bar{x}_i / \Sigma n_i = 29.125$
$\Sigma n_i(\bar{x}_i - \bar{\bar{x}})^2 = 9(32.22-29.125)^2 + 10(28.10-29.125)^2 + 13(27.08-29.125)^2 = 216.122$
$\Sigma df_i s_i^2 = 8(45.944) + 9(5.211) + 12(5.577) = 481.378$

H_o: $\mu_1 = \mu_2 = \mu_3$
H_1: at least one mean is different
$\alpha = .05$
C.R. F > $F^2_{29,.05} = 3.3277$
calculations:

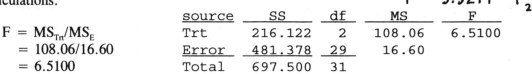

source	SS	df	MS	F
Trt	216.122	2	108.06	6.5100
Error	481.378	29	16.60	
Total	697.500	31		

F = MS_{Trt}/MS_E
  = 108.06/16.60
  = 6.5100

conclusion:
 Reject H_o; there is sufficient evidence to conclude that at least one mean is different.

9. The completed ANOVA table appears below.

source	SS	df	MS	F
Trt	2.17	2	1.085	.1542
Error	112.57	16	7.036	
Total	114.74	18		

Complete the ANOVA table as follows.
(1) Since $SS_{Trt} + 112.57 = 114.74$, it follows that $SS_{Trt} = 114.74 - 112.57 = 2.17$.
(2) Since there were three groups, $k = 3$ and $df_{Trt} = k - 1 = 2$.
(3) Since $n_1 = 5$ and $n_2 = 7$ and $n_3 = 7$, $df_E = \Sigma df_i = 4 + 6 + 6 = 16$.
(4) $MS_{Trt} = SS_{Trt}/df_{Trt} = 2.17/2 = 1.085$.
(5) $MS_E = SS_E/df_E = 112.57/16 = 7.036$. [STORE this, do not round it off.]
(6) $F = MS_{Trt}/MS_E = 1.085/7.036 = .1542$
NOTE: The text could also have asked, "What would be the variance of all 19 scores taken as one large group?" The answer is $s^2 = SS_{Tot}/df_{Tot} = 114.74/18 = 6.374$.

10. The completed ANOVA table appears below.

source	SS	df	MS	F
Trt	21.34	3	7.113	5.6631
Error	144.45	115	1.256	
Total	165.79	118		

Complete the ANOVA table by starting as follows.
(1) Because there are 4 groups, $df_{Trt} = 3$.
(2) Because the individual df's are 11,24,49,31, $df_{Error} = 11+24+49+31 = 115$.

11. The following preliminary values are identified and/or calculated, with only the values in column A and the total column different from those in exercise 3.

	A	B	C	D	total
n	6	10	8	5	29
Σx	1479	74	147	92	1792
Σx^2	1961341	684	2951	1710	1966686
$\bar{x}$	246.500	7.400	18.375	18.400	61.793
s^2	319353.500	15.156	35.696	4.300	

$\bar{\bar{x}} = \Sigma n_i \bar{x}_i / \Sigma n_i$
$= [6(246.500) + 10(7.400) + 8(18.375) + 5(18.400]/(6 + 10 + 8 + 5) = 1792/29$
$= 61.793$ [NOTE: This must always agree with the $\bar{x}$ in the "total" column.]
$\Sigma n_i(\bar{x}_i - \bar{\bar{x}})^2 = 6(246.500 - 61.793)^2 + 10(7.400 - 61.793)^2 + 8(18.375 - 61.793)^2$
$+ 5(18.400 - 61.793)^2$
$= 258781.784$
$\Sigma df_i s_i^2 = 5(31935.500) + 9(15.156) + 7(35.696) + 4(4.300) = 1597170.976$

$H_0: \mu_A = \mu_B = \mu_C = \mu_D$
$H_1:$ at least one mean is different
$\alpha = .01$
C.R. $F > F_{25,.01}^3 = 4.6755$
calculations:

source	SS	df	MS	F
Trt	258781.784	3	86260.595	1.3502
Error	1597170.976	25	63886.839	
Total	1855952.760	28		

$F = MS_{Trt}/MS_E$
$= 86260/63886$
$= 1.3502$

conclusion:
Do not reject H_0; there is not sufficient evidence to conclude that at least one mean is different.

The results are affected by the outlier in what may be an unexpected manner. Because the sample mean of group 1 increases dramatically, making the groups appear "more different" than before, one might expect even stronger evidence to reject H_o than the original calculated $F = 11.6743$. Instead, however, the calculated F drops to $F = 1.3502$ and the conclusion becomes that the groups are <u>not</u> significantly different at all. In general, the effect of an outlier cannot be predicted with certainty. In this case, the MS_E (the estimate of the unexplained variability [or inherent variance] within the groups) increased to the point that the differences between the sample means were no longer significant -- they were consistent with the differences expected between such small samples each having such a large variance.

NOTE: Complete the ANOVA table as follows:
(1) Enter $SS_{Trt} = \Sigma n_i(\bar{x}_i-\bar{\bar{x}})^2$ and $SS_E = \Sigma df_i s_i^2$ values from the preliminary calculations.
(2) Enter $df_{Trt} = k-1$ and $df_E = \Sigma df_i = \Sigma(n_i-1) = \Sigma n_i - k$.
(3) Add the SS and df columns to find SS_{Tot} and df_{Tot}. [The df_{Tot} must equal $\Sigma n_i - 1$.]
(4) Calculate $MS_{Trt} = SS_{Trt}/df_{Trt}$ and $MS_E = SS_E/df_E$.
(5) Calculate $F = MS_{Trt}/MS_E$.
As a final check, calculate s^2 (i.e., the variance of all the scores in one large group) two different ways as indicated below. If these answers agree, the problem is probably correct.
* from the "total" column in the table for the preliminary calculations:
$$s^2 = [n\Sigma x^2 - (\Sigma x)^2]/[n(n-1)]$$
$$= [29(1966686)-(1792)^2]/[29(28)] = 53822630/812 = 66284.0$$
* from the "total" row of the ANOVA table
$$s^2 = SS_{Tot}/df_{Tot}$$
$$= 1855952.760/28 = 66284.0$$

12. The only requirement is that all the temperatures be measured on the same scale. Whether that scale is Fahrenheit or Celsius will not affect the test statistic. Since the F statistic is the ratio of two variances, the numerator and denominator are unaffected by additive changes to the data and are both multiplied by the square of multiplicative changes to the data.

11-4 Two-Way ANOVA

NOTE: The formulas and principles in this section are logical extensions of the previous sections.
$SS_{Row} = \Sigma n_i(\bar{x}_i-\bar{\bar{x}})^2$ for i=1,2,3... [for each row]
$SS_{Col} = \Sigma n_j(\bar{x}_j-\bar{\bar{x}})^2$ for j=1,2,3... [for each column]
$SS_{Tot} = \Sigma(x-\bar{\bar{x}})^2$ [for all the x's]
When there is only one observation per cell...
the unexplained variation is
$SS_E = SS_{Tot} - SS_{Row} - SS_{Col}$
and there is not enough data to measure interaction.
When there is more than one observation per cell...
the unexplained variation (i.e., the failure of items in the same cell to respond the same) is
$SS_E = \Sigma(x-\bar{x}_{ij})^2$ [for each cell -- i,e., for each i,j (row,col) combination]
and the interaction sum of squares is
$SS_{Int} = SS_{Tot} - SS_{Row} - SS_{Col} - SS_E$.
Since the data will be analyzed from statistical software packages, however, the above formulas need not be used by hand.

1. a. $MS_{Int} = 1263$
   b. $MS_E = 695$
   c. $MS_{Star} = 350$
   d. $MS_{MPAA} = 14$

2. H_o: there is no star rating/MPAA rating interaction
H_1: there is star rating/MPAA rating interaction
$\alpha = .05$ [or another appropriate value]
C.R. $F > F_{8,.05}^3 = 4.0662$
calculations:
  $F = MS_{Int}/MS_E$
    $= 1263/695 = 1.8173$

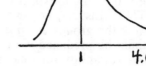

conclusion:
  Do not reject H_o; there is not sufficient evidence to conclude that there is a star rating/MPAA rating interaction.
The evidence does not support a conclusion that movies of different star ratings tend to have different relationships between MPAA ratings and lengths.

3. H_o: $\mu_P = \mu_F = \mu_G = \mu_E$
H_1: at least one mean is different
$\alpha = .05$ [or another appropriate value]
C.R. $F > F_{8,.05}^3 = 4.0662$
calculations:
  $F = MS_{Star}/MS_E$
    $= 350/695 = .5036$

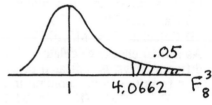

conclusion:
  Do not reject H_o; there is not sufficient evidence to conclude that at least one mean is different.
The evidence does not support a conclusion that movies of different star ratings tend to have different lengths.

4. H_o: $\mu_{G/PG/PG\text{-}13} = \mu_R$
H_1: the means are different
$\alpha = .05$ [or another appropriate value]
C.R. $F > F_{8,.05}^1 = 5.3177$
calculations:
  $F = MS_{MPAA}/MS_E$
    $= 14/695 = .0201$

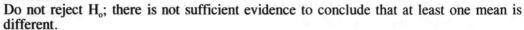

conclusion:
  Do not reject H_o; there is not sufficient evidence to conclude that the means are different.
The evidence does not support a conclusion that movies of different MPAA ratings tend to have different lengths.

5. H_o: $\mu_{G/PG/PG\text{-}13} = \mu_R$
H_1: the means are different
$\alpha = .05$
C.R. $F > F_{3,.05}^1 = 10.128$
calculations:
  $F = MS_{MPAA}/MS_E$
    $= 0/266 = 0$

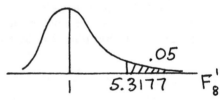

conclusion:
  Do not reject H_o; there is sufficient evidence to conclude that the means are different.

6. H_o: $\mu_P = \mu_F = \mu_G = \mu_E$
  H_1: at least one mean is different
  $\alpha = .05$
  C.R. $F > F^3_{3,.05} = 9.2766$
  calculations:
    $F = MS_{Star}/MS_E$
    $= 153/266 = .5752$
  conclusion:
    Do not reject H_o; there is not sufficient evidence to conclude that at least one mean is different.

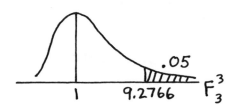

7. H_o: $\mu_A = \mu_B = \mu_C = \mu_D$
  H_1: at least one mean is different
  $\alpha = .05$
  C.R. $F > F^3_{6,.05} = 4.7571$
  calculations:
    $F = MS_{Typist}/MS_E$
    $= 7.333/.417 = 17.585$
  conclusion:
    Reject H_o; there is sufficient evidence to conclude that at least one mean is different.

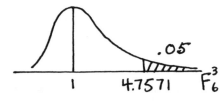

8. H_o: $\mu_I = \mu_{II} = \mu_{III}$
  H_1: at least one mean is different
  $\alpha = .05$
  C.R. $F > F^2_{6,.05} = 5.1433$
  calculations:
    $F = MS_{WP}/MS_E$
    $= 60.083/.417 = 144.0839$
  conclusion:
    Reject H_o; there is sufficient evidence to conclude that at least one mean is different.

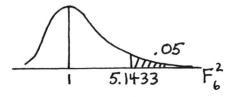

9. As indicated by the Minitab output given **[in the Student Manual]**, reversing the roles of the rows and columns changes the order in which the results are presented but does not change the numerical values identified with the MPAA ratings or the star ratings.

10. The analysis of variance technique tests for differences between means by comparing various variances. Since adding a constant to each score does not affect any variability, no values in the ANOVA summary table are changed at all.

11. The Minitab output at the right was obtained from the original data of exercises 1-4 and 9 as indicated. Multiplying each value in the table by 10 multiplies the entries in the df, SS and MS columns by $10^2 = 100$. [As the entries as given to the nearest whole number, there are discrepancies caused by rounding.]

```
MTB > LET C1=10*C1
MTB > TWOWAY C1 C2 C3

ANALYSIS OF VARIANCE   C1

SOURCE         DF         SS          MS
row             1       1406        1406
column          3     104919       34973
INTERACTION     3     379019      126340
ERROR           8     555950       69494
TOTAL          15    1041294
```

NOTE: All F values remain unchanged. Testing the hypothesis of exercise 3, for example,
    $F = MS_{Star}/MS_E = 34973/69494 = .5033$
agrees, except for .0003 discrepancy caused by rounding, with the result given in that exercise.

12. The group means for the n=5 movies in each of the 6 cells of Table 11-4 are as follows.

		STAR RATING			
		fair	good	excellent	
MPAA RATING	G/PG/PG-13	102.4	99.4	134.6	112.1
	R	115.2	115.2	112.2	114.2
		108.8	107.7	123.4	

a. Adding 100 minutes to each G/PG/PG-13 film rated excellent yields the desired effect.

		STAR RATING			
		fair	good	excellent	
MPAA RATING	G/PG/PG-13	102.4	99.4	234.6	145.4
	R	115.2	115.2	112.2	114.2
		108.8	107.7	173.4	

b. Adding 100 minutes to each G/PG/PG-13 film yields the desired effect.

		STAR RATING			
		fair	good	excellent	
MPAA RATING	G/PG/PG-13	202.4	199.4	234.6	212.1
	R	115.2	115.2	112.2	114.2
		158.8	157.7	173.4	

c. Adding 100 minutes to each film rated excellent yields the desired effect.

		STAR RATING			
		fair	good	excellent	
MPAA RATING	G/PG/PG-13	102.4	99.4	234.6	145.4
	R	115.2	115.2	212.2	147.5
		108.8	107.7	223.4	

Review Exercises

1. The following preliminary values are identified and/or calculated.

	A	B	C	D
n	5	5	5	5
Σx	265	295	245	325
Σx^2	14083	17415	12015	21193
$\bar{x}$	53	59	49	65
s^2	9.5	2.5	2.5	17.0

$k = 4$ $\bar{\bar{x}} = \Sigma\bar{x}_i/k$
$n = 5$ $= 56.5$
$s_{\bar{x}}^2 = \Sigma(\bar{x}_i-\bar{\bar{x}})^2/(k-1)$ $s_p^2 = \Sigma s_i^2/k$
$= 147/3 = 49$ $= 31.5/4 = 7.875$

H_o: $\mu_A = \mu_B = \mu_C = \mu_D$
H_1: at least one mean is different
$\alpha = .05$
C.R. $F > F_{16,.05}^3 = 3.2389$
calculations:
$F = ns_{\bar{x}}^2/s_p^2$
$= 5(49)/7.875 = 31.1111$

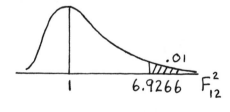

conclusion:
  Reject H_o; there is sufficient evidence to conclude that at least one mean is different.

NOTE: Since $n_A=n_B=n_C=n_D$, this exercise was completed using the techniques and format of section 11-2 for equal sample sizes. It could also have been completed using the techniques and format of section 11-3 for unequal sample sizes. The method for unequal sample sizes works for all one-way ANOVA problems; the procedure given for equal sample sizes is a short-cut of that method that takes advantage of the algebraic simplification occurring when all the n's are equal.

2. Refer to the NOTE above. The following preliminary values are identified and/or calculated.

	A	B	C	total
n	5	5	5	15
Σx	42	39	49	130
Σx^2	374	311	487	1172
$\bar{x}$	8.4	7.8	9.8	8.667
s^2	5.3	1.7	1.7	

$k = 3$ $\bar{\bar{x}} = \Sigma\bar{x}_i/k$
$n = 5$ $= 8.667$
$s_{\bar{x}}^2 = \Sigma(\bar{x}_i-\bar{\bar{x}})^2/(k-1)$ $s_p^2 = \Sigma s_i^2/k$
$= 2.106/2 = 1.053$ $= 8.7/3 = 2.9$

H_o: $\mu_A = \mu_B = \mu_C$
H_1: at least one mean is different
$\alpha = .01$
C.R. $F > F_{12,.01}^2 = 6.9266$
calculations:
$F = ns_{\bar{x}}^2/s_p^2$
$= 5(1.053)/2.9 = 1.8161$
conclusion:
  Do not reject H_o; there is not sufficient evidence to conclude that at least one mean is different.

3. The following preliminary values are identified and/or calculated.

	A	B	C	total
n	5	5	7	17
Σx	.49	.37	.33	1.19
Σx^2	.0483	.0279	.0159	.0921
$\bar{x}$	.0980	.0740	.0471	.0700
s^2	.0000700	.0001300	.0000571	

$\bar{\bar{x}} = \Sigma n_i \bar{x}_i / \Sigma n_i$
$= [5(.0980) + 5(.0740) + 7(.0471)]/(5 + 5 + 7) = 1.19/17$
$= .0700$ [NOTE: This must always agree with the $\bar{x}$ in the "total" column.]
$\Sigma n_i(\bar{x}_i - \bar{\bar{x}})^2 = 5(.0980 - .0700)^2 + 5(.0740 - .0700)^2 + 7(.0471 - .0700)^2 = .007657$
$\Sigma df_i s_i^2 = 4(.0000700) + 4(.0001300) + 6(.0000571) = .001143$

H_o: $\mu_A = \mu_B = \mu_C$
H_1: at least one mean is different
$\alpha = .05$
C.R. $F > F_{14,.05}^2 = 3.7389$
calculations:

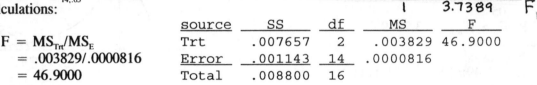

source	SS	df	MS	F
Trt	.007657	2	.003829	46.9000
Error	.001143	14	.0000816	
Total	.008800	16		

$F = MS_{Trt}/MS_E$
$= .003829/.0000816$
$= 46.9000$

conclusion:
Reject H_o; there is sufficient evidence to conclude that at least one mean is different.

NOTE: Complete the ANOVA table as follows:
(1) Enter $SS_{Trt} = \Sigma n_i(\bar{x}_i - \bar{\bar{x}})^2$ and $SS_E = \Sigma df_i s_i^2$ values from the preliminary calculations.
(2) Enter $df_{Trt} = k-1$ and $df_E = \Sigma df_i = \Sigma(n_i-1) = \Sigma n_i - k$.
(3) Add the SS and df columns to find SS_{Tot} and df_{Tot}. [The df_{Tot} must equal $\Sigma n_i - 1$.]
(4) Calculate $MS_{Trt} = SS_{Trt}/df_{Trt}$ and $MS_E = SS_E/df_E$.
(5) Calculate $F = MS_{Trt}/MS_E$.
As a final check, calculate s^2 (i.e., the variance of all the scores in one large group) two different ways as indicated below. If these answers agree, the problem is probably correct.
   * from the "total" column in the table for the preliminary calculations:
   $s^2 = [n\Sigma x^2 - (\Sigma x)^2]/[n(n-1)]$
   $= [17(.0921)-(1.19)^2]/[17(16)] = .1496/272 = .00055$
   * from the "total" row of the ANOVA table
   $s^2 = SS_{Tot}/df_{Tot}$
   $= .008800/16 = .00055$

4. The following preliminary values are identified and/or calculated. Working with the rounded $\bar{x}$ and s^2 given in the text will not yield exact answers. As always, the numbers shown below are not necessarily the precise values stored in the calculator and employed within the problem.

	A	B	C	total
n	4	5	3	12
Σx	87	128	74	289
Σx^2	1899	3300	1828	7027
$\bar{x}$	21.75	25.60	24.67	24.08
s^2	2.25	5.80	1.33	

$\bar{\bar{x}} = \Sigma n_i \bar{x}_i / \Sigma n_i = 24.08$
$\Sigma n_i(\bar{x}_i - \bar{\bar{x}})^2 = 4(21.75 - 24.08)^2 + 5(25.60 - 24.08)^2 + 3(24.67 - 24.08)^2 = 34.30$
$\Sigma df_i s_i^2 = 3(2.25) + 4(5.80) + 2(1.33) = 32.62$

H_o: $\mu_A = \mu_B = \mu_C$
H_1: at least one mean is different
$\alpha = .05$
C.R. $F > F^2_{9,.05} = 4.2565$
calculations:

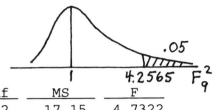

	source	SS	df	MS	F
$F = MS_{Trt}/MS_E$	Trt	34.30	2	17.15	4.7322
$= 17.15/3.624$	Error	32.62	9	3.624	
$= 4.7322$	Total	66.92	11		

conclusion:
    Reject H_o; there is sufficient evidence to conclude that at least one mean is different.

5. H_o: there is no engine-size/transmission-type interaction
   H_1: there is engine-size/transmission-type interaction
   $\alpha = .05$
   C.R. $F > F^2_{6,.05} = 5.1433$
   calculations:
       $F = MS_{Int}/MS_E$
       $= .6/11.3 = .0531$

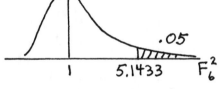

conclusion:
    Do not reject H_o; there is not sufficient evidence to conclude that there is engine-size/transmission-type interaction.

6. H_o: $\mu_{1.5} = \mu_{2.2} = \mu_{2.5}$
   H_1: at least one mean is different
   $\alpha = .05$
   C.R. $F > F^2_{6,.05} = 5.1433$
   calculations:
       $F = MS_{Size}/MS_E$
       $= 21.6/11.3 = 1.9115$

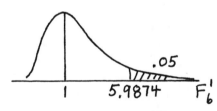

conclusion:
    Do not reject H_o; there is not sufficient evidence to conclude that at least one mean is different.

7. H_o: $\mu_A = \mu_M$
   H_1: the means are different
   $\alpha = .05$
   C.R. $F > F^1_{6,.05} = 5.9874$
   calculations:
       $F = MS_{Trans}/MS_E$
       $= 40.3/11.3 = 3.5664$

conclusion:
    Do not reject H_o; there is not sufficient evidence to conclude that the means are different.

Chapter 12

Statistical Process Control

12-2 Control Charts for Variation and Mean

NOTE: In this section, n = number of observations per sample subgroup
k = number of sample subgroups

1. $\bar{\bar{x}} = \Sigma\bar{x}/k = 99.860/20 = 4.993$

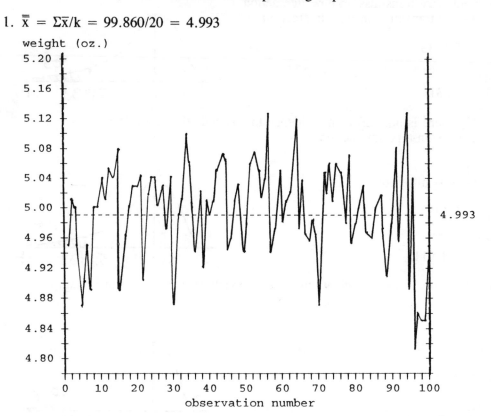

No; there does not appear to be a pattern suggesting that the process is not within statistical control.

2. $\bar{R} = \Sigma R/k = 2.48/20 = .124$ $\qquad$ $LCL = D_3\bar{R} = 0(.124) = 0$
$\qquad\qquad\qquad\qquad\qquad\qquad\qquad\qquad$ $UCL = D_4\bar{R} = 2.114(.124) = .262$

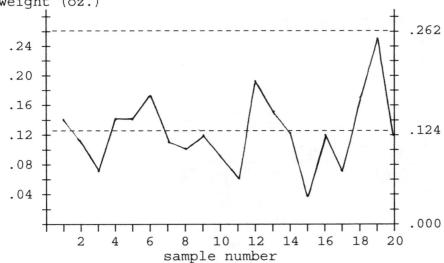

3. $\bar{\bar{x}} = \Sigma\bar{x}/k = 99.860/20 = 4.993$
$\bar{R} = \Sigma R/k = 2.48/20 = .124$
$LCL = \bar{\bar{x}} - A_2\bar{R} = 1.993 - (.577)(.124) = 4.993 - .072 = 4.921$
$UCL = \bar{\bar{x}} + A_2\bar{R} = 1.993 + (.577)(.124) = 4.993 + .072 = 5.065$

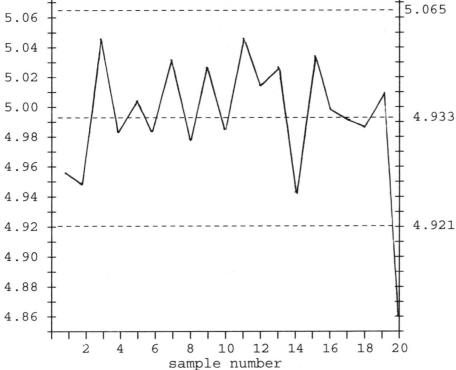

The process requires correction.

4. $\bar{\bar{x}} = \Sigma\bar{x}/k = 352.25/24 = 14.68$

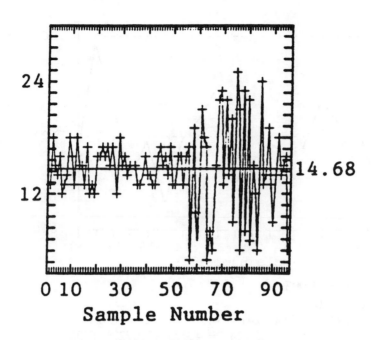

The process is not within statistical control. There is increasing variation.

5. $\bar{R} = \Sigma R/k = 179/24 = 7.458$ $\text{LCL} = D_3\bar{R} = 0(7.458) = 0$
$\text{UCL} = D_4\bar{R} = 2.282(7.458) = 17.109$

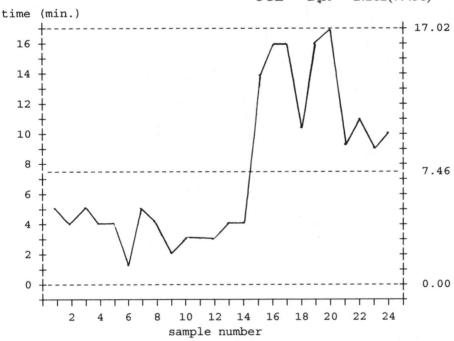

The process variation is not within statistical control. There is a shift up.

6. $\bar{\bar{x}} = \Sigma\bar{x}/k = 352.25/24 = 14.68$
$\bar{R} = \Sigma R/k = 179/24 = 7.458$
$LCL = \bar{\bar{x}} - A_2\bar{R} = 14.68 - (.729)(7.458) = 14.68 - 5.44 = 9.24$
$UCL = \bar{\bar{x}} + A_2\bar{R} = 14.68 + (.729)(7.458) = 14.68 + 5.44 = 20.11$

The process is in statistical control.

7. This chart requires a full sheet and appears by itself on the next page.

8. $\bar{R} = \Sigma R/k = 14.958$
$LCL = D_3\bar{R} = 0(14.958) = 0$
$UCL = D_4\bar{R} = 2.004(14.958) = 29.98$

The process is in statistical control.

7. $\bar{\bar{x}} = \Sigma\bar{x}/k = 20.306$

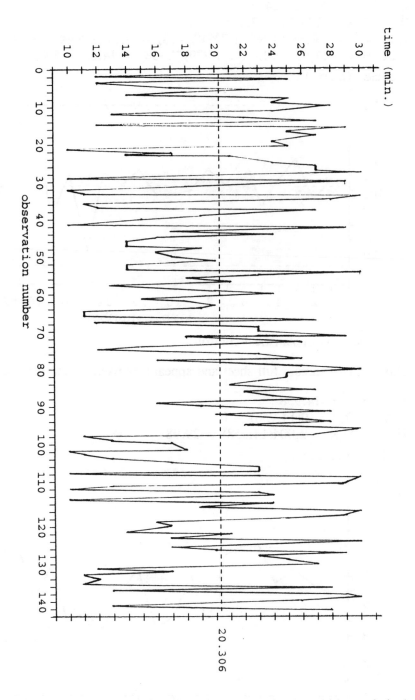

No; there does not appear to be a pattern suggesting that the process is not within statistical control.

8. This chart appears on the previous page.

9. $\bar{\bar{x}} = \Sigma\bar{x}/k = 20.306$
$\bar{R} = \Sigma R/k = 14.958$
$LCL = \bar{\bar{x}} - A_2\bar{R} = 20.306 - (.483)(14.958) = 20.306 - 7.225 = 13.08$
$UCL = \bar{\bar{x}} + A_2\bar{R} = 20.306 + (.483)(14.958) = 20.306 + 7.225 = 27.53$

time (min.)

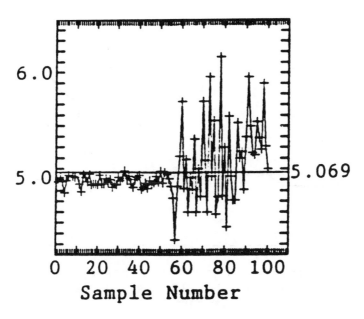

The process is working well. If there were a downward trend, then the delivery times would be being improved and no correction should be made.

10. $\bar{\bar{x}} = \Sigma\bar{x}/k = 101.380/20 = 5.069$

The process is not in control. There is increasing variability and an upward drift.

11. $\bar{R} = \Sigma R/k = 10.50/20 = .525$
    $LCL = D_3\bar{R} = 0(.525) = 0$
    $UCL = D_4\bar{R} = 2.114(.525) = 1.110$

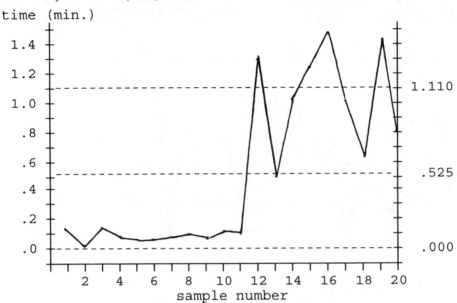

The process variation is not within statistical control. There is a shift up, and there are points above the upper control limit.

12. $\bar{\bar{x}} = \Sigma\bar{x}/k = 101.380/20 = 5.069$
    $\bar{R} = \Sigma R/k = 10.50/20 = .525$
    $LCL = \bar{\bar{x}} - A_2\bar{R} = 5.069 - (.577)(.525) = 5.069 - .303 = 5.372$
    $UCL = \bar{\bar{x}} + A_2\bar{R} = 5.069 + (.577)(.525) = 5.069 + .303 = 4.766$

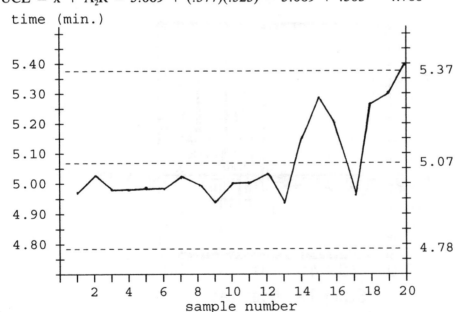

The process is not in statistical control. There is an upward trend and violation of the upper control limit.

13. a. $\bar{s} = \Sigma s/k = 2.8979/20 = .145$
   $LCL = B_3\bar{s} = 0(.145) = 0$
   $UCL = B_4\bar{s} = 2.089(.145) = .303$

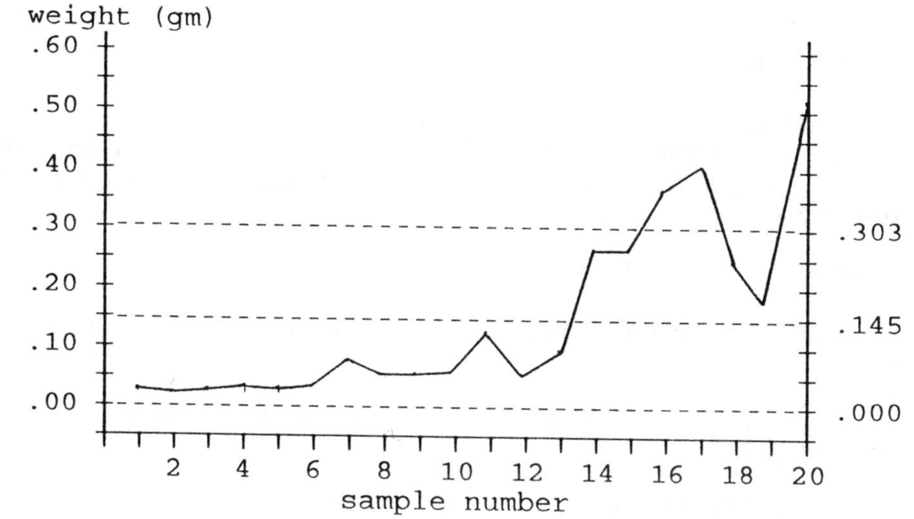

weight (gm)

As concluded by the original method, the process is out of statistical control.

b. $\bar{\bar{x}} = \Sigma\bar{x}/k = 5.7098$ [given in the text]
   $\bar{s} = \Sigma s/k = 2.8979/20 = .145$
   $LCL = \bar{\bar{x}} - A_3\bar{s} = 5.7098 - (1.427)(.145) = 5.7098 - .2068 = 5.503$
   $UCL = \bar{\bar{x}} + A_3\bar{s} = 5.7098 + (1.427)(.145) = 5.7098 + .2068 = 5.917$

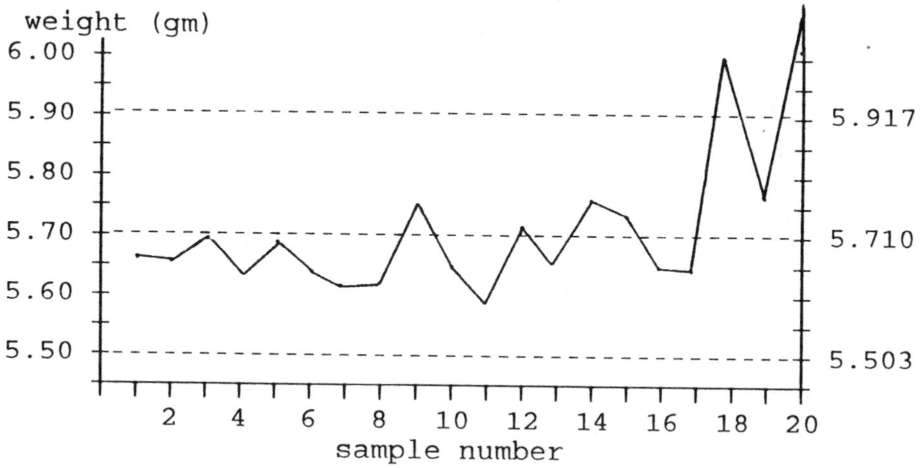

weight (gm)

As concluded by the original method, the process is out of statistical control.

14. For eight consecutive points all to lie above or below the mean, the first point can be anywhere and the next seven points each have probability .5 of "matching" it. The probability that all seven match to produce eight consecutive points as desired is $(.5)^7 = .0078125$.

12-3 Control Charts for Attributes

1. This process is within statistical control. Since the first third of the sample means are generally less than the overall mean, the middle third are generally more than the overall mean, and the final third are generally less than the overall mean, however, one may wish to check future analyses to see whether such a patter tends to repeat itself.

2. This process is within statistical control.

3. This process is out of statistical control. There is an upward trend, and there is a point above the upper control limit.

4. This process is out of statistical control. There are points lying beyond the control limits and there appears to be an upward trend.

5. $\bar{p} = (\Sigma x)/(\Sigma n) = (3+2+...+5)/(21)(300) = 108/6300 = .0171$
   $\sqrt{\bar{p}\cdot\bar{q}/n} = \sqrt{(.017)(.983)/300} = .00749$
   $LCL = \bar{p} - 3\sqrt{\bar{p}\cdot\bar{q}/n} = .0171 - 3(.00749) = .0171 - .0225 = 0$ [since it cannot be negative]
   $UCL = \bar{p} + 3\sqrt{\bar{p}\cdot\bar{q}/n} = .0171 + 3(.00749) = .0171 + .0225 = .0396$
   NOTE: The 21 sample proportions are: .010 .007 .013 .023 .010 .050 .060 .007 .020 .013 .010 .017 .013 .020 .017 .007 .013 .010 .020 .003 .017

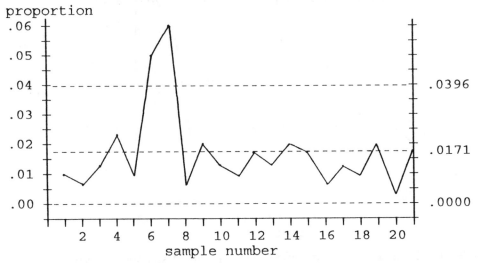

The process is out of statistical control because there are points above the upper control limit. If there were temporary employees on duty on precisely the days the process was out of statistical control, then either the temporary employees need to be better trained/screened or temporary employees should not be hired -- perhaps vacations could be staggered, output could be reduced, or regular employees could put in overtime.

6. The process is within statistical control.

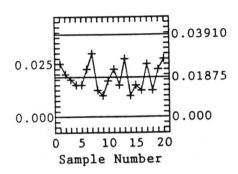

7. $\bar{p} = (\Sigma x)/(\Sigma n) = (4+3+...+5)/(18)(120) = 89/2160 = .0412$
$\sqrt{\bar{p}\cdot\bar{q}/n} = \sqrt{(.041)(.959)/120} = .0181$
$LCL = \bar{p} - 3\sqrt{\bar{p}\cdot\bar{q}/n} = .0412 - 3(.0181) = .0412 - .0544 = 0$ [since it cannot be negative]
$UCL = \bar{p} + 3\sqrt{\bar{p}\cdot\bar{q}/n} = .0412 + 3(.0181) = .0412 + .0544 = .0956$
NOTE: The 18 sample proportions are: .033 .025 .025 .017 .042 .033 .025 .133 .050 .025
.025 .017 .042 .058 .067 .033 .050 .042

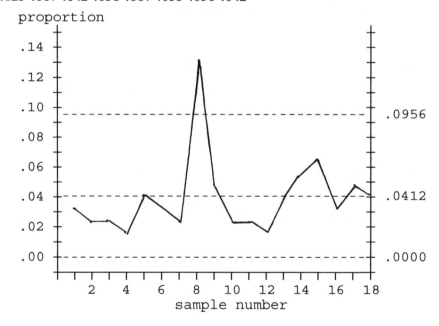

The process is out of statistical control because there is a point above the upper control limit.
Since it is an isolated point, the company should examine the circumstances associated with that
one specific time period (to look for a specific correctable cause -- e.g., a new person on duty)
before ordering a general shutdown and/or retooling.

8. The process is within statistical control.

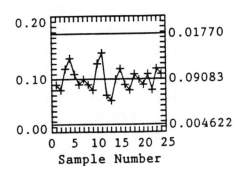

9. $\bar{p} = (\Sigma x)/(\Sigma n) = (10+8+\ldots+11)/(20)(400) = 150/8000 = .01875$

    a. $\hat{p} = .01875$
       $\alpha = .05$
       $\hat{p} \pm z_{.025}\sqrt{\hat{p}\hat{q}/n}$
       $.01875 \pm 1.960\sqrt{(.01875)(.98125)/8000}$
       $.01875 \pm .00297$
       $.0158 < p < .0217$

    b. original claim: $p \leq .01$ [normal approximation to the binomial, use z]
       H_o: $p \leq .01$
       H_1: $p > .01$
       $\alpha = .05$
       C.R. $z > z_{.05} = 1.645$
       calculations:
          $z_{\hat{p}} = (\hat{p} - \mu_{\hat{p}})/\sigma_{\hat{p}}$
             $= (.01875 - .01)/\sqrt{(.01)(.99)/8000}$
             $= .00875/.00111$
             $= 7.866$
       conclusion:
       Reject H_o; there is sufficient evidence to conclude that $p > .01$.

10. The charts are identical except for the labels. Since the labels on the np chart are n = 200 times the labels on the chart for p, the center line is $200(.016) = 3.20$ and the upper control limit is $200(.04262) = 8.52$.

11. $\bar{p} = (\Sigma x)/(\Sigma n) = .05$
In both parts (a) and (b), the center line occurs at .05.

a. $n = 100$
$\sqrt{\bar{p} \cdot \bar{q}/n} = \sqrt{(.05)(.95)/100} = .0218$
$LCL = \bar{p} - 3\sqrt{\bar{p} \cdot \bar{q}/n} = .05 - 3(.0218) = .05 - .0654 = 0$ [since it cannot be negative]
$UCL = \bar{p} + 3\sqrt{\bar{p} \cdot \bar{q}/n} = .05 + 3(.0218) = .05 + .0654 = .1154$

b. $n = 300$
$\sqrt{\bar{p} \cdot \bar{q}/n} = \sqrt{(.05)(.95)/300} = .0126$
$LCL = \bar{p} - 3\sqrt{\bar{p} \cdot \bar{q}/n} = .05 - 3(.0126) = .05 - .0378 = .0123$
$UCL = \bar{p} + 3\sqrt{\bar{p} \cdot \bar{q}/n} = .05 + 3(.0126) = .05 + .0378 = .0878$

c. The lower and upper control limits are closer to the center line in part (b). This has the advantage of being better able (i.e., on the basis of less deviance from the long run average) to detect when the process is out of statistical control, but it has the disadvantage of requiring the examination of a larger sample size. The chart in part (b) would be better able to detect a shift from 5% to 10% because the larger sample size would cause less fluctuation about the 5% or 10% long run averages and make the shift more noticeable.

Review Exercises

1. $\bar{\bar{x}} = \Sigma\bar{x}/k = 5028/20 = 2.514$

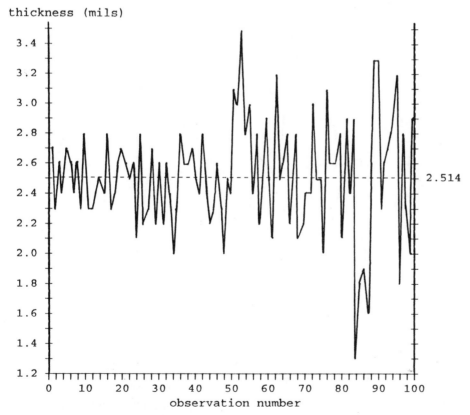

It appears that there __may__ be an increase in variation.

2. The process is out of statistical control because there is a point beyond the upper control limit and there appears to be an upward trend.

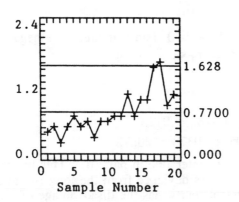

3. $\bar{\bar{x}} = \Sigma\bar{x}/k = 5028/20 = 2.514$
   $\bar{R} = \Sigma R/k = 154/20 .77$
   $LCL = \bar{\bar{x}} - A_2\bar{R} = 2.514 - (.577)(.77) = 2.514 - .444 = 2.070$
   $UCL = \bar{\bar{x}} + A_2\bar{R} = 2.514 + (.577)(.77) = 2.514 + .444 = 2.958$

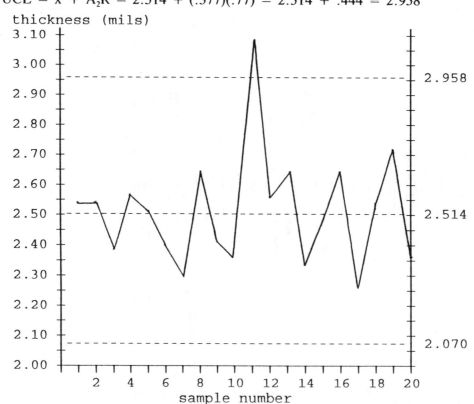

The process is out of statistical control because there is a point above the upper control limit and requires correction.

4. While there is one point below the lower control limit, some would say that having fewer defects than expected is not really a problem. Since the unusually low number was followed by the highest number on the chart, it might be appropriate to investigate whether one shift could have some how passed its defects on to the next one.

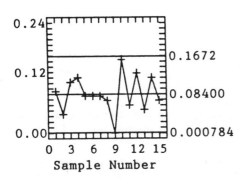

5. $\bar{p} = (\Sigma x)/(\Sigma n) = (2+6+\ldots+4)/(24)(50) = 145/1200 = .1208$

$\sqrt{\bar{p}\cdot\bar{q}/n} = \sqrt{(.1208)(.8792)/50} = .0461$

LCL $= \bar{p} - 3\sqrt{\bar{p}\cdot\bar{q}/n} = .1208 - 3(.0461) = .1208 - .1383 = 0$ [since it cannot be negative]

UCL $= \bar{p} + 3\sqrt{\bar{p}\cdot\bar{q}/n} = .1208 + 3(.0461) = .1208 + .1383 = .2591$

NOTE: The 24 sample proportions are: .04 .12 .06 .08 .22 .20 .18 .24 .04 .06 .08 .04 .06 .10 .08 .12 .24 .20 .22 .24 .10 .04 .06 .08

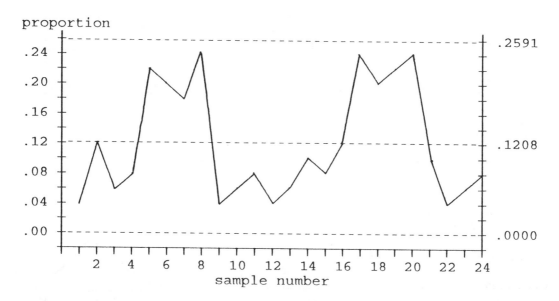

The process is out of statistical control because there is a cyclical pattern; corrective action should be taken.

Chapter 13

Nonparametric Statistics

13-2 Sign Test

NOTE: Table A-7 gives only x_L, the <u>lower</u> critical value for the sign test. Accordingly, the text lets x be the <u>smaller</u> of the number of $+$'s or the number of $-$'s and warns the user to use common sense to avoid concluding the reverse of what the data indicates.

An alternative approach maintains the natural agreement between the alternative hypothesis and the critical region and is consistent with the logic and notation of parametric tests. Let x <u>always</u> be the number of $+$'s. The problem's symmetry means that the upper critical value is $x_U = n - x_L$.

Since this alternative approach builds directly on established patterns and provides insight into rationale of the sign test, its C.R. and calculations and picture [notice that if H_o is true then $\mu_x = n/2$] for each exercise are given to the right of the method using only lower critical values.

1. Let Humorous be group 1.
   claim: median difference $= 0$

pair	A	B	C	D	E	F	G	H
H-S	+	-	+	+	+	+	+	+

   n = 8 +'s and -'s

   H_o: median difference $= 0$
   H_1: median difference $\neq 0$
   $\alpha = .05$
   C.R. $x \leq x_{L,8,.025} = 0$ <u>OR</u>   C.R. $x \leq x_{L,8,.025} = 0$
   $\qquad\qquad\qquad\qquad\qquad\qquad\qquad x \geq x_{U,8,.025} = 8\text{-}0 = 8$

   calculations: calculations:
   $\quad$ x = 1 $\quad$ x = 7
   conclusion:
   $\quad$ Do not reject H_o; there is not sufficient evidence to conclude that median difference $\neq 0$.

2. Let MicroAir be group 1.
   claim: median difference $= 0$

pair	A	B	C	D	E	F	G	H	I	J	K
M-F	+	+	+	+	0	+	+	-	+	+	+

   n = 10 +'s and -'s

   H_o: median difference $= 0$
   H_1: median difference $\neq 0$
   $\alpha = .05$
   C.R. $x \leq x_{L,10,.025} = 1$ <u>OR</u>   C.R. $x \leq x_{L,10,.025} = 1$
   $\qquad\qquad\qquad\qquad\qquad\qquad\qquad x \geq x_{U,10,.025} = 10\text{-}1 = 9$

   calculations: calculations:
   $\quad$ x = 1 $\quad$ x = 9
   conclusion:
   $\quad$ Reject H_o; there is sufficient evidence to conclude that median difference $\neq 0$ (in fact, median difference > 0).
   Since MicroAir takes significantly longer to process requests, choose Flight Services.

3. Let Before be group 1.
   claim: median difference = 0

pair	A	B	C	D	E	F	G	H	I	J
B-A	+	+	+	-	+	+	0	+	+	+

   n = 9 +'s and -'s

   H_0: median difference = 0
   H_1: median difference ≠ 0
   α = .05
   C.R. $x \leq x_{L,9,.025} = 1$ <u>OR</u>   C.R. $x \leq x_{L,9,.025} = 1$
                                               $x \geq x_{U,9,.025} = 9-1 = 8$

   calculations: calculations:
   x = 1   x = 8
   conclusion:
   Reject H_0; there is sufficient evidence to conclude that median difference ≠ 0 (in fact, median difference > 0).
   We can be 95% certain that the drug lowers systolic blood pressure.

4. Let Before be group 1.
   claim: median difference > 0

pair	A	B	C	D	E	F	G	H
B-A	-	+	+	+	+	+	+	+

   n = 8 +'s and -'s

   H_0: median difference ≤ 0
   H_1: median difference > 0
   α = .05
   C.R. $x \leq x_{L,8,.05} = 1$ <u>OR</u>   C.R. $x \geq x_{U,8,.05} = 8-1 = 7$
   calculations: calculations:

   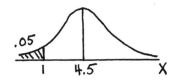

   x = 1   x = 7
   conclusion:
   Reject H_0; there is sufficient evidence to conclude that median difference > 0 -- i.e., the pain scores are higher before hypnosis.
   Yes; hypnosis does appear to be effective in reducing pain.
   NOTE: When using the lower-tail approach above at the left in a one-tailed test, check whether any result in the critical region is in the direction of the alternative hypothesis -- i.e., do not automatically reject H_0 because the smaller of the number of +'s or -'s is in the critical region.

5. Let Before be group 1.
   claim: median difference < 0

pair	1	2	3	4	5	6	7	8	9	10
B-A	-	-	+	-	0	-	-	-	-	-

   n = 9 +'s and -'s

   H_0: median difference ≥ 0
   H_1: median difference < 0
   α = .05
   C.R. $x \leq x_{L,9,.05} = 1$ <u>OR</u>   C.R. $x \leq x_{L,9,.05} = 1$
   calculations: calculations:
   x = 1   x = 1
   conclusion:
   Reject H_0; there is sufficient evidence to conclude that median difference < 0.
   Yes; we can be 95% certain that the new course is effective. The Dean of Curriculum should support the proposal.

6. Let the pre's be group 1.
   claim: median difference = 0

   ```
   pre - post | +  0  0  0  +  +  +  +  0  +
   n = 6 +'s and -'s
   ```

   H_o: median difference = 0
   H_1: median difference $\neq$ 0
   α = .05
   C.R. $x \leq x_{L,6,.025} = 0$ <u>OR</u>   C.R. $x \leq x_{L,6,.025} = 0$
   $\phantom{C.R. x \leq x_{L,6,.025} = 0 OR C.R. } x \geq x_{U,6,.025} = 6\text{-}0 = 6$

   calculations: calculations:
     $x = 0$ $x = 6$
   conclusion:
     Reject H_o; there is sufficient evidence to conclude that median difference $\neq$ 0 (in fact,
     median difference > 0 -- i.e., the pre weights are heavier).
   Since the pre weights are heavier, we conclude there is a weight loss during training.

7. Let Test I be group 1.
   claim: median difference = 0

   ```
   pair | 1  2  3  4  5  6  7  8  9 10
   I-II |  -  -  -  +  -  -  -  -  -  -
   n = 10 +'s and -'s
   ```

   H_o: median difference = 0
   H_1: median difference $\neq$ 0
   α = .05
   C.R. $x \leq x_{L,10,.025} = 1$ <u>OR</u>   C.R. $x \leq x_{L,10,.025} = 1$
   $\phantom{C.R. x \leq x_{L,10,.025} = 1 OR C.R. } x \geq x_{U,10,.025} = 10\text{-}1 = 9$

   calculations: calculations:
     $x = 1$ $x = 9$
   conclusion:
     Reject H_o; there is sufficient evidence to conclude that median difference $\neq$ 0 (in fact,
     median difference < 0).
   We can be 95% certain that the tests give different results (and that Test I tends to give lower
     scores). Further research is necessary to determine which of the tests best measures IQ.

   NOTE: This test may not be the most appropriate one for the intended purpose. The sign test
   measures whether the overall median difference between the two tests is 0, and <u>not</u> whether the
   two tests give the same results for individuals. If the median difference were 0 and there were
   approximately half +'s and half -'s, the tests could still be very different. It could be, for
   example, that one test grossly overestimates the IQ's of half the students and grossly
   underestimates the IQ's of the other half -- a distinct possibility if the test is sexually or
   ethnically biased so as to help half the students and hurt the others.

8. Let Before be group 1.

claim: median difference < 0 (assuming "affect" to mean "improve")

```
B-A | - - - - - - + 0 + -
n = 9 +'s and -'s
```

H_o: median difference ≥ 0
H_1: median difference < 0
$\alpha = .05$

C.R. $x \leq x_{L,9,.05} = 1$ **OR**   C.R. $x \leq x_{L,9,.05} = 1$
calculations: calculations:
  x = 2   x = 2

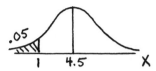

conclusion:

Do not reject H_o; there is not sufficient evidence to conclude that median difference < 0.
The course appears to be helpful, but we cannot be 95% certain that the observed
improvements were not merely chance fluctuations in performance.

9. Let Sitting be group 1.

claim: median difference = 0

pair	1	2	3	4	5	6	7	8	9	10
Si-Su	+	+	+	+	+	+	+	+	+	+

n = 10 +'s and -'s

H_o: median difference $= 0$
H_1: median difference $\neq 0$
$\alpha = .05$

C.R. $x \leq x_{L,10,.025} = 1$ **OR**   C.R. $x \leq x_{L,10,.025} = 1$
                                    $x \geq x_{U,10,.025} = 10-1 = 9$
calculations: calculations:
  x = 0   x = 10

conclusion:

Reject H_o; there is sufficient evidence to conclude that median difference $\neq 0$ (in fact,
median difference > 0).

10. Let Before be group 1.

claim: median difference = 0

```
B-A | + + + + - + + + + 0 + + +
n = 12 +'s and -'s
```

H_o: median difference $= 0$
H_1: median difference $\neq 0$
$\alpha = .05$ [or another appropriate value]

C.R. $x \leq x_{L,12,.025} = 2$ **OR**   C.R. $x \leq x_{L,12,.025} = 2$
                                      $x \geq x_{U,12,.025} = 12-2 = 10$
calculations: calculations:
  x = 1   x = 11

conclusion:

Reject H_o; there is sufficient evidence to conclude that median difference $\neq 0$ (in fact,
median difference > 0 -- i.e., the anxiety scores are higher before the program).

11. Let favoring Covariant be a success.
    claim: $p > .5$

    7 +'s   3 -'s
    n = 10 +'s or -'s

    H_o: $p \le .5$
    H_1: $p > .5$
    $\alpha = .05$
    C.R. $x \le x_{L,10,.05} = 1$ **OR**   C.R. $x \ge x_{U,10,.05} \doteq 10\text{-}1 = 9$
    calculations: calculations:
    , $x = 3$ $x = 7$
    conclusion:
        Do not reject H_o; there is not sufficient evidence to conclude that $p > .5$.

12. Let scores above 40 be +'s.
    claim: median ≥ 40

    32 +'s   43 -'s
    n = 75 +'s or -'s
    Since n > 25, use z with
    $\mu_x = n/2 = 75/2 = 37.5$
    $\sigma_x = \sqrt{n}/2 = \sqrt{75}/2 = 4.330$

    H_o: median ≥ 0
    H_1: median < 0
    $\alpha = .05$
    C.R. $z < -z_{.05} = -1.645$ **OR**   C.R. $z < -z_{.05} = -1.645$
    calculations: calculations:
        $x = 32$ $x = 32$
        $z_x = [(x+.5)-\mu_x]/\sigma_x$ $z_x = [(x+.5)-\mu_x]/\sigma_x$
            $= [32.5 - 37.5]/4.330$ $= [32.5 - 37.5]/4.330$
            $= -5/4.330 = -1.155$ $= 5/4.330 = -1.155$
    conclusion:
        Do not reject H_o; there is not sufficient evidence to conclude that median < 40.

13. Let the post-training scores be group 1 (i.e., let those with improved grades be +'s).
    claim: median improvement > 0

    59 +'s   36 -'s   5 0's
    n = 95 +'s or -'s
    Since n > 25, use z with
    $\mu_x = n/2 = 95/2 = 47.5$
    $\sigma_x = \sqrt{n}/2 = \sqrt{95}/2 = 4.873$

    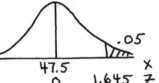

    H_o: median improvement ≤ 0
    H_1: median improvement > 0
    $\alpha = .05$
    C.R. $z < -z_{.05} = -1.645$ **OR**   C.R. $z > z_{.05} = 1.645$
    calculations: calculations: [See the NOTE following the exercise.]
        $x = 36$ $x = 59$
        $z_x = [(x+.5)-\mu_x]/\sigma_x$ $z_x = [(x-.5)-\mu_x]/\sigma_x$
            $= [36.5 - 47.5]/4.873$ $= [58.5 - 47.5]/4.873$
            $= -11/4.873 = -2.257$ $= 11/4.863 = 2.257$
    conclusion:
        Reject H_o; there is sufficient evidence to conclude that median improvement > 0.

NOTE: The correction for continuity is a conservative adjustment intending to make less likely a false rejection of H_o by shifting the x value .5 units toward the middle. When x is the smaller of the number of +'s or the number of -'s, this always involves replacing x with x+.5. In the alternative approach, x is replaced with either x+.5 or x-.5 according to which one shifts the value toward the middle -- i.e., with x+.5 when $x < \mu_x = (n/2)$, and with x-.5 when $x > \mu_x = (n/2)$.

14. Let positive increases be +'s.
claim: median increase > 0

```
    32 +'s        18 -'s
    n = 50 +'s or -'s
```
Since n > 25, use z with
$\mu_x = n/2 = 50/2 = 25$
$\sigma_x = \sqrt{n}/2 = \sqrt{50}/2 = 3.536$

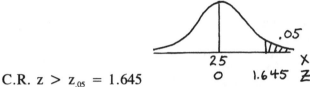

H_o: median increase ≤ 0
H_1: median increase > 0
$\alpha = .05$
C.R. $z < -z_{.05} = -1.645$ OR   C.R. $z > z_{.05} = 1.645$
calculations: calculations: [See the NOTE preceding the exercise.]
   $x = 18$ $x = 36$
   $z_x = [(x+.5)-\mu_x]/\sigma_x$ $z_x = [(x-.5)-\mu_x]/\sigma_x$
       $= [18.5 - 25]/3.536$ $= [31.5 - 25]/3.536$
       $= -6.5/3,536 = -1.838$ $= 6.5/3.536 = 1.838$
conclusion:
Reject H_o; there is sufficient evidence to conclude that median increase > 0.

15. Let those greater than the hypothesized median be +'s.
claim: median $= 0$

```
    36 +'s        22 -'s        2 0's
    n = 58 +'s or -'s
```
Since n > 25, use z with
$\mu_x = n/2 = 58/2 = 29$
$\sigma_x = \sqrt{n}/2 = \sqrt{58}/2 = 3.808$

H_o: median $= 0$
H_1: median $\neq 0$
$\alpha = .01$
C.R. $z < -z_{.005} = -2.575$ OR   C.R. $z < -z_{.005} = -2.575$
                                   $z > z_{.005} = 2.575$

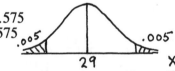

calculations: calculations:
   $x = 22$ $x = 36$
   $z_x = [(x+.5)-\mu_x]/\sigma_x$ $z_x = [(x-.5)-\mu_x]/\sigma_x$
       $= [22.5 - 29]/3.808$ $= [35.5 - 29]/3.808$
       $= -6.5/3.808$ $= 6.5/3.808$
       $= -1.707$ $= 1.707$
conclusion:
Do not reject H_o; there is not sufficient evidence to conclude that median $\neq 40$.

16. Let those with increased reaction times be $+$'s.
claim: median change $= 0$

    22 +'s   6 -'s   2 0's
    n = 28 +'s or -'s

Since $n > 25$, use z with
    $\mu_x = n/2 = 28/2 = 14$
    $\sigma_x = \sqrt{n}/2 = \sqrt{28}/2 = 2.646$

H_o: median change $= 0$
H_1: median change $\neq 0$
$\alpha = .01$
C.R. $z < -z_{.005} = -2.575$ **OR**   C.R. $z < -z_{.005} = -2.575$
                                             $z > z_{.005} = 2.575$

calculations: calculations:
    x = 6   x = 22
    $z_x = [(x+.5)-\mu_x]/\sigma_x$ $z_x = [(x-.5)-\mu_x]/\sigma_x$
      $= [6.5 - 14]/2.835$ $= [21.5 - 14]/2.646$
      $= -7.5/2.646 = -2.858$ $= 7.5/2.646 = 2.835$

conclusion:
    Reject H_o; there is sufficient evidence to conclude that median change $\neq 0$ (in fact, median change > 0 -- alcohol increases reaction times).
Yes; it appears the companies are justified in insuring drinkers at higher rates.

17. Let the expensive pistol scores be group 1.
claim: median difference $= 0$

    24 +'s   16 -'s
    n = 40 +'s or -'s

Since $n > 25$, use z with
    $\mu_x = n/2 = 40/2 = 20$
    $\sigma_x = \sqrt{n}/2 = \sqrt{40}/2 = 3.162$

H_o: median difference $= 0$
H_1: median difference $\neq 0$
$\alpha = .05$
C.R. $z < -z_{.025} = -1.960$ **OR**   C.R. $z < -z_{.025} = -1.960$
                                               $z > z_{.025} = 1.960$

calculations: calculations:
    x = 16   x = 24
    $z_x = [(x+.5)-\mu_x]/\sigma_x$ $z_x = [(x-.5)-\mu_x]/\sigma_x$
      $= [16.5 - 20]/3.162$ $= [23.5 - 20]/3.162$
      $= -3.5/3.162 = -1.107$ $= 3.5/3.162 = 1.107$

conclusion:
    Do not reject H_o; there is not sufficient evidence to conclude that median difference $\neq 0$.
Yes; on the basis of this test the less expensive pistol should probably be purchased.

18. Let amounts above 640 be $+$'s.
claim: median ≥ 640
    + + 0 + - - - - - - - - - -
    n = 14 +'s and -'s

H_o: median $\geq$ 640
H_1: median $<$ 640
α = .05
C.R. x $\leq$ $x_{L,14,.05}$ = 3   OR   C.R. x $\leq$ $x_{L,14,.05}$ = 3
calculations: calculations:
    x = 3   x = 3

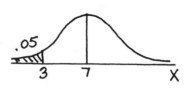

conclusion:
    Reject H_o; there is sufficient evidence to conclude that median $<$ 640.

19. Let those greater than the hypothesized median be +'s.
    claim: median = 5.670   Since n $>$ 25, use z with

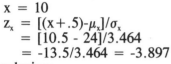

    10 +'s   38 -'s   2 0's μ_x = n/2 = 48/2 = 24
    n = 48 +'s or -'s σ_x = $\sqrt{n}$/2 = $\sqrt{48}$/2 = 3.464

H_o: median = 5.670
H_1: median $\neq$ 5.670
α = .05 [or another appropriate value]
C.R. z $<$ -$z_{.025}$ = -1.960   OR   C.R. z $<$ -$z_{.025}$ = -1.960
                                                      z $>$ $z_{.025}$ = 1.960

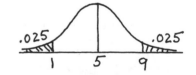

calculations: calculations:
    x = 10   x = 10
    z_x = [(x+.5)-μ_x]/σ_x z_x = [(x+.5)-μ_x]/σ_x
        = [10.5 - 24]/3.464   = [10.5 - 24]/3.464
        = -13.5/3.464 = -3.897   = -13.5/3.464 = -8.897
conclusion:
    Reject H_o; there is sufficient evidence to conclude that median $\neq$ 5.670 (in fact,
    median $<$ 5.670).

20. Let those with greater 8:00 am temperatures be +'s.
    claim: median difference = 0
    - - - - + + - 0 - - -
    n = 10 +'s and -'s

H_o: median difference = 0
H_1: median difference $\neq$ 0
α = .05
C.R. x $\leq$ $x_{L,10,.025}$ = 1   OR   C.R. x $\leq$ $x_{L,10,.025}$ = 1
                                                  x $\geq$ $x_{U,9,.025}$ = 9
calculations: calculations:
    x = 2   x = 2
conclusion:
    Do not reject H_o; there is not sufficient evidence to conclude that median difference $\neq$0.

21. The n=15 scores arranged in order are: 1 2 3 6 7 8 8 9 10 11 14 17 23 25 30.
    From Table A-7, k = 3.
    The 95% confidence interval is x_{k+1} $<$ median $<$ x_{n-k}
                                    x_4 $<$ median $<$ x_{12}
                                    6 $<$ median $<$ 17

NOTE: From the list of ordered scores and the fact that k = 3, it is apparent that one would
reject a hypothesized median of 6 or less or of 17 or more. As in the parametric cases, the
1-α confidence interval is composed of precisely those values not rejected in a two-tailed test at
the α level.

22. Let scores above 100 be +'s.
claim: median $\geq$ = 100

    40 +'s   60 -'s   21 0's

using the usual method: discarding the zeros
    n = 100 +'s or -'s
Since n > 25, use z with
    μ_x = n/2 = 100/2 = 50
    σ_x = $\sqrt{n}$ /2 = $\sqrt{100}$/2 = 5.000

H_o: median $\geq$ 0
H_1: median < 0
α = .05
C.R. z < $-z_{.05}$ = -1.645
calculations:
    x = 40
    z_x = $[(x+.5)-\mu_x]/\sigma_x$
        = [40.5 - 50]/5.000 = -9.5/5.000 = -1.900
conclusion:
    Reject H_o; there is sufficient evidence to conclude that median < 100.

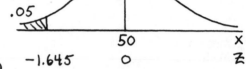

using the second method: counting half the zeros each way (and discarding the odd zero)
50 +'s   70 -'s
n = 120 +'s or -'s
Since n > 25, use z with
    μ_x = n/2 = 120/2 = 60
    σ_x = $\sqrt{n}$ /2 = $\sqrt{120}$/2 = 5.477

H_o: median $\geq$ 0
H_1: median < 0
α = .05
C.R. z < $-z_{.05}$ = -1.645
calculations:
    x = 50
    z_x = $[(x+.5)-\mu_x]/\sigma_x$
        = [50.5 - 60]/5.477 = -9.5/5.477 = -1.734
conclusion:
    Reject H_o; there is sufficient evidence to conclude that median < 100.

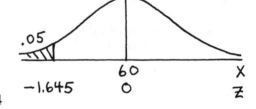

using the third method: counting the zeros in a one-tailed test to favor H_o
61 +'s   60 -'s
n = 161 +'s or -'s
Since n > 25, use z with
    μ_x = n/2 = 121/2 = 60.5
    σ_x = $\sqrt{n}$ /2 = $\sqrt{121}$/2 = 5.500

H_o: median $\geq$ 0
H_1: median < 0
α = .05
C.R. z < $-z_{.05}$ = -1.645
calculations:
    x = 60
    z_x = $[(x+.5)-\mu_x]/\sigma_x$
        = [60.5 - 60.5]/5.500 = 0/5.500 = 0
conclusion:
    Do not reject H_o; there is not sufficient evidence to conclude that median < 100.

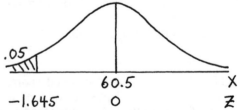

23. Let having high blood pressure be a success (i.e., those with high blood pressure are +'s).
claim: $p \geq .5$

```
n-50 +'s (the minority of the sample)
   50 -'s (the majority of the sample)
 n = # of +'s or -'s
```

Since $n > 25$, use z with
$\mu_x = n/2$
$\sigma_x = \sqrt{n}/2$

H_o: $p \geq .50$
H_1: $p < .50$
$\alpha = .01$
C.R. $z < -z_{.01} = -2.327$
calculations:
$x = n - 50$

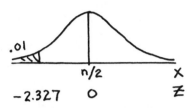

$z_x = [(x+.5)-\mu_x]/\sigma_x$	multiply by σ_x
$z_x\sigma_x = (x+.5)-\mu_x$	substitute the given values
$-2.327 \cdot \sqrt{n}/2 = n - 50 + .5 - (n/2)$	multiply by 2
$-2.327 \cdot \sqrt{n} = 2n - 100 + 1 - n$	simplify
$-2.327 \cdot \sqrt{n} = n - 99$	square
$5.415n = n^2 - 198n + 9801$	simplify
$0 = n^2 - 203.415n + 9801$	use quadratic formula: $a = 1$
$n = [-(-203.415) \pm \sqrt{(-203.415)^2 - 4(1)(9801)}]/2(1)$	$b = -203.415$
$= [203.415 \pm \sqrt{2173.633}]/2$	$c = 9801$
$= [203.415 \pm 46.662]/2$	
$= 250.037/2$ or $156.793/2$	
$= 125.02$ or 78.40	

If $n \leq 78.40$, the 50 persons without high blood pressure form a large enough majority to allow the conclusion that $p < .50$. Since n must be a whole number, 78 is the largest value n can assume.
NOTE: If $n \geq 125.02$, the 50 persons without high blood pressure would form a small enough minority to allow the conclusion that $p > .50$.

24. Use the binomial table for $p=.5$ and reject for lower-tailed x values summing to less than .03.
For $1 \leq n \leq 5$, $P(x=0) > .03$
For $6 \leq n \leq 8$, $P(x \leq 0) \leq .03$ but $P(x \leq 1) > .03$
For $9 \leq n \leq 11$, $P(x \leq 1) \leq .03$ but $P(x \leq 2) > .03$
For $12 \leq n \leq 13$, $P(x \leq 2) \leq .03$ but $P(x \leq 3) > .03$
For $14 \leq n \leq 15$, $P(x \leq 3) \leq .03$ but $P(x \leq 4) > .03$
In summary, the desired critical values are as follows.

```
n: 1 2 3 4 5 6 7 8 9 10 11 12 13 14 15
x: * * * * * 0 0 0 1  1  1  2  2  3  3
```

13-3 Wilcoxon Signed-Ranks Test for Two Dependent Samples

NOTE: Table A-8 gives only T_L, the <u>lower</u> critical value for the signed-ranks test. Accordingly, the text lets T be the <u>smaller</u> of the sum of positive ranks or the sum of the negative ranks and warns the user to use common sense to avoid concluding the reverse of what the data indicates.

An alternative approach maintains the natural agreement between the alternative hypothesis and the critical region and is consistent with the logic and notation of parametric tests. Let T <u>always</u> be the sum of the positive ranks. By symmetry, the upper critical value is $T_U = \Sigma R - T_L$.

Since this alternative approach builds directly on established patterns and provides insight into rationale of the signed-ranks test, its C.R. and calculations and picture [notice that if H_o is true then $\mu_T = \Sigma R/2$] for each exercise are given to the right of the method using only lower critical values.

1. Let Humorous be group 1.
   claim: the populations have the same distribution

pair	A	B	C	D	E	F	G	H
H-S	2.3	-1.0	1.7	1.6	1.4	2.5	1.3	2.6
R	6	-1	5	4	3	7	2	8

   n = 8 non-zero ranks
   $\Sigma R- = 1$
   $\Sigma R+ = 35$
   $\Sigma R = 36$ [check: $\Sigma R = n(n+1)/2 = 8(9)/2 = 36$]

   H_o: the populations have the same distribution
   H_1: the populations have different distributions
   $\alpha = .01$
   C.R. $T \leq T_{L,8,.005} = 0$ <u>OR</u>   C.R. $T \leq T_{L,8,.005} = 0$
                                                      $T \geq T_{U,8,.005} = 36-0 = 36$
   calculations: calculations:
   $T = 1$ $T = 35$

   conclusion:
   Do not reject H_o; there is not sufficient evidence to conclude that the populations have different distributions

2. Let MicroAir be group 1.
   claim: the populations have the same distribution

pair	A	B	C	D	E	F	G	H	I	J	K
M-F	3	3	4	1	0	5	9	-1	3	7	13
R	4	4	6	1.5	-	7	9	-1.5	4	8	10

   n = 10 non-zero ranks
   $\Sigma R- = 1.5$
   $\Sigma R+ = 53.5$
   $\Sigma R = 55$ [check: $\Sigma R = n(n+1)/2 = 10(11)/2 = 55$]

   H_o: the populations have the same distribution
   H_1: the populations have different distributions
   $\alpha = .05$
   C.R. $T \leq T_{L,10,.025} = 8$ <u>OR</u>   C.R. $T \leq T_{L,10,.025} = 8$
                                                      $T \geq T_{U,10,.025} = 55-8 = 47$
   calculations: calculations:
   $T = 1.5$ $T = 53.5$

   conclusion:
   Reject H_o; there is sufficient evidence to conclude that the populations have different distributions (in fact, the first population has larger scores).

3. Let Before be group 1.
   claim: the populations have the same distribution

pair	A	B	C	D	E	F	G	H	I	J
B-A	2	14	17	-7	17	12	0	15	33	16
R	1	4	7.5	-2	7.5	3	0	5	4	6

   n = 9 non-zero ranks
   $\Sigma R-$ = 2
   $\Sigma R+$ = 43
   ΣR = 45   [check: $\Sigma R = n(n+1)/2 = 9(10)/2 = 45$]

   H_o: the populations have the same distribution
   H_1: the populations have different distributions
   α = .05
   C.R. $T \leq T_{L,9,.025} = 6$ OR   C.R. $T \leq T_{L,9,.025} = 6$
                                              $T \geq T_{U,9,.025} = 45-6 = 39$
   calculations: calculations:
      T = 2   T = 43
   conclusion:
      Reject H_o; there is sufficient evidence to conclude that the populations have different
      distributions (in fact, the Before group has higher scores).

NOTE: This manual follows the text and the directions to the exercises of this section by using "the
populations have the same distribution" as the null hypothesis. To be more precise, the signed-
rank test doesn't test "distributions" but tests the "location" (i.e., central tendency -- as opposed to
variation) of distributions. The test discerns whether one group taken as a whole tends to have
higher or lower scores than another group taken as a whole. The test does not discern whether one
group is more variable than another. This distinction is made clear in the wording of the
conclusion when rejecting H_o in a two-tailed test (as in exercise #3 above) and when using a one-
tailed test (as in exercise #5 below). For further insight into the limitations of the test and its
insensitivity to differences in variation, see the NOTE accompanying exercise #7.
   In addition, each exercise uses a minus sign preceding ranks associated with negative
differences. While the ranks themselves are not negative, the use of the minus sign helps to
organize the information.

4. Let Before be group 1.
   claim: population 1 has higher scores

pair	A	B	C	D	E	F	G	H
H-S	-.2	4.1	1.6	1.8	3.2	2.0	2.9	9.6
R	-1	7	2	3	6	4	5	8

   n = 8 non-zero ranks
   $\Sigma R-$ = 1
   $\Sigma R+$ = 35
   ΣR = 36   [check: $\Sigma R = n(n+1)/2 = 8(9)/2 = 36$]

   H_o: the populations have the same distribution
   H_1: population 1 has higher scores
   α = .01
   C.R. $T \leq T_{L,8,.01} = 2$ OR   C.R. $T \geq T_{U,8,.01} = 36-2 = 34$
   calculations: calculations:
      T = 1   T = 35
   conclusion:
      Reject H_o; there is sufficient evidence to conclude that the population 1 has higher scores.

5. Let Before be group 1.
   claim: the Before population has lower scores

pair	1	2	3	4	5	6	7	8	9	10
B-A	-36	-50	15	-2	0	-22	-36	-6	-26	-13
R	-7.5	-9	4	-1	0	-5	-7.5	-2	-6	-3

   $n = 9$ non-zero ranks
   $\Sigma R- = 41$
   $\Sigma R+ = 4$
   $\Sigma R = 45$ [check: $\Sigma R = n(n+1)/2 = 9(10)/2 = 45$]

   H_0: the populations have the same distribution
   H_1: the Before population has lower scores
   $\alpha = .01$
   C.R. $T \leq T_{L,9,.01} = 3$ OR   C.R. $T \leq T_{L,9,.01} = 3$
   calculations: calculations:
   $T = 4$ $T = 4$

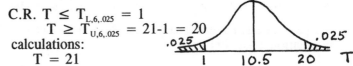

   conclusion:
   Do not reject H_0; there is not sufficient evidence to conclude that the Before population has lower scores.
   Since he cannot be 95% certain that the new course is effective, the Dean of Curriculum may not wish to support the proposal.

   NOTE: In general, the signed-rank test of this section is a more powerful test (i.e., better able to detect departures from H_0) than the sign test of the previous section. Accordingly, one can often reject H_0 with the signed-rank test but not with the sign test, or one can often find the results farther into the region of rejection with the signed-rank test than with the sign test. While it is mathematically possible to reject H_0 with the sign test but not with the signed-rank test, this exercise is not a proper illustration of that phenomenon -- since H_0 was rejected at the .05 level in the previous section and the .01 level was employed here. At the .05 level, the C.R. is $T \leq 6$ and H_0 is rejected as it was with the sign test in the previous section.

6. Let Pre be group 1.
   claim: the populations have the same distribution

pair	A	B	C	D	E	F	G	H	I	J	K
Pr-Po	5	0	0	0	0	8	1	1	4	0	1
R	5	-	-	-	-	6	2	2	4	-	2

   $n = 6$ non-zero ranks
   $\Sigma R- = 0$
   $\Sigma R+ = 21$
   $\Sigma R = 21$ [check: $\Sigma R = n(n+1)/2 = 6(7)/2 = 21$]

   H_0: the populations have the same distribution
   H_1: the populations have different distributions
   $\alpha = .05$
   C.R. $T \leq T_{L,6,.025} = 1$ OR   C.R. $T \leq T_{L,6,.025} = 1$
                                          $T \geq T_{U,6,.025} = 21-1 = 20$
   calculations: calculations:
   $T = 0$ $T = 21$
   conclusion:
   Reject H_0; there is sufficient evidence to conclude that the populations have different distributions (in fact, the Pre-training group has higher weights).

7. Let Test I be group 1.
   claim: the populations have the same distribution

pair	1	2	3	4	5	6	7	8	9	10
I-II	-7	-9	-2	4	-4	-4	-5	-7	-7	-4
R	-8	-10	-1	3.5	-3.5	-3.5	-6	-8	-8	-3.5

   n = 10 non-zero ranks
   ΣR- = 51.5
   ΣR+ = 3.5
   ΣR = 55.0   [check: ΣR = n(n+1)/2 = 10(11)/2 = 55]

   H_o: the populations have the same distribution
   H_1: the populations have different distributions
   α = .01
   C.R. $T \leq T_{L,10,.005} = 3$ OR   C.R. $T \leq T_{L,10,.005} = 3$
                                          $T \geq T_{U,10,.005} = 55-3 = 52$
   calculations: calculations:
     T = 3.5   T = 3.5

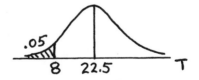

   conclusion:
     Do not reject H_o; there is not sufficient evidence to conclude that the populations have different distributions.

   NOTE: This test may not be the most appropriate one for the intended purpose. The signed-rank test measures overall differences between the groups and **not** whether the two tests give the same results for individuals. If one test gave half the students higher scores and the other half of the students scores that were lower by the same amounts, then ΣR- would equal ΣR+ (so we could not reject H_o) but the distributions would be very different. This happens if one test either (a) "enhances" the scores so that the top half of the students score higher than their true IQ's and the bottom half of the students score lower than their true IQ's or (b) fails to differentiate at all so that everyone is pulled toward the middle -- then the top half of the students score lower than their true IQ's and the bottom half of the students score higher than their true IQ's.

8. Let Before be group 1.
   claim: population 1 has lower scores

pair	A	B	C	D	E	F	G	H	I	J
B-A	-36	-8	-32	-4	-9	-29	13	0	2	-9
R	-9	-3	-8	-2	-4.5	-7	6	-	1	-4.5

   n = 9 non-zero ranks
   ΣR- = 38
   ΣR+ = 7
   ΣR = 45   [check: ΣR = n(n+1)/2 = 9(10)/2 =45]

   H_o: the populations have the same distribution
   H_1: population 1 has lower scores
   α = .05
   C.R. $T \leq T_{L,9,.05} = 8$ OR   C.R. $T \leq T_{L,9,.05} = 8$
   calculations: calculations:
     T = 7   T = 7
   conclusion:
     Reject H_o; there is sufficient evidence to conclude that population 1 has lower scores -- i.e., that the school is effective.

9. Let Sitting be group 1.
   claim: the populations have the same distribution

pair	1	2	3	4	5	6	7	8	9	10
Si-Su	.99	1.60	.98	.82	1.01	1.54	.21	.70	1.67	1.32
R	5	9	4	3	6	8	1	2	10	7

   n = 10 non-zero ranks
   $\Sigma R- = 0$
   $\Sigma R+ = 55$
   $\Sigma R = 55$ [check: $\Sigma R = n(n+1)/2 = 10(11)/2 = 55$]

   H_o: the populations have the same distribution
   H_1: the populations have different distributions
   $\alpha = .05$
   C.R. $T \leq T_{L,10,.025} = 8$ OR   C.R. $T \leq T_{L,10,.025} = 8$
   $\qquad\qquad\qquad\qquad\qquad\qquad\qquad T \geq T_{U,10,.025} = 55-8 = 47$

   calculations: calculations:
   $T = 0$ $T = 55$

   conclusion:
   Reject H_o; there is sufficient evidence to conclude that the populations have different distributions (in fact, the sitting positions have larger capacity scores).

10. Let Before be group 1.
   claim: the populations have the same distribution

pair	A	B	C	D	E	F	G	H	I	J	K	L	M
B-A	21	33	2	16	-1	12	4	8	16	0	18	18	15
R	11	12	2	7.5	-1	5	3	4	7.5	-	9.5	9.5	6

   n = 12 non-zero ranks
   $\Sigma R- = 1$
   $\Sigma R+ = 77$
   $\Sigma R = 78$ [check: $\Sigma R = n(n+1)/2 = 12(13)/2 = 78$]

   H_o: the populations have the same distribution
   H_1: the populations have different distributions
   $\alpha = .05$
   C.R. $T \leq T_{L,12,.025} = 14$ OR   C.R. $T \leq T_{L,12,.025} = 14$
   $\qquad\qquad\qquad\qquad\qquad\qquad\qquad T \geq T_{U,12,.025} = 78-14 = 64$

   calculations: calculations:
   $T = 1$ $T = 77$

   conclusion:
   Reject H_o; there is sufficient evidence to conclude that the populations have different distributions (in fact, the Before group has higher scores).

diff	rank
0	0
0	0
0	0
1	1
3	2.5
-3	-2.5
5	4.5
-5	-4.5
-6	-6
-8	-7
10	8
12	10
12	10
-12	-10
14	12.5
14	12.5
16	14.5
16	14.5
18	16
19	17.5
19	17.5
23	19
26	20
27	21.5
-27	-21.5
29	23
30	24
33	25.5
33	25.5
35	27
38	28
40	29
42	30
44	31
47	32
52	33.5
52	33.5
59	35
72	36

11. The manual system is group 1.
The 39 differences are listed in order at the right side of the page.

   a. claim: the populations have the same distribution
      n = 36 non-zero ranks (n > 30, use z approximation for test)
      $\Sigma R- = 51.5$
      $\Sigma R+ = 614.5$
      $\Sigma R = 666.0$ [check: $\Sigma R = n(n+1)/2 = 36(37)/2 = 666$]
   H_o: the populations have the same distribution
   H_1: the populations have different distributions
   $\alpha = .01$
   C.R. $z < z_{.005} = -2.575$ OR   C.R. $z < z_{.005} = -2.575$
   $\qquad\qquad\qquad\qquad\qquad\qquad\qquad\qquad z > z_{.005} = 2.575$

   calculations: calculations:
      T = 51.5   T = 614.5
      $z = (T - \mu_T)/\sigma_T$ $z = (T - \mu_T)/\sigma_T$
      $\quad = (51.5 - 333)/\sqrt{36(37)(73)/24}$ $\quad = (614.5 - 333)/63.651$
      $\quad = -281.5/63.651$ $\quad = 281.5/63.651$
      $\quad = -4.423$ $\quad = 4.423$
   conclusion:
      Reject H_o; there is sufficient evidence to conclude that the
      populations have different distributions (in fact, the manual
      system tends to produce longer times).

   b. claim: the scanner population has lower scores
      n = 36 non-zero ranks (n > 30, use z approximation for test)
      $\Sigma R- = 51.5$
      $\Sigma R+ = 614.5$
      $\Sigma R = 666.0$ [check $\Sigma R = n(n+1)/2 = 36(37)/2 = 666$
   H_o: the populations have the same distribution
   H_1: the manual population has higher scores
   $\alpha = .01$
   C.R. $z < -z_{.01} = -2.327$ OR   C.R. $z > z_{.01} = 2.327$
   calculations: calculations:
      T = 51.5   T = 614.5
      $z = (T - \mu_T)/\sigma_T$ $z = (T - \mu_T)/\sigma_T$
      $\quad = (51.5 - 333)/\sqrt{36(37)(73)/24}$ $\quad = (614.5 - 333)/63.651$
      $\quad = -281.5/63/651$ $\quad = 281.5/63.651$
      $\quad = -4.423$ $\quad = 4.423$
   conclusion:
      Reject H_o:; there is sufficient evidence to conclude that the
      manual population has higher scores -- i.e., that the scanner
      population has lower scores.

12. Since T is defined as the lower of $\Sigma R-$ and $\Sigma R+$, its smallest possible value is 0. Its largest
   value occurs when $\Sigma R- = \Sigma R+ = (1/2)(\Sigma R) = (1/2)[n(n+1)/2] = n(n+1)/4$.
   a. 0 and 8(9)/4 = 18
   b. 0 and 10(11)/4 = 27.5
   c. 0 and 50(51)/4 = 637.5

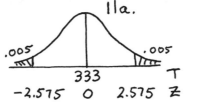

11a.

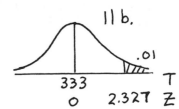

11 b.

13. For n = 100 non-zero ranks, use the z approximation with

$\mu_T = n(n+1)/4$
$\quad = 100(101)/4 = 2525$
$\sigma_T = \sqrt{[n(n+1)(2n+1)/24]}$
$\quad = \sqrt{[100(101)(201)/24]} = 290.84$

For $\alpha = .05$ in a two-tailed test, the critical z in 1.960.
The critical T (i.e., the smaller of $\Sigma R-$ and $\Sigma R+$) is found by solving as follows.

$z = (T - \mu_T)/\sigma_T < -1.96$
$(T - 2525)/290.84 < -1.96$
$\quad\quad T - 2525 < -570.05$
$\quad\quad\quad\quad T < 1954.95$

With the assumption of no ties in ranks, the C.R. is T ≤ 1954. NOTE: In general, however, assigning the average of the tied ranks would introduce (1/2)'s -- but not (1/3)'s, (1/4)'s, etc., since the average of n consecutive integers is always a either a whole number (if n is odd) or a half number (if n is even). The proper critical value for T would then be 1954.5.

14. The 50 differences obtained by subtracting 10,000 from each score, and the signed ranks of the differences are as follows.

-1910	-890	7810	2350	-6330		-19	-10	49	22	-46
4800	100	16580	7330	5970		40	1.5	50	48	44
-1200	1860	-2230	1550	2430		-11	18	-21	-15	23
780	3260	-4970	220	1430		8	31	-41	3	13
3490	1600	3520	-2530	-5490		33	16	34	-25	-42
4310	4760	3410	-5520	-2550		38	39	32	-43	-26
-2460	-6750	630	-3600	330		-24	-47	6	-35	4
-1840	510	-690	2700	-100		-17	5	-7	27	-1.5
-2800	-3830	2010	6200	1450		-29	-37	20	45	14
-1230	-860	-3180	-2720	-3610		-12	-9	-30	-28	-36

n = 100
$\Sigma R- = 611.5$
$\Sigma R+ = 663.6$
$\Sigma R = 1275.0$ [check: (50)(51)/2 = 1275

For n = 50 non-zero ranks, use the z approximation with

$\mu_T = n(n+1)/4$
$\quad = 50(51)/4 = 637.5$
$\sigma_T = \sqrt{[n(n+1)(2n+1)/24]}$
$\quad = \sqrt{[50(51)(101)/24]} = 103.59$

Let T = 611.5, the smaller of the sums.

H_o: median = 10,000
H_1: median ≠ 10,000
$\alpha = .05$
C.R. $z < -z_{.025} = -1.960$
$\quad\quad z >$ [not needed, since T is the smaller of $\Sigma R-$ and $\Sigma R+$]

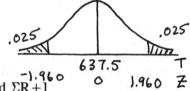

calculations:

$z = (T - \mu_T)/\sigma_T$
$\quad = (611.5 - 637.5)/103.59 = -26/103.59 = -.251$

conclusion: Do not reject H_o, there is not sufficient evidence to conclude that median ≠ 10,000.

13-4 Wilcoxon Rank-Sum Test for Two Independent Samples

1. Below are the scores (in order) for each group. The group listed first is considered group 1.

CA	R	MD	R
329	4	307	1
330	5	317	2
331	6	325	3
343	8	332	7
370	14	345	9
397	17	351	10
411	21	354	11.5
420	22	354	11.5
424	23	361	13
438	25	378	15
441	26	379	16
446	27	400	18
448	28	409	19
452	29	410	20
459	30	427	24
	285		180.0

$n_1 = 15 \qquad \Sigma R_1 = 285$
$n_2 = 15 \qquad \Sigma R_2 = 180$

$n = \Sigma n = 30 \qquad \Sigma R = 465$

check: $\Sigma R = n(n+1)/2$
$= 30(31)/2$
$= 465$

$R = \Sigma R_1 = 285$

$\mu_R = n_1(n+1)/2$
$= 15(31)/2$
$= 232.5$

$\sigma_R^2 = n_1 n_2 (n+1)/12$
$= (15)(15)(31)/12$
$= 581.25$

H_o: the populations have the same distribution
H_1: the populations have different distributions
$\alpha = .05$
C.R. $z < -z_{.025} = -1.96$
    $z > z_{.025} = 1.96$
calculations:
$z_R = (R - \mu_R)/\sigma_R$
$= (285 - 232.5)/\sqrt{581.25}$
$= 52.5/24.109$
$= 2.178$

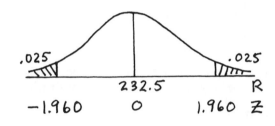

conclusion:
    Reject H_o; there is sufficient evidence to conclude that the populations have different distributions (in fact, population 1 has larger salaries).
On the basis of salaries alone, California is preferred. In practice, one also must consider directly related factors (e.g., cost of living) and subjective factors (e.g., quality of living).

NOTE: As in the previous section, the manual follows the wording in the text and tests the hypothesis that "the populations have the same distribution" with the understanding that the test detects only differences in location and not differences in variability.

2. Below are the scores (in order) for each group. The group listed first is considered group 1.

Mult	R	One	R
.6	1	4.3	7.5
2.1	2	4.7	9
2.2	3	5.7	12
2.8	4	5.9	13.5
3.5	5	6.0	15.5
3.7	6	6.2	18.5
4.3	7.5	6.3	21
5.2	10	6.8	22
5.6	11	7.1	23.5
5.9	13.5	7.2	25
6.0	15.5	7.4	26.5
6.2	18.5	7.4	26.5
6.2	18.5	7.9	29
6.2	18.5	8.2	30
7.1	23.5	8.8	32
7.8	28	8.9	33
8.7	31	9.3	35
9.2	34	9.4	36
9.6	37	10.2	38
11.2	39	11.5	40.5
11.5	40.5		494.0
11.8	42		
13.3	43		
14.3	44		
	496		

$n_1 = 24$ $\Sigma R_1 = 496$
$n_2 = 20$ $\Sigma R_2 = 494$

$n = \Sigma n = 44$ $\Sigma R = 990$

check: $\Sigma R = n(n+1)/2$
$= 44(45)/2$
$= 990$

$R = \Sigma R_1 = 496$

$\mu_R = n_1(n+1)/2$
$= 24(45)/2$
$= 540$

$\sigma_R^2 = n_1 n_2 (n+1)/12$
$= (24)(20)(45)/12$
$= 1800$

H_o: the populations have the same distribution
H_1: the populations have different distributions
$\alpha = .05$
C.R. $z < -z_{.025} = -1.96$
     $z > z_{.025} = 1.96$
calculations:
$z_R = (R - \mu_R)/\sigma_R$
$= (496 - 540)/\sqrt{1800} = -44/42.426 = -1.037$

conclusion:
Do not reject H_o; there is not sufficient evidence to conclude that the populations have different distributions.

3. Below are the scores (in order) for each group. The group listed first is considered group 1.

O-C	R	Con	R
210	1	334	7.5
287	2	349	11
288	3	402	12
304	4	413	14
305	5	429	15
308	6	445	16
334	7.5	460	18.5
340	9	476	20.5
344	10	483	21
407	13	501	22
455	17	519	23
463	19	594	24
	96.5		203.5

$n_1 = 12$ $\Sigma R_1 = 96.5$
$n_2 = 12$ $\Sigma R_2 = 203.5$

$n = \Sigma n = 24$ $\Sigma R = 300.0$

check: $\Sigma R = n(n+1)/2$
$= 24(25)/2$
$= 300$

$R = \Sigma R_1 = 96.5$

$\mu_R = n_1(n+1)/2$
$= 12(25)/2 = 150$

$$\sigma_R^2 = n_1 n_2 (n+1)/12$$
$$= (12)(12)(25)/12 = 300$$

H_o: the populations have the same distribution
H_1: the populations have different distributions
$\alpha = .01$
C.R. $z < -z_{.005} = -2.575$
$\quad\; z > z_{.005} = 2.575$
calculations:
$\quad z_R = (R - \mu_R)/\sigma_R$
$\quad\quad = (96.5 - 150)/\sqrt{300}$
$\quad\quad = -53.5/17.321$
$\quad\quad = -3.089$

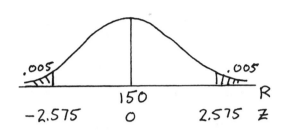

conclusion:
   Reject H_o; there is sufficient evidence to conclude that the populations have different
   distributions (in fact, population 1 has smaller volumes).
Based on this result, we can be 99% confident that there are biological factors related to
   obsessive-compulsive disorders.

4. Below are the scores (in order) for each group. The group listed first is considered group 1.

Rt A	R	Rt B	R
72	1	76	2
83	3	87	4
91	5	95	8.5
92	6	96	10
93	7	97	11
95	8.5	99	13
98	12	102	15.5
100	14	104	17
102	15.5	106	18
108	19.5	114	22
108	19.5	115	23
112	21	116	24
117	25	121	26
128	27	132	28
	184		222

$n_1 = 14 \quad\quad \Sigma R_1 = 184$
$n_2 = 14 \quad\quad \Sigma R_2 = 222$
_____   _____
$n = \Sigma n = 28 \quad\quad \Sigma R = 406$

check: $\Sigma R = n(n+1)/2$
$\quad\quad\quad\; = 28(29)/2$
$\quad\quad\quad\; = 406$

$R = \Sigma R_1 = 184$

$\mu_R = n_1(n+1)/2$
$\quad\; = 14(29)/2$
$\quad\; = 203$

$\sigma_R^2 = n_1 n_2 (n+1)/12$
$\quad = (14)(14)(29)/12$
$\quad = 473.67$

H_o: the populations have the same distribution
H_1: the populations have different distributions
$\alpha = .05$
C.R. $z < -z_{.025} = -1.96$
$\quad\; z > z_{.025} = 1.96$
calculations:
$\quad z_R = (R - \mu_R)/\sigma_R$
$\quad\quad = (184 - 203)/\sqrt{473.67}$
$\quad\quad = -19/21.764 = -.873$

conclusion:
   Do not reject H_o; there is not sufficient evidence to conclude that the populations have
   different distributions.
No; it does not appear that one route has any advantage over the other.

5. Below are the scores (in order) for each group. The group listed first is considered group 1.

Beer	R	Liqr	R
.129	1	.182	9
.146	2	.185	10
.148	3	.190	12.5
.152	4	.205	15
.154	5	.220	17
.155	6	.224	18
.164	7	.225	19.5
.165	8	.226	20.5
.187	11	.227	21
.190	12.5	.234	22
.203	14	.241	23
.212	16	.247	24
	89.5	.253	25
		.257	26
			261.5

$n_1 = 12$ $\Sigma R_1 = 89.5$
$n_2 = 14$ $\Sigma R_2 = 261.5$

$n = \Sigma n = 26$ $\Sigma R = 351.0$

check: $\Sigma R = n(n+1)/2$
$= 26(27)/2$
$= 351$

$R = \Sigma R_1 = 89.5$

$\mu_R = n_1(n+1)/2$
$= 12(27)/2$
$= 162$

$\sigma^2_R = n_1 n_2 (n+1)/12$
$= (12)(14)(27)/12$
$= 378$

H_o: the populations have the same distribution
H_1: the populations have different distributions
$\alpha = .05$
C.R. $z < -z_{.025} = -1.96$
     $z > z_{.025} = 1.96$
calculations:
$z_R = (R - \mu_R)/\sigma_R$
$= (89.5 - 162)/\sqrt{378}$
$= -72.5/19.442$
$= -3.729$

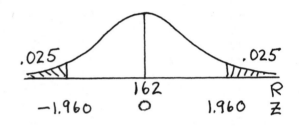

conclusion:
   Reject H_o; there is sufficient evidence to conclude that the populations have different distributions (in fact, population 1 has lower BAC levels).
Based on these results, it appears that the liquor drinkers have a higher BAC level and are (presumably) more dangerous.

6. Below are the scores (in order) for each group. The group listed first is considered group 1.

with	R	w/o	R
141	8	111	1
154	12	124	2
157	13.5	127	3
157	13.5	133	4
163	17	135	5
165	18	137	6
168	19	138	7
173	21	144	9
178	22	148	10
181	23.5	151	11
189	25.5	159	15
192	27	162	16
200	28	170	20
201	29	181	23.5
219	30	189	25.5
	307		158

$n_1 = 15$ $\Sigma R_1 = 307$
$n_2 = 15$ $\Sigma R_2 = 158$

$n = \Sigma n = 30$ $\Sigma R = 666$

check: $\Sigma R = n(n+1)/2$
$= 30(31)/2$
$= 465$

$R = \Sigma R_1 = 307$

$\mu_R = n_1(n+1)/2$
$= 15(31)/2$
$= 232.5$

$\sigma^2_R = n_1 n_2 (n+1)/12$
$= (15)(15)(31)/12 = 581.25$

H_0: the populations have the same distribution
H_1: the populations have different distributions
$\alpha = .05$
C.R. $z < -z_{.025} = -1.96$
$\quad\quad z > z_{.025} = 1.96$
calculations:
$\quad z_R = (R - \mu_R)/\sigma_R$
$\quad\quad = (307 - 232.5)/\sqrt{581.25}$
$\quad\quad = 74.5/24.109 = 3.090$

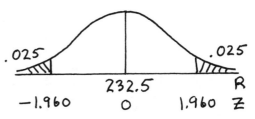

conclusion:
Reject H_0; there is sufficient evidence to conclude that the populations have different distributions (in fact, population 1 has higher).
Yes; the course appears to be effective and should continue to be offered.

7. Below are the scores (in order) for each group. The group listed first is considered group 1.

Exp	R	Cont	R
21.37	1	54.22	5
34.26	2	70.02	13
44.09	3	70.70	14
44.71	4	74.01	18
57.29	6	94.23	24
59.77	7	99.08	25
59.78	8	104.06	26
60.83	9	111.26	27
64.05	10	118.43	29
66.87	11	118.58	30
69.95	12	119.89	31
72.14	15	120.76	32
73.46	16	121.67	33
74.27	17	121.70	34
75.38	19	122.80	35
76.59	20	138.27	36
80.03	21		412
82.25	22		
92.72	23		
117.80	28		
	254		

$n_1 = 20$ $\quad \Sigma R_1 = 254$
$n_2 = 16$ $\quad \Sigma R_2 = 412$

$n = \Sigma n = 36 \quad\quad \Sigma R = 666$

check: $\Sigma R = n(n+1)/2$
$\quad\quad\quad = 36(37)/2$
$\quad\quad\quad = 666$

$R = \Sigma R_1 = 254$

$\mu_R = n_1(n+1)/2$
$\quad\quad = 20(37)/2$
$\quad\quad = 370$

$\sigma_R^2 = n_1 n_2 (n+1)/12$
$\quad\quad = (20)(16)(37)/12$
$\quad\quad = 986.67$

H_0: the populations have the same distribution
H_1: the populations have different distributions
$\alpha = .05$
C.R. $z < -z_{.025} = -1.96$
$\quad\quad z > z_{.025} = 1.96$
calculations:
$\quad z_R = (R - \mu_R)/\sigma_R$
$\quad\quad = (254 - 370)/\sqrt{986.67}$
$\quad\quad = -116/31.411 = -3.693$
conclusion:
Reject H_0; there is sufficient evidence to conclude that the populations have different distributions (in fact, population 1 has lower scores).

8. Below are the scores (in order) for each group. The group listed first is considered group 1.

placebo	R	calcium	R
96.5	1	102.7	2
104.8	4	104.4	3
106.1	5	106.6	6
107.2	7	108.0	8
108.5	9	109.6	10
113.6	12	113.2	11
116.3	14.5	114.7	13
118.1	16.5	116.3	14.5
120.4	18	118.1	16.5
122.5	21	120.9	19
123.1	22	121.4	20
124.6	25	123.4	23
128.8	27	124.3	24
	182	127.7	26
		129.1	28
			224

$$n_1 = 13 \qquad \Sigma R_1 = 182$$
$$n_2 = 15 \qquad \Sigma R_2 = 224$$

$$n = \Sigma n = 28 \qquad \Sigma R = 406$$

$$\text{check: } \Sigma R = n(n+1)/2$$
$$= 28(29)/2$$
$$= 406$$

$$R = \Sigma R_1 = 182$$

$$\mu_R = n_1(n+1)/2$$
$$= 13(29)/2$$
$$= 188.5$$

$$\sigma_R^2 = n_1 n_2 (n+1)/12$$
$$= (13)(15)(29)/12$$
$$= 471.25$$

H_o: the populations have the same distribution
H_1: the populations have different distributions
$\alpha = .05$
C.R. $z < -z_{.025} = -1.96$
    $z > z_{.025} = 1.96$
calculations:

$$z_R = (R - \mu_R)/\sigma_R$$
$$= (182 - 188.5)/\sqrt{471.25}$$
$$= -6.5/21.708 = -.299$$

conclusion:
Do not reject H_o; there is not sufficient evidence to conclude that the populations have different distributions of blood pressure levels.

9. Below are the scores (in order) for each group. The group listed first is considered group 1.

Male	R	Female	R
12.27	2	6.66	1
12.63	3	13.88	4
17.98	8	13.96	5
18.06	9	14.34	6
19.54	12	15.89	7
19.84	13	18.71	10
20.20	15	19.20	11
22.12	16	20.15	14
22.99	17	23.90	19
23.01	18	29.85	23
23.93	20	31.13	24
25.63	21		124
25.73	22		
32.20	25		
32.56	26		
39.53	27		
	254		

$$n_1 = 16 \qquad \Sigma R_1 = 254$$
$$n_2 = 11 \qquad \Sigma R_2 = 124$$

$$n = \Sigma n = 27 \qquad \Sigma R = 378$$

$$\text{check: } \Sigma R = n(n+1)/2$$
$$= 27(28)/2$$
$$= 378$$

$$R = \Sigma R_1 = 254$$

$$\mu_R = n_1(n+1)/2$$
$$= 16(28)/2$$
$$= 224$$

$$\sigma_R^2 = n_1 n_2 (n+1)/12$$
$$= (16)(11)(28)/12$$
$$= 410.67$$

H_o: the populations have the same distribution
H_1: the populations have different distributions
$\alpha = .05$
C.R. $z < -z_{.025} = -1.96$
    $z > z_{.025} = 1.96$
calculations:
$z_R = (R - \mu_R)/\sigma_R$
   $= (254 - 224)/\sqrt{410.67}$
   $= 30/20.265 = 1.480$

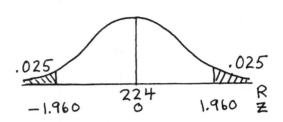

conclusion:
Do not reject H_o; there is not sufficient evidence to conclude that the populations have different distributions.

10. Below are the scores (in order) for each group. The group listed first is considered group 1.

E to D	R	D to E	R
7.10	1	26.63	11
16.32	2	26.68	12
20.60	3	27.24	14
21.06	4	27.62	15
21.13	5	29.34	21
21.96	6	29.49	22
24.23	7	30.20	24
24.64	8	30.26	25
25.49	9	32.34	29
26.43	10	32.54	30
26.69	13	33.53	34
27.85	16	33.62	35
28.02	17	34.02	36
28.71	18	35.32	37
28.89	19	35.91	38
28.90	20	42.91	41
30.02	23		424
30.29	26		
30.72	27		
31.73	28		
32.83	31		
32.86	32		
33.31	33		
38.81	39		
39.29	40		
	437		

$n_1 = 25$ $\qquad \Sigma R_1 = 437$
$n_2 = 16$ $\qquad \Sigma R_2 = 424$

$n = \Sigma n = 41$ $\qquad \Sigma R = 861$

check: $\Sigma R = n(n+1)/2$
            $= 41(42)/2$
            $= 861$

$R = \Sigma R_1 = 437$

$\mu_R = n_1(n+1)/2$
    $= 25(42)/2$
    $= 525$

$\sigma_R^2 = n_1 n_2(n+1)/12$
    $= (25)(16)(42)/12$
    $= 1400$

H_o: the populations have the same distribution
H_1: the populations have different distributions
$\alpha = .05$
C.R. $z < -z_{.025} = -1.96$
    $z > z_{.025} = 1.96$
calculations:
$z_R = (R - \mu_R)/\sigma_R$
   $= (437 - 525)/\sqrt{1400}$
   $= -88/37.417 = -2.352$

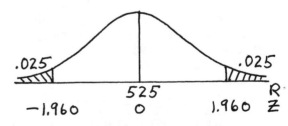

conclusion:
Reject H_o; there is sufficient evidence to conclude that the populations have different distributions (in fact, population 1 has lower scores).

11. Below are the scores (in order) for each group. The group listed first is considered group 1.

red	R	yel	R
.817	1	.822	2
.844	3	.845	4
.864	6	.861	5
.870	7	.890	10
.871	8	.890	11
.876	9	.892	12
.896	13	.905	14
.909	15	.912	17
.913	19	.912	17
.920	22.5	.912	17
.921	24	.914	20.5
.931	30.5	.914	20.5
.933	32	.920	22.5
.941	35	.922	25
.952	36	.925	26
.957	37	.927	27
.958	38.5	.929	28
	336.5	.930	29
		.931	30.5
		.935	33
		.936	34
		.958	38.5
		.971	40
		.986	41
			524.5

$n_1 = 17$ $\qquad \Sigma R_1 = 336.5$
$n_2 = 24$ $\qquad \Sigma R_2 = 524.5$

$n = \Sigma n = 41$ $\qquad \Sigma R = 861.0$

check: $\Sigma R = n(n+1)/2$
$\qquad\qquad = 41(42)/2$
$\qquad\qquad = 861$

$R = \Sigma R_1 = 336.5$

$\mu_R = n_1(n+1)/2$
$\qquad = 17(42)/2$
$\qquad = 357$

$\sigma_R^2 = n_1 n_2(n+1)/12$
$\qquad = (17)(24)(42)/12$
$\qquad = 1428$

H_o: the populations have the same distribution
H_1: the populations have different distributions
$\alpha = .05$ [or another appropriate value]
C.R. $z < -z_{.025} = -1.96$
$\quad\;\; z > z_{.025} = 1.96$
calculations:
$\quad z_R = (R - \mu_R)/\sigma_R$
$\qquad = (336.5 - 357)/\sqrt{1428}$
$\qquad = -20.5/37.789 = -.542$
conclusion:

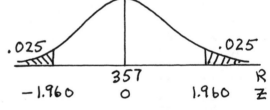

Do not reject H_o; there is not sufficient evidence to conclude that the populations have different distributions.

12. Let the males be group 1: $n_1 = 59$, $\Sigma R_1 = 2030.5$, $n_2 = 11$, $\Sigma R_2 = 454.5$, $n = 70$
$\quad \mu_R = n_1(n+1)/2 = 59(71)/2 = 2094.5$, $\sigma_R^2 = n_1 n_2(n+1)/12 = 59(11)(71)/12 = 3839.9$
H_o: the populations have the same distribution
H_1: the populations have different distributions
$\alpha = .05$ [or another appropriate value]
C.R. $z < -z_{.025} = -1.96$
$\quad\;\; z > z_{.025} = 1.96$
calculations:
$\quad z_R = (R - \mu_R)/\sigma_R$
$\qquad = (2030.5 - 2094.5)/\sqrt{3839.9}$
$\qquad = -64.0/61.967 = -1.033$
conclusion:

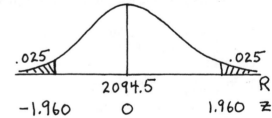

Do not reject H_o; there is not sufficient evidence to conclude that the populations have different distributions.

13. a. The format of the previous exercises produces the following. Consider the A's group 1.

A	R	B	R
1	16		
2	17		
3	18		
4	19		
5	20		
6	21		
7	22		
8	23		
9	24		
10	25		
11	26		
12	27		
13	28		
14	29		
15	30		
120	345		

$n_1 = 15 \qquad \Sigma R_1 = 120$
$n_2 = 15 \qquad \Sigma R_2 = 345$

$n = \Sigma n = 30 \qquad \Sigma R = 465$

check: $\Sigma R = n(n+1)/2$
$\qquad\qquad = 30(31)/2$
$\qquad\qquad = 465$

$R = \Sigma R_1 = 120$

$\mu_R = n_1(n+1)/2$
$\quad = 15(31)/2$
$\quad = 232.5$

$\sigma_R^2 = n_1 n_2(n+1)/12$
$\quad = (15)(15)(31)/12$
$\quad = 581.25$

H_o: the populations have the same distribution
H_1: the populations have different distributions
$\alpha = .05$
C.R. $z < -z_{.025} = -1.96$
$\qquad z > z_{.025} = 1.96$
calculations:
$z_R = (R - \mu_R)/\sigma_R$
$\quad = (120 - 232.5)/\sqrt{581.25}$
$\quad = -112.5/24.109 = -4.666$

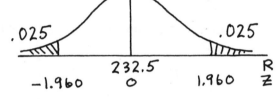

conclusion:
Reject H_o; there is sufficient evidence to conclude that the populations have different distributions (in fact, population A has lower scores).

NOTE: These are the most one-sided results possible for population A having lower scores than population B -- every score from A was lower than the lowest score from B. The statistical conclusion (i.e., rejecting H_o) is the one demanded by common sense. With these sample sizes it is not possible to have a calculated z less than -4.666.

b. The format of the previous exercises produces the following. Consider the A's group 1.

A	R	B	R
1	2		
3	4		
5	6		
7	8		
9	10		
11	12		
13	14		
15	16		
17	18		
19	20		
21	22		
23	24		
25	26		
27	28		
29	30		
225	240		

$n_1 = 15 \qquad \Sigma R_1 = 225$
$n_2 = 15 \qquad \Sigma R_2 = 240$

$n = \Sigma n = 30 \qquad \Sigma R = 465$

check: $\Sigma R = n(n+1)/2$
$\qquad\qquad = 30(31)/2$
$\qquad\qquad = 465$

$R = \Sigma R_1 = 225$

$\mu_R = n_1(n+1)/2$
$\quad = 15(31)/2$
$\quad = 232.5$

$$\sigma_R^2 = n_1 n_2 (n+1)/12$$
$$= (15)(15)(31)/12 = 581.25$$

H_o: the populations have the same distribution
H_1: the populations have different distributions
$\alpha = .05$
C.R. $z < -z_{.025} = -1.96$
    $z > z_{.025} = 1.96$
calculations:
$$z_R = (R - \mu_R)/\sigma_R$$
$$= (225 - 232.5)/\sqrt{581.25}$$
$$= -7.5/24.109 = -.311$$
conclusion:

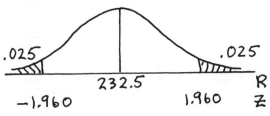

.025   .025
           232.5   R
−1.960   1.960   Z

Do not reject H_o; there is not sufficient evidence to conclude that the populations have different distributions.

NOTE: These results represent an alternating inter-mixing of A scores and B scores. The statistical conclusion (i.e., not rejecting H_o) is the one demanded by common sense.

c. Parts (a) and (b) represent two possible extreme data patterns. In part (a), all the low scores in one group and all the high scores in the other; in part (b), the ordered scores alternate between the two groups. Common sense demands that the hypothesis of identical population distributions be rejected in part (a) and not be rejected in part (b). The statistical conclusions support the demands of common sense.

d. The format of the previous exercises produces the following. Consider the A's group 1.

A	R	B	R
16	2		
17	2		
18	3		
19	4		
20	5		
21	6		
22	7		
23	8		
24	9		
25	10		
26	11		
27	12		
28	13		
29	14		
30	15		
345	120		

$n_1 = 15$ $\Sigma R_1 = 345$
$n_2 = 15$ $\Sigma R_2 = 120$
———————   ———————
$n = \Sigma n = 30$ $\Sigma R = 465$

check: $\Sigma R = n(n+1)/2$
                $= 30(31)/2$
                $= 465$

$R = \Sigma R_1 = 345$

$\mu_R = n_1(n+1)/2$
      $= 15(31)/2$
      $= 232.5$

$\sigma_R^2 = n_1 n_2 (n+1)/12$
      $= (15)(15)(31)/12$
      $= 581.25$

H_o: the populations have the same distribution
H_1: the populations have different distributions
$\alpha = .05$
C.R. $z < -z_{.025} = -1.96$
    $z > z_{.025} = 1.96$
calculations:
$$z_R = (R - \mu_R)/\sigma_R$$
$$= (345 - 232.5)/\sqrt{581.25}$$
$$= 112.5/24.109 = 4.666$$
conclusion:

.025   .025
           232.5   R
1.960   0   1.960   Z

Reject H_o; there is sufficient evidence to conclude that the populations have different distributions (in fact, population A has higher scores).

Compared to part (a), the R value changes from 120 to 345 -- the μ_R and σ_R, which depend only on the sample sizes, remain the same. This has the effect of reversing the sign on the calculated z statistic.

NOTE: Part (c) is the "mirror image" of part (a) -- these are the most one-sided results possible for population A having higher scores than population B -- every score from A was higher than the highest score from B. The statistical conclusion (i.e., rejecting H_o) is the one demanded by common sense. With these sample sizes it is not possible to have a calculated z greater than 4.666.

e. The format of the previous exercises produces the following. Consider the A's group 1.

A	R	B	R
	30		16
	2		17
	3		18
	4		19
	5		20
	6		21
	7		22
	8		23
	9		24
	10		25
	11		26
	12		27
	13		28
	14		29
	15		1
	149		316

$n_1 = 15 \qquad \Sigma R_1 = 149$
$n_2 = 15 \qquad \Sigma R_2 = 316$

$n = \Sigma n = 30 \qquad \Sigma R = 465$

check: $\Sigma R = n(n+1)/2$
$= 30(31)/2$
$= 465$

$R = \Sigma R_1 = 149$

$\mu_R = n_1(n+1)/2$
$= 15(31)/2$
$= 232.5$

$\sigma_R^2 = n_1 n_2(n+1)/12$
$= (15)(15)(31)/12$
$= 581.25$

H_o: the populations have the same distribution
H_1: the populations have different distributions
$\alpha = .05$
C.R. $z < -z_{.025} = -1.96$
$\qquad z > z_{.025} = 1.96$
calculations:
$z_R = (R - \mu_R)/\sigma_R$
$\quad = (149 - 232.5)/\sqrt{581.25}$
$\quad = -83.5/24.109$
$\quad = -3.463$

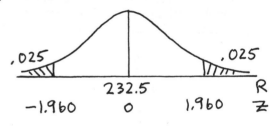

conclusion:

Reject H_o; there is sufficient evidence to conclude that the populations have different distributions (in fact, population A has lower scores).

Compared to part (a), the R value changes from 120 to 149 -- the μ_R and σ_R, which depend only on the sample sizes, remain the same. This has the effect of moving the calculated z statistic away from its negative limit of -4.667 to a less dramatic (but still significant) -3.463.

NOTE: These are strange results. Every score from A except one was lower than the second lowest score from B -- and that one exception from A was the highest score of all, while the lowest score from B was the lowest score of all. Common sense suggests that the populations from which the samples come are highly skewed -- that the A scores are basically low with a very small percentage of high scores, and that the B scores are basically high with a very small percentage of low scores. This being the case, the scores from A are generally lower than the scores from B and we hope the statistical test would so conclude -- as indeed it did.

14. The denominators of the Mann-Whitney and Wilcoxon statistics are the same.
The numerator of the Mann-Whitney statistic is
$$U - n_1n_2/2 = [n_1n_2 + n_1(n_1+1)/2 - R] - n_1n_2/2$$
$$= n_1n_2/2 + n_1(n_1+1)/2 - R$$
$$= (n_1/2)(n_2+n_1+1) - R1$$
$$= -[R - n_1(n_1+n_2+1)/2]$$
$$= -[\text{numerator of Wilcoxon statistic}]$$

15. With $n_1 = 2$ and $n_2 = 2$ (so that $n = n_1 + n_2 = 4$), the test of hypothesis is the same as in the
previous exercises except that the samples are not large enough to use the normal
approximation and the critical region must be based on $R = \Sigma R_1$ instead of z. Let the A's be
considered group 1.

    H_o: the populations have the same distribution
    H_1: the populations have different distributions
    $\alpha = .05$
    C.R. $R < ?$
           $R > ?$

a. There are $4!/2!2! = 6$ possible arrangements of 2 scores from A and 2 scores from B.
Those 6 arrangements and their associated $R = \Sigma R_1$ values are as follows.

```
      rank
   1  2  3  4 │ R
   A  A  B  B │ 3
   A  B  A  B │ 4
   A  B  B  A │ 5
   B  A  A  B │ 5
   B  A  B  A │ 6
   B  B  A  A │ 7
```

b. If H_o is true and populations A and B are identical, then there is no reason to expect any of
the arrangements over another and each of the 6 arrangements is equally likely. The
possible values of R and the probability associated with value is therefore as given below.

R	P(R)	NOTE: R·P(R)	R^2·P(R)
3	1/6	3/6	9/6
4	1/6	4/6	16/6
5	2/6	10/6	50/6
6	1/6	6/6	36/6
7	1/6	7/6	49/6
	1	30/6	160/6

$\mu_R = \Sigma R \cdot P(R)$
$= 30/6 = 5$
$\sigma_R^2 = \Sigma R^2 \cdot P(R) - (\mu_R)^2$
$= 160/6 - 5^2$
$= 10/6$

NOTE: The fact that each of the arrangements is equally likely can be proven using
conditional probability formulas for selecting A's and B's from a finite population of 2 A's
and 2 B's as follows.
$$P(A_1A_2B_3B_4) = P(A_1)\cdot P(A_2)\cdot P(B_3)\cdot P(B_4) = (2/4)\cdot(1/3)\cdot(2/2)\cdot(1/1) = 4/24 = 1/6$$
$$P(A_1B_2A_3B_4) = P(A_1)\cdot P(B_2)\cdot P(A_3)\cdot P(B_4) = (2/4)\cdot(2/3)\cdot(1/2)\cdot(1/1) = 4/24 = 1/6$$
$$P(A_1B_2B_3A_4) = P(A_1)\cdot P(B_2)\cdot P(B_3)\cdot P(A_4) = (2/4)\cdot(2/3)\cdot(1/2)\cdot(1/1) = 4/24 = 1/6$$
$$P(B_1A_2A_3B_4) = P(B_1)\cdot P(A_2)\cdot P(A_3)\cdot P(B_4) = (2/4)\cdot(2/3)\cdot(1/2)\cdot(1/1) = 4/24 = 1/6$$
$$P(B_1A_2B_3A_4) = P(B_1)\cdot P(A_2)\cdot P(B_3)\cdot P(A_4) = (2/4)\cdot(2/3)\cdot(1/2)\cdot(1/1) = 4/24 = 1/6$$
$$P(B_1B_2A_3A_4) = P(B_1)\cdot P(B_2)\cdot P(A_3)\cdot P(A_4) = (2/4)\cdot(1/3)\cdot(2/2)\cdot(1/1) = 4/24 = 1/6$$
In addition, the μ_R and σ_R^2 values calculated above using the probability formulas agree with
the formulas
$$\mu_R = n_1(n+1)/2 = 2(5)/2 = 5$$
$$\sigma_R^2 = n_1n_2(n+1)/12 = (2)(2)(5)/12 = 20/12 = 10/6$$
given in this section. Since the distribution -- as evidenced by the P(R) values -- is not
normal, however, we cannot use these values to convert to z scores for use with Table A-2.

c. The P(R) values indicate that the smallest α at which the test can be performed is the 2/6 = .333 level with

    C.R. $R \leq 3$
          $R \geq 7$

as illustrated at the right -- using discrete bars instead of a continuous normal distribution. Even with the smallest or largest possible R values (i.e., R = 3 or R = 7) it would not be possible to reject H_o using $\alpha = .05$.

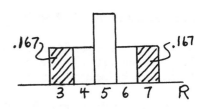

16. With $n_1 = 3$ and $n_2 = 3$ (so that $n = n_1 + n_2 = 6$), the test of hypothesis is the same as in the previous exercises except that the samples are not large enough to use the normal approximation and the critical region must be based on $R = \Sigma R_1$ instead of z. Let the A's be considered group 1.

    H_o: the populations have the same distribution
    H_1: the populations have different distributions
    $\alpha = .05$
    C.R. $R < ?$
          $R > ?$

a. There are 6!/3!3! = 20 possible arrangements of 2 scores from A and 2 scores from B. Those 20 arrangements and their associated $R = \Sigma R_1$ values are as follows.

rank 1 2 3 4 5 6	R
A A A B B B	6
A A B A B B	7
A A B B A B	8
A A B B B A	9
A B A A B B	8
A B B A A B	10
A B A B B A	10
A B B A B A	11
A B B B A A	12
A B A B A B	9
B A B A B A	12
B A A A B B	9
B A A B A B	10
B A B A A B	11
B A A B B A	11
B A B B A A	13
B B A A A B	12
B B A A B A	13
B B A B A A	14
B B B A A A	15

R	P(R)	
6	1/20	= .05
7	1/20	= .05
8	2/20	= .10
9	3/20	= .15
10	3/20	= .15
11	3/20	= .15
12	3/20	= .15
13	2/20	= .10
14	2/20	= .05
15	1/20	= .05
	1.00	

b. If H_o is true and populations A and B are identical, then there is no reason to expect any of the arrangements over another and each of the 20 arrangements is equally likely. The possible values of R and the probability associated with value is therefore as given above.

c. Yes. The above table indicates that a .10 level test that there is no difference between treatments A and B may be defined by C.R. $R \leq 6$
                                                $R \geq 15$

13-5 Kruskal-Wallis Test

NOTE: As in the previous sections, the manual follows the wording in the text and tests the hypothesis that "the populations have the same distribution" with the understanding that the test detects only differences in location and not differences in variability.

1. Below are the scores (in order) for each group. The group listed first is group 1, etc.

18-20	R	21-29	R	30+	R
97.1	2	97.9	6	97.4	1
97.7	5	98.2	8.5	97.5	3.5
98.0	7	98.2	8.5	97.5	3.5
98.4	10	99.0	14	98.6	12.5
98.5	11	99.6	15	98.6	12.5
	35		52.0		33.0

$n_1 = 5 \quad R_1 = 35$
$n_2 = 5 \quad R_2 = 52$
$n_3 = 5 \quad R_3 = 33$

$n = \Sigma n = 15 \quad \Sigma R = 120$

check:
$\Sigma R = n(n+1)/2$
$= 15(16)/2$
$= 120$

H_o: the populations have the same distribution
H_1: the populations have different distributions
$\alpha = .05$
C.R. $H > \chi^2_{2,.05} = 5.991$

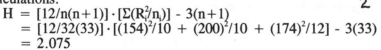

calculations:
$H = [12/n(n+1)] \cdot [\Sigma(R_i^2/n_i)] - 3(n+1)$
$= [12/15(16)] \cdot [(35)^2/5 + (52)^2/5 + (33)^2/5] - 3(16)$
$= [.05] \cdot [1003.6] - 48 = 2.180$

conclusion:
Do not reject H_o; there is not sufficient evidence to conclude that the populations have different distributions.

2. Below are the scores (in order) for each group. The group listed first is group 1, etc.

D&T	R	MA	R	KG&L	R
65	3	61	2	55	1
70	4.5	72	8	71	6
70	4.5	75	12	72	8
75	12	78	20	72	8
76	14	80	22	73	10
77	17	80	22	75	12
77	17	81	25	77	17
81	25	84	28	77	17
82	27	85	29	77	17
90	30	96	32	80	22
	154		200	81	25
				92	31
					174

$n_1 = 10 \quad R_1 = 154$
$n_2 = 10 \quad R_2 = 200$
$n_3 = 12 \quad R_3 = 174$

$n = \Sigma n = 32 \quad \Sigma R = 528$

check:
$\Sigma R = n(n+1)/2$
$= 32(33)/2$
$= 528$

H_o: the populations have the same distribution
H_1: the populations have different distributions
$\alpha = .05$
C.R. $H > \chi^2_{2,.05} = 5.991$
calculations:
$H = [12/n(n+1)] \cdot [\Sigma(R_i^2/n_i)] - 3(n+1)$
$= [12/32(33)] \cdot [(154)^2/10 + (200)^2/10 + (174)^2/12] - 3(33)$
$= 2.075$

conclusion:
Do not reject H_o; there is not sufficient evidence to conclude that the populations have different distributions.

3. Below are the scores (in order) for each group. The group listed first is group 1, etc.

T	R	L	R	E	R
25.9	1	27.0	4	26.3	3
26.0	2	27.6	7	27.1	5
27.3	6	28.1	8	29.2	13
28.2	19	28.8	11	29.8	14
28.5	10	29.0	12	30.0	15
	28		42		50

$n_1 = 5 \quad R_1 = 28$
$n_2 = 5 \quad R_2 = 42$
$n_3 = 5 \quad R_3 = 50$

$n = \Sigma n = 15 \quad \Sigma R = 120$

check:
$\Sigma R = n(n+1)/2$
$= 15(16)/2$
$= 120$

H_o: the populations have the same distribution
H_1: the populations have different distributions
$\alpha = .05$
C.R. $H > \chi^2_{2,.05} = 5.991$
calculations:

$H = [12/n(n+1)] \cdot [\Sigma(R_i^2/n_i)] - 3(n+1)$
$= [12/15(16)] \cdot [(28)^2/5 + (42)^2/5 + (50)^2/5] - 3(16)$
$= [.05] \cdot [1009.6] - 48 = 2.480$

conclusion:
   Do not reject H_o; there is not sufficient evidence to conclude that the populations have different distributions.

4. Below are the ranks (in the original order) for each group. The group listed first is group 1, etc.

2-Par	1-Par	Trans
13	7	1
14	4	8
11.5	6	9.5
9.5	11.5	2
15	4	4
63	32.5	24.5

$n_1 = 5 \quad R_1 = 63$
$n_2 = 5 \quad R_2 = 32.5$
$n_3 = 5 \quad R_3 = 24.5$

$n = \Sigma n = 15 \quad \Sigma R = 120.0$

check:
$\Sigma R = n(n+1)/2$
$= 15(16)/2$
$= 120$

H_o: the populations have the same distribution
H_1: the populations have different distributions
$\alpha = .05$
C.R. $H > \chi^2_{2,.05} = 5.991$
calculations:

$H = [12/n(n+1)] \cdot [\Sigma(R_i^2/n_i)] - 3(n+1)$
$= [12/15(16)] \cdot [(63)^2/5 + (32.5)^2/5 + (24.5)^2/5] - 3(16)$
$= 8.255$

conclusion:
   Reject H_o; there is sufficient evidence to conclude that the populations have different distributions of measures of social adjustment.

5. Below are the scores (in order) for each group. The group listed first is group 1, etc.

1	R	2	R	3	R	4	R	5	R
2.9	9.5	2.7	2	2.8	5.5	2.7	2	2.8	5.5
2.9	9.5	3.2	19.5	2.8	5.5	2.7	2	3.1	15
3.0	12	3.3	24	2.8	5.5	2.9	9.5	3.5	31.5
3.1	15	3.4	28	3.2	19.5	2.9	9.5	3.7	36
3.1	15	3.4	28	3.3	24	3.2	19.5	4.1	44
3.1	15	3.6	34	3.3	24	3.2	19.5	4.1	44
3.1	15	3.8	39	3.5	31.5	3.3	24	4.2	46.5
3.7	36	3.8	39	3.5	31.5	3.3	24		222.5
4.7	36	4.0	42	3.5	31.5	3.4	28		
3.9	41	4.1	44	3.8	39		138.0		
4.2	46.5	4.3	48		217.5				
	250.5		347.5						

$n_1 = 11 \quad R_1 = 250.5$
$n_2 = 11 \quad R_2 = 347.5$
$n_3 = 10 \quad R_3 = 217.5$
$n_4 = 9 \quad R_4 = 138$
$n_5 = 7 \quad R_5 = 222.5$

$n = \Sigma n = 48 \quad \Sigma R = 1176.0$

check:
$\Sigma R = n(n+1)/2$
$= 48(49)/2$
$= 1176$

H_0: the populations have the same distribution
H_1: the populations have different distributions
$\alpha = .10$
C.R. $H > \chi^2_{4,.10} = 7.779$
calculations:

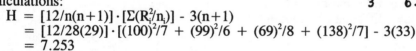

$$H = [12/n(n+1)] \cdot [\Sigma(R_i^2/n_i)] - 3(n+1)$$
$$= [12/48(49)] \cdot [(250.5)^2/11 + (347.5)^2/11 + (217.5)^2/10 + (138)^2/9 + (222.5)^2/7] - 3(49)$$
$$= [.00510] \cdot [30601.356] - 147 = 9.129$$

conclusion:

Reject H_0; there is sufficient evidence to conclude that the populations have different distributions.

6. Below are the ranks (in the original order) for each group. The group listed first is group 1, etc.

GA	VT	TX	OR
11	21	9	25
5	7	4	17
23	13	12	24
8	16	18	26
19	15	1	10
6	27	2	22
28	99	3	14
100		20	138
		69	

$n_1 = 7 \quad R_1 = 100$
$n_2 = 6 \quad R_2 = 99$
$n_3 = 8 \quad R_3 = 69$
$n_4 = 7 \quad R_4 = 138$

$n = \Sigma n = 28 \quad \Sigma R = 406$

check:
$\Sigma R = n(n+1)/2$
$= 28(29)/2$
$= 406$

H_0: the populations have the same distribution
H_1: the populations have different distributions
$\alpha = .10$
C.R. $H > \chi^2_{3,.10} = 6.251$
calculations:

$$H = [12/n(n+1)] \cdot [\Sigma(R_i^2/n_i)] - 3(n+1)$$
$$= [12/28(29)] \cdot [(100)^2/7 + (99)^2/6 + (69)^2/8 + (138)^2/7] - 3(33)$$
$$= 7.253$$

conclusion:

Reject H_0; there is sufficient evidence to conclude that the populations have different distributions.

7. Below are the scores for each group. The group listed first is group 1, etc.

1	R	4	R	7	R
20	19	18	14.5	21	21
18	14.5	20	19	16	9
27	24	19	17	14	6
17	11	12	3.5	23	22
18	14.5	9	1	17	11
13	5	18	14.5	17	11
24	23	28	25	20	19
15	7.5	10	2		99
	118.5	15	7.5		
		12	3.5		
			107.5		

$n_1 = 8 \quad R_1 = 118.5$
$n_2 = 10 \quad R_2 = 107.5$
$n_3 = 7 \quad R_3 = 99.0$

$n = \Sigma n = 25 \quad \Sigma R = 325.0$

check:
$\Sigma R = n(n+1)/2$
$= 25(26)/2$
$= 325$

H_o: the populations have the same distribution
H_1: the populations have different distributions
$\alpha = .05$
C.R. $H > \chi^2_{2,.05} = 5.991$

calculations:

$H = [12/n(n+1)] \cdot [\Sigma(R_i^2/n_i)] - 3(n+1)$
$= [12/25(26)] \cdot [(118.5)^2/8 + (107.5)^2/10 + (99)^2/7] - 3(26)$
$= [.0185] \cdot [4311.049] - 78 = 1.589$

conclusion:

Do not reject H_o; there is not sufficient evidence to conclude that the populations have different distributions.

8. Below are the scores for each group. The group listed first is group 1, etc.

1	R	4	R	7	R
.50	11.5	.55	13	3.04	23
1.00	18.5	.46	9.5	1.09	21
1.05	20	.94	17	.23	1
.42	7	.29	4	1.00	18.5
.84	15	.26	3	6.20	25
.33	5.5	.33	5.5	.46	9.5
.90	16	1.70	22	3.20	24
.83	14	.50	11.5		122.0
	107.5	.43	8		
		.25	2		
			95.5		

$n_1 = 8 \quad R_1 = 107.5$
$n_2 = 10 \quad R_2 = 95.5$
$n_3 = 7 \quad R_3 = 122.0$

$n = \Sigma n = 25 \quad \Sigma R = 325.0$

check:
$\Sigma R = n(n+1)/2$
$= 25(26)/2$
$= 325$

H_o: the populations have the same distribution
H_1: the populations have different distributions
$\alpha = .05$
C.R. $H > \chi^2_{2,.05} = 5.991$

calculations:

$H = [12/n(n+1)] \cdot [\Sigma(R_i^2/n_i)] - 3(n+1)$
$= [12/25(26)] \cdot [(107.5)^2/8 + (95.5)^2/10 + (122)^2/7] - 3(26)$
$= 4.76 =$

conclusion:

Do not reject H_o; there is not sufficient evidence to conclude that the populations have different distributions.

9. Below are the scores for each group. The group listed first is group 1, etc.

1	R	4	R	7	R
1.9	13.5	2.8	23	2.7	22
2.4	21	1.8	11.5	1.9	13.5
1.5	3.5	3.2	24	1.3	1
1.6	6	2.1	16.5	1.7	9
1.6	6	1.4	2	2.2	19
1.5	3.5	2.1	16.5	2.0	15
1.7	9	4.2	25	2.2	19
2.2	19	1.7	9		98.5
	81.5	1.8	11.5		
		1.6	6		
			145.0		

$n_1 = 8 \quad R_1 = 81.5$
$n_2 = 10 \quad R_2 = 145.0$
$n_3 = 7 \quad R_3 = 98.5$

$n = \Sigma n = 25 \quad \Sigma R = 325.0$

check:
$\Sigma R = n(n+1)/2$
$= 24(25)/2$
$= 325$

H_o: the populations have the same distribution
H_1: the populations have different distributions
$\alpha = .05$
C.R. $H > \chi^2_{2,.05} = 5.991$
calculations:

$H = [12/n(n+1)] \cdot [\Sigma(R_i^2/n_i)] - 3(n+1)$
  $= [12/25(26)] \cdot [(81.5)^2/8 + (145)^2/10 + (98.5)^2/7] - 3(26)$
  $= [.0217] \cdot [3381.6] - 78 = 1.732$

conclusion:
   Do not reject H_o; there is not sufficient evidence to conclude that the populations have different distributions.

10. Below are the scores for each group. The group listed first is group 1, etc.

1	R	4	R	7	R
1.9	14	2.8	22	3.0	23
2.4	20.5	1.8	12	1.8	12
1.5	4.5	3.2	24	1.1	1
1.6	7	2.1	16.5	1.4	2.5
1.6	7	1.4	2.5	2.3	19
1.5	4.5	2.1	16.5	2.0	15
1.7	9.5	4.2	25	2.4	20.5
2.2	18	1.7	9.5		93.0
	85.0	1.8	12		
		1.6	7		
			147.0		

$n_1 = 8 \quad R_1 = 85.0$
$n_2 = 10 \quad R_2 = 147.0$
$n_3 = 7 \quad R_3 = 93.0$

$n = \Sigma n = 25 \qquad \Sigma R = 325.0$

check:
$\Sigma R = n(n+1)/2$
   $= 24(25)/2$
   $= 325$

H_o: the populations have the same distribution
H_1: the populations have different distributions
$\alpha = .05$
C.R. $H > \chi^2_{2,.05} = 5.991$
calculations:

$H = [12/n(n+1)] \cdot [\Sigma(R_i^2/n_i)] - 3(n+1)$
  $= [12/25(26)] \cdot [(85)^2/8 + (147)^2/10 + (93)^2/7] - 3(26)$
  $= 1.377$

conclusion:
   Do not reject H_o; there is not sufficient evidence to conclude that the populations have different distributions.
   Compared to exercise #9, the increased variability in the zone 7 scores has increased the problem's uncertainty and lowered the calculated H statistic from 1.732 to 1.377.

11. Below are the scores (in order) for each group. The group listed first is group 1, etc.

below 2750	R	2750- 2999	R	3000- 3499	R	above 3500	R
27	9	24	2	27	9	21	1
29	20	26	5	28	15	25	3
30	23.5	27	9	28	15	26	5
30	23.5	28	15	28	15	26	5
31	26.5	28	15	29	20	27	9
33	29	28	15	30	23.5	27	9
34	30	28	15	30	23.5		32
35	31	29	20		121.0		
32	32	31	26.5				
	224.5	32	28				
			150.5				

$n_1 = 9 \quad R_1 = 224.5$
$n_2 = 10 \quad R_2 = 150.5$
$n_3 = 7 \quad R_3 = 121$
$n_4 = 6 \quad R_4 = 32$

$n = \Sigma n = 32 \qquad \Sigma R = 528.0$

check:
$\Sigma R = n(n+1)/2$
   $= 32(33)/2$
   $= 528$

H_o: the populations have the same distribution
H_1: the populations have different distributions
$\alpha = .05$
C.R. $H > \chi^2_{3,.05} = 7.815$

calculations:
$$H = [12/n(n+1)] \cdot [\Sigma(R_i^2/n_i)] - 3(n+1)$$
$$= [12/32(33)] \cdot [(224.5)^2/9 + (150.5)^2/10 + (121)^2/7 + (32)^2/6] - 3(33)$$
$$= [.01136] \cdot [10127.29] - 99 = 16.083$$

conclusion:
Reject H_o; there is sufficient evidence to conclude that the populations have different distributions.

12. Below are the ranks (in the original order) for each group. The group listed first is group 1, etc.

red	ora	yel	bro	tan	gre
13	47.5	25	11	86	30
83	19	55.5	6	74.4	50
72.5	70	38	67	20	7
3	55.5	93	8	82	28.5
86	64	45.5	21.5	86	78.5
45.5	44	25	76.5	17.5	33
88.5	28.5	42	80.5	99	15.5
62	69	97	61	64	1
14	42	54	80.5	25	71
15.5	439.5	57	72.5	78.5	314.5
47.4		27	52.5	32	
31		10	95	664.5	
40		88.5	76.5		
9		58	52.5		
59.5		50	94		
36		42	74.5		
17.5		38	23		
723.5		67	34.5		
		4	67		
		12	100		
		38	50		
		64	5		
		34.5	21.5	92	
		59.5	98	90	
		1124.5	96	84	
			91	2	
				1783.5	

$n_1 = 17 \quad R_1 = 723.5$
$n_2 = 9 \quad R_2 = 439.5$
$n_3 = 24 \quad R_3 = 1124.5$
$n_4 = 30 \quad R_4 = 1783.5$
$n_5 = 11 \quad R_5 = 664.5$
$n_6 = 9 \quad R_6 = 314.5$

$n = \Sigma n = 100 \qquad \Sigma R = 5050.0$

check:
$$\Sigma R = n(n+1)/2$$
$$= 100(101)/2$$
$$= 5050$$

H_o: the populations have the same distribution
H_1: the populations have different distributions
$\alpha = .05$
C.R. $H > \chi^2_{5,.05} = 11.071$

calculations:
$$H = [12/n(n+1)] \cdot [\Sigma(R_i^2/n_i)] - 3(n+1)$$
$$= [12/100(101)] \cdot [(723.5)^2/17 + (439.5)^2/9 + (1124.5)^2/24 + (1783.5)^2/30$$
$$+ (664.5)^2/11 + (314.5)^2/9] - 3(101)$$
$$= 8.408$$

conclusion:
Do not reject H_o; there is not sufficient evidence to conclude that the populations have different distributions.
No; the results do not indicate a problem that requires corrective action.

13. a. $H = [12/n(n+1)] \cdot [\Sigma(R_i^2/n_i)] - 3(n+1)$ where $n = \Sigma n_i = 8(6) = 48$

$= [12/48(49)] \cdot [\Sigma(R_i^2/6)] - 3(49)$

$= [1/196] \cdot (1/6)[\Sigma(R_i^2)] - 147$

$= [1/1176] \cdot [R_1^2 + R_2^2 + R_3^2 + R_4^2 + R_5^2 + R_6^2 + R_7^2 + R_8^2] - 147$

b. Since adding or subtracting a constant to each score does not affect the order of the scores, their ranks and the calculated H statistic are not affected.

c. Since multiplying or dividing each score by a positive constant does not affect the order of the scores, their ranks and the calculated H statistic are not affected.

14. For 15 total scores in 3 groups of 5 each, the sum of the ranks is $15(16)/2 = 120$.
The smallest H results when $R_1 = R_2 = R_3 = 40$.

$H = [12/n(n+1)] \cdot [\Sigma(R_i^2/n_i)] - 3(n+1)$

$= [12/(15)(16)] \cdot [40^2/5 + 40^2/5 + 40^2/5] - 3(16) = 0$

The largest H occurs when there is maximum separation of the groups.

$R_1 = 1 + 2 + 3 + 4 + 5 = 15$

$R_2 = 6 + 7 + 8 + 9 + 10 = 40$

$R_3 = 11 + 12 + 13 + 14 + 15 = 65$

$H = [12/n(n+1)] \cdot [\Sigma(R_i^2/n_i)] - 3(n+1)$

$= [12/(15)(16)] \cdot [15^2/5 + 40^2/5 + 65^2/5] - 3(16) = 12.5$

15. NOTE: Be careful when counting the number of tied ranks; in addition to the easily recognized ".5's," there are 3 12's, 3 16's and 3 23's. The following table organizes the calculations.

rank	t	$T = t^3 - t$
4.5	2	6
7.5	2	6
9.5	2	6
12	3	24
16	3	24
19.5	2	6
23	3	24
29.5	4	60
32.5	2	6
34.5	2	6
37.5	2	6
40.5	2	6
48.5	2	6
		186

correction factor:

$1 - \Sigma T/(n^3-n) = 1 - 186/(60^3-60)$

$= 1 - 186/215940$

$= 1 - .000861$

$= .999139$

The original calculated test statistic is $H = 3.489$.
The corrected calculated test statistic is $H = 3.489/.999139 = 3.492$

16. The algebra is tedious and the outline of the proof is sketched below.
The key useful fact relating $n_1 + n_2 = n$ to R_1 and R_2 is

$R_1 + R_2 = n(n+1)/2$ which implies $(n+1) = 2(R_1 + R_2)/n$ and $(n+1)^2 = 4(R_1 + R_2)^2/n^2$.

Also, let $D = n_1 n_2 (n_1 + n_2 + 1)/12 = n_1 n_2 (n+1)/12$.

(1) $H = [12/n(n+1)][R_1^2/n_1 + R_2^2/n_2] - 3(n+1)$

$= [12/n(n+1)][(n_2 R_1^2 + n_1 R_2^2)/n_1 n_2 - n(n+1)^2/4]$

$= [12/n_1 n_2 (n+1)][(n_2 R_1^2 + n_1 R_2^2)/n - n_1 n_2 (n+1)^2/4]$

$= [1/D][(n_2 R_1^2 + n_1 R_2^2)/n - n_1 n_2 (R_1 + R_2)^2/n^2]$

$= [1/D][A]$

(2) $z = [1/\sqrt{D}][R_1 - n_1(n+1)/2]$
$\quad\quad = [1/\sqrt{D}][R_1 - n_1(R_1+R_2)/n]$
$\quad\quad = [1/\sqrt{D}][(n_2R_1 - n_1R_2)/n]$
$\quad z^2 = [1/D][(n_2R_1 - n_1R_2)^2/n^2]$
$\quad\quad = [1/D][B]$

For calculated values, algebra shows that A = B and completes the proof that values, $z^2 = H$.
For critical values, tables show that $(z_{\alpha/2})^2 = \chi^2_{1,\alpha}$ for any α.
$\quad$ For example, $(z_{.025})^2 = (1.96)^2 = 3.841 = \chi^2_{1,.05}$

13-6 Rank Correlation

1. a. Since n ≤ 30, use Table A-9. CV: $r_s = \pm.450$
   b. Since n > 30, use Formula 13-1. CV: $r_s = \pm1.960/\sqrt{49} = \pm.280$
   c. Since n > 30, use Formula 13-1. CV: $r_s = \pm2.327/\sqrt{39} = \pm.373$

2. a. Since n ≤ 30, use Table A-9. CV: $r_s = \pm.689$
   b. Since n > 30, use Formula 13-1. CV: $r_s = \pm2.05/\sqrt{81} = \pm.228$
   c. Since n > 30, use Formula 13-1. CV: $r_s = \pm2.575/\sqrt{49} = \pm.368$

NOTE: This manual calculates $d = R_x - R_y$, thus preserving the sign of d. This convention means
Σd must equal 0 and provides a check for the assigning and differencing of the ranks.

3. The following table summarizes the calculations.

x	R_x	y	R_y	d	d^2
63	2	43	5	-3	9
68	3	44	6	-3	9
71	5	39	4	1	1
55	1	30	3	-2	4
70	4	28	2	2	4
75	6	20	1	5	25
				0	52

$r_s = 1 - [6(\Sigma d^2)]/[n(n^2-1)]$
$\quad = 1 - [6(52)]/[6(35)]$
$\quad = 1 - 1.486$
$\quad = -.486$

4. The following table summarizes the calculations.

x	R_x	y	R_y	d	d^2
28	1.5	16	2	-0.5	0.25
28	1.5	17	3	-1.5	2.25
35	3	12	1	2	4
37	4	19	4	0	0
40	5	20	5	0	0
				0.0	6.50

$r_s = 1 - [6(\Sigma d^2)]/[n(n^2-1)]$
$\quad = 1 - [6(6.5)]/[5(24)]$
$\quad = 1 - .325$
$\quad = .675$

5. The following table summarizes the calculations.

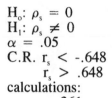

R_x	R_y	d	d^2
2	5	-3	9
6	2	4	16
3	3	0	0
3	8	-3	9
7	10	-3	9
10	9	1	1
9	1	8	64
8	7	1	1
4	6	-2	4
1	4	-3	9
		0	122

$$r_s = 1 - [6(\Sigma d^2)]/[n(n^2-1)]$$
$$= 1 - [6(122)]/[10(99)]$$
$$= 1 - .739$$
$$= .261$$

$H_o: \rho_s = 0$
$H_1: \rho_s \neq 0$
$\alpha = .05$
C.R. $r_s < -.648$
$r_s > .648$
calculations:
$r_s = .261$
conclusion:

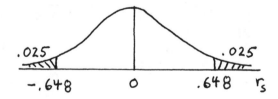

Do not reject H_o; there is not sufficient evidence to conclude that $\rho_s \neq 0$.

6. The following table summarizes the calculations.

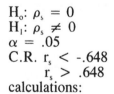

R_x	R_y	d	d^2
2	5	-3	9
7	2	5	25
6	3	3	9
4	8	-4	16
5	10	-5	25
8	9	-1	1
9	1	8	64
10	7	3	9
3	6	-3	9
1	4	-3	9
		0	176

$$r_s = 1 - [6(\Sigma d^2)]/[n(n^2-1)]$$
$$= 1 - [6(176)]/[10(99)]$$
$$= 1 - 1.067$$
$$= -.067$$

$H_o: \rho_s = 0$
$H_1: \rho_s \neq 0$
$\alpha = .05$
C.R. $r_s < -.648$
$r_s > .648$
calculations:
$r_s = -.067$
conclusion:
Do not reject H_o; there is not sufficient evidence to conclude that $\rho_s \neq 0$.

7. The following table summarizes the calculations.

R_x	R_y	d	d^2
1	4	-3	9
2	3	-1	1
7	5	2	4
4	6	-2	4
6	7	-1	1
8	8	0	0
3	2	1	1
5	1	4	16
		0	36

$$r_s = 1 - [6(\Sigma d^2)]/[n(n^2-1)]$$
$$= 1 - [6(36)]/[8(63)]$$
$$= 1 - .429$$
$$= .571$$

H_o: $\rho_s = 0$
H_1: $\rho_s \neq 0$
$\alpha = .05$
C.R. $r_s < -.738$
$\quad r_s > .738$
calculations:
$\quad r_s = .571$
conclusion:

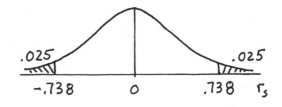

Do not reject H_o; there is not sufficient evidence to conclude that $\rho_s \neq 0$.

8. The following table summarizes the calculations.

R_x	R_y	d	d^2
4	1	3	9
3	5	-2	4
5	6	-1	1
6	2	4	16
7	7	0	0
8	8	0	0
2	3	-1	1
1	4	-3	9
		0	40

$$r_s = 1 - [6(\Sigma d^2)]/[n(n^2-1)]$$
$$= 1 - [6(40)]/[8(63)]$$
$$= 1 - .476$$
$$= .524$$

H_o: $\rho_s = 0$
H_1: $\rho_s \neq 0$
$\alpha = .05$
C.R. $r_s < -.738$
$\quad r_s > .738$
calculations:
$\quad r_s = .524$
conclusion:

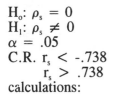

Do not reject H_o; there is not sufficient evidence to conclude that $\rho_s \neq 0$.

9. The following table summarizes the calculations.

x	R_x	y	R_y	d	d^2
107	10	111	10.5	-0.5	0.25
96	5	97	3	2	4
103	8	116	12	-4	16
90	3	107	8	-5	25
96	5	99	4	1	1
113	12	111	10.5	1.5	2.25
86	1	85	1	0	0
99	7	108	9	-2	4
109	11	102	6	5	25
105	9	105	7	2	4
96	5	100	5	0	0
89	2	93	2	0	0
				0.0	81.50

$$r_s = 1 - [6(\Sigma d^2)] / [n(n^2-1)]$$
$$= 1 - [6(81.5)] / [12(143)]$$
$$= 1 - .285$$
$$= .715$$

H_o: $\rho_s = 0$
H_1: $\rho_s \neq 0$
$\alpha = .05$
C.R. $r_s < -.591$
    $r_s > .591$
calculations:
    $r_s = .715$
conclusion:

Reject H_o; there is sufficient evidence to conclude that $\rho_s \neq 0$ (in fact, $\rho_s > 0$).

NOTE: Exercise 9 as presented is decidedly different from all the other exercises in this section, and the above test may be open to challenge. When finding the correlation between x and y, it makes no difference which variable carries which designation -- BUT THE PROBLEM MUST BE WELL-DEFINED AND INTERNALLY CONSISTENT. If x is "height" and y is "weight," for example, that designation must be kept for all pairs and one cannot rank one person's height with the other weights. Similarly if x is "test 1" and y is "test 2." Here, however, x and y are assumed to be "twin 1" and "twin 2" -- but how is it determined which twin is which? The problem could be made well-defined and internally consistent by declaring, for example, that x is the "older twin" -- but this was not stated in the exercise. Interchanging which twin is x and which is y for twins 1,2,4,9 produces the results and the same data lead to the <u>opposite conclusion!</u> As in all the other exercises, there should be a clear and consistent determination of what is x and what is y. Even if the choice of x and y was made randomly, that fact should be reported with the data.

x	R_x	y	R_y	d	d^2
111	11	107	8	3	9
97	5	96	4	1	1
103	8	116	12	-4	16
107	10	90	2	8	64
96	3.5	99	5	-1.5	2.25
113	12	111	11	1	1
86	1	85	1	0	0
99	6	108	9	-3	9
102	7	109	10	-3	9
105	9	105	7	2	4
96	3.5	100	6	-2.5	6.25
89	2	93	3	-1	1
				0.0	122.50

$$r_s = 1 - [6(\Sigma d^2)] / [n(n^2-1)]$$
$$= 1 - [6(122.5)] / [12(143)]$$
$$= 1 - .428$$
$$= .572$$

10. The following table summarizes the calculations.

x	R_x	y	R_y	d	d^2
91	5	4.56	2	3	9
92	6	6.48	6	0	0
82	2	5.99	5	-3	9
85	3	7.92	8	-5	25
87	4	5.36	4	0	0
80	1	3.32	1	0	0
94	7	7.32	7	0	0
97	8	5.27	3	5	25
				0	68

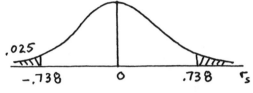

$$r_s = 1 - [6(\Sigma d^2)]/[n(n^2-1)]$$
$$= 1 - [6(68)]/[8(63)]$$
$$= 1 - .810$$
$$= .190$$

$H_0: \rho_s = 0$
$H_1: \rho_s \neq 0$
$\alpha = .05$
C.R. $r_s < -.738$
        $r_s > .738$
calculations:
    $r_s = .190$
conclusion:

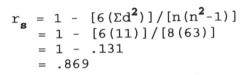

Do not reject H_0; there is not sufficient evidence to conclude that $\rho_s \neq 0$.

11. The following table summarizes the calculations.

x	R_x	y	R_y	d	d^2
.27	1	2	2.5	-1.5	2.25
1.41	3	3	4.5	-1.5	2.25
2.19	5.5	3	4.5	1	1
2.83	7	6	8	-1	1
2.19	5.5	4	6	-0.5	0.25
1.81	4	2	2.5	1.5	2.25
.85	2	1	1	1	1
3.05	8	5	7	1	1
				0.0	11.00

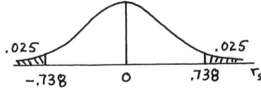

$$r_s = 1 - [6(\Sigma d^2)]/[n(n^2-1)]$$
$$= 1 - [6(11)]/[8(63)]$$
$$= 1 - .131$$
$$= .869$$

$H_0: \rho_s = 0$
$H_1: \rho_s \neq 0$
$\alpha = .05$
C.R. $r_s < -.738$
        $r_s > .738$
calculations:
    $r_s = .869$
conclusion:

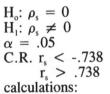

Reject H_0; there is sufficient evidence to conclude that $\rho_s \neq 0$ (in fact, $\rho_s > 0$).

12. The following table summarizes the calculations.

x	R_x	y	R_y	d	d^2
2.41	1	2	2.5	-1.5	2.25
7.57	4	3	4.5	-0.5	0.25
9.55	7	3	4.5	2.5	6.25
8.82	6	6	8	-2	4
8.72	5	4	6	-1	1
6.96	3	2	2.5	0.5	0.25
6.83	2	1	1	1	1
11.42	8	5	7	1	1
				0.0	16.00

$r_s = 1 - [6(\Sigma d^2)]/[n(n^2-1)]$
$\quad = 1 - [6(16)]/[8(63)]$
$\quad = 1 - .190$
$\quad = .810$

$H_o: \rho_s = 0$
$H_1: \rho_s \neq 0$
$\alpha = .05$
C.R. $r_s < -.738$
$\quad\;\; r_s > .738$
calculations:
$\quad r_s = .869$
conclusion:
   Reject H_o; there is sufficient evidence to conclude that $\rho_s \neq 0$ (in fact, $\rho_s > 0$).

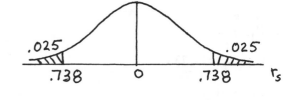

13. The following table summarizes the calculations.

x	R_x	y	R_y	d	d^2
17.2	1	.19	3	-2	4
43.5	6	.20	4.5	1.5	2.25
30.7	4	.26	8	-4	16
53.1	8	.16	1	7	49
37.2	5	.24	7	-2	4
21.0	2	.20	4.5	-2.5	6.25
27.6	3	.18	2	1	1
46.3	7	.23	6	1	1
				0.0	83.50

$r_s = 1 - [6(\Sigma d^2)]/[n(n^2-1)]$
$\quad = 1 - [6(83.5)]/[8(63)]$
$\quad = 1 - .994$
$\quad = .006$

$H_o: \rho_s = 0$
$H_1: \rho_s \neq 0$
$\alpha = .05$
C.R. $r_s < -.738$
$\quad\;\; r_s > .738$
calculations:
$\quad r_s = .006$
conclusion:
   Do not reject H_o; there is not sufficient evidence to conclude that $\rho_s \neq 0$.

14. The following table summarizes the calculations.

x	R_x	y	R_y	d	d^2
167	5	4.23	5	0	0
191	7	4.69	7	0	0
112	1	3.21	1	0	0
129	3	3.47	3	0	0
140	4	3.72	4	0	0
173	6	4.45	6	0	0
119	2	3.36	2	0	0
				0	0

$$r_s = 1 - [6(\Sigma d^2)]/[n(n^2-1)]$$
$$= 1 - [6(0)]/[7(48)]$$
$$= 1 - .000$$
$$= 1.000$$

$H_o: \rho_s = 0$
$H_1: \rho_s \neq 0$
$\alpha = .05$
C.R. $r_s < -.786$
       $r_s > .786$
calculations:
    $r_s = 1.000$
conclusion:
    Reject H_o; there is sufficient evidence to conclude that $\rho_s \neq 0$ (in fact, $\rho_s > 0$).

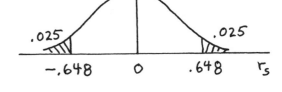

15. The following table summarizes the calculations.

x	R_x	y	R_y	d	d^2
17	6	73	8	-2	4
21	8	66	4	4	16
11	2.5	64	3	-0.5	0.25
16	5	61	2	3	9
15	4	70	6	-2	4
11	2.5	71	7	-4.5	20.25
24	9	90	10	-1	1
27	10	68	5	5	25
19	7	84	9	-2	4
8	1	52	1	0	0
				0.0	83.50

$$r_s = 1 - [6(\Sigma d^2)]/[n(n^2-1)]$$
$$= 1 - [6(83.5)]/[10(99)]$$
$$= 1 - .506$$
$$= .494$$

$H_o: \rho_s = 0$
$H_1: \rho_s \neq 0$
$\alpha = .05$
C.R. $r_s < -.648$
       $r_s > .648$
calculations:
    $r_s = .494$
conclusion:
    Do not reject H_o; there is not sufficient evidence to conclude that $\rho_s \neq 0$.

16. The following table summarizes the calculations.

x	R_x	y	R_y	d	d^2
11.6	8	13.1	8	0	0
8.3	7	10.6	6	1	1
3.6	5	10.1	5	0	0
.6	1	4.4	2	-1	1
6.9	6	11.5	7	-1	1
2.5	3	6.6	4	-1	1
2.4	2	3.6	1	1	1
2.6	4	5.3	3	1	1
				0	6

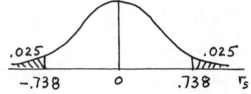

$$r_s = 1 - [6(\Sigma d^2)]/[n(n^2-1)]$$
$$= 1 - [6(6)]/[8(63)]$$
$$= 1 - .071$$
$$= .929$$

$H_o: \rho_s = 0$
$H_1: \rho_s \neq 0$
$\alpha = .05$
C.R. $r_s < -.738$
    $r_s > .738$
calculations:
    $r_s = .929$
conclusion:

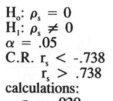

Reject H_o; there is sufficient evidence to conclude that $\rho_s \neq 0$ (in fact, $\rho_s > 0$).

17. The following table summarizes the calculations.

x	R_x	y	R_y	d	d^2
.65	6	14.7	7	-1	1
.55	4	12.3	5	-1	1
.72	7	14.6	6	1	1
.83	8	15.1	8	0	0
.57	5	5.0	4	1	1
.51	3	4.1	2.5	0.5	0.25
.43	2	3.8	1	1	1
.37	1	4.1	2.5	-1.5	2.25
				0.0	7.50

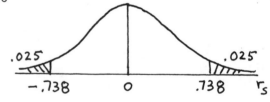

$$r_s = 1 - [6(\Sigma d^2)]/[n(n^2-1)]$$
$$= 1 - [6(7.5)]/[8(63)]$$
$$= 1 - .089$$
$$= .911$$

$H_o: \rho_s = 0$
$H_1: \rho_s \neq 0$
$\alpha = .05$
C.R. $r_s < -.738$
    $r_s > .738$
calculations:
    $r_s = .911$
conclusion:

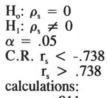

Reject H_o; there is sufficient evidence to conclude that $\rho_s \neq 0$ (in fact, $\rho_s > 0$).

18. The following table summarizes the calculations.

x	R_x	y	R_y	d	d^2
-62	2	21	7	-5	25
-41	5	13	4	1	1
-36	6	12	3	3	9
-26	8	3	1	7	49
-33	7	6	2	5	25
-56	3	22	8	-5	25
-50	4	14	5	-1	1
-66	1	19	6	-5	25
				0	160

$$r_s = 1 - [6(\Sigma d^2)]/[n(n^2-1)]$$
$$= 1 - [6(160)]/[8(63)]$$
$$= 1 - 1.905$$
$$= -.905$$

$H_o: \rho_s = 0$
$H_1: \rho_s \neq 0$
$\alpha = .05$
C.R. $r_s < -.738$
     $r_s > .738$
calculations:
   $r_s = -.905$
conclusion:
   Reject H_o; there is sufficient evidence to conclude that $\rho_s \neq 0$ (in fact, $\rho_s < 0$).

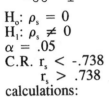

19. The following table summarizes the calculations.

x	R_x	y	R_y	d	d^2
3522	1	.20	1	0	0
3597	2	.22	2	0	0
4171	7	.23	3	4	16
4258	8	.29	4	4	16
3993	4	.31	5	-1	1
3971	3	.33	7.5	-4.5	20.25
4042	5	.33	7.5	-2.5	6.25
4053	6	.32	6	0	0
				0.0	59.50

$$r_s = 1 - [6(\Sigma d^2)]/[n(n^2-1)]$$
$$= 1 - [6(59.5)]/[8(63)]$$
$$= 1 - .708$$
$$= .292$$

$H_o: \rho_s = 0$
$H_1: \rho_s \neq 0$
$\alpha = .05$
C.R. $r_s < -.738$
     $r_s > .738$
calculations:
   $r_s = .292$
conclusion:
   Do not reject H_o; there is not sufficient evidence to conclude that $\rho_s \neq 0$.

20. The following table summarizes the calculations.

x	R_x	y	R_y	d	d^2
1364	7	652	7	0	0
53	1	20	1	0	0
457	4	163	3	1	1
717	5	210	5	0	0
384	3	175	4	-1	1
1432	8	821	8	0	0
365	2	143	2	0	0
1626	9	1098	9	0	0
840	6	459	6	0	0
				0	2

$$r_s = 1 - [6(\Sigma d^2)]/[n(n^2-1)]$$
$$= 1 - [6(2)]/[9(80)]$$
$$= 1 - .017$$
$$= .983$$

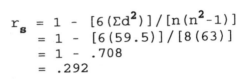

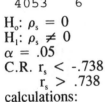

$H_o: \rho_s = 0$
$H_1: \rho_s \neq 0$
$\alpha = .05$
C.R. $r_s < -.683$
     $r_s > .683$
calculations:
     $r_s = .983$
conclusion:
     Reject H_o; there is sufficient evidence to conclude that $\rho_s \neq 0$ (in fact, $\rho_s > 0$).

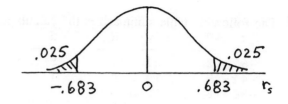

21. The following table summarizes the calculations.

x	R_x	y	R_y	d	d^2
65.5	2	12.6	3	-1	1
68.3	9	36.4	11	-2	4
73.1	17	13.6	4	13	169
71.8	15	43.2	13	2	4
67.3	5	32.4	9	-4	16
70.2	14	34.8	10	4	16
68.7	11	54.6	15	-4	16
68.4	10	11.6	2	8	64
64.8	1	157.8	17	-16	256
70.0	12.5	20.2	7	5.5	30.25
68.0	7	19.4	6	1	1
66.8	4	28.4	8	-4	16
70.0	12.5	89.2	16	-3.5	12.25
65.7	3	11.4	1	2	4
72.0	16	18.8	5	11	121
68.1	8	48.8	14	-6	36
67.6	6	40.4	12	-6	36
					802.50

$r_s = 1 - [6(\Sigma d^2)] / [n(n^2-1)]$
$\quad = 1 - [6(802.5)] / [17(288)]$
$\quad = 1 - .983$
$\quad = .017$

$H_o: \rho_s = 0$
$H_1: \rho_s \neq 0$
$\alpha = .05$
C.R. $r_s < -.490$
     $r_s > .490$
calculations:
     $r_s = .017$
conclusion:
     Do not reject H_o; there is not sufficient evidence to conclude that $\rho_s \neq 0$.

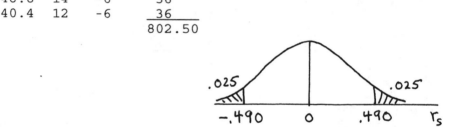

22. The following table summarizes the calculations.

x	R_x	y	R_y	d	d^2
34	3.5	29.6	5	-1.5	2.25
34	3.5	7.4	1	2.5	6.25
34	3.5	11.2	3	0.5	0.25
42	7	33.6	7	0	0
34	3.5	7.8	2	1.5	2.25
34	3.5	26.8	4	-0.5	0.25
42	8	49.0	9	1	1
45	9	47.4	8	-1	1
34	3.5	29.8	6	-2.5	6.25
				0.0	19.50

$r_s = 1 - [6(\Sigma d^2)] / [n(n^2-1)]$
$\quad = 1 - [6(19.5)] / [9(80)]$
$\quad = 1 - .1625$
$\quad = .8375$

H_o: $\rho_s = 0$
H_1: $\rho_s \neq 0$
$\alpha = .05$
C.R. $r_s < -.683$
　　 $r_s > .683$
calculations:
　 $r_s = .8375$

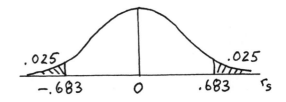

conclusion:
　Reject H_o; there is sufficient evidence to conclude that $\rho_s \neq 0$ (in fact, $\rho_s > 0$).
NOTE: The above formula for r_s yields an exact answer only if there are no ties. Application of formula 9-1 on the ranks gives an exact answer of $r_s = .807$.

23. The following table summarizes the calculations.

x	R_x	y	R_y	d	d^2
2948	18	189.1	19.5	-1.5	2.25
3536	27	205.2	29	-2	4
3472	26	194.8	24	2	4
2782	12	189.1	19.5	-7.5	56.25
3766	30	178.7	5	25	625
4367	32	225.1	32	0	0
2649	6	183.4	13.5	-7.5	56.25
2526	3	182.3	12	-9	81
2665	7	183.4	13.5	-6.5	42.25
3374	25	199.3	25	0	0
2863	17	184.8	15.5	1.5	2.25
3010	20	184.8	15.5	4.5	20.25
2779	11	179.0	7	4	16
3315	23	201.7	27	-4	16
2788	14	181.2	10.5	3.5	12.25
3290	22	201.6	26	-4	16
2360	2	170.9	2	0	0
2775	10	179.6	8	2	4
2619	5	178.9	6	-1	1
3253	21	193.1	21	0	0
1650	1	147.4	1	0	0
3628	28	205.1	28	0	0
3784	31	212.4	31	0	0
2602	4	177.0	4	0	0
2717	9	187.9	18	-9	81
2992	19	194.4	23	-4	16
3354	24	193.7	22	2	4
3697	29	205.5	30	-1	1
2784	13	181.2	10.5	2.5	6.25
2804	15	186.9	17	-2	4
2823	16	176.3	3	13	169
2682	8	180.7	9	-1	1
				0.0	1241.00

$$r_s = 1 - [6(\Sigma d^2)]/[n(n^2-1)]$$
$$= 1 - [6(1241)]/[32(1023)]$$
$$= 1 - .227$$
$$= .773$$

since n > 30,
CV: $\pm z_{\alpha/2}/\sqrt{n-1} = \pm 1.960/\sqrt{31}$
　　　　 $= \pm .352$

H_o: $\rho_s = 0$
H_1: $\rho_s \neq 0$
$\alpha = .05$
C.R. $r_s < -.352$
  $r_s > .352$
calculations:
  $r_s = .773$
conclusion:
  Reject H_o; there is sufficient evidence to conclude that $\rho_s \neq 0$ (in fact, $\rho_s > 0$).

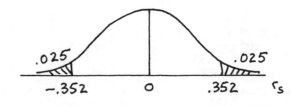

24. There are n=32 pairs of values. The weight and highway mpg ranks as as follows.

```
wgt: 18     27   26   12   30   32    6       3      7   25   17   20   11   23   14   22
mpg: 26.5    9    9   20    1    3   23.5   26.5   30   20    9   15   15   15   15   15

wgt:  2    10    5   21    1   28   31    4    9   19   24     29   13   15   16    8
mpg: 29     2  23.5 23.5  32    5    5    9   20    5  23.5    9   15   28   15   31
```
$\Sigma d^2 = 8815.5$ $r_s = 1 - [6(8815.5)]/[32(1023)] = 1 - 1.616 = -.616$
Since n>30, CV $= \pm 1.96\sqrt{31} = \pm.352$.

H_o: $\rho_s = 0$
H_1: $\rho_s \neq 0$
$\alpha = .05$
C.R. $r_s < -.352$
  $r_s > .352$
calculations:
  $r_s = -.616$
conclusion:
  Reject H_o; there is sufficient evidence to conclude that $\rho_s \neq 0$ (in fact, $\rho_s < 0$).
  NOTE: The above formula for r_s yields an exact answer only if there are no ties. Application of formula 9-1 on the ranks gives an exact answer of $r_s = -.630$.

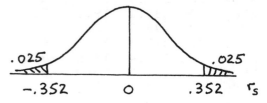

25. a. Excluding ties, the six possible rankings for the second judge are 123 132 213 231 312 321.

b. The following table organizes the calculations.

2nd judge	d	d^2	Σd^2	$r_s = 1 - [6(\Sigma d^2)]/[n(n^2-1)]$
123	0,0,0	0,0,0	0	$1 - [6 \cdot 0]/[3 \cdot 8] = 1 - 0 = 1$
132	0,-1,1	0,1,1	2	$1 - [6 \cdot 2]/[3 \cdot 8] = 1 - .5 = .5$
213	-1,1,0	1,1,0	2	$1 - [6 \cdot 2]/[3 \cdot 8] = 1 - .5 = .5$
231	-1,-1,2	1,1,4	6	$1 - [6 \cdot 6]/[3 \cdot 8] = 1 - 1.5 = -.5$
312	-2,1,1	4,1,1	6	$1 - [6 \cdot 6]/[3 \cdot 8] = 1 - 1.5 = -.5$
321	-2,0,2	4,0,4	8	$1 - [6 \cdot 8]/[3 \cdot 8] = 1 - 2 = -1$

c. $P(r_s > .9) = P(r_s = 1) = 1/6 = .167$

26. a. $t_{6,.025} = 2.365$; $r_s^2 = (2.447)^2/[(2.447)^2 + 6] = .499$, $r_s = \pm.707$
b. $t_{13,.025} = 2.160$; $r_s^2 = (2.160)^2/[(2.160)^2 + 13] = .264$, $r_s = \pm.514$
c. $t_{28,.025} = 2.048$; $r_s^2 = (2.048)^2/[(2.048)^2 + 28] = .130$, $r_s = \pm.361$
d. $t_{28,.005} = 2.763$; $r_s^2 = (2.763)^2/[(2.763)^2 + 28] = .214$, $r_s = \pm.463$
e. $t_{6,.005} = 3.707$; $r_s^2 = (3.707)^2/[(3.707)^2 + 6] = .696$, $r_s = \pm.834$

27. a. For the pictured data, r_s would give a better measure of the relationship than would r. Since the relationship is non-linear, the linear correlation coefficient r cannot accurately measure its strength. It is true, however, that there is a positive correlation -- i.e., smaller x's are associated with smaller y's and larger x's are associated with larger y's. The rank correlation coefficient r_s measures precisely such an association -- independent of the magnitude of the increases (i.e., whether the increase is linear, quadratic or irregular).

b. When dealing with n ranks, $\Sigma R = n(n+1)/2$ -- this is always true (i.e., whether or not there are ties in the ranks) and may be used as a check to see if the ranks have been assigned correctly. It is also true that $\Sigma R^2 = n(n+1)(2n+1)/6$ whenever there are no ties in ranks, and that ΣR^2 is less than that whenever there are ties. The shortcut formula

$r_s = 1 - [6(\Sigma d^2)]/[n(n^2-1)]$

was derived using $\Sigma R^2 = n(n+1)(2n+1)/6$ (i.e., assuming there were no ties). If x is ranked from low to high to produce the usual R_x values and y is ranked from high to low, each y rank will be $(n+1) - R_y$ -- where R_y represents the usual low to high ranking.
  Instead of $d = R_x - R_y$,
  we now have $d = R_x - [(n+1) - R_y]$.
If there are no ties, algebra and the above formulas for ΣR and ΣR^2 can be used to show the resulting r_s will be precisely the negative of the r_s obtained by ranking both variables from low to high. Whether or not there are ties, the "true" formula always produces that result -- viz., that ranking one of the variables from high to low reverses the sign of r_s.

c. Refer to the comments and notation in part (b).
  If both researchers rank from low to high, then $d = R_x - R_y$;
    if they both rank from high to low, then $d = [(n+1) - R_x] - [(n+1) - R_y] = R_y - R_x$.
  While the signed d's are negatives of each other. the d^2 values and r_s remain unchanged.

d. Since the two-tailed critical value for $\alpha = .10$ places .05 in each tail, it may be used for a one-tailed test at the .05 level. The requested test is as follows.

$H_o: \rho_s \leq 0$
$H_1: \rho_s > 0$
$\alpha = .05$
C.R. $r_s > .564$
calculations:
  $r_s = .855$
conclusion:
  Reject H_o; there is sufficient evidence to conclude that $\rho_s > 0$.

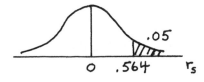

13-7 Runs Test for Randomness

NOTE: In each exercise, the item that appears first in the sequence is considered to be of the first type and its count is designated by n_1.

1. $n_1 = 7$ (# of A's)
   $n_2 = 2$ (# of B's)
   G = 2 (# of runs -- one run of A's, followed by one run of B's)
   CV: 1,6 (from Table A-10)

2. $n_1 = 10$ (# of A's)
   $n_2 = 10$ (# of B's)
   G = 9 (# of runs -- one run of A's, followed by one run of B's)
   CV: 6,16 (from Table A-10)

3. a. In increasing order, the values are 3 7 7 8 9 10 12 16 18 20
   and the median is $(9 + 10)/2 = 9.5$

   b. The original sequence 3 8 7 7 9 12 10 16 20 18 becomes the following B's and A's.
   B B B B B A A A A A

   c. $n_1 = 5$ $n_2 = 5$ $G = 2$

   d. CV: 2,10

   e. Since $G \le 2$, we conclude there is not a random sequence of values above and below the median.

4. a. In increasing order, the values are 2 2 2 3 4 4 4 5 6 7 8 9 9 12
   and the median is $(4 + 5)/2 = 4.5$

   b. The original sequence becomes B B B B B B A A B A A A A A

   c. $n_1 = 7$ $n_2 = 7$ $G = 4$

   d. CV: 3,13

   e. Since $3 \le G \le 13$, we fail to reject the hypothesis of randomness above and below the median.

5. Since $n_1 = 6$ and $n_2 = 24$, use the normal approximation.

$$\mu_G = 2n_1n_2/(n_1+n_2) + 1$$
$$= 2(6)(24)/30 + 1 = 10.6$$
$$\sigma_G^2 = [2n_1n_2(2n_1n_2-n_1-n_2)]/[(n_1+n_2)^2(n_1+n_2-1)]$$
$$= [2(6)(24)(258)]/[(30)^2(29)] = 2.847$$

H_o: the sequence is random
H_1: the sequence is not random
$\alpha = .05$
C.R. $z < -z_{.025} = -1.960$
      $z > z_{.025} = 1.960$
calculations:
   $G = 12$
   $z_G = (G - \mu_G)/\sigma_G$
      $= (12 - 10.6)/\sqrt{2.847}$
      $= 1.4/1.687 = .830$

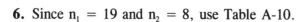

conclusion:
   Do not reject H_o; there is not sufficient evidence to conclude that the sequence is not random.

6. Since $n_1 = 19$ and $n_2 = 8$, use Table A-10.

H_o: the sequence is random
H_1: the sequence is not random
$\alpha = .05$
C.R. $G \le 7$
      $G \ge 17$
calculations:
   $G = 11$
conclusion:
   Do not reject H_o; there is not sufficient evidence to conclude that the values do not occur in a random sequence.

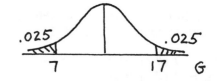

7. When the scores are arranged in order, the median is found to be $(.14 + .15)/2 = .145$.
In the usual above-below notation, the original sequence may be written as
   A A A A A A A A A B A B B B B B B B B B
Since $n_1 = 10$ and $n_2 = 10$, use Table A-10.

H_o: values above and below the median occur in a random sequence
H_1: values above and below the median do not occur in a random sequence
$\alpha = .05$
C.R. $G \leq 6$
     $G \geq 16$
calculations:
   $G = 4$

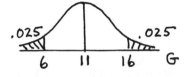

conclusion:
   Reject H_o; there is sufficient evidence to conclude that the values above and below the median do not occur in a random sequence (in fact, values on the same side of the median tend to occur in groups).

8. $\bar{x} = 134/17 = 7.9$.
   A B A A B A B A A A A A B B A A A
   Since $n_1 = 12$ and $n_2 = 5$, use Table A-10.

H_o: the sequence is random
H_1: the sequence is not random
$\alpha = .05$
C.R. $G \leq 4$
     $G \geq 12$
calculations:
   $G = 9$

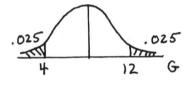

conclusion:
   Do not reject H_o; there is not sufficient evidence to conclude that the values do not occur in a random sequence.

9. When the scores are arranged in order, the median is found to be $(812 + 856)/2 = 834$.
In the usual above-below notation, the original sequence may be written as
   A A A A A A A A B B B B B B B B
Since $n_1 = 8$ and $n_2 = 8$, use Table A-10.

H_o: values above and below the median occur in a random sequence
H_1: values above and below the median do not occur in a random sequence
$\alpha = .05$
C.R. $G \leq 4$
     $G \geq 14$
calculations:
   $G = 2$

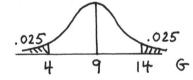

conclusion:
   Reject H_o; there is sufficient evidence to conclude that the values above and below the median do not occur in a random sequence (in fact, values on the same side of the median tend to occur in groups).

10. median $= (264.32 + 264.11)/2 = 264.215$
A A A A A A B B B B B B
Since $n_1 = 6$ and $n_2 = 6$, use Table A-10.

H_o: values above and below the median occur in a random sequence
H_1: values above and below the median do not occur in a random sequence
$\alpha = .05$
C.R. $G \leq 3$
  $G \geq 11$
calculations:
 $G = 2$
conclusion:

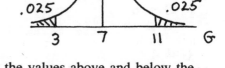

Reject H_o; there is sufficient evidence to conclude that the values above and below the median do not occur in a random sequence (in fact, values on the same side of the median tend to occur in groups).

11. The sequence in Y's and O's is as follows.
 Y Y Y Y Y Y O Y O O Y Y Y Y O
 O O O O O O O O O O O O O Y Y
 Y O O O O Y Y O Y O O O O O O
Since $n_1 = 17$ and $n_2 = 28$, use the normal approximation.
 $\mu_G = 2n_1n_2/(n_1+n_2) + 1$
  $= 2(17)(28)/45 + 1 = 22.156$
 $\sigma_G^2 = [2n_1n_2(2n_1n_2-n_1-n_2)]/[(n_1+n_2)^2(n_1+n_2-1)]$
  $= [2(17)(28)(907)]/[(45)^2(44)] = 9.691$

H_o: the sequence is random
H_1: the sequence is not random
$\alpha = .05$
C.R. $z < -z_{.025} = -1.960$
  $z > z_{.025} = 1.960$
calculations:
 $G = 12$
 $z_G = (G - \mu_G)/\sigma_G$
  $= (12 - 22.156)/\sqrt{9.691}$
  $= -10.156/3.113 = -3.262$
conclusion:

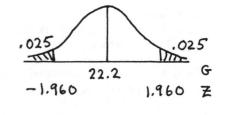

Reject H_o; there is sufficient evidence to conclude that the sequence is not random (in fact, cases from the same age level tend to occur in groups).

12. Since $n_1 = 20$ and $n_2 = 20$, use Table A-10.

H_o: the sequence is random
H_1: the sequence is not random
$\alpha = .05$
C.R. $G \leq 14$
  $G \geq 28$
calculations:
 $G = 8$
conclusion:

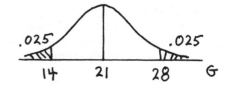

Reject H_o; there is sufficient evidence to conclude that the values do not occur in a random sequence (in fact, like values tend to occur in groups).

13. Since $n_1 = 50$ and $n_2 = 37$, use the normal approximation.

$\mu_G = 2n_1n_2/(n_1+n_2) + 1$
$\quad = 2(50)(37)/87 + 1 = 43.529$
$\sigma_G^2 = [2n_1n_2(2n_1n_2-n_1-n_2)]/[(n_1+n_2)^2(n_1+n_2-1)]$
$\quad = [2(50)(37)(3613)]/[(87)^2(86)] = 20.537$

H_0: the sequence is random
H_1: the sequence is not random
$\alpha = .05$
C.R. $z < -z_{.025} = -1.960$
$\quad z > z_{.025} = 1.960$
calculations:
$\quad G = 48$
$\quad z_G = (G - \mu_G)/\sigma_G$
$\quad\quad = (48 - 43.529)/\sqrt{20.54}$
$\quad\quad = 4.471/4.532 = .987$

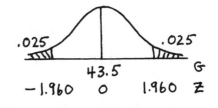

conclusion:
Do not reject H_0; there is not sufficient evidence to conclude that the sequence is not random.

14. Define July 1 to be the middle of the year.
A A A A A A B B B B B B
Since $n_1 = 6$ and $n_2 = 6$, use Table A-10.

H_0: dates after and before the middle occur in a random sequence
H_1: dates after and before the middle do not occur in a random sequence
$\alpha = .05$
C.R. $G \leq 3$
$\quad G \geq 11$
calculations:
$\quad G = 2$

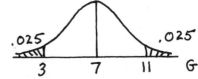

conclusion:
Reject H_0; there is sufficient evidence to conclude that the values after and before the middle do not occur in a random sequence (in fact, dates on the same side of the middle tend to occur in groups).

15. The sequence in O's and E's is as follows.
O E O O O E E O O O E O O O O E O E E E
E E E O O E O E O O O E E E E E O O O O
E O O O O O O O O E O E E E O O E O E E
O O E O E O E O E E E E E E E E E E O O
E E E E E O E E E O O E E O O O E E O O
Since $n_1 = 49$ and $n_2 = 51$, use the normal approximation.

$\mu_G = 2n_1n_2/(n_1+n_2) + 1$
$\quad = 2(49)(51)/100 + 1 = 50.98$
$\sigma_G^2 = [2n_1n_2(2n_1n_2-n_1-n_2)]/[(n_1+n_2)^2(n_1+n_2-1)]$
$\quad = [2(49)(51)(4898)]/[(100)^2(99)] = 24.727$

H_o: the sequence is random
H_1: the sequence is not random
$\alpha = .05$
C.R. $z < -z_{.025} = -1.960$
      $z > z_{.025} = 1.960$
calculations:
  $G = 43$
  $z_G = (G - \mu_G)/\sigma_G$
      $= (43 - 50.98)/\sqrt{24.73}$
      $= -7.98/4.973 = -1.605$
conclusion:
  Do not reject H_o; there is not sufficient evidence to conclude that the sequence is not random.

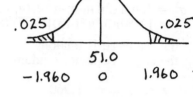

16. There are 49 digits above 4.5 and 51 digits above 4.5
Since $n_1 = 49$ and $n_2 = 51$, use the normal approximation.
  $\mu_G = 2n_1 n_2/(n_1 + n_2) + 1$
      $= 2(49)(51)/100 + 1 = 50.98$
  $\sigma_G^2 = [2n_1 n_2(2n_1 n_2 - n_1 - n_2)]/[(n_1 + n_2)^2(n_1 + n_2 - 1)]$
      $= [2(49)(51)(4898)]/[(100)^2(99)] = 24.727$

H_o: the sequence is random
H_1: the sequence is not random
$\alpha = .05$
C.R. $z < -z_{.025} = -1.960$
      $z > z_{.025} = 1.960$
calculations:
  $G = 54$
  $z_G = (G - \mu_G)/\sigma_G$
      $= (54 - 50.98)/\sqrt{24.73}$
      $= 3.02/4.973 = .607$
conclusion:
  Do not reject H_o; there is not sufficient evidence to conclude that the sequence is not random.

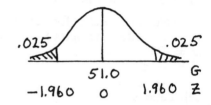

17. The minimum possible number of runs is $G = 2$ and occurs when all the A's are together and all the B's are together (e.g., A A B B).
The maximum possible number of runs is $G = 4$ and occurs when the A's and B's alternate (e.g., A B A B).
Because the critical region for $n_1 = n_2 = 2$ is
        C.R. $G \leq 1$
              $G \geq 6$
the null hypothesis of the sequence being random can never be rejected at the .05 level. Very simply, this means that it is not possible for such a small sample to provide 95% certainty that a non-random phenomenon is occurring.

18. For $n_1 = 20$ and $n_2 = 20$ using the normal approximation,

$\mu_G = 2n_1n_2/(n_1+n_2) + 1$
$\quad = 2(20)(20)/40 + 1 = 21$
$\sigma_G^2 = [2n_1n_2(2n_1n_2-n_1-n_2)]/[(n_1+n_2)^2(n_1+n_2-1)]$
$\quad = [2(20)(20)(760)]/[(40)^2(39)] = 9.74$
$\sigma_G = 3.12$

Consider the following test.

H_o: the sequence is random
H_1: the sequence is not random
$\alpha = .05$
C.R. $z < -z_{.025} = -1.960$
$\quad\ z > z_{.025} = 1.960$
calculations:

$z_G = (G - \mu_G)/\sigma_G$
$\pm 1.96 = (G - \mu_G)/\sigma_G$
$\quad\ G = \mu_G \pm 1.96\sigma_G$
$\quad\quad = 21 \pm 1.96(3.12)$
$\quad\quad = 21 \pm 6.12$
$\quad\quad = 14.88, 27.12$ [the critical values in terms of G]

Since G must be an integer, $G < 14.88$ and $G > 27.12$ agree with the $G \leq 14$ and $G \geq 28$ of Table A-10.

19. a. and b. The 84 sequences and the number of runs in each are as follows.

AAABBBBBB-2	BAAABBBBB-3	BBAAABBBB-3	BBBAAABBB-3	BBBBAAABB-3	BBBBBAAAB-3
AABABBBBB-4	BAABABBBB-5	BBAABABBB-5	BBBAABABB-5	BBBBAABAB-5	BBBBBAABA-4
AABBABBBB-4	BAABBABBB-5	BBAABBABB-5	BBBAABBAB-5	BBBBAABBA-4	
AABBBABBB-4	BAABBBABB-5	BBAABBBAB-5	BBBAABBBA-4		
AABBBBABB-4	BAABBBBAB-5	BBAABBBBA-4			
AABBBBBAB-4	BAABBBBBA-4				
AABBBBBBA-3					BBBBBBAAA-2
ABAABBBBB-4	BABAABBBB-5	BBABAABBB-5	BBBABAABB-5	BBBBABAAB-5	BBBBBABAA-4
ABABABBBB-6	BABABABBB-7	BBABABABB-7	BBBABABAB-7	BBBBABABA-6	
ABABBABBB-6	BABABBABB-7	BBABABBAB-7	BBBABABBA-6		
ABABBBABB-6	BABABBBAB-7	BBABABBBA-6			
ABABBBBAB-6	BABABBBBA-6				
ABABBBBBA-5					
ABBAABBBB-4	BABBAABBB-5	BBABBAABB-5	BBBABBAAB-5	BBBBABBAA-4	
ABBABABBB-6	BABBABABB-7	BBABBABAB-7	BBBABBABA-6		
ABBABBABB-6	BABBABBAB-7	BBABBABBA-6			
ABBABBBAB-6	BABBABBBA-6				
ABBABBBBA-5					
ABBBAABBB-4	BABBBAABB-5	BBABBBAAB-5	BBBABBBAA-4		
ABBBABABB-6	BABBBABAB-7	BBABBBABA-6			
ABBBABBAB-6	BABBBABBA-6				
ABBBABBBA-5					
ABBBBAABB-4	BABBBBAAB-5	BBABBBBAA-4			
ABBBBABAB-6	BABBBBABA-6				
ABBBBABBA-4					
ABBBBBAAB-4	BABBBBBAA-4				
ABBBBBABA-5					
ABBBBBBAA-3					

c. Below is the distribution for G, the number of runs, found from the above sequences.

G	P(G)	
2	2/84 =	.023
3	7/84 =	.083
4	21/84 =	.250
5	24/84 =	.286
6	20/84 =	.238
7	10/84 =	.229
	84/84 =	1.000

d. Based on the above distribution, a two-tailed test at the .05 level (that places .025 or less in each tail) has

C.R. $G \le 2$
$G \ge 8$

(for which it will never be possible to reject by being in the upper tail).

e. The critical region in part (d) agrees exactly with Table A-10.

f. From the distribution in part (c)...   From Formula 13-2...

G	P(G)	G·P(G)
2	2/84	4/84
3	7/84	21/84
4	21/84	84/84
5	24/84	120/84
6	20/84	120/84
7	10/84	70/84
		419/84

$\mu_G = \Sigma G \cdot P(G)$
$= 419/84$

$\mu_G = 2n_1 n_2 / (n_1 + n_2) + 1$
$= 2(3)(6)/(9) + 1$
$= 5$
$[= 420/84]$

NOTE: In most large-sample formulas in this chapter, the given μ and σ are exact and the distribution is approximately normal. In this section, the given μ and σ are approximations that improve as n_1 and n_2 increase -- and even for the small samples here the error is only 1/84.

20. a.
ABBBBBBBBBAA
AABBBBBBBBBA
BAAABBBBBBBB
BBAAABBBBBBB
BBBAAABBBBBB
BBBBAAABBBBB
BBBBBAAABBBB
BBBBBBAAABBB
BBBBBBBAAABB
BBBBBBBBAAAB

If G=3, the sequence must begin and end with the same letter and all of the other letter must be grouped together without interruption in the interior of the sequence. There are 10 such sequences.

b. P(G=3) = 10/220 = .045

c. No; a two tailed test has only .025 in each tail and .045 > .025.

d. For n_1 and $n_2 = 9$, the lower critical region is $G \le 2$.

e. For $n_1 = 3$ and $n_2 = 9$ using the normal approximation,
$\mu_G = 2n_1 n_2 / (n_1 + n_2) + 1$
$= 2(3)(9)/12 + 1 = 5.5$
$\sigma_G^2 = [2n_1 n_2 (2n_1 n_2 - n_1 - n_2)]/[(n_1 + n_2)^2 (n_1 + n_2 - 1)]$
$= [2(3)(9)(42)]/[(12)^2(11)] = 1.43$
$\sigma_G = 1.197$

Consider the following test.

H_o: the sequence is random
H_1: the sequence is not random
$\alpha = .05$
C.R. $z < -z_{.025} = -1.960$
  $z > z_{.025} = 1.960$
calculations:
  $z_G = (G - \mu_G)/\sigma_G$
 $\pm 1.96 = (G - \mu_G)/\sigma_G$
  $G = \mu_G \pm 1.96\sigma_G$
   $= 5.5 \pm 1.96(1.197)$
   $= 5.5 \pm 2.35$
   $= 3.15, 7.85$ [the critical values in terms of G]

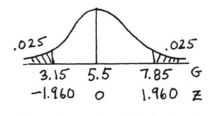

Since G must be an integer, $G < 3.15$ and $G > 7.85$ corresponds to $G \leq 3$ and $G \geq 7$ -- which does not agree with Table A-10 and the correct answer because the normal approximation is not appropriate for $n_1 = 3$ and $n_2 = 9$,

Review Exercises

1. Since $n_1 = 20$ and $n_2 = 5$, use Table A-10.

H_o: the sequence is random
H_1: the sequence is not random
$\alpha = .05$ [only value in Table A-10]
C.R. $G \leq 5$
  $G \geq 12$
calculations:
 $G = 5$
conclusion:

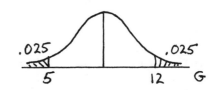

Reject H_o; there is sufficient evidence to conclude that the sequence is not random (in fact, similar types tend to be grouped together).

2. Using B-A: $- + + + - + - - + + - +$
$n = 12$ +'s and -'s
$x = 5$
C.R. $x \leq 2$
do not reject

3. The following table summarizes the calculations.

R_x	R_y	d	d^2
1	4	-3	9
4	2	2	4
3	1	2	4
5	6	-1	1
6	3	3	9
2	7	-5	25
7	5	2	4
		0	56

$r_s = 1 - [6(\Sigma d^2)]/[n(n^2-1)]$
 $= 1 - [6(56)]/[7(48)]$
 $= 1 - 1$
 $= 0$

H_0: $\rho_s = 0$
H_1: $\rho_s \neq 0$
$\alpha = .05$
C.R. $r_s < -.786$
$\qquad r_s > .786$
calculations:
$\qquad r_s = 0$
conclusion:
$\qquad$ Do not reject H_0; there is not sufficient evidence to conclude that $\rho_s \neq 0$.

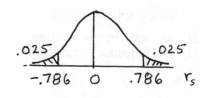

4. When the scores in each precinct are ordered, the ranks are as follows.
$\qquad$ Precinct 1: 1 3 4 7 7 9 10 11 12.5 15 17 20
$\qquad$ Precinct 2: 2 5 7 12.5 14 16 18 19 21 22 23 24
R = 116.5; z = (116.5 - 150)/17.32 = -1.934
C.R. z < -1.960 or z > 1.960
do not reject

5. Below are the scores (in order) for each group.

1	R	2	R	3	R
160	1	165	3	162	2
172	4	174	6	175	7
173	5	176	9	177	11.5
176	9	180	16.5	179	14.5
176	9	181	18	187	23
177	11.5	184	20	195	27.5
178	13	186	22	210	31
179	14.5	190	24	215	32
180	16.5	192	26	216	33
183	19	195	27.5	220	34
185	21	200	29	222	35
191	25	201	30		250.5
	148.5		231.0		

$n_1 = 12 \qquad R_1 = 148.5$
$n_2 = 12 \qquad R_2 = 231.0$
$n_3 = 11 \qquad R_3 = 250.5$

$n = \Sigma n = 35 \qquad \Sigma R = 630.0$

check:
$\Sigma R = n(n+1)/2$
$\qquad = 35(36)/2$
$\qquad = 630$

H_0: the populations have the same distribution
H_1: the populations have different distributions
$\alpha = .05$
C.R. $H > \chi^2_{2,.05} = 5.991$
calculations:
$\qquad H = [12/n(n+1)] \cdot [\Sigma(R_i^2/n_i)] - 3(n+1)$
$\qquad\quad = [12/35(36)] \cdot [(148.5)^2/12 + (231)^2/12 + (250.5)^2/11] - 3(36)$
$\qquad\quad = [.00952] \cdot [11989] - 108$
$\qquad\quad = 6.181$
conclusion:
$\qquad$ Reject H_0; there is sufficient evidence to conclude that the populations have different distributions.

6. Using A-B, the signed ranks are -5 -1 -2.5 2.5 -6 -9 -7 4 -8
n = 9 non-zero differences
T = $\Sigma R+$ = 6.5
C.R. T $\leq$ 6
do not reject

7. Let Test A be group 1.
claim: median difference = 0

pair	1	2	3	4	5	6	7	8	9
H-S	-	-	-	+	-	-	-	+	-

$n = 9$ +'s and -'s

H_o: median difference = 0
H_1: median difference $\neq$ 0
$\alpha = .05$
C.R. $x \leq x_{L,9,.025} = 1$ **OR**   C.R. $x \leq x_{L,9,.025} = 1$
$\phantom{C.R. x \leq x_{L,9,.025} = 1 OR }$ $x \geq x_{U,9,.025} = 9-1 = 8$

calculations: calculations:
   $x = 2$ $x = 2$

conclusion:
   Do not reject H_o; there is not sufficient evidence to conclude that median difference $\neq$ 0.

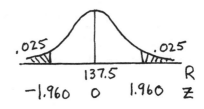

8. Using B-A: - - + - - - 0 0
$n = 6$ +'s and -'s
$x = 1$
C.R. $x \leq 0$
do not reject

9. Below are the scores (in order) for each group. The group listed first is considered group 1.

Ora	R	Wes	R
14	4	5	1
26	6	7	2
39	9.5	10	3
60	16	17	5
62	17	27	7
63	18	35	8
66	19	39	9.5
70	20	40	11
75	21	48	12
79	23	49	13
86	24	50	14
	177.5	54	15
		78	22
			122.5

$n_1 = 11$ $\Sigma R_1 = 177.5$
$n_2 = 13$ $\Sigma R_2 = 122.5$

$n = \Sigma n = 24$ $\Sigma R = 300.0$

check: $\Sigma R = n(n+1)/2$
$ = 24(25)/2$
$ = 300$

$R = \Sigma R_1 = 177.5$

$\mu_R = n_1(n+1)/2$
$ = 11(25)/2$
$ = 137.5$

$\sigma_R^2 = n_1 n_2(n+1)/12$
$ = (11)(13)(25)/12$
$ = 297.917$

H_o: the populations have the same distribution
H_1: the populations have different distributions
$\alpha = .05$
C.R. $z < -z_{.025} = -1.960$
$ z > z_{.025} = 1.960$
calculations:
   $z_R = (R - \mu_R)/\sigma_R$
   $ = (177.5 - 137.5)/\sqrt{297.917}$
   $ = 40/17.260 = 2.317$
conclusion:
   Reject H_o; there is sufficient evidence to conclude that the populations have different distributions (in fact, the Orange County scores tend to be higher).

10. $x/n = 1594/22$ giving the sequence B B A A A A A A A A A A A B B B B A B B B B A
$n_1 = 10$, $n_2 = 12$; $G = 6$
C.R. $G \le 7$ or $G \ge 17$
reject the hypothesis of randomness (in favor of scores above/below the mean being grouped together)

11. The following table summarizes the calculations.

R_x	R_y	d	d^2
1	1	0	0
3	2	1	1
2	3	-1	1
7	8	-1	1
6	6	0	0
10	7	3	9
14	14	0	0
15	15	0	0
11	4	7	49
9	10	-1	1
12	13	-1	1
4	5	-1	1
8	9	-1	1
5	11	-6	36
13	12	1	1
		0	102

$$r_s = 1 - [6(\Sigma d^2)]/[n(n^2-1)]$$
$$= 1 - [6(102)]/[15(224)]$$
$$= 1 - .182$$
$$= .818$$

H_o: $\rho_s = 0$
H_1: $\rho_s \ne 0$
$\alpha = .05$
C.R. $r_s < -.525$
    $r_s > .525$
calculations:
    $r_s = .818$
conclusion:
    Reject H_o; there is sufficient evidence to conclude that $\rho_s \ne 0$ (in fact, $\rho_s > 0$).

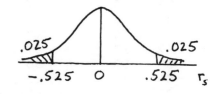

12. Since $\Sigma R_1 = \Sigma R_2 = \Sigma R_3$ with equal sample sizes, $H = 0$.
C.R. $H > 5.991$
do not reject

13. Let Before be group 1.
claim: the populations have the same distribution

pair	1	2	3	4	5	6	7	8	9	10	11	12	13	14
I-II	-2	0	-4	-4	-1	2	-4	-3	0	-1	-5	-2	-5	-3
R	-4	0	-9	-9	-1.5	4	-9	-6.5	0	-1.5	-11.5	-4	-11.5	-6.5

$n = 12$ non-zero ranks
$\Sigma R- = 74.0$
$\Sigma R+ = 4.0$
$\Sigma R = 78.0$ [check: $\Sigma R = n(n+1)/2 = 12(13)/2 = 78$]

H_o: the populations have the same distribution
H_1: the populations have different distributions
$\alpha = .05$
C.R. $T \leq T_{L,12,.025} = 14$ OR   C.R. $T \leq T_{L,12,.025} = 14$
                                                      $T \geq T_{U,12,.025} = 78\text{-}14 = 64$
calculations: calculations:
   $T = 4$ $T = 4$

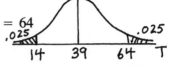

conclusion:
   Reject H_o; there is sufficient evidence to conclude that the populations have different distributions (in fact, the Before scores tend to be lower).

14. Rank the scores to obtain $\Sigma d^2 = 18.5$ and $r_s = .959$
C.R. $r_s < -.545$ or $r_s > .545$
reject the claim of no rank correlation and conclude the consumptions are positively correlated.

15. Let Before be group 1.
   claim: median difference $= 0$

pair	1	2	3	4	5	6	7	8	9	10	11	12	13	14
H-S	-	0	-	-	-	+	-	-	0	-	-	-	-	-

   $n = 12$ +'s and -'s

H_o: median difference $= 0$
H_1: median difference $\neq 0$
$\alpha = .05$
C.R. $x \leq x_{L,12,.025} = 2$ OR   C.R. $x \leq x_{L,12,.025} = 2$
                                                   $x \geq x_{U,12,.025} = 12\text{-}2 = 10$
calculations: calculations:
   $x = 1$ $x = 1$

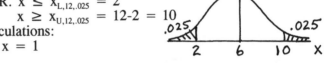

conclusion:
   Reject H_o; there is sufficient evidence to conclude that median difference $\neq 0$ (in fact, median difference < 0 -- i.e., the Before scores tend to be lower).

16. Below are the scores (in order) for each group. The group listed first is group 1, etc.

paper	R	radio	R	none	R
845	11	811	10	612	3
907	15	782	8	574	2
639	4	749	7	539	1
883	14	863	12	641	5
806	9	872	13	666	6
	53		50		17

$n_1 = 5 \quad R_1 = 53$
$n_2 = 5 \quad R_2 = 50$
$n_3 = 5 \quad R_3 = 17$

$n = \Sigma n = 15 \quad \Sigma R = 120$

check:
$\Sigma R = n(n+1)/2$
$= 15(16)/2$
$= 120$

H_o: the populations have the same distribution
H_1: the populations have different distributions
$\alpha = .05$
C.R. $H > \chi^2_{2,.05} = 5.991$

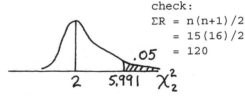

calculations:
   $H = [12/n(n+1)] \cdot [\Sigma(R_i^2/n_i)] - 3(n+1)$
   $= [12/15(16)] \cdot [(53)^2/5 + (50)^2/5 + (17)^2/5] - 3(16) = 7.980$
conclusion:
   Reject H_o; there is sufficient evidence to conclude that the populations have different distributions.

17. Below are the scores (in order) for each group. The group listed first is considered group 1.

MO	R	CS	R
23.00	1	29.00	10
24.50	2	30.99	12
24.75	3	32.00	14
26.00	4	32.99	16
27.00	5	33.00	17
27.98	6	33.98	18
27.99	7	33.99	19
28.15	8	34.79	20
29.99	10	35.79	21
29.99	10	37.75	22
31.50	13	38.99	30
32.75	15		192
	84		

$n_1 = 12$ $\Sigma R_1 = 84$
$n_2 = 11$ $\Sigma R_2 = 192$

$n = \Sigma n = 23$ $\Sigma R = 276$

check: $\Sigma R = n(n+1)/2$
$= 23(24)/2$
$= 276$

$R = \Sigma R_1 = 84$

$\mu_R = n_1(n+1)/2$
$= 12(24)/2$
$= 144$
$\sigma_R^2 = n_1 n_2 (n+1)/12$
$= (11)(12)(24)/12$
$= 264$

H_o: the populations have the same distribution
H_1: the populations have different distributions
$\alpha = .05$
C.R. $z < -z_{.025} = -1.960$
$z > z_{.025} = 1.960$
calculations:
$z_R = (R - \mu_R)/\sigma_R$
$= (84 - 144)/\sqrt{264}$
$= -60/16.248 = -3.693$
conclusion:
Reject H_o; there is sufficient evidence to conclude that the populations have different distributions (in fact, the mail order prices are lower).

18. $n_1 = 20$, $n_2 = 10$; $G = 5$
C.R. $G \leq 9$ or $G \geq 20$
reject the hypothesis of randomness and conclude that males and females tended to be grouped together

19. Below are the scores (in order) for each group.

1	R	2	R	3	R
195	3	187	1	193	2
198	4	210	5	212	6
223	9	222	8	215	7
240	12	238	11	231	10
251	13	256	15	252	14
	41		40	260	16
				267	17
					72

$n_1 = 5$ $R_1 = 41$
$n_2 = 5$ $R_2 = 40$
$n_3 = 7$ $R_3 = 72$

$n = \Sigma n = 17$ $\Sigma R = 153$

check:
$\Sigma R = n(n+1)/2$
$= 17(18)/2$
$= 153$

H_o: the populations have the same distribution
H_1: the populations have different distributions
$\alpha = .01$
C.R. $H > \chi^2_{2,.01} = 9.210$
calculations:

$\begin{aligned}
H &= [12/n(n+1)] \cdot [\Sigma(R_i^2/n_i)] - 3(n+1) \\
&= [12/17(18)] \cdot [(41)^2/5 + (40)^2/5 + (72)^2/7] - 3(18) \\
&= [.0392] \cdot [1396.77] - 54 = .775
\end{aligned}$

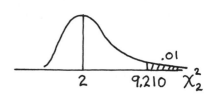

conclusion:
Do not reject H_o; there is not sufficient evidence to conclude that the populations have different distributions.

20. Use the Wilcoxon rank sum test.
$n_1 = 13$, $n_2 = 17$, $R = 163$
$z = (163 - 201.5)/23.894 = -1.611$
for α .05, CV $= \pm 1.960$
do not reject the hypothesis of equality

21. Let Judge A be group 1.
claim: the populations have the same distribution

pair	1	2	3	4	5	6	7	8
I-II	-.8	.7	-1.6	-.1	.7	.9	1.0	0
R	-4	2.5	-7	-1	2.5	5	6	0

n = 7 non-zero ranks
$\Sigma R- = 12.0$
$\Sigma R+ = 16.0$
$\Sigma R = 28.0$ [check: $\Sigma R = n(n+1)/2 = 7(8)/2 = 28$]

H_o: the populations have the same distribution
H_1: the populations have different distributions
$\alpha = .05$

C.R. $T \le T_{L,7,.025} = 2$ OR   C.R. $T \le T_{L,7,.025} = 2$
$T \ge T_{U,7,.025} = 28-2 = 26$

calculations: calculations:
$T = 12$ $T = 16$

conclusion:
Do not reject H_o; there is not sufficient evidence to conclude that the populations have different distributions.

NOTE: This test may not be the most appropriate one for the intended purpose. The signed-rank test measures overall differences between the judges and <u>not</u> whether the two judges give the same results for individuals. If one judge gave half the students higher scores and the other half of the students scores that were lower by the same amounts, then $\Sigma R-$ would equal $\Sigma R+$ (so we could not reject H_o) but the distributions would be very different. Thus the test measures whether the overall typical scores are the same for the two judges and not whether the two judges agreed on a contestant-by-contestant basis.

22. Let the right eye be group 1.
claim: the populations have the same distribution

pair	A	B	C	D	E	F	G	H
R-L	-.5	-.4	-.2	-2.3	-2.5	-3.3	2.3	-2.6
R	-3	-2	-1	-4.5	-6	-8	4.5	-7

n = 8 non-zero ranks
$\Sigma R- = 31.5$
$\Sigma R+ = 4.5$
$\Sigma R = 36.0$ [check: $\Sigma R = n(n+1)/2 = 8(9)/2 = 36$]

H_o: the populations have the same distribution
H_1: the populations have different distributions
$\alpha = .05$
C.R. $T \leq T_{L,8,.025} = 4$ OR   C.R. $T \leq T_{L,8,.025} = 4$
                                     $T \geq T_{U,8,.025} = 36-4 = 32$

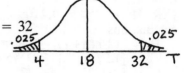

calculations: calculations:
  $T = 4.5$ $T = 4.5$
conclusion:
    Do not reject H_o; there is not sufficient evidence to conclude that the populations have different distributions

23. Since $n_1 = 25$ and $n_2 = 8$, use the normal approximation.
    $\mu_G = 2n_1n_2/(n_1+n_2) + 1$
        $= 2(25)(8)/33 + 1$
        $= 13.121$
    $\sigma_G^2 = [2n_1n_2(2n_1n_2-n_1-n_2)]/[(n_1+n_2)^2(n_1+n_2-1)]$
        $= [2(25)(8)(367)]/[(33)^2(32)]$
        $= 4.212$

H_o: the sequence is random
H_1: the sequence is not random
$\alpha = .05$
C.R. $z < -z_{.025} = -1.960$
    $z > z_{.025} = 1.960$
calculations:
  $G = 5$
  $z_G = (G - \mu_G)/\sigma_G$
     $= (5 - 13.121)/\sqrt{4.212}$
     $= -8.121/2.052 = -3.957$

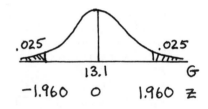

conclusion:
    Reject H_o; there is sufficient evidence to conclude that the sequence is not random (in fact, the N and Y responses tend to occur in groups).

24. The following table summarizes the calculations.

beer	R_x	wine	R_y	d	d^2
32.3	5	3.1	7	-2	4
29.4	3	4.4	8	-5	25
35.3	7	2.3	5	2	4
34.9	6	1.7	4	2	4
29.9	4	1.4	3	1	1
28.7	2	1.2	1.5	0.5	0.25
26.8	1	1.2	1.5	-0.5	0.25
41.4	8	3.0	6	2	4
					42.50

$r_s = 1 - [6(\Sigma d^2)]/[n(n^2-1)]$
  $= 1 - [6(42.5)]/[8(63)]$
  $= 1 - .506$
  $= .494$

H_o: $\rho_s = 0$
H_1: $\rho_s \neq 0$
$\alpha = .05$
C.R. $r_s < -.738$
    $r_s > .738$
calculations:
  $r_s = .494$

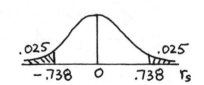

conclusion:
    Do not reject H_o; there is not sufficient evidence to conclude that $\rho_s \neq 0$.